Bioavailability and Analysis of Vitamins in Foods

Bioavailability and Analysis of Vitamins in Foods

G.F.M. Ball

CHAPMAN & HALL
London · Weinheim · New York · Tokyo · Melbourne · Madras

Published by Chapman & Hall, 2–6 Boundary Row, London SE1 8HN

Chapman & Hall, 2–6 Boundary Row, London SE1 8HN, UK

Chapman & Hall GmbH, Pappelallee 3, 69469 Weinheim, Germany

Chapman & Hall USA, 115 Fifth Avenue, New York, NY 10003, USA

Chapman & Hall Japan, ITP-Japan, Kyowa Building, 3F, 2-2-1 Hirakawacho, Chiyoda-ku, Tokyo 102, Japan

Chapman & Hall Australia, 102 Dodds Street, South Melbourne, Victoria 3205, Australia

Chapman & Hall India, R. Seshadri, 32 Second Main Road, CIT East, Madras 600 035, India

First edition 1998

© 1998 G.F.M. Ball

Typeset in Palatino 10/12 by Pure Tech India Ltd, Pondicherry
Printed in the United Kingdom by T.J. International, Padstow, Cornwall

ISBN 0 412 78090 9

A catalogue record for this book is available from the British Library

Library of Congress Catalog Card Number: 97-68950

♾ Printed on permanent acid-free text paper, manufactured in accordance with ANSI/NISO Z39.48-1992 and ANSI/NISO Z39.48-1984 (Permanence of Paper).

Contents

Contents

Preface

Every country in the world is concerned with the nutritional status of its population and in utilizing its natural food resources in the most effective way possible. Surveys based on food intakes and food compositional data are being conducted with the object of establishing recommended intakes of vitamins. These recommendations are constantly being changed as new knowledge comes to light.

Analytical techniques using physicochemical and microbiological methods have been largely developed to determine the total vitamin content of a food commodity or diet using the most rigorous extraction method commensurate with the stability of the vitamin. The extraction procedures frequently involve prolonged heating of suitably prepared food samples at extremes of pH to liberate vitamins from chemically bound forms in the food matrix or to remove a preponderance of fat from fatty foods. For several vitamins the data obtained by these means grossly overestimate the nutritional value of the food because the human digestive system fails to liberate bound vitamin forms for subsequent absorption by the intestine. This statement is borne out by reports of vitamin deficiency in situations where the dietary supply of vitamin is adequate on the basis of conventional analysis. Various research laboratories are directing their effort toward the estimation of bioavailable vitamin, i.e. the proportion of vitamin in the food which is available for utilization by the body. So far, few data have been published and there are many gaps in the knowledge required to interpret experimental results.

The main purpose of this book is to discuss the important factors that influence the bioavailability of vitamins in foods. The physiological processes of digestion and absorption are given special attention because the intestine is highly selective in allowing nutrients to enter the body proper. Other factors include the composition of the diet, the effects of alcohol and drugs, age and state of health. Food processing can result in losses of the more labile vitamins and sometimes the bioavailability of the remaining amount of vitamin is also reduced. On the other hand, domestic cooking can increase the bioavailability of vitamins from certain foods.

Vitamin assays in foods are carried out for a variety of purposes: to implement regulatory enforcement; to check compliance with contract specifications and nutrient labelling regulations; to provide quality assurance for supplemented products; to provide data for food composition tables; to study changes in vitamin content attributable to food processing, packaging and storage; and to assess the effects of geographical, environmental and seasonal conditions. In addition, new varieties of food plants, including cereal grains, are frequently being introduced. The book surveys methods of determining vitamins in foods, with emphasis on the extraction procedures employed. Rather than using heat and strong acid or alkali to estimate the total (potential) vitamin content, a few methods use enzymatic digestion of the food matrix to estimate the available vitamin.

As to the scope of the book, the first chapter describes the functional anatomy of the small intestine and the physiology of absorption as a foundation to understanding the specific absorption mechanisms for the individual vitamins discussed in the later chapters. Chapter 2 outlines the principles of the various physicochemical, microbiological and biospecific techniques encountered in vitamin analysis and includes high-performance liquid chromatographic (HPLC) methods for determining two or three vitamins concurrently or simultaneously. HPLC and other methods for determining individual vitamins are described in the respective chapters which follow.

I would like to express my sincere thanks to the many people who found the time to respond to my requests for information and to those who encouraged me in my endeavours.

George F. M. Ball
Windsor, Berkshire
England

1

Physiological aspects of vitamin bioavailability

1.1 THE ROLE OF VITAMINS IN HUMAN NUTRITION

1.1.1 Introduction

Vitamins are a group of organic compounds which are essential in very small amounts for the normal functioning of the human body. They have widely varying chemical and physiological functions and are broadly distributed in natural food sources. Thirteen vitamins are recognized in human nutrition and these may be conveniently classified, according to their solubility, into two groups. The fat-soluble vitamins are represented by vitamins A, D, E and K; also included are the 50 or so carotenoids that possess varying degrees of vitamin A activity. The water-soluble vitamins comprise vitamin C and the members of the vitamin B group, namely thiamin (vitamin B$_1$), riboflavin (vitamin B$_2$), niacin, vitamin B$_6$, pantothenic acid, biotin, folate and vitamin B$_{12}$. This simple classification reflects to some extent the bioavailability of the vitamins, as the solubility affects their mode of intestinal absorption and their uptake by tissues. The solubility properties also have a direct bearing on the analytical methods employed in vitamin assays.

For many of the vitamins, biological activity is attributed to a number of structurally related compounds known as vitamers. The vitamers pertaining to a particular vitamin display, in most cases, similar qualitative biological properties to one another, but, because of subtle differences in their chemical structures, exhibit varying degrees of potency. Provitamins are vitamin precursors, i.e. naturally occurring substances which are not themselves vitamins, but which can be converted by normal body metabolism into vitamins.

It is often stated that vitamins cannot be produced in the body and must, therefore, be supplied in the diet. This statement is valid for some of the vitamins, but is not strictly true for others. For example, vitamin D can be formed in the skin upon adequate exposure to ultraviolet radiation; vitamin K is normally produced in sufficient amounts by intestinal bacteria; and niacin can be synthesized *in vivo* from an amino acid precursor, tryptophan. With the possible exception of vitamins D and K, vitamins must be supplied by the diet because they cannot be produced in adequate amounts by the human body. Plants have the ability to synthesize most of the vitamins and serve as primary sources of these dietary essentials.

1.1.2 Nutritional vitamin deficiency

Several of the B-group vitamins serve as coenzymes for enzymes that function in the catabolism of foodstuffs to produce energy for the organism. A typical coenzyme consists of a protein (apoenzyme) to which is attached the vitamin. The vitamin portion of the coenzyme is usually responsible for the attachment of the enzyme to the substrate. If the specific vitamin is not available to form the coenzyme, the sequence of chemical changes in the metabolic process cannot proceed and the product whose change is blocked accumulates in the tissues: alternatively, metabolism is diverted to another pathway. For some B-group vitamins, deficiency results in a biochemical defect which is manifested as a disease with characteristic symptoms. Subclinical deficiency and marginal deficiency are synonymous terms used to describe conditions in individuals who are not clinically nutrient deficient, but who appear to be close to it. An alternative and perhaps better term proposed by Herbert (1990) is 'early negative nutrient balance', which is used when laboratory measurements indicate that an individual is losing more of a nutrient than is being absorbed.

Development of deficiency

By reference to the sequence of events in the development of vitamin deficiency, Pietrzik (1985) emphasized the importance of preventing functional metabolic disturbances that can evolve into overt clinical symptoms. This sequence can be subdivided into six stages as follows.

- **Stage 1** Body stores of the vitamin are progressively depleted. A decreased vitamin excretion in the urine is often the first sign. Normal blood levels are maintained by homeostatic mechanisms in the very early stages of deficiency.
- **Stage 2** The urinary excretion of the vitamin is further decreased and vitamin concentrations in the blood and other tissues are lowered. A diminished concentration of vitamin metabolites might also be observed.
- **Stage 3** There are changes in biochemical parameters such as low concentrations of the vitamin in blood, urine and tissues, and a low activity of vitamin-dependent enzymes or hormones. Immune response might also be reduced. Nonspecific subclinical symptoms such as general malaise, insomnia, loss of appetite and other mental changes appear.
- **Stage 4** The biochemical changes become more severe and morphological or functional disturbances are observed. These disturbances might be corrected by vitamin dosing in therapeutic amounts within a relatively short time or vitamin supplementation in amounts of (or exceeding) the recommended dietary allowances over a longer period. Malformation of cells is reversible at this stage.
- **Stage 5** The classical clinical symptoms of vitamin deficiency will appear. Anatomical alterations characterized by reversible damage of tissues might be cured in general by hospitalization of the patient. In most cases there are deficiencies of several nutrients and a complicated dietetic and therapeutic regimen has to be followed.
- **Stage 6** The morphological and functional disturbances will become irreversible, finally leading to death in extreme cases.

From the health point of view, it is proposed that the borderline vitamin deficiency is represented by the transition from the third to the fourth stage.

Causes of deficiency

The causes of nutritional vitamin deficiency are any one or combination of the following: inadequate ingestion, inadequate absorption, inadequate utilization, increased requirement, increased excretion and increased destruction (Herbert and Das, 1994).

Effects of dietary fibre on absorption of nutrients
Dietary fibre refers to the unavailable carbohydrate provided by plant cell wall material and comprises principally nonstarch polysaccharides, which include cellulose, and lignin. Diets that contain processed foods may contribute noncellulosic polysaccharides such as gums, mucilages and algal polysaccharides, many of which are structurally similar to components of the plant cell wall (Southgate, 1982).

The consumption of diets high in dietary fibre has profound effects on the physical properties of the intestinal contents. Dietary fibre is an adsorbent with water-holding capacity, catalytic surfaces and cation exchange properties. Soluble polysaccharides leach out from the plant cell wall structures and analogous polysaccharides used as food additives become fully hydrated during passage through the stomach and proximal small intestine. The resultant increase in viscosity affects absorption. Also, the sheer bulk of dietary fibre can entrap micronutrients and affect the mixing of the intestinal contents with consequential effects on absorption. These effects seem to delay the absorption of glucose and other water-soluble organic nutrients, but do not cause malabsorption (Southgate, 1993).

The binding of bile salts to various food fibre sources and isolated fibre components has been repeatedly demonstrated by various *in vitro* techniques (Vahouny, 1982). There is further *in vitro* evidence to suggest that certain types of dietary fibre may either interfere with the formation of mixed micelles in the intestinal lumen or effectively alter the normal diffusion and accessibility of micellar lipids to the absorptive surface of the intestinal mucosa. Such events could compromise the absorption of lipids, including the fat-soluble vitamins.

Effects of alcohol on nutritional status
In many societies alcoholic beverages are considered to be part of the basic food supply. Alcohol is also consumed for its mood-altering effects and is thereby a psychoactive drug. Under both circumstances, a large intake of alcohol has profound effects on nutritional status. Because alcohol (ethanol) is rich in energy but otherwise empty in nutritional value, the excessive consumption of alcoholic beverages promotes the development of primary malnutrition by displacing other nutrients in the diet (Lieber, 1988). Causes of vitamin deficiencies in alcoholism are decreased dietary intake, decreased intestinal absorption and alterations in vitamin metabolism.

1.1.3 Recommended daily allowances

Although it is not possible experimentally to determine the precise human requirement for the various vitamins, certain minimum quantities of each vitamin have been shown to be necessary for the maintenance of good health. Recommended Dietary Allowances (RDA) have been published in the United States for vitamins A, D, E and K, thiamin, riboflavin, niacin, vitamin B_6, folate, vitamin B_{12} and vitamin C (National Research Council, 1989). Other countries and international bodies have compiled similar recommendations. In the United Kingdom data for Reference Nutrient Intakes (RNI) have been published by the Department of Health and Social Security (Department of Health, 1991). RDA and RNI are nutritional standards which have the same statistical basis and are defined as the amount of the nutrient which is adequate to ensure that the requirements of all essentially healthy people are met.

The addition of vitamins to a particular processed food is intended to provide a specific proportion of the RDA or RNI. General principles for the addition of nutrients to foods have been established by the Codex Alimentarius Commission (1987) and the United States Food and Drug Administration (1987).

1.1.4 Vitamin enhancement of foods

The terms restoration, fortification, enrichment, standardization and nutrification have been used (Anon., 1989) to describe various ways of enhancing the vitamin content of foods.

Restoration involves the replacement, in full or in part, of vitamin losses incurred during processing. The addition of vitamins A and D to skimmed milk powder and of vitamin D to evaporated milk are examples of non-legislative vitamin restoration in the UK. Other examples are the replacement of B-group vitamins in flour to compensate for the losses incurred in the milling of cereals to low extraction rates, and the addition of vitamin C to instant potato.

Fortification refers to the addition of vitamins to foods that represent ideal carriers for a particular vitamin, but which do not necessarily contain that vitamin naturally. It is especially carried out to fulfil the role of a food in the diet. Thus margarine is fortified with vitamin A in many countries to replace the vitamin A that is lost from the diet when margarine is substituted for butter. Vitamin D is added at higher levels than found in butter as a public health measure since the extra is considered necessary for the population as a whole.

Enrichment refers to the addition of vitamins above the initial natural levels to make a product more marketable.

Standardization refers to additions to compensate for natural fluctuations in vitamin content. For instance, milk and butter are subject to seasonal variations in their vitamin A and D contents, and hence these vitamins are added to some dairy products to maintain constant levels.

Nutrification means the addition of vitamins to formulated or fabricated foods marketed as meal replacers.

Vitamins are also added to perform specific processing functions. For instance, β-carotene (the principal source of dietary vitamin A) is added to products such as pasta, margarine, cakes and processed cheeses to impart colour. Vitamin E and vitamin C (in the form of ascorbyl palmitate) can be used as antioxidants to stabilize pure oils and fats, including margarines. Ascorbic acid is used for a variety of purposes in food processing, such as a reducing agent involved in the formation of the cured meat colour in the curing of bacon and ham.

1.2 GENERAL PHYSIOLOGY OF DIGESTION AND ABSORPTION

1.2.1 General concepts

Knowledge of the physiology of digestion and absorption of the vitamins is important towards understanding the utilization and conservation of these micronutrients. Digestion refers to the chemical and physical modifications which render ingested food constituents absorbable by the small intestine. Absorption is the process by which the products of digestion pass out of the digestive tract into the cells lining it, and from there into the bloodstream or lymph. In the healthy human, the absorption of foodstuffs is very efficient during passage through the small intestine. After a mixed meal all of the carbohydrate, about 95% of the fat and 90% of the protein are absorbed. A complex meal may be fully digested and absorbed in 3 hours.

The gastrointestinal tract, although containing many enzymes, is metabolically less active than the liver. Its metabolic functions are therefore directed primarily toward delivering ingested nutrients to the liver for further processing. Blood from the intestine drains into the superior mesenteric vein, which unites with the splenic vein to form the portal vein. The portal vein conveys nonlipid products of digestion from the intestine to the liver, together with products of erythrocyte destruction from the spleen. All absorbed nonlipid material passes through the liver and nutrients are extracted by the liver cells for storage or metabolic processing. The liver also detoxifies drugs and noxious chemicals. Detoxified blood mixed with hepatic arterial blood pours into the hepatic veins and returns to the heart via the inferior vena cava. The blood is then sent to the lungs, and the oxygenated blood returns to the heart for redistribution to all parts of the body.

The liver produces bile, which is secreted continuously into the gallbladder, whence it is discharged intermittently into the duodenum. Bile is necessary for the emulsification of ingested fats. Many nutrients, including vitamin D, folate and vitamin B_{12} and also bile salts, are subject to an enterohepatic circulation through their excretion in bile and subsequent reabsorption by the small intestine. The liver also produces copious amounts of lymph to sustain the lymphatic system. Absorbed lipids and lipid-soluble substances in the form of chylomicrons are transported via the lymphatic system and thoracic ducts to the subclavian veins and into the systemic circulation, bypassing the liver.

The gastrointestinal tract is depicted diagrammatically in Figure 1.1. Digestion commences as soon as the food enters the mouth. Mastication helps to break up large particles of food, whilst also mixing food with saliva, which acts as a lubricant and contains the diastatic enzyme ptyalin. In the stomach, the churning action and the presence of hydrochloric acid and pepsin converts the food bolus into a liquid chyme. Most absorption takes place and is completed in the small intestine, which is about 6 m in length and 3 cm in diameter in the adult. The small intestine consists of three portions, starting with the duodenum, which is only about 30 cm long. The jejunum accounts for the remaining two-fifths of

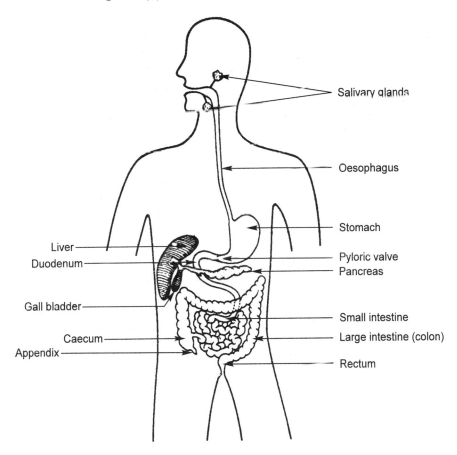

Figure 1.1 The gastrointestinal tract.

the small intestine, and the ileum the remaining three-fifths. On arrival at the duodenum, the acid chyme is buffered by the bicarbonate in pancreatic juice and bile. Brunner's glands in the duodenum produce an alkaline secretion containing mucus. The enzymes secreted into the duodenum by the pancreas can, together with the bile, initiate and carry through the whole digestive process. The final stages of digestion take place on the luminal surface or within the cytoplasm of the cells lining the small intestine. The amount of enzymes produced by the intestinal glands (crypts of Lieberkühn) is almost negligible and the small quantities of disaccharidases and dipeptidases are of cellular origin. Another intestinal enzyme, alkaline phosphatase, is bound primarily to the brush-border membrane of epithelial cells and is also present in soluble form in the lumen. The various stages of digestion are coordinated by the action of the nervous system, endocrine system and circulatory system.

The lumen of the upper gastrointestinal tract is characterized by a wide range of pH values. Fordtran and Locklear (1966) demonstrated postprandial pH values in humans of 1.8–3.4 in the stomach, 6.8–7.8 in the lower small intestine, and intermediate values (3.5–7) in the duodenum and proximal jejunum.

When absorption has been accomplished, the jejunum and ileum are actively involved in the regulation of electrolyte and fluid balance. The primary functions of the large intestine (colon) are to secrete mucus and to absorb water, glucose and inorganic salts.

1.2.2 Functional anatomy of the small intestine, liver and kidney

Small intestine

The intestinal wall comprises four principal layers as shown in Figure 1.2. The outermost serosal layer is covered with peritoneum and contains blood vessels, lymphatics and nerve fibres which pass through it to the other layers. The next layer, the muscularis externa, consists of an inner band of circularly orientated and an outer band of longitudinally orientated smooth muscle fibres arranged in a spiral fashion. The muscularis externa facilitates peristalsis and, by churning, aids in mixing the chyme with the digestive enzymes. The third layer, the submucosa, comprises course areolar connective tissue and contains plexuses of blood vessels and lymphatics. The innermost layer, the mucosa, consists of an epithelium, lubricated by mucus, resting upon a basement membrane. This in turn is supported by a layer of loose, areolar connective tissue termed the lamina propria, which is attached to a thin sheet of smooth muscle, the muscularis mucosa.

To allow a maximum rate of absorption, the intestinal mucosa shows specialization that increases its surface area to $c.300\,m^2$. Firstly, the

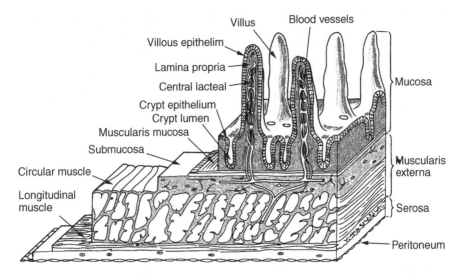

Figure 1.2 Diagrammatic structure of the small intestinal wall (sectional view).

mucosa, with a core of submucosa, is thrown into permanent concentric folds called plicae circulares which protrude into the intestinal lumen. The folds are up to 1 cm in height and any one fold may extend two-thirds or more around the circumference of the intestine. Rarely do the folds completely encircle the lumen. The plicae are most frequent in the distal duodenum and proximal jejunum. Thereafter, they become less frequent and are rare in the distant ileum. As well as increasing the absorptive area, the plicae circulares retard the passage of chyme, allowing more time for effective digestion and absorption.

The absorptive area is greatly increased by the presence of tightly packed projections called villi, each about 1 mm long, giving the mucosal surface a velvet-like appearance. In the human, the villi are numerous in the duodenum and jejunum, but fewer in the ileum. In the proximal duodenum they are broad, ridge-like structures, changing to tall, leaf-like villi in the distal duodenum/proximal jejunum. Thereafter, they gradually transform into short finger-like processes in the distal jejunum and ileum. The villus is the functional absorptive unit of the small intestine. Contained within the lamina propria core of each villus is a capillary network with a supplying arteriole and draining venule. A blind-ending lymphatic vessel (lacteal) in the centre of each villus drains into a plexus of collecting vesicles in the submucosa. The lacteal is surrounded by smooth muscle elements that extend from the muscularis mucosa. During digestion the muscularis mucosa contracts, causing the villi to sway and contract intermittently. These movements and contractions may have the effect of 'milking' the lacteals, forcing the lymph and absorbed nutrients (fats) into the lymphatic vesicles. Each villus is covered by an epithelium composed of a single layer of columnar absorptive cells (enterocytes) interspersed occasionally with mucus-secreting goblet cells. The epithelium is supported by a basement membrane formed by a layer of fine fibrillar material composed of collagen and mucopolysac-charides. Undifferentiated cells in the epithelium of the intestinal glands (crypts) proliferate and migrate upwards to replace enterocytes and other cells that are continuously desquamating from the tip of the villus into the intestinal lumen. The migrating cells become differentiated as they move up the villus.

The essential features of an enterocyte that concern its absorptive function are shown in Figure 1.3. The luminal cell 'brush border' consists of minute projections of the apical plasma membrane called microvilli,

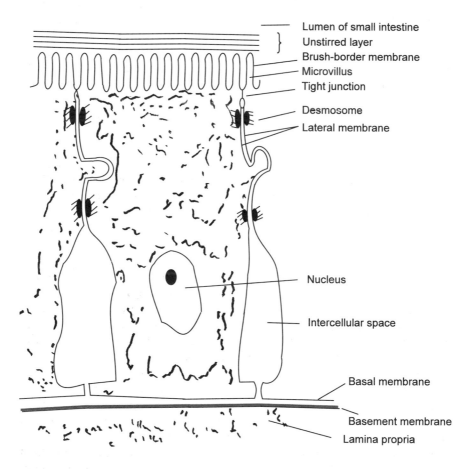

Figure 1.3 Longitudinal section of an enterocyte showing the membranes and junction points.

which are packed at a density of about $200\,000/mm^2$ in the human jejunum. The luminal surfaces of the microvilli are covered with a fuzzy coat of glycoprotein-rich filaments known as the glycocalyx. The glycocalyx may play an important role in terminal digestion by providing an extracellular site for brush-border enzymes. The fuzzy coat cannot be removed by washing with mucolytic or proteolytic agents, and hence it is not an accumulation of adsorbed mucus secreted by goblet cells. Rather, the glycocalyx is a dynamic component of enterocytes, requiring an intact cell for its synthesis and maintenance (Ito, 1969).

Just below the level of the bases of the microvilli, each cell is tightly bound to the adjacent cells by a junctional complex involving an exceptionally close association of the lateral plasma membranes. These tight junctions prevent macromolecules from entering the villus by intercellular channels. Immediately beneath the tight junctions and at other points on the lateral plasma membrane are looser intercellular connections called desmosomes. The apical parts of adjacent cells are closely interlocked by cytoplasmic processes but the basal parts are separated at intervals by intercellular canaliculi. These channels become greatly dilated to form intercellular spaces when ions and water are being actively absorbed.

Liver

The liver is divided into right and left lobes of approximately equal size and subdivided into a large number of lobules. It is covered by a fibroconnective tissue capsule from which thin connective tissue septa enter the matrix. Each lobule contains a central vein and portal canals at its periphery. Portal canals are surrounded by small amounts of fibroconnective tissue and contain a branch of the hepatic artery, a branch of the portal vein, and a bile ductule. Lymphatic vessels within the portal canal produce lymph which passes indirectly into the lymphatic capillaries. The matrix of the liver is composed of parenchymal cells (hepatocytes) arranged in a series of branching and anastomosing perforated plates. These form a sponge-like three-dimensional lattice in which the spaces (sinusoids) are filled with blood. At the surface adjacent to a sinusoidal blood space, the hepatocyte is separated from the sinusoidal lining by a narrow perisinusoidal space called the space of Disse (Figure 1.4). At this surface the plasma membrane of the hepatocyte is covered by numerous long microvilli. Bile produced by the hepatocytes passes into narrow channels (bile canaliculi), which are simply spaces between adjacent hepatocytes and lead to interlobular bile ductules.

In addition to its parenchymal cells, the liver contains various kinds of nonparenchymal cells; namely endothelial cells that line the sinusoidal spaces, Kupffer cells, pit cells and stellate cells (Blomhoff and Wake,

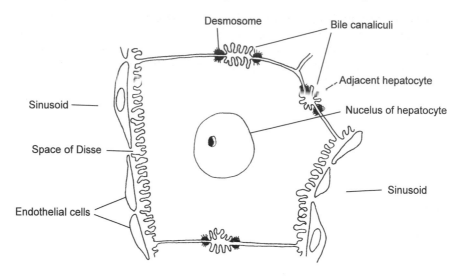

Figure 1.4 Longitudinal section of a hepatocyte. The cell in this section shows five surfaces: two adjacent to sinusoids and separated from the endothelium by the perisinusoidal space, and three opposed to other hepatocytes. Irregular microvilli protrude into the perisinusoidal space. On the other three surfaces, three bile canaliculi are illustrated, each limited laterally by desmosomes.

1991). The endothelial cells are characterized by the presence of numerous large fenestrations and by the absence of a supporting basal lamina. These fenestrations, along with spaces between adjacent endothelial cells, allow the plasma of the sinusoidal blood free access to the perisinusoidal space, thus permitting easy exchange of metabolites between hepatocytes and the blood. Stellate cells are the major vitamin A-storing cells in the liver; they line the space of Disse and comprise 5–15% of total liver cells.

Cirrhosis of the liver, common in chronic alcoholics, is caused by alcohol poisoning at such frequent intervals that the hepatocytes cannot recover fully between bouts of drinking. When this happens, fibroblasts grow in place of the hepatocytes and the liver becomes irreversibly clogged with connective tissue.

Kidney

The kidney plays an essential role in the conservation of water-soluble vitamins by selectively reabsorbing these essential nutrients from the glomerular filtrate and returning them to the blood. Reabsorption takes place in the proximal convoluted tubule, a region of the nephron which is composed of a single layer of epithelial cells with a brush border of microvilli at the luminal surface. The tubules are enveloped by blood capillaries whose walls are highly permeable. The outer membrane of the tubular epithelial cell rests on a basement membrane and is invaginated to form a labyrinth of basal channels. Adjacent membranes of tubule cells are separated by intercellular spaces, and fluid circulates through these spaces and basal channels. This fluid bathes the cells of the proximal convoluted tubule and the surrounding capillary network.

1.2.3 Biological barriers encountered during the absorption of nutrients

The enterocyte constitutes the only barrier of physiological significance controlling the movement of solutes from the intestinal lumen to the blood capillaries and lacteals in the villus.

The environment immediately adjacent to the mucous membrane of the small intestine is different to the environment in the bulk phase of the lumen. Convection of solutes does not take place in the membrane-localized microenvironment and no amount of peristaltic mixing can affect it. The pH at the luminal surface is two units lower than that of the bulk luminal phase and varies less than ±0.5 units despite large pH variations in the intestinal chyme. The existence of this acidic microenvironment or 'unstirred water layer' was originally proposed by pharmacologists in 1957 to account for the absorption of weakly acidic or basic drugs. According to pH-partition theory, weak acids or bases can only cross the lipoidal cell membrane in the nonionized form by the process of non-ionic diffusion. The degree of solute ionization depends upon the pH of the medium and the solute's pK_a value.

Shiau *et al.* (1985) showed that the formation of the acidic microclimate is most likely due to the presence of mucin which covers the entire surface of the epithelium. Mucopolysaccharides possess a wide range of ionizable groups and hence mucin is an ampholyte. If the luminal chyme is of low pH, the ampholyte is positively charged and so it repels additional hydrogen ions entering the microclimate. If, on the other hand, the chyme is alkaline, the ampholyte becomes negatively charged and retains hydrogen ions within the microclimate. In this manner, the mucin layer functions as a restrictive barrier for hydrogen ions diffusing in and out of the microclimate.

Essentially, mammalian cell membranes are composed of a bilayer of phospholipid covered on each surface by a layer of porous protein. Most water-soluble molecules can permeate the protein layers without hindrance but, being lipid-insoluble, they cannot pass through the phospholipid bilayer. However, the bilayer is occasionally perforated by a mosaic of tiny channels of porous protein that bring the interstitial and cellular fluids into continuity. Water-soluble molecules that are small enough to enter these channels can therefore traverse the entire width of the membrane.

The enterocyte possesses two types of plasma membrane: (1) the brush-border membrane containing disaccharidases and other digestive enzymes but not Na$^+$–K$^+$-dependent ATPase, and (2) the basolateral membrane containing the ATPase but not digestive enzymes. The ATPase is the enzyme responsible for the continuous pumping of sodium ions from the cell into the interstitial fluid in exchange for the intake of potassium ions, using ATP (adenosine triphosphate) as the metabolic source of energy. The brush-border membrane is wider and has a higher protein-to-lipid ratio than the basolateral membrane. The latter membrane contains occasional gaps which are large enough to permit passage of chylomicrons and other lipoproteins from the epithelial intercellular space to the lamina propria. The morphological and functional differences between the two types of membrane are important in the active transport of sugars, amino acids and certain water-soluble vitamins. Enterocytes take up nutrients not only from the lumen across the brush-border membrane, but also from the bloodstream across the basolateral membrane.

The thin walls of the blood capillaries in the villus are composed of endothelial cells that contain a number of structurally recognizable pathways. These include circular openings (fenestrations), some of which are provided with an aperture or diaphragm. The fenestrations tend to face the intestinal lumen and facilitate the rapid uptake of nonlipid nutrients which have traversed the villus epithelium.

The walls of the lacteals are composed of endothelial cells that have no fenestrations. However, gaps of sufficient size to permit passage of chylomicrons are present between some adjacent endothelial cells.

1.2.4 Movement of solutes across cell membranes

The absorption of water-soluble vitamins involves movement of the vitamin molecules from the intestinal lumen across the brush-border membrane into the cytoplasm of the enterocyte, and exit from the cytoplasm across the basolateral membrane into the serosa. The principal mechanisms involved will be defined to avoid ambiguity.

The term 'transport' as used in this text refers to the mediated transfer of a solute across a biological membrane. It excludes simple diffusion, which is nonmediated. Mediated transfer involves mobile carrier proteins that carry solutes from one side of the membrane to the other. Specificity is imparted by the tertiary and quaternary structures of the protein molecule. Only if a solute's spatial configuration fits into the protein carrier will the carrier convey it across the membrane. At physiological concentrations, the movement of several vitamins (thiamin, riboflavin, pantothenic acid, biotin and vitamin C) across the membranes takes place by carrier-mediated transport.

The interaction of a solute with its carrier is characterized by saturation at high solute concentrations, stereospecificity and competition with structural analogues. These properties are shared by the interaction of a substrate and an enzyme and, because of these similarities, the carrier may be considered as a specialized membrane enzyme. The relationship between substrate concentration and the rate of an enzyme-catalysed reaction can be used to describe the interaction of a substrate with its carrier. In Figure 1.5 V_{max} represents the maximum velocity attained when the substrate concentration is increased to a very high level. V_{max} is proportional both to the total enzyme concentration and to the velocity with which the enzyme catalyses transformation of substrate molecules once it has combined with them. The Michaelis constant, K_m, can be defined as the substrate concentration required to elicit half the maximum velocity. K_m therefore describes the affinity of the substrate for its

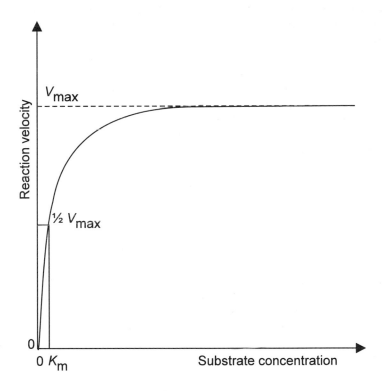

Figure 1.5 Relationship between substrate concentration and the rate of an enzyme-catalysed reaction.

binding site on the carrier ('enzyme'). When two substrates are transported on the same carrier, the observed K_m for a particular substrate will be an 'apparent K_m' and will depend on the concentration of the second substrate (Kimmich, 1981).

The movement of solutes across cell membranes can take place by simple diffusion, facilitated diffusion and various types of active transport.

Simple diffusion

This is due to the random movement of molecules that ultimately results in their even distribution. Molecules move along a concentration gradient from a higher concentration to a lower concentration until a state of equilibrium is reached. The rate of diffusion depends on the size and concentration of the solutes, and on temperature and pressure. In the case of niacin and vitamin B_6, a favourable gradient for cellular uptake is maintained by intracellular metabolism of the vitamin to a nontransportable form. Simple diffusion exhibits linearity between solute concentration and rate of transport (i.e. no saturation kinetics), no dependence on metabolic energy, no evidence of stereospecificity, and no competition with structural analogues.

Facilitated diffusion

This is a carrier-mediated transport mechanism in which the energy is provided by the concentration gradient that exists between the inside and the outside of the cell. As the name implies, the carrier facilitates the diffusion of the solute to the other side of the membrane. No metabolic energy from the cell is required. Solutes can only be moved from a region of higher concentration to one of a lower concentration, and therefore facilitated diffusion does not lead to concentrative accumulation of the solute.

Active transport

When the net movement of a solute across a membrane takes place against a concentration gradient (uphill transport) it is called active transport, and this requires a source of metabolic energy from the cell. The energy can be supplied directly by the progress of a cellular chemical reaction, in which case the transport system is called primary active transport. Alternatively, the energy source can be supplied indirectly through a coupling to the flux of some other solute down its electrochemical gradient; in this case the transport system is called secondary active transport.

1.2.5 Experimental techniques for studying absorption and transmembrane transport mechanisms

Much of the knowledge of the process of intestinal absorption is based on experimental data obtained using everted intestinal sacs, intestinal perfusion techniques and intestinal loops. Information about transmembrane transport has been largely obtained using two experimental approaches: (1) isolated or cultured intact enterocytes, and (2) membrane vesicles prepared from either the mucosal or serosal aspect of the intestinal epithelium.

Everted intestinal sacs

This *in vitro* technique, introduced by Wilson and Wiseman (1954), provides information about the incorporation of the test solute into the intestinal tissue. The preparation consists of a short length of rat intestine turned inside out so that the epithelial brush border is on the outer surface. The sac is filled with a physiological solution, sufficient in volume to distend the wall, and an oxygen bubble is introduced. The sac is tied at both ends and suspended in a well-oxygenated incubation medium maintained at 37 °C. The radioactively labelled solute under investigation is placed in the large volume of incubation medium and the solute taken up is measured both in the cells and in the physiological solution within the sac. The preparation should remain viable for at least 10 minutes, allowing both net absorption and tissue uptake to be studied under physiological conditions. A number of adjacent portions of intestine from the same animal can be studied simultaneously.

Intestinal perfusion techniques

These *in vivo* techniques are based on the principle of luminal disappearance of the test solute. They allow measurement of absorption only under steady state conditions.

Studies with experimental animals

In this method a short length (segment) of the intestine of an experimental animal (usually the rat) is perfused *in situ* without interrupting the blood or lymph supply. Absorption studies therefore take into account normal hormonal and nervous endogenous factors as well as normal blood and lymphatic circulations.

   In a procedure described by Said, Hollander and Katz (1984), rats are anaesthetized with ether and the abdomen is opened by a midline incision. An inflow catheter is inserted into the proximal end of the jejunum and an outflow catheter is introduced 50 cm distally. The intestinal segment is flushed with phosphate buffer to remove any residual intestinal contents and then flushed with air. The segment is replaced in the peritoneal cavity and the abdomen is surgically closed. The rat is allowed to awaken and is placed in a restraining cage. The test solutions containing the labelled solute of interest (a ^{14}C-labelled folate compound) and ^{3}H-labelled inulin as a nonabsorbable marker are infused via the inflow catheter using a perfusion pump. After a period of equilibration, the outflow is collected over a specified period of time. Aliquots of the effluent are analysed for radioactivity by double isotope counting and calculating technique. The rate of solute absorbed into the circulation is calculated from its rate of disappearance from the perfusate. Absorption results are expressed in terms of dry intestinal length after correction for changes in volume indicated by the nonabsorbable marker concentration.

Clinical studies

A technique described by Bailey *et al.* (1984) uses a triple-lumen perfusion tube which is inserted through the mouth of a human subject into a specific segment of the intestine. The tube is constructed by cementing together three tubes, 2 mm in diameter, one of which is shorter than the other two. After an overnight fast, the perfusion tube is swallowed by the subject and the tube is correctly positioned with the aid of a fluoroscope. The radioactive test substance along with a nonabsorbable marker (polyethylene glycol) is introduced into the proximal portion of the intestinal segment through the shorter 2 mm tube. Intestinal fluid is then aspirated from the distal end of the longer segment and analysed for radioactivity and the nonabsorbable marker.

Intestinal loops

This *in vivo* technique requires the opening of the abdomen of an experimental animal under anaesthesia and isolating several equal lengths of intestine by ligatures without obstruction of major blood vessels. The test solute along with a nonabsorbable marker is introduced into each segment and after the predetermined time the segments are detached. The luminal content of each segment is collected for analysis of the test solute remaining and the nonabsorbable marker.

Intestinal loops allow determination of initial rates of disappearance, which is useful in kinetic studies. One disadvantage of the method lies in the small volumes (particularly in rats) of luminal solutions that are injected into the segments (0.1 ml/cm segment of rat intestine). This small volume does not allow for control of pH and the initial concentration falls rapidly, complicating long-term studies (Selhub, Dhar and Rosenberg, 1983).

Isolated or cultured cell preparations

These preparations allow determination of initial rates of solute uptake and provide information on the intracellular accumulation and metabolic fate of the solute. Cell culture systems appear to offer significant advantages regarding heterogeneity of cell populations, cell viability and other factors compared with freshly isolated cells (Bowman, McCormick and Rosenberg, 1989).

Membrane vesicles

Membrane vesicles allow solute transport properties to be investigated independently in the brush-border and basolateral membranes without the complications of cellular metabolism or intracellular compartmentation. Brush-border membrane vesicles can be prepared by homogenizing isolated enterocytes or mucosal scrapings and precipitating non-brush-border particulate matter with $10\,mM\ Ca^{2+}$ or Mg^{2+}. The brush borders, because of their greater density of negative surface charges, are not precipitated and remain in the supernatant fluid. Centrifugation of this fluid provides a pellet of membrane vesicles, 95% of which have the same membrane orientation as in the intact enterocyte (i.e. they are 'the right side out'). Basolateral membrane vesicles can be obtained from a source of isolated enterocytes by differential centrifugation followed by separation on a density gradient. Vesicles of brush-border and basolateral membranes can be prepared simultaneously using the technique of free-flow electrophoresis to isolate them (Murer and Kinne, 1980).

1.3 DIGESTION, ABSORPTION AND TRANSPORT OF DIETARY FAT – A VEHICLE FOR THE ABSORPTION OF FAT-SOLUBLE VITAMINS

Absorption of the fat-soluble vitamins takes place mainly in the proximal jejunum and depends on the proper functioning of the digestion and absorption of dietary fat. The efficiency of absorption of fat-soluble vitamins parallels that of fat absorption and is affected by the nature of the lipid component of the diet.

The fat content of a typical Western diet is composed mainly of triglycerides (storage lipids) accompanied by smaller amounts of phospholipids (membrane lipids) and sterols. The diet contributes 1–2 g of phospholipid each day, mainly as phosphatidylcholine, but 10 times this amount enters the small intestinal lumen in the bile. The major sterol in the Western diet is cholesterol, which is predominantly, but not exclusively, of animal origin. Most dietary cholesterol is present as the free sterol, with only 10–15% as cholesteryl ester. Plant sterols, such as β-sitosterol, account for 20–25% of total dietary sterol (Tso, 1994).

The digestion of triglycerides begins in the stomach and involves different lipases secreted by the salivary glands and the gastric mucosa. The stomach is also the major site for emulsification of dietary fat. The coarse lipid emulsion, on entering the duodenum, is emulsified into smaller globules by the detergent action of bile salts aided by the churning action of the intestine. The adsorption of bile salts on to the surface of the fat globules increases the lipolytic activity of pancreatic lipases which hydrolyse triglycerides to β-monoglycerides and free fatty acids. A co-factor called colipase is present in pancreatic juice and is required for lipase activity when bile salt is present. During its detergent action, bile salts exist in a monomolecular solution. Above a critical concentration of bile salts, the bile constituents (bile salts, phospholipid and cholesterol) form aggregates called micelles, in which the polar ends of the molecules are orientated towards the surface and the nonpolar portion forms the interior. The β-monoglycerides and free fatty acids are sufficiently polar to combine with the micelles to form mixed micelles. These are stable water-soluble structures which can dissolve fat-soluble vitamins and other hydrophobic compounds in their oily interior.

In the jejunum the mixed micelles are able to convey the fat-soluble vitamins across the mucin layer to the brush border of the enterocyte, where they are absorbed along with other lipids. The bile salts are left behind to be actively reabsorbed in the distal ileum, whence they return to the liver to be recycled via the gall-bladder. Within the enterocyte, triglycerides are resynthesized and packaged, together with fat-soluble vitamins, in the form of chylomicrons. These structures are composed of lipid droplets coated with a stabilizing layer of phospholipid and cholesterol and, in turn, with a specific protein. The chylomicrons pass out of the basal membrane of the enterocyte and enter the central lacteal of the villus. From there they pass into the larger lymphatic channels draining the intestine, into the thoracic duct, and ultimately into the systemic circulation.

On entering the bloodstream, the chylomicrons are attacked by lipoprotein lipase, which hydrolyses the triglycerides they carry. At the same time, there is a selective exchange of lipids and apolipoproteins with high density lipoproteins (HDL). Within a few minutes the large triglyceride-rich chylomicrons are thus converted into much smaller, triglyceride-depleted lipoproteins, the chylomicron remnants. Chylomicron remnants contain apolipoprotein E that they acquire both from HDL in circulation and from hepatocytes. Apolipoprotein E mediates the binding of chylomicron remnants to lipoprotein receptors embedded in the cell membranes of the liver or other tissues and the remnants are taken up into these cells by pinocytosis.

Formation of chylomicrons depends on the *de novo* synthesis of certain apolipoproteins in the small intestine. Treatment of experimental animals with inhibitors of protein synthesis such as cycloheximide or puromycin markedly decreases triglyceride absorption; the enterocytes become packed with triglyceride, which cannot be released into lymph as chylomicrons (Sabesin and Isselbacher, 1965). The human genetic disorder abetalipoproteinaemia is characterized by steatorrhoea and a histological picture of lipid-filled enterocytes, identical to that seen in cycloheximide-treated animals (Glickman *et al.*, 1979). It is commonly believed that abetalipoproteinaemia is caused by the inability to synthesize apolipoprotein B. However, this lipoprotein has been shown to be synthesized by the enterocytes of abetalipoproteinaemic patients and, furthermore, the apolipoprotein B gene is not abnormal in such patients (Tso, 1994). Thus the aetiology of abetalipoproteinaemia may be a defect in the association of intracellular lipid with apolipoprotein B, a prerequisite for the normal packaging of chylomicrons. In support of this suggestion, Wetterau *et al.* (1992) demonstrated that the microsomal triglyceride transfer protein, believed to participate in chylomicron formation, is lacking in abetalipoproteinaemic patients.

1.4 TRANSPORT OF GLUCOSE IN THE INTESTINE AND KIDNEY – A MODEL FOR THE ABSORPTION OF WATER-SOLUBLE VITAMINS

The intestinal and renal transport of glucose are discussed here because they have been well studied and the postulated mechanisms have given rise to useful models, which can be applied to studying the transport of certain water-soluble vitamins. The various models for glucose transport are extensions of Crane's sodium-gradient hypothesis for energy coupling (Crane, 1962, 1965). A model based on the mechanisms discussed in this text is illustrated in Figure 1.6.

Intestinal absorption

Soon after eating a meal the high concentration of glucose in the intestinal lumen may allow uptake to take place by simple diffusion. After a while, the luminal concentration of glucose becomes lower than the concentration in the enterocyte or in the intercellular space, and therefore the glucose must be absorbed against a concentration gradient. The uphill influx of glucose across the brush-border membrane is coupled to the passage of sodium ions (Na^+) and occurs by virtue of a Na^+–glucose carrier (cotransporter) located in this membrane. The carrier possesses specific binding sites for both D-glucose and Na^+; there are about 10^6 carriers per enterocyte. D-Galactose, having an almost identical structure to D-glucose, shares the same carrier site, but glucose has a greater affinity for the carrier and is transported at a greater rate. Sodium-dependent glucose transport across the brush-border membrane is rather specifically

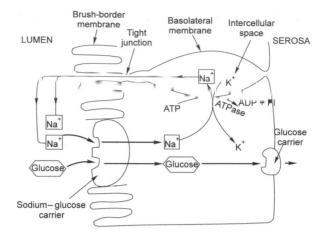

Figure 1.6 Schematic representation of carrier-mediated D-glucose transport in mammalian intestine. Two molecules of Na$^+$ bind per molecule of D-glucose at the brush-border membrane and energy is provided by the basolateral Na$^+$–K$^+$-dependent ATPase. ATP, adenosine triphosphate; ADP, adenosine diphosphate; Pi, inorganic phosphate.

inhibited by phloridzin (phlorizin), a β-glucoside which competes with D-glucose for the binding site on the carrier but is not itself transported.

The membrane potential plays an essential role in the sodium-coupled transport of glucose across the brush-border membrane. Transport occurs as a result of decreasing carrier affinity for glucose between the two surfaces of the membrane. At the outer surface the membrane potential initially promotes binding of Na$^+$ to a site deep within the carrier protein molecule, thereby activating the carrier and maximizing its affinity for glucose. The carrier binds sodium at a stoichiometric ratio of two Na$^+$ ions to a single glucose molecule; thus the free carrier and the loaded carrier bear a net charge of opposite sign. The negative membrane potential drives the cationic Na$^+$–glucose–carrier complex to the inner surface of the membrane where the molecular conformation of the complex alters to a lower energy state and Na$^+$ and glucose are released into the cytoplasm of the enterocyte. The free anionic carrier bearing a single net negative charge is then driven back to the outer surface of the membrane, where it once again becomes activated and initiates another shuttle.

The energy required for the uphill transport of glucose across the brush-border membrane is provided by the electrochemical potential gradient for sodium across this barrier. This energy source has two components: an electrical potential difference of about 40 mV across the brush-border membrane (cell interior negative) and a sodium concentration gradient. Both the electrical and chemical components are established by the continuous extrusion of sodium out of the enterocyte by the action of the basolateral sodium pump. The transport mechanism is therefore secondary active transport because it is indirectly linked to a source of metabolic energy, i.e. the energy derived from the hydrolysis of ATP during the extrusion of sodium. If the sodium pump is blocked by the action of ouabain, a specific inhibitor of Na^+–K^+-dependent ATPase, uphill glucose transport is abolished.

The sodium-coupled entry of glucose and the sodium extrusion mechanism are electrogenic; that is, they both generate electric currents at their respective brush-border and basolateral membranes. The electrogenic entry of sodium results in a marked depolarization of the brush-border membrane, i.e. the cell interior becomes less negative with respect to the luminal surface. Because the intercellular tight junction of the intestinal epithelium has a high conductivity for sodium ions, Na^+ easily equilibrates between the extracellular media on the basolateral and luminal sides of the enterocyte. By virtue of this rapid paracellular pathway for the flow of Na^+, the current generated at one of the limiting membranes can influence the potential difference across the other. This electrical coupling facilitates an electrochemical feedback between the two membranes (Hopfer, 1977).

The downhill efflux of glucose from the cytoplasm into the intercellular space is attributed partly to simple diffusion through the relatively porous basolateral membrane, but mostly to facilitated diffusion through this membrane. The facilitated diffusion mechanism is mediated by a sodium-independent carrier which can transport D-fructose as well as D-glucose and D-galactose. Although the carrier has a low affinity for glucose, its high capacity ensures that this sugar does not accumulate within the enterocyte under physiological conditions.

The role of glucagon (Debnam and Sharp, 1993) and cAMP (adenosine 3′,5′-cyclic monophosphate) (Sharp and Debnam, 1994) in the control of glucose transport across the brush-border and basolateral membranes of rat jejunal enterocytes has been studied. Glucagon, a pancreatic hormone, increases sodium–glucose cotransport across the brush-border membrane, apparently by causing a decrease in sodium permeability of the membrane. This action of glucagon is linked to elevated levels of the 'second messenger' cAMP. A stimulatory action of cAMP on glucose transport across the brush-border and basolateral membranes of isolated enterocytes was demonstrated and appeared to affect the V_{max} of the transport system rather than the K_m (affinity). The experimental results suggested that cAMP is an intracellular mediator of the actions of glucagon on intestinal glucose uptake.

Adaptive regulation

The increased requirement for all nutrients during pregnancy, lactation and other conditions is met primarily by an increase in the absorptive capacity of the small intestine. The enhanced absorption is achieved through an increase in the height of villi and in the number of villi per unit surface area and also through an increase in intestinal length. This nonspecific effect may be mediated by hormones.

Changes in specific transport systems depend on dietary availability or body stores of their substrates. This adaptation can be rapid (24 h or less) and is not associated with morphological adaptations in the intestine. Dietary carbohydrate increases the rate of intestinal glucose transport with an increase in V_{max} but no change in K_m. Starvation reduces the rate of glucose transport with a decrease in V_{max} – again without a change in K_m (Karasov and Diamond, 1983). An increase in V_{max} with no significant change in K_m indicates that the regulatory mechanism involves an increase in the number of carriers, and not an increased affinity of the carrier for its substrate. This effect could result from the synthesis of more carrier proteins in existing enterocytes or the production of new enterocytes with a higher density of carriers. The signal to a transporter system may be the luminal concentration of the transported substrate itself.

Diamond and Karasov (1987) predicted that intestinal vitamin transporters would prove to be suppressed by high intakes of their dietary substrates. The reason behind their prediction was as follows. The transporter is most needed at low dietary substrate levels for any essential trace nutrient that has a fixed daily requirement but that does not contribute significantly to the body's energy needs. At high dietary levels the fixed requirement could be extracted by fewer transporters or even by simple diffusion. Since the synthesis of a transport protein involves biosynthetic energy, transport synthesis should be suppressed if the energy costs exceed the benefits that the transporter provides.

The prediction of suppressed transport at high dietary intakes has proved to be true for ascorbic acid, biotin and thiamin, but not for pantothenic acid, for which transporter activity is independent of dietary levels (Ferraris and Diamond, 1989). These authors noted that the ratio of carrier-mediated uptake to the diffusional (nonsaturable) uptake component at a concentration equal to the transporter's K_m is high for the three regulated vitamins but low for the nonregulated vitamin. It appears that intestinal transporters are regulated only if they make the dominant contribution to uptake.

Renal reabsorption

The role of the kidney in maintaining whole-body glucose homeostasis is to conserve or excrete glucose, and the kidney fulfils a similar role in the body conservation of water-soluble vitamins. The amount of glucose excreted depends on the amount of glucose filtered by the glomerulus and the amount subsequently reabsorbed by the proximal convoluted tubule. The amount of glucose in the glomerular filtrate depends on the plasma glucose concentration and the glomerular filtration rate. As this amount increases, the ability of the renal tubules to reabsorb glucose reaches a maximum (the renal threshold) and the excess glucose that is not reabsorbed is excreted in the urine.

Glucose uptake at the brush-border membrane of the tubule cells takes place by a secondary active transport mechanism that is similar, if not identical, to that described for the intestinal uptake of glucose. There is molecular evidence for the existence of two renal Na^+–glucose cotransporters (Pajor, Hirayama and Wright, 1992). A 'high-affinity, low-capacity' carrier protein with a $2:1$ Na^+/glucose stoichiometry is found in the outer medulla and appears to be identical in amino acid sequence to the intestinal Na^+–glucose cotransporter. The second transporter, a 'low-affinity, high-capacity' carrier protein with a $1:1$ Na^+/glucose stoichiometry, is found in the outer cortex.

Transport of glucose across the basolateral membrane of the tubule cell takes place by sodium-independent facilitated diffusion (Thorens, Lodish and Brown, 1990)

1.5 BIOAVAILABILITY OF VITAMINS

1.5.1 General concepts

The term bioavailability, as applied to vitamins in foods, may be defined as the proportion of the quantity of vitamin ingested that undergoes intestinal absorption and utilization by the body. Utilization encompasses transport of the absorbed vitamin to the tissues, cellular uptake and conversion to a form which can fulfil some biochemical function. The word 'available' is key here – the vitamin may also be metabolized within the cell to a nonfunctional form for subsequent excretion or simply stored within the cell for future use. Leklem (1986) stressed that while a given amount of a vitamin may be absorbed and enter the circulation, tissues may not take up the vitamin because they have reached saturation. In this case, a certain proportion of the circulating vitamin would be excreted and appear unavailable if urinary excretion or tissue accumulation are the criteria being used to assess bioavailability. Thus any definition must be viewed as an operational definition within the context of the method used to determine bioavailability.

The bioavailability of a vitamin should not be confused with the vitamin content of a food. For example, food processing can result in the loss of a labile vitamin, but the bioavailability of the remaining amount of vitamin may or may not be altered. Thus nutrient stability and bioavailability are clearly distinct, but equally important issues.

Many factors influence the absorption of a vitamin from a particular food or diet. Malabsorption may occur in the presence of gastrointestinal disorders or disease. The nutritional status of an individual with respect to a particular vitamin is implicated through adaptive regulation of intestinal vitamin transport. Absorption depends on the chemical form and physical state in which the vitamin exists within the food matrix. These properties may be influenced by the effects of food processing and cooking, particularly in the case of niacin, vitamin B_6 and folate. Vitamins that exist in bound forms in foods exhibit lower efficiencies of digestion and absorption compared with their unbound forms ingested, for example, in tablet form. Certain dietary components can retard or enhance a vitamin's absorption, therefore the composition of the diet is an important consideration. Other ingested substances such as alcohol and drugs may interfere with the physiological mechanisms of absorption. Because bioavailability is influenced by such a diverse range of interacting parameters, the amount of available vitamin in a diet or individual food can vary considerably.

1.5.2 Methods for determining vitamin bioavailability

For the determination of absolute or true bioavailability, it is necessary first to measure the total amount of the vitamin present in the food or diet by means of a microbiological or physicochemical assay. Bioavailability is then calculated by dividing the concentration of available vitamin (as determined by animal or human bioassay) by the total concentration of the vitamin. (Bioassays and the term 'available' are discussed in the following chapter.) Determination of the total vitamin content requires an extraction procedure capable of liberating all bound forms of the vitamin. This, in practice, is very difficult to achieve without destroying the more labile vitamins. Consequently, almost all methods for determining vitamin bioavailability in foods yield a measurement of relative bioavailability. This is the observed response obtained when the animal is fed the test food or diet relative to the response predicted for 100% bioavailability as indicated by a dose–response curve obtained using a pure vitamin standard.

Rats or chicks have been used extensively as experimental animals, but these animal studies are now thought to have relatively little value in predicting bioavailability for humans. This is because of problems such as intestinal synthesis of water-soluble vitamins, coprophagy (faecal recycling) in rats, the need to feed diets in dry form, and some metabolic differences between animals and humans. The main emphasis nowadays in the field of nutrient bioavailability has turned to the use of protocols with human subjects in order to avoid the uncertain relevance of animal models.

The relative bioavailability of the vitamin under investigation is usually calculated using the slope-ratio assay (Littell, Lewis and Henry, 1995). The data obtained after statistical analysis have to be interpreted on the basis of the vitamin's chemistry as well as the physiology and biochemistry of utilization and function. The range of bioavailability values reported for an experimental treatment group can be very wide, making the average value practically meaningless. Because the bioavailability of a nutrient depends on the conditions of the assessment (feeding regimen, diet composition, etc.) the results must be accompanied by the experimental protocol employed.

Methods for assessing vitamin bioavailability have been discussed by Gregory (1988) and some aspects of these methods are briefly discussed here. Results tend to be highly variable and the best estimate is probably obtained by combining the results from several different methods.

Balance studies

In a balance study, human subjects are fed a diet containing a known amount of the test nutrient, and the difference between what is ingested and what is recovered in the faeces represents apparent absorption. No correction is made for endogenous loss which occurs through intestinal secretions and mucosal cell sloughing. The balance method does not give information on the utilization of the nutrient in the body – this can only be implied. The method is not valid for B-group vitamins because microflora in the colon interfere by synthesizing and metabolizing these vitamins. It has, however, been successfully applied to fat-soluble vitamins.

Dose–effect relationships

A widely used approach for determining vitamin bioavailability is to examine the dose–effect relationship based on a physiological response. The response criterion employed should specifically reflect the utilization of the vitamin of interest as, for example, direct measurement of the vitamin or one of its metabolites in blood plasma or tissues, or any functional measurement, such as enzyme activity. The use of multiple dosage levels in both test and reference diets permits statistical analysis by slope-ratio methods.

When assessing plasma response in human subjects, multiple blood samples should be taken for at least 8 h to permit an evaluation of the area under the plasma concentration versus time curve. Such an approach provides greater pharmacokinetic validity in the experimental design and interpretation (Gregory, 1988).

Urinary excretion

Measurement of the urinary excretion of water-soluble vitamins in humans is a frequently used index of the bioavailability of these vitamins. At a given level of ingestion, an increase in urinary excretion of the vitamin in question is indicative of an increase in intestinal absorption which, in turn, indicates an increase in that vitamin's bioavailability. This approach assumes that tissue saturation of the vitamin exists, that vitamin absorbed in excess of need is excreted in the urine, and that no metabolic processes are occurring which lower the need for the vitamin (Yu and Kies, 1993). Studies using chronic administration of a basal diet containing the test food are preferred to single-dose protocols because lower amounts of the test food in each meal are required to elicit a measurable response. The results of extended protocols reflect the actual extent of utilization of a vitamin and are less subject to differences in the rate of absorption than short-term studies (Gregory, 1988).

Use of isotopes

Radioactive isotopes have been used extensively as tracers to study nutrient metabolism in animals and humans. Foods are labelled with the isotope, either intrinsically or extrinsically, and the fate of the labelled nutrient following ingestion is monitored in the body. The nutrient or its metabolites can be specifically determined in blood, tissues, urine and faeces, allowing the detailed study of absorption, distribution, metabolism and excretion.

Intrinsic labelling involves introducing the labelled nutrient into the plant or animal food source during growth and development, resulting in incorporation of label into the tissues. It has the advantage that the labelled nutrient undergoes biological distribution and metabolism similar to that of the endogenous nutrient. Plant foods can be labelled intrinsically by adding the isotopically labelled nutrient to the nutrient solution of hydroponically grown plants. The use of short-term applications of the isotope with growing plants, for instance by stem-injection or foliar application, results in the preferential intrinsic labelling of tissues undergoing rapid growth. In this event, distribution of isotope within the total plant tissue may not represent that of the endogenous nutrient (Ammerman, 1995). Animal tissue may be intrinsically labelled by adding the labelled nutrient to the diet or administering by gavage (Janghorbani and Young, 1982).

Extrinsic labelling involves mixing the isotopically labelled nutrient with the food just before consumption and is far simpler and less expensive than using intrinsically labelled food. Interpretation depends on the assumption that the labelled nutrient equilibrates with all pools of the endogenous nutrient in the food.

Although there is no solid evidence of harmful effects from the ionizing radiation which arises from the doses used, there is justifiable reluctance to administer radioactive isotopes to humans. Stable isotopes provide an alternative and have been used extensively for studying the bioavailability of trace elements. They emit no radiation and can therefore be used safely as tracers in human studies. Furthermore, they do not undergo decay and hence there is no time constraint on the experimental design. Stable isotopes suitable for labelling include deuterium (^{2}H), ^{13}C, ^{15}N and ^{18}O.

Stable isotopes, unlike radioisotopes, are universally present in biological tissues. Therefore, when designing a stable isotope experiment it is necessary to achieve a sufficient isotope enrichment in the food of interest so that the added tracer can be quantified with the necessary degree of precision. Deuterium is generally the isotope of choice in vitamin bioavailability studies because routes of labelling can often be devised for incorporating multiple ^{2}H atoms per molecule, thereby increasing the molecular mass in proportion to the number of atoms incorporated. Determination of the isotopic enrichment is performed by tandem gas chromatography–mass spectrometry with selected-ion monitoring (Bier and Matthews, 1982).

The application of stable isotopes to the study of vitamin bioavailability is at an early stage of development and stable-isotopically labelled vitamin compounds are not commercially available. In addition, several problems need to be overcome. Although the potential exists for food enrichment with labelled vitamins, there are practical limitations in achieving sufficient labelling. The amount of stable isotope required for accurate analysis far exceeds the amount of tracer used in radioisotopic studies. It may add significantly to the total quantity of the vitamin, in which case it can not be considered a tracer.

REFERENCES

Ammerman, C.B. (1995) Methods for estimation of mineral bioavailability. In *Bioavailability of Nutrients for Animals. Amino Acids, Minerals, and Vitamins* (eds C.B. Ammerman, D.H. Baker and A.J. Lewis), Academic Press, New York, pp. 83–94.

Anon. (1989) *Nutritional Enhancement of Food*, IFST Technical Monograph No. 5, Institute of Food Science & Technology (UK), London.

Bailey, L.B., Cerda, J.J. Bloch, B.S. *et al.* (1984) Effect of age on poly- and mono-glutamyl folacin absorption in human subjects. *J. Nutr.*, **114**, 1770–6.

Bier, D.M. and Matthews, D.E. (1982) Stable isotope tracer methods for in vivo investigations. *Fed. Proc.*, **41**, 2679–85.

Blomhoff, R. and Wake, K. (1991) Perisinusoidal stellate cells of the liver: important roles in retinol metabolism and fibrosis. *FASEB, J.*, **5**, 271–7.

Bowman, B.B., McCormick, D.B. and Rosenberg, I.H. (1989) Epithelial transport of water-soluble vitamins. *Annu. Rev. Nutr.*, **9**, 187–99.

Codex Alimentarius Commission (1987) *General Principles for the Addition of Essential Nutrients to Foods, Alinorm 87/26*, Appendix 5, Food and Agriculture Organization, Rome.

Crane, R.K. (1962) Hypothesis for mechanism of intestinal active transport of sugars. *Fed. Proc.*, 21, 891–5

Crane, R.K. (1965) Na⁺ dependent transport in the intestine and other animal tissues. *Fed. Proc.*, **24**, 1000–6.

Debnam, E.S. and Sharp, P.A. (1993) Acute and chronic effects of pancreatic glucagon on sugar transport across the brush-border and basolateral membranes of rat jejunal enterocytes. *Exp. Physiol.*, **78**, 197–207.

Department of Health (1991) *Dietary Reference Values for Food Energy and Nutrients for the United Kingdom*. Report on Health and Social Subjects, No. 41, HM Stationery Office, London.

Diamond, J.M. and Karasov, W.H. (1987) Adaptive regulation of intestinal nutrient transporters. *Proc. Natl Acad. Sci. USA*, **84**, 2242–5.

Ferraris, R.P. and Diamond, J.M. (1989) Specific regulation of intestinal nutrient transporters by their dietary substrates. *Ann. Rev. Physiol.*, **51**, 125–41.

Fordtran, J.S. and Locklear, T.W. (1966) Ionic constituents and osmolality of gastric and small-intestinal fluids after eating. *Am. J. Dig. Dis.*, **11**, 503–21.

Glickman, R.M., Green, P.H.R., Lees, R.S. *et al.* (1979) Immunofluorescence studies of apolipoprotein B in intestinal mucosa – absence in abetalipoproteinemia. *Gastroenterology*, **76**, 288–92.

Gregory, J.F. III (1988) Recent developments in methods for the assessment of vitamin bioavailability. *Food Technol.*, **42**(10), 230, 233, 235, 237–8.

Herbert, V. (1990) Development of human folate deficiency. In *Folic Acid Metabolism in Health and Disease* (eds M.F. Picciano, E.L.R. Stokstad and J.F. Gregory III), Wiley-Liss, Inc., New York, pp. 195–210.

Herbert, V. and Das, K.C. (1994) Folic acid and vitamin B₁₂. In *Modern Nutrition in Health and Disease*, 8th edn, Vol. 1 (eds M.E. Shils, J.A. Olson and M. Shike), Lea & Febiger, Philadelphia, pp. 402–25.

Hopfer, U. (1977) Isolated membrane vesicles as tools for analysis of epithelial transport. *Am. J. Physiol.*, **233**, E445–9.

Ito, S. (1969) Structure and function of the glycocalyx. *Fed. Proc.*, **28**, 12–25.

Janghorbani, M. and Young, V.R. (1982) Advances in the use of stable isotopes of minerals in human studies. *Fed. Proc.*, **41**, 2702–8.

Karasov, W.H. and Diamond, J.M. (1983) Adaptive regulation of sugar and amino acid transport by vertebrate intestine. *Am. J. Physiol.*, **245**, G443–62.

Kimmich, G.A. (1981) Intestinal absorption of sugar. In *Physiology of the Gastrointestinal Tract*, Vol. 2 (ed. L.R. Johnson), Raven Press, New York, pp. 1035–61.

Leklem, J. (1986) Bioavailability of vitamins: applications to human nutrition. In *Food and Agricultural Research Opportunities to Improve Human Nutrition* (eds A.R. Doberenz, J.A. Milner and B.S. Schweigert), College of Human Resources, University of Delaware, Newark, Delaware, pp. A56–71.

Lieber, C.S. (1988) The influence of alcohol on nutritional status. *Nutr. Rev.*, **46**, 241–54.

Littell, R.C., Lewis, A.J. and Henry, P.R. (1995) Statistical evaluation of bioavailability assays. In *Bioavailability of Nutrients for Animals. Amino Acids, Minerals, and Vitamins* (eds C.B. Ammerman, D.H. Baker and A.J. Lewis), Academic Press, New York, pp. 5–33.

Murer, H. and Kinne, R. (1980) The use of isolated membrane vesicles to study epithelial transport processes. *J. Membrane Biol.*, **55**, 81–95.

National Research Council (1989) *Recommended Dietary Allowances*, 10th edn, National Academy Press, Washington, D.C.

Pajor, A.N., Hirayama, B.A. and Wright, E.M. (1992) Molecular evidence for two renal Na$^+$/glucose cotransporters. *Biochim. Biophys. Acta*, **1106**, 216–20.

Pietrzik, K. (1985) Concept of borderline vitamin deficiencies. *Int. J. Vitam. Nutr. Res.*, Suppl. 27, 61–73.

Sabesin, S.M. and Isselbacher, K.J. (1965) Protein synthesis inhibition mechanism for the production of impaired fat absorption. *Science*, **147**, 1149–51.

Said, H.M., Hollander, D. and Katz, D. (1984) Absorption of 5-methyltetrahydrofolate in rat jejunum with intact blood and lymphatic vessels. *Biochim. Biophys. Acta*, **775**, 402–8.

Selhub, J., Dhar, G.J. and Rosenberg, I.H. (1983) Gastrointestinal absorption of folates and antifolates. *Pharmac. Ther.*, **20**, 397–418.

Sharp, P.A. and Debnam, E.S. (1994) The role of cyclic AMP in the control of sugar transport across the brush-border and basolateral membranes of rat jejunal enterocytes. *Exp. Physiol.*, **79**, 203–14.

Shiau, Y.-F., Fernandez, P., Jackson, M.J. and McMonagle, S. (1985) Mechanisms maintaining a low-pH microclimate in the intestine. *Am. J. Physiol.*, **248**, G608–17.

Southgate, D.A.T. (1982) Definitions and terminology of dietary fiber. In *Dietary Fiber in Health and Disease* (eds G. V. Vahouny and D. Kritchevsky), Plenum Press, New York, pp. 1–7.

Southgate, D.A.T. (1993) Effects of dietary fibre on the bioavailability of nutrients. In *Bioavailability '93. Nutritional, Chemical and Food Processing Implications of Nutrient Availability.* Conference proceedings, Part 1 (ed. U. Schlemmer), Ettlingen, May 9–12, Bundesforschungsanstalt für Ernährung, pp. 128–37.

Thorens, B., Lodish, H.F. and Brown, D. (1990) Differential localization of two glucose transporter isoforms in rat kidney. *Am. J. Physiol.*, **259**, C286–94.

Tso, P. (1994) Intestinal lipid absorption. In *Physiology of the Gastrointestinal Tract*, 3rd edn (ed. L.R. Johnson), Raven Press, New York, pp. 1867–1907.

United States Food and Drug Administration (1987) *Code of Federal Regulations*, Title 21, *Nutritional Quality Guidelines for Foods*, Part 104.5, Food and Drug Administration, Washington, D.C.

Vahouny, G.V. (1982) Dietary fibers and intestinal absorption of lipids. In *Dietary Fiber in Health and Disease* (eds G. V. Vahouny and D. Kritchevsky), Plenum Press, New York, pp. 203–27.

Wetterau, J.R., Aggerbeck, L.P., Bouma, M.E. *et al.* (1992) Absence of microsomal triglyceride transfer protein in individuals with abetalipoproteinemia. *Science*, **258**, 999–1001.

Wilson, T.H. and Wiseman, G. (1954) The use of sacs of everted small intestine for the study of the transference of substances from the mucosal to the serosal surface. *J. Physiol.*, **123**, 116–25.

Yu, B.H. and Kies, C. (1993) Niacin, thiamin, and pantothenic acid bioavailability to humans from maize bran as affected by milling and particle size. *Plant Foods Human Nutr.*, **43**, 87–95.

2

Laboratory procedures and some of the analytical techniques used in vitamin determinations

2.1 ASSESSMENT OF VITAMIN ACTIVITY IN A FOOD COMMODITY OR DIET

2.1.1 Bioassays

In the context of human and animal nutrition, vitamins are physiologically active substances. To assess the nutritional value of a food commodity or diet with respect to a particular vitamin it is necessary to determine the vitamin's biological activity. Biological activity refers to the ability of a nutrient to fulfil a specific biological function or metabolic requirement. The only direct means of determining biological activity is a bioassay based on biological function. Such assays were the means by which the vitamins were originally recognized. They are still required today to measure the biopotency of new vitamin derivatives and food supplements, as well as validating new nonbiological assay methods.

For many bioassay procedures it is not practical or ethical to use human subjects, therefore animal models (mainly rats or chicks) are used instead. Although animal bioassays provide a measure of the available vitamin, they may not extrapolate to human nutrition. Aside from metabolic differences between animals and humans, animal bioassays require the use of vitamin-depleted animals and semi-purified basal rations. Growth assays use only immature animals. In addition, bioassays are subject to ambiguity because of the potential for altered biosynthesis or utilization of vitamins by intestinal microflora. Stimulation of the response of animals consuming test diets could occur if vitamins synthesized by the microflora were utilized, either by direct absorption in the lower intestinal tract or, in the case of rodents, indirectly by coprophagy.

The result obtained from a bioassay should ideally reflect the combined *in vivo* response to all vitamin-active compounds and vitamin precursors present in the test sample in relation to their biological activities. Such a result can only be achieved using a bioassay that measures a true biological response. This may be either the overall growth response or a specific physiological response such as calcification of bone induced by vitamin D. Bioassays based on biological function depend on a series of sequential physiological events: intestinal absorption, plasma transport, tissue uptake, metabolism and, ultimately, biological function at the target cells. They are, therefore, an index of 'available' vitamin – a term which is not synonymous with 'bioavailability'.

Bioassays may be either curative or prophylactic. In a typical bioassay of the curative type, rats or chicks are fed a specially formulated basal diet, which provides for all their nutritional requirements except for the vitamin under assay. When a decline in body weight or the appearance of a specific vitamin deficiency disorder is observed, the animals are divided into treatment groups and fed either the test food sample or a standard preparation of the vitamin incorporated into the basal diet. The extent to which the growth retardation or deficiency disorder has been cured is then estimated, and from this estimate the vitamin content of the test sample can be calculated. In prophylatic bioassays the animals receive the test sample and standard vitamin preparation at the beginning of the assay, and their ability to prevent the deficiency disorder is compared.

Bioassays based on biological function are the most difficult to conduct, as well as being lengthy and expensive. Consequently, alternative bioassays have been developed that are simpler and less expensive. Among these latter assays, elevation of plasma vitamin levels depends on the extent of intestinal absorption and the efficiency of plasma transport. Liver-storage assays are somewhat more specific because they depend on tissue storage of the vitamin. However, neither of these assays accounts for biological function at the cellular level, therefore they do not reflect the true physiological vitamin activity of the test sample. Bioassays of this type have utility if the vitamin in the test sample and the standard possess similar biological activity and are administered to the experimental animal in an identical manner. The results are difficult to interpret correctly if the sample contains several vitamers of varying potency, as is the case with vitamin E.

2.1.2 *In vitro* analytical techniques

Nowadays, most vitamin determinations are performed using microbiological assays or physicochemical methods of analysis. These may be referred to collectively as *in vitro* techniques. Microbiological assays, like animal bioassays, measure the combined response of the active substances, and take into account utilization at the cellular level. However, they do not account for the complexities of mammalian digestion and absorption. In addition, the assay organisms frequently respond to molecular fragments of vitamins that are inactive in mammals. Physicochemical assays permit the quantification of the principal substances that are responsible for the biological activity but, in many cases, they do not account for all of the active substances.

In vitro techniques require prior extraction of the vitamins from the food matrix in order to facilitate their measurement. Often, the extraction process is designed to liberate vitamin which is chemically bound to other dietary constituents in the food matrix, thus providing an assay value representing the total amount of vitamin present. For some vitamins, such as niacin, this approach often leads to a serious overestimation of the amount of biologically available vitamin in the food. In recent years less rigorous, more selective sample extraction procedures have been proposed, which simulate (as far as possible) mammalian digestion, in attempts to estimate the biologically available vitamin content. Assays using such extraction procedures provide a better estimate of the sample's nutritional value than assays using rigorous nonselective extraction procedures.

2.2 OBJECTIVES AND ANALYTICAL APPROACH

The analytical approach depends on what needs to be measured in order to provide the required information to fulfil the objective. For example, the chemist in the food manufacturing industry needs to know how much of the vitamin that was added in fortification is actually present in the finished product, as vitamin losses may be inevitable as a result of processing. In practice, this requirement is comparatively simple, as the vitamin in its parent form will be added as a stabilized preparation that can be readily extracted in a measurable quantity. In routine quality control work the fortified product will be reasonably consistent from batch to batch, and therefore no unforseen problems associated with the food matrix should be encountered once a suitable analytical method has been established. The quality control manager will want a quick and reliable result on a regular basis and with minimum cost per analysis, therefore the method should be simple and rapid with good repeatability (i.e. within-laboratory precision).

The investigator who needs to determine the nutritional value of a food commodity or diet is faced with a more demanding analytical challenge. Most of the vitamins occur naturally in trace amounts and in a variety of chemical forms (vitamers) and some are chemically bound to proteins or carbohydrates in the food matrix. Moreover, the matrix will vary enormously in composition if different diets are being studied.

The general analytical procedure (apart from animal or human bioassays) for determining vitamins in foods can be broken down into six main stages: sampling, extraction, clean-up, measurement, calculation of results and interpretation of data. Other factors to be considered are maintenance of sample integrity, storage of the sample pending analysis and preparation of the sample for analysis.

Taylor (1986) has discussed the role of collaborative interlaboratory studies in the validation of analytical methods and Hamaker (1986) has explained the statistical principles involved.

2.3 LABORATORY FACILITIES

All operations with vitamin solutions and vitamin-containing materials should be carried out in subdued light or in low-actinic amber glassware. A dedicated laboratory fitted with gold fluorescent tubes which exclude radiation wavelengths of less than 500 nm is the ideal. Microbiological manipulations should be performed well away from the sample preparation area, preferably in a separate room of structural simplicity to facilitate cleaning. An essential requirement is a readily available supply of fresh glass-distilled water; deionized water is not suitable for folate and vitamin B_{12} assays (Pearson, 1967). All glassware should be kept exclusively for microbiological assays, and a meticulous washing-up regime is essential to ensure that the glassware is scrupulously clean before use. Several of the water-soluble vitamins are rapidly destroyed by alkali, and glassware should be acid-washed before rinsing to prevent inactivation of vitamins by surface films of alkaline detergents.

2.4 SAMPLING

The International Union of Pure and Applied Chemistry (IUPAC) has prepared a document intended to furnish concepts, terms and definitions in the field of analytical chemistry with the emphasis on sampling (Horwitz, 1990).

All too often the analyst has little influence over either the design or implementation of sampling. Nevertheless, the analyst must have a perception of the uncertainty involved in sampling, since this determines to a large extent the variability that can be tolerated in the analysis. If, for example, the sampling error is more than two-thirds of the total error, then a reduction in analytical error is only of marginal importance.

A major problem encountered in food analysis is the heterogeneity of foods. Pomeranz and Meloan (1978) referred to a macroheterogeneity (among various units of a lot) and microheterogeneity (within various parts of a unit). The latter is especially important in vitamin assays. Thus, thiamin and nicotinic acid are concentrated in the aleurone and scutellum tissues of the wheat kernel, whilst the starchy endosperm is relatively devoid of these vitamins. The uneven distribution of fat in meat makes it necessary to express analytical values on a fat-free basis. Heterogeneity may also be a problem in vitamin-enriched foods such as breakfast cereals, in which the distribution of vitamins applied by spray-drying is frequently far from uniform. It is evident that the 'concentration' of a nutrient is actually a mean value with an associated uncertainty. Therefore, before a food sample can be analysed, consideration must be taken to obtain a representative sample in order to reduce this uncertainty to a minimum. Regardless of the accuracy of the analysis, if the samples are not representative, the data may be worthless.

The objective of sampling from a nutritional standpoint is to ensure that the laboratory sample is representative of the consignment or lot. A single truly representative subsample from the consignment can only be obtained if the consignment is homogeneous. This is rarely the case and it is usually necessary to select a number of individual portions or units (increments) in an attempt to produce a representative sample. This selection is most reliably made using simple random sampling, which is based on mathematical probability and is not synonymous with haphazard sampling (Kratochvil and Taylor, 1981; Garfield, 1989). Using statistical tables of random numbers, any portion or unit of the consignment has an equal chance of being selected. The increments may be either analysed individually and the results averaged, or they may be combined to form a single composite sample. The analysis of individual increments, selected in accordance with a properly designed statistical plan, provides information about between-sample variability (i.e. the homogeneity of the lot) at the expense of additional analytical effort. The number of samples required for estimating the mean and variability of a particular lot depends upon the homogeneity of the lot, the size of the lot, and the risk one is willing to take on reaching a wrong conclusion (Elkins and Dudek, 1985). The less homogeneous the lot and the larger the lot, the more samples are required, but lot homogeneity is more important than lot size in determining the number of samples required. The statistical concepts involved in designing a sampling plan have been discussed by the Association of Vitamin Chemists, Inc. (1966) and summarized by Pomeranz and Meloan (1978).

2.5 EXTRACTION AND CLEAN-UP PROCEDURES USED IN PHYSICOCHEMICAL, MICROBIOLOGICAL AND BIOSPECIFIC METHODS

2.5.1 Extraction procedures

The appropriate method of extraction depends upon the following criteria: the analytical information required; the nature of the food matrix; the form in which the vitamin occurs naturally or is added (different bound forms of vitamins are often found in meat, plant and dairy products); the nature and relative amounts of potentially interfering substances; the stability of the vitamin towards heat and extremes of pH; and the selectivity and specificity of the analytical method to be used.

Acid hydrolysis

Thiamin and riboflavin are effectively liberated from their parent coenzyme molecules by autoclaving food samples at 121 °C and 100 kPa

(15 lb/in^2) steam pressure for 30 min with 0.1 N (0.1 M) hydrochloric acid (HCl). This treatment denatures proteins and hydrolyses starch to soluble sugars. It also liberates sufficient amounts of free fatty acids to cause an interference in microbiological assays. Acid hydrolysis is also used for the extraction of niacin, vitamin B$_6$ and biotin, the hydrolysis conditions depending upon the type of food.

Alkaline hydrolysis

Alkaline hydrolysis (saponification) effectively removes the preponderance of triglycerides from fatty food samples, whilst also releasing the fat-soluble vitamins from lipoprotein complexes and other bound forms. Saponification is conventionally carried out by refluxing the suitably prepared sample with a mixture of ethanol and 60% w/v aqueous potassium hydroxide (KOH) solution in the presence of pyrogallol as an antioxidant for 30 min. The procedure can be used to extract vitamins A, D and E, but it is not expedient for vitamin K, which is rapidly decomposed in alkaline media. The hydrolysis reaction attacks ester linkages, releasing the fatty acids from glycerides and phospholipids, and also from esterified sterols and carotenoids. It also breaks down chlorophylls into small water-soluble fractions and dissolves any gelatin that might have been present in the vitamin premix added to supplemented foods.

The amount of ethanolic KOH solution required depends on the fat content of the sample. A rough guide is to use 5 ml of 60% w/v aqueous KOH and 15 ml of ethanol per 1 g of fat (Reynolds, 1985). A slow stream of nitrogen is introduced into the saponification flask via a side-arm at the start and end of the process. A nitrogen flow is not necessary during the actual refluxing because a blanket of alcohol vapour prevents aerial oxidation during boiling. Rapid cooling after saponification is important. The liberation of the unstable retinol and α-tocopherol from their relatively stable esters during saponification demands protective measures against light and oxygen throughout the subsequent procedure.

The sterols, carotenoids, fat-soluble vitamins, etc., which constitute the unsaponifiable fraction, are extractable from the saponification digest by liquid–liquid extraction using a water-immiscible organic solvent, after adding water to the digest to facilitate the separation of the aqueous and organic phases. The fatty acids, which are precipitated as their potassium salts (soaps), and the glycerol are not extractable under alkaline conditions. Multiple extractions are necessary to ensure a quantitative transference of the vitamin of interest in accordance with partition theory. The combined solvent extracts are dried over anhydrous sodium sulphate and carefully evaporated to dryness under vacuum in a rotary evaporator. During rotary evaporation the flask is rapidly filled with inert solvent vapour, but it is recommended to introduce nitrogen before the end of evaporation. The residue is redissolved in a small volume of a suitable solvent for analysis or further purification.

The nicotinic acid in mature cereal grains is liberated from its chemically bound forms by autoclaving with calcium hydroxide solution

Enzymatic digestion

Enzymatic hydrolysis using lipase is an effective alternative to saponification for removing triglycerides, and has been applied to the extraction of vitamin K, which is alkali-labile. Proteinaceous samples, such as meat, can be digested with papain to dissolve the proteins that have been denatured during previous acid hydrolysis. Starchy foods such as breakfast cereals can be digested with a diastatic enzyme before saponification so as to avoid the formation of lumps (Bui, 1987).

For the determination of thiamin, it is necessary to hydrolyse thiamin phosphate esters with an enzyme preparation containing phosphatase activity. Pantothenic acid and folate, which are unstable towards extremes of pH, can be released from bound forms using selected enzymes.

The commercial enzyme preparations used are crudely purified natural materials and may contain appreciable amounts of vitamins. Blank determinations must be run in parallel with the extracted samples to allow correction for the vitamin content of the enzyme preparation. The enzyme must be added in sufficient quantity to overcome the effects of naturally occurring inhibitors.

Direct solvent extraction

The fat-soluble vitamins can be extracted from the food matrix without chemical change using a solvent system that is capable of effectively penetrating the tissues and breaking lipoprotein bonds.

The Röse-Gottlieb method is particularly suitable for extracting the total fat from milk products and infant formulas. It entails treatment of the reconstituted milk sample with ammonia solution and alcohol in the cold, and extraction with a diethyl ether–petroleum ether mixture. The alcohol precipitates the protein, which dissolves in the ammonia, allowing the fat to be extracted with the mixed ethers. The method is suitable for the extraction of vitamins A and D, but not for vitamins E and K, which are labile under alkaline conditions.

Supercritical fluid extraction

Supercritical fluid extraction (SFE) has been used for many years as a large-scale extraction process in the food industry, the most widely known application being the decaffeination of green coffee beans. The use of supercritical fluids in analytical chemistry began with supercritical fluid chromatography (SFC), which has been practised since the mid-1960s as a subcategory of gas chromatography. Analytical SFE only began to develop separately from SFC in the mid-1980s. Early SFE instruments were add-on features offered with SFC systems, and interfaced directly to the chromatographic column. More recently, the development of SFE independent of SFC has produced a new generation of instrumentation designed for routine analytical sample preparation.

Principle

If a gas is compressed, it will at some pressure liquefy. However, if the gas is held above its critical temperature, while being simultaneously compressed to a pressure exceeding its critical pressure, it will not liquefy, but instead will become what is called a 'supercritical fluid'. Supercritical fluids behave rather like dense gases in that they take the shape of and fill their container. They are also easily compressible at moderate pressures and low temperatures. Their physical properties are somewhere between liquids and gases. They have densities and solvating power similar to liquid organic solvents, but, like gases, have extremely high diffusivities and very low viscosities.

The unique properties of supercritical fluids make them particularly suitable for extracting compounds from solids or semi-solid food samples. Their low viscosity and absence of surface tension allow them to penetrate a matrix very rapidly, and their solvating power enables them to dissolve the solutes. The high diffusion coefficients of solutes in the supercritical fluid media permit rapid mass transfer out of the matrix. The solvating power of a supercritical fluid is directly related to its density, with density a function of pressure. Therefore, stepping up the pressure will increase the solvating power, and this provides the means by which the extraction can be optimized.

Many supercritical fluids are gases at ambient conditions. For carbon dioxide (CO_2), the most widely used extractant for SFE, the critical temperature is 31 °C and the critical pressure is 73 atm (Majors, 1991). The phase diagram for CO_2 (Figure 2.1) shows the critical point at which the two-phase (gas and liquid) system undergoes an abrupt transition to form the single supercritical fluid phase. The comparatively low critical temperature and pressure of supercritical CO_2 makes it a practical solvent as far as instrumental design is concerned. Other desirable characteristics of supercritical CO_2 are that it is nontoxic, nonflammable, noncorrosive, nonreactive, colourless and odourless. Cryogenic-grade CO_2 is not pure enough for most SFE applications, the purity being particularly important when highly sensitive detectors are used in the

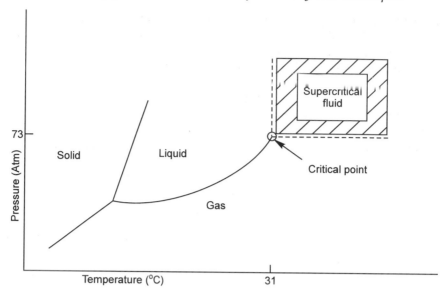

Figure 2.1 Phase diagram for carbon dioxide. CO_2 becomes supercritical at 1050 psi (73 atm) and 31 °C.

analytical step. High-purity SFE-grade CO_2 is commercially available with a helium headspace to increase tank pressure.

Supercritical CO_2 is an excellent extraction medium for nonpolar compounds such as alkanes and terpenes, but unfortunately it does not have sufficient solvating power at typical working pressure (80–600 atm) to quantitatively extract solutes that are quite polar (Hawthorne, 1990). The extraction of more-polar analytes can be achieved by adding to the supercritical CO_2 a small percentage of a polar organic modifier, such as methanol. It is possible to selectively extract classes of compounds from a given sample by varying the pressure or temperature of the supercritical CO_2, by varying the modifier concentration, or by a combination of both.

Instrumentation

In a typical instrumental configuration a high pressure pump is used to deliver the supercritical fluid at a constant controllable pressure to the extraction vessel, which is placed in an oven or heating block to maintain the vessel at a temperature above the critical temperature of the supercritical fluid. During the extraction, the soluble compounds are partitioned from the bulk sample matrix into the supercritical fluid, then swept through a flow restrictor into a collecting or trapping device that is at ambient pressure. The depressurized supercritical fluid (now a gas for CO_2) is vented, and the extracted compounds are retained. This system, in which the supercritical fluid is continuously flowing through the extraction vessel and the analyte is collected continuously, is referred to as dynamic SFE. SFE may also be performed in the static mode by pressurizing the extraction vessel, and extracting the sample with no outflow of the supercritical fluid. After a set period of time, a valve is opened to allow the soluble compounds to be swept into the collection device. In a third method, the recirculating mode, the same fluid is pumped through the sample repeatedly then, after some time, it is pumped to the collection device. Of the three approaches, the dynamic method seems preferable because the supercritical fluid is constantly renewed during the extraction (Hawthorne, 1990).

Either syringe pumps or reciprocating pumps can be used, and both types meet the flow and pressure requirements. Syringe pumps are ideal for delivering unmodified supercritical fluids, their only disadvantage being their limited volume. Modified supercritical fluids are most commonly delivered using two reciprocal pumps, one delivering the supercritical fluid and the other, with proportioning valves, delivering the modifier.

SFE can be employed either as an off-line method, in which a 'stand-alone' extraction instrument is used to collect the sample extract for subsequent analysis, or an on-line method, in which the extraction instrument is coupled directly to an analytical chromatographic instrument. Off-line SFE is inherently simpler to perform and allows the extract to be analysed repeatedly if required.

In early off-line SFE systems, the extracted components were collected in an open vessel containing a few millilitres of a liquid solvent, but this technique results in a loss of trace analytes due to aerosol formation on depressurization, even with cryogenic cooling (Hawthorne, 1990). The analytical trap is a better technique, in which the depressurization takes place across a cryogenically cooled column packed with a solid adsorbent. The extract is concentrated at the column head and the analyte is backflushed with a suitable solvent. This technique serves as an additional experimental parameter for increased selectivity and clean-up.

On-line SFE is usually coupled either with capillary gas chromatography (GC) or SFC; reports of on-line coupling with high-performance liquid chromatography (HPLC) are rare. SFE–SFC offers prospects for the extraction and determination of the fat-soluble vitamins in food. The SFE–SFC coupling has been achieved by flowing the extract through a cold injector loop, by cryogenically trapping the extract upon an adsorbent column for subsequent backflushing on to the SFC column, or by quantitatively transferring the extract directly on to the chromatographic column. The elimination of sample handling between extraction and chromatographic analysis reduces the possibility for loss and degradation of the analyte, thereby improving the precision and accuracy of the determination. There is also the potential to achieve maximum sensitivity by direct transfer of the extract onto the chromatographic column. A major drawback of on line SFE is the danger of introducing bulk amounts of interfering co-extractants into the chromatographic system, which could exceed its analytical capacity and possibly ruin the SFC column.

Applications
Applications for the extraction of fat-soluble vitamins from foods using supercritical CO_2 include the extraction of carotenes from sweet potatoes (Spanos, Chen and Schwartz, 1993) and vegetables (Marsili and Callahan, 1993), tocopherols (vitamin E) from wheat germ (Saito *et al.*, 1989) and phylloquinone (vitamin K) from infant formula (Schneiderman *et al.*, 1988).

2.5.2 Clean-up procedures

The extracts prepared by treatment of the test material may require some form of clean-up before the vitamins can be measured. The extent of the clean-up depends upon the ratio of analyte to interfering substances, and also upon the sensitivity and selectivity of the analytical technique. Colorimetric, titrimetric and spectrophotometric techniques have rather poor specificity and require extensive sample clean-up. Direct fluorimetric techniques are rather more specific, while chromatographic techniques, particularly HPLC, have relatively high specificity. Microbiological assays are subject to interference from many substances, although these can usually be diluted out owing to the high sensitivity of such techniques. It is frequently necessary, however, to remove lipoidal material prior to performing a microbiological assay.

Nonchromatographic techniques

Deproteinization
Treatment of food homogenates or acid hydrolysates with trichloroacetic acid, perchloric acid or other deproteinizing agent in the cold denatures and precipitates the proteins, which can then be removed by filtration or centrifugation. An alternative method, applicable to milk and milk-based products such as yoghurt and infant formulas, is to precipitate the protein by acidification with dilute HCl at room temperature, followed by filtration. The optimum pH range is between 4.0 and 4.6 (the isoelectric point of casein); below pH 4.0 the precipitated particles are too fine to facilitate a rapid and effective filtration (Woodrow, Torrie and Henderson, 1969).

Precipitation of sterols

In fat-soluble vitamin assays, the bulk of the sterols can be removed from the unsaponifiable fraction of the sample by precipitation from a freezing methanolic solution, followed by filtration.

Solvent extraction

Before determining B-group vitamins by microbiological assay, it is often necessary to remove the fat from high-fat samples (e.g. meat, eggs, cheese, wheat germ, maize, soya beans) by means of Soxhlet extraction with either diethyl ether or light petroleum ether. This step also prevents the formation of oily emulsions, which may hinder complete extraction of the vitamin to be determined.

Dialysis

This is a size-exclusion technique which has found useful application in isolating pantothenic acid from proteins following enzymatic digestion of autoclaved food samples (Walsh, Wyse and Hansen, 1979). In this simple process, the food digest–enzyme mixture is transferred with water washings into a short length of semipermeable dialysis tubing, 1 cm in diameter, which is tied securely at both ends and immersed in a relatively large volume (60–150 ml) of cold (5 °C) distilled water. Large molecules (molecular weight > 10 000) are retained, but smaller molecules, including free pantothenic acid, diffuse through the tubing into the more dilute surrounding medium, which is stirred continuously in the cold to maintain the concentration difference. After 8 h, all of the pantothenic acid will have diffused out of the food digest–enzyme mixture, and the water extract can be used for the assay.

Open-column chromatography

Liquid–solid (adsorption) chromatography using gravity-flow glass columns dry-packed with magnesia, alumina or silica gel have been used to purify sample extracts in the determination of vitamins D and K. Such columns enable separations directly comparable with those obtained by thin-layer chromatography to be carried out rapidly on a preparative scale.

Ion exchange chromatography has been utilized to purify sample extracts in HPLC assays for thiamin, niacin, vitamin B_6, folate and vitamin C.

Gel permeation chromatography

High-pressure gel permeation chromatography, employing up to four 300×7.8 mm internal diameter µStyragel 100 Å columns connected in series, has been used in vitamin A, D and E assays to remove trigly-cerides in the analysis of oils and margarine, breakfast cereals and infant formulas by HPLC (Landen, 1980, 1982, 1985; Landen and Eiten-miller, 1979). µStyragel 100 Å is a semi-rigid gel composed of 10 µm particles of polystyrene cross-linked with divinylbenzene and has an average pore size of c.40 angstroms (1 Å $= 10^{-10}$ m) (Vivilecchia et al., 1977).

Solid-phase extraction

Solid-phase extraction, a refinement of open-column chromatography, uses small disposable prepacked cartridges to facilitate rapid clean-up of sample extracts prior to analysis by GC and HPLC (Majors, 1986). The full range of silica-based stationary phases encountered in HPLC column packings is commercially available. The average silica particle size is typically 40 µm (Bond Elut and Bakerbond) or 60 µm (Sep-Pak) and allows easy elution under low pressure. The small mean pore diameter (typically 60 Å) excludes proteins of molecular weight higher than 15 000–20 000, and the large surface area (typically 500–600 m^2/g) confers a high sample-loading capacity. The successive conditioning, loading, washing and elution of solid-phase extraction cartridges are carried out by a step change in solvent strength using the smallest possible volumes of solvent. The cartridges may be operated under positive pressure using a hand-held syringe or under negative pressure using a vacuum manifold. The latter technique is preferable, as multiple samples can be processed simultaneously, and the solvent flow rates can be precisely controlled with the aid of a vacuum gauge.

Purification of the sample extract can be effected in two ways, after first conditioning the sorbent with an appropriate solvent in order to solvate the functional groups of the stationary phase. In the sample clean-up mode, a stationary phase is selected which has a very high affinity for the analyte and little or no affinity for the matrix, therefore the sorbent retains the analyte and unwanted material passes through. After loading the sample, the cartridge is washed with an appropriate solvent to remove further unwanted material, and the analyte is finally eluted with the minimum volume of a solvent that is just strong enough to displace it from the sorbent. This technique provides the opportunity for trace enrichment, in which a large volume of dilute sample is passed through the cartridge, and the enriched sample can be displaced with a small volume of solvent.

In the matrix removal mode, the sample extract is simply passed through the cartridge. Unwanted material will be retained, while the analyte will pass through the sorbent. This strategy is usually chosen when the analyte is present in high concentration.

2.6 GAS CHROMATOGRAPHY

2.6.1 General aspects

Principle

Gas chromatography (GC) is a technique for separating volatile substances by passing a stream of inert gas (mobile phase) over a film of nonvolatile organic liquid (stationary phase) contained within a heated column. The sample extract is introduced at the column inlet, and the volatile components are retained as a result of selective molecular interactions between the components and the stationary phase. Separation is effected as a consequence of the numerous sorption–desorption cycles that take place as the volatile sample material passes through the column. Detection of the separated components is carried out by continuous monitoring of the gaseous column effluent.

Column technology

GC may be carried out using either packed-bed (packed) columns or capillary columns. In packed columns the stationary phase is coated onto microparticles of a porous support material (diatomaceous earth) contained in a glass column. The surfaces of untreated diatomaceous earth are covered with silanol ($\equiv$Si—OH) groups, which function as active sites in promoting severe peak tailing of polar compounds. The diatomaceous earth is therefore deactivated before coating with stationary phases of low or intermediate polarity by treatment with dimethyldichlorosilane (DMCS) or similar silylating reagent. Typical packed column dimensions are 2–4 mm internal diameter (i.d.) and 1.5 m up to 3 m or 4 m length.

In modern capillary columns the stationary phase is chemically bonded to the inner surface of a fused-silica capillary, leaving a lumen extending throughout the length of the column. Separation of sample components is governed solely by the rate of mass transfer in the stationary phase, and the absence of a support leads to an improved inertness. The standard capillary columns are of 0.25 or 0.32 mm i.d. and 10 m up to 50 m or even 100 m in length, with a film thickness of up to 1 μm (typically 0.25 μm). The column is coated on the outside with a protective layer of polyimide resin to impart mechanical strength and flexibility.

Capillary columns provide superior separation efficiencies compared with packed columns, and the sharper peaks facilitate a more accurate integration, as well as a greatly improved detectability. Another feature of capillary columns is a low stationary phase bleed, which is an advantage in temperature-programmed separations. On the other hand, packed columns offer the advantages of high sample capacity, sample ruggedness, ease of operation and low cost.

Capillary columns require special injection systems to cope with the limited sample capacity. Since 1983, wide-bore, thick-film capillary columns have become available which bridge the gap between capillary and packed column GC. The combination of increased i.d. and increased film thickness permits a significant increase in the sample capacity, allowing the use of a simple on-column injector (as used for packed columns) with its inherent ease of operation and quantitative assay. The column i.d. is most commonly 0.53 mm, and cross-linked nonpolar polysiloxane phases can be coated in thicknesses of 1–8 μm in 1 μm increments. The immobilized stationary phase is non-extractable, which allows the column to be flushed with pure solvents to remove contaminants, nonvolatiles and pyrolysis products. A 10 m length wide-bore capillary column approximates the sample capacity and separation efficiency of a 1.8 m × 2 mm i.d. packed column with a 3–5% phase loading at the same operating conditions (Wiedemer, McKinley and Rendl, 1986). However, when the carrier gas flow rate is optimized for the capillary tubing, the wide-bore column produces far superior separations, with an increased speed of analysis. The low pressure drop across the column offers a very practical way to increase the efficiency several-fold, simply by increasing the column length.

Detectors

The detector monitors the gaseous column effluent and measures compositional variations attributable to eluted components. The output signal is produced in response to the mass or concentration of the eluted compound passing through the detector. The linear range refers to the range of solute concentration over which the detector response is linear, from the minimum detectable level (two or three times the noise level) to the upper concentration, which produces a departure from linearity of about 5%. Different compounds can produce signals of varying magnitude for a given concentration, so the signal observed needs to be corrected by applying a response factor to obtain accurate ratios of components in a mixture. For a mass-sensitive detector, such as the flame ionization detector, the response factor is expressed as the ratio of the peak area to the mass of solute injected. Response factors are frequently determined using an internal standard as a marker. A known amount of the internal standard is added to a standard mixture of all the components of interest, and the response factors relative to the internal standard are calculated. The same amount of internal standard is added to each sample, and the peak areas are corrected by dividing the areas by the response factors.

Three types of GC detector employed for water-soluble vitamin determinations are ionization detectors; they are the flame ionization detector, the nitrogen–phosphorus detector and the electron capture detector. A flame photometric detector has also been employed. Interfacing a gas chromatograph with a mass spectrometer (GC–MS) provides information on the molecular weight and structure of the compound, in addition to its quantitative detection.

Derivatization techniques

A prerequisite for a sample to be analysed by GC is that the sample components are sufficiently volatile without decomposing under the conditions of separation. For the water-soluble vitamins the required volatility and stability are achieved either by making a suitable degradation product or preparing a chemical derivative by reaction with a suitable reagent. The most widely used derivatization technique is silylation, whereby a trimethylsilyl (TMS) group is introduced into a wide variety of organic compounds containing —OH, —COOH, —SH, —NH$_2$ and =NH groups. The replacement of active hydrogen by the silyl group reduces the polarity of the compound and decreases the possibility of hydrogen bonding, with the resultant increase in volatility. Furthermore, stability is enhanced by reduction in the number of reactive sites. The trifluoroacetate (TFA) derivative offers the advantage of compatibility with the highly sensitive electron capture detector (Ahuja, 1976).

2.6.2 Applications

Fat-soluble vitamins

Vitamin E
During the 1960s, GC, using packed columns, was widely applied to the determination of vitamin E in foods (Ball, 1988). Sample saponification was necessary, followed in most cases by further purification to remove interfering sterols. The introduction of HPLC in the 1970s led to the demise of GC applications, but the development of fused silica capillary columns has promoted a revival. Capillary columns facilitate the separation of the trimethylsilyl ethers of all eight vitamin E vitamers, and enable the tocopherols and major plant sterols in margarines to be determined simultaneously (Slover *et al.*, 1985). Marks (1988) was able to determine the tocopherols in deodorizer distillate (a product obtained mainly from the processing of crude soybean oil) without the need for saponification. The advantages of using a capillary column over a packed column for this application are summarized in Table 2.1.

A major problem encountered in the gas chromatographic analysis of animal-derived foods is cholesterol, which accompanies vitamin E in the

Table 2.1 Summary of advantages of using capillary GC for the assay of deodorizer distillate vs packed-column GC (Reproduced with permission from Marks, 1988, © *J. Am. Oil Chem. Soc.*, **65**(12), 10, 36–9, American Oil Chemist's Society.)

Item	Capillary	Packed column
Sample preparation	5–10 min	2–3 h
Chromatography	32 min, temp programmed	40 min, isothermal
Resolution	98% resolution of β and γ tocopherol isomers	No resolution between β and γ isomers
Interferences	None	Possible interferences with all tocopherols and internal standard
Accuracy	99%+	Unknown
Precision	±1.1% RSD[a]	±3.2% RSD[a]
Total assay time	c.40 min	c.4 h

[a] Relative standard deviation.

unsaponifiable fraction. Even using capillary columns, the TMS ethers of cholesterol and α-tocopherol are poorly resolved from one another. However, this separation problem can be overcome by forming the heptafluorobutyryl esters (Ulberth, 1991).

In a method developed by Kmostak and Kurtz (1993) for determining supplemental vitamin E acetate in animal feed premixes, 1 g samples were heated at 70 °C for 10 min with 1 ml of a solution of 1% acetic acid in ether and 10 ml of hexane to break down the premix matrix. The extraction mixture was shaken for 20 min and then centrifuged, after which the hexane layer was passed through 2 g of deactivated alumina for clean-up. After two further extractions with hexane, the combined hexane fractions were evaporated just to dryness under a gentle stream of nitrogen. The residue was taken up with internal standard solution (squalene in hexane) and injected onto a 12.5 m CPSil 5CB capillary column. In this method the vitamin E was not derivatized.

Water-soluble vitamins

Thiamin
Attempts to chromatograph TMS and other derivatives of thiamin have been unsuccessful because the derivatives are not sufficiently volatile at normal gas chromatographic operating temperatures, whereas they pyrolyse above 250 °C (Velíšek and Davídek, 1986). The ability of sulphite treatment to quantitatively split the thiamin molecule into substituted pyrimidine and thiazole moieties has been utilized in GC methods for determining thiamin in food samples (Echols *et al.*, 1983; Echols, Miller and Thompson, 1985; Echols, Miller and Foster, 1986; Velíšek *et al.*, 1986). The

thiazole compound possesses the necessary attributes of being soluble in organic solvents and sufficiently volatile to permit its direct determination by GC. The general analytical procedure entails acid and enzyme hydrolysis of the sample, cleavage of thiamin by bisulphite, extraction of the thiazole compound by chloroform, and GC. The flame ionization detector provides adequate sensitivity for the analysis of food samples containing relatively high concentrations of thiamin: the nitrogen– phosphorus detector is about 1000 times more sensitive and is more selective.

Niacin

At least one GC method has been reported (Tanaka *et al.*, 1989) for determining nicotinamide added illegally to fresh meat to stabilize the red colour. There appear to be no GC applications for determining naturally occurring niacin in foods.

Vitamin B_6

A GC method with electron capture detection has been reported for determining the trifluoroacetate (TFA) derivatives of vitamin B_6 vitamers and applied to hydrolysed extracts of enriched white bread, nonfat dry milk and peas (Lim *et al.*, 1982).

Pantothenic acid

A technique for determining pantothenic acid is to chromatograph the pantoyl lactone formed from pantothenic acid by acid hydrolysis (Tarli, Benocci and Neri, 1971). This approach is applicable to foodstuffs because the hydrolysis reaction liberates the lactone from the free and bound pantothenic acid in the food matrix with a recovery of $> 95\%$ (Velíšek and Davídek, 1986). Davídek *et al.* (1985) applied the technique to the analysis of fresh beef liver, spray-dried egg yolk, soybean flour, whole-grain wheat flour and dried bakers' yeast. On comparison of the GC results with results obtained by the currently accepted microbiological method, no significant difference was found at the 95% probability level, and the correlation coefficient was $r = 0.975$. The GC method was considered to be more reliable than the microbiological method, as it was more rapid and simpler and showed higher reproducibility.

2.7 SUPERCRITICAL FLUID CHROMATOGRAPHY

2.7.1 General aspects

Principle

Supercritical fluid chromatography (SFC) combines some of the best properties of GC and HPLC, thus creating a new and complementary technique. SFC has greater resolving power than HPLC because the high diffusivities of supercritical fluids relative to liquids improves the analyte mass transfer between the mobile and stationary phase. Also, the low gas-like viscosities of supercritical fluids result in lower pressure drops across chromatographic columns than observed in HPLC, thus allowing faster mobile phase flow rates than is possible with liquids. The lower pressure drop also allows longer columns to be used in SFC than in HPLC, giving higher theoretical-plate numbers. The resolving power of SFC is not as good as that of capillary GC. However, whereas the carrier gas in GC has no effect upon selectivity, the mobile phase in SFC has the solvating power of a liquid, so the different solubilities of individual sample components provide selectivity. Another advantage of SFC over GC is the low operating temperature when supercritical CO_2 is used, which permits the analysis of thermolabile compounds. Increasing the pressure with time (pressure programming) increases the solvating power of the supercritical fluid thereby causing the elution of compounds of increasingly higher molecular weight, analogous to temperature programming in GC. In summary, the supercritical fluid viscosity and diffusion properties allow fast efficient analytical separations to be performed (as in the case of GC) but the solvating ability provides selectivity (as in HPLC).

Instrumentation

A capillary SFC system consists of three basic components: a high-pressure pump that delivers the mobile phase under supercritical pressures; a precision oven equipped with injector, column and detector; and a microprocessor-based control system. The required low flow rate (several microlitres per minute) and the need for a pulseless fluid supply make syringe pumps the best choice. In the injector there is a pneumatically driven internal sample loop and a four-port rotary microvalve for introducing the sample. An adaptor facilitates split injections, and fast valve switching is used in the nonsplit mode. Microprocessor control and a high-pressure transducer combine to provide accurate mobile phase pressure delivery from 1 to 400 atm/min and pressure ramping.

The standard detector is the flame ionization detector – the universal detector most commonly used in GC – but selective GC detectors can also be installed. The UV/visible and fluorescence spectrophotometric detectors commonly employed in HPLC are also compatible with SFC after modification for high pressure and high temperature operation. One of the major advantages of SFC over HPLC is that interfacing with mass spectrometry is much more straightforward. SFC can also be interfaced with Fourier transform infrared spectroscopy (FTIR) detection, which, like mass spectrometry, provides direct structural information on a separated compound. Carbon dioxide has intense infra-red absorptions which severely limit the range over which spectroscopic data can be collected, but supercritical xenon is an ideal mobile phase for on-line SFC–FTIR (Healey, Jenkins and Poliakoff, 1989).

Modern SFC instruments facilitate changing the density of the super-critical fluid by programmed pressure ramping, and so have adjustable solvating power. By selection of appropriate columns and pressure ramp-ing, which may be viewed as somewhat analogous to gradient elution in HPLC, it is possible to separate complex mixtures at moderate tempera-tures using pure supercritical CO_2 as the mobile phase. In order to elute highly polar solutes, it may be necessary to add a polar modifier such as methanol to the supercritical CO_2.

The column

Both open tubular capillary and packed columns have been used exten-sively in SFC. Capillary columns are constructed of highly deactivated fused silica with internal diameters of 100 μm or less, and are coated internally with an immobilized polysiloxane-based stationary phase. Modern packed columns contain microparticulate silica, which may be covered with a bonded stationary phase such as an octadecyl (C_{18}) hydrocarbon.

Capillary columns, being highly permeable, exhibit very little pressure drop per unit length, allowing long column lengths to be used to produce high efficiency separations, and exploiting the full potential of pressure ramping. They generally produce sharper chromatographic peaks com-pared with packed columns, resulting in a higher detector sensitivity. Capillary columns are usually used for the analysis of complex mixtures, but they suffer from a low sample capacity, which may hinder the quantification of trace components.

Compared with capillary columns, standard packed columns provide faster analyses and a higher sample capacity, but the pressure drop per unit length can be great owing to their low permeability. This causes excessive differences in fluid densities along the column, and hence substantial variations of solute retention, since mobile phase solvating power is directly proportional to density. As a result, significant losses of resolution may be observed.

Micropacked columns containing large spherical (average diameter $= 113$ μm) porous siliceous particles coated with stationary phase have a low packing density and are therefore more permeable than standard packed SFC columns. Micropacked columns permit analyses with accept-able efficiency, resolution and analysis time, but the pressure drop is similar to that of a capillary column (Ibáñez, Herraiz and Reglero, 1993; Ibáñez *et al.*, 1993).

2.7.2 Applications

Fat-soluble vitamins

White *et al.* (1988) compared the utility of two bonded phase capillary columns, DB-5 and DB-WAX (Carbowax 20M), for separating a group of seven fat-soluble vitamins.

Snyder, Taylor and King (1993) characterized the tocopherol (vitamin E) content of a commercial antioxidant formulation and a deodorizer distillate using capillary SFC without the need for sample pretreatment. The separation and quantification of α-, β-, γ- and δ-tocopherols was achieved using an SB-Octyl-50 (Dionex) column. Common plant sterols eluted after the tocopherols (Figure 2.2).

The advantages of using two in-line coupled micropacked SFC columns loaded with different stationary phases (SE-54 and Carbowax 20) for fat-soluble vitamin analysis was evaluated and some experimental variables were optimized (Ibáñez *et al.*, 1995).

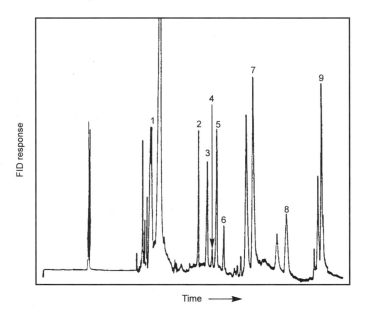

Figure 2.2 SFC of the tocopherols present in a deodorizer distillate. Peaks: (1) fatty acids; (2) squalene; (3) δ-tocopherol; (4) β-tocopherol; (5) γ-tocopherol; (6) α-tocopherol; (7) phytosterols; (8) diglycerides; (9) triglycerides. Capillary column, 10 m × 100 μm i.d. SB-Octyl-50 (Dionex); carrier fluid, SFC-grade carbon dioxide; conditions, density/pressure programme and inverse temperature programme; detector, flame ionization; injection volume, 200 nl. (Reproduced with permission from Snyder, Taylor and King, 1993 © *J. Am. Oil. Chem. Soc.*, **70** (4), 349–354, American Oil Chemist's Society.)

2.8 HIGH-PERFORMANCE LIQUID CHROMATOGRAPHY

2.8.1 General aspects

Principle

In HPLC, a small volume (typically 10–100 µl) of the suitably prepared sample extract is applied to a column packed with a microparticulate material, whose surfaces constitute the stationary phase, and the sample components are eluted under high pressure with a liquid mobile phase. Detection of the separated components is achieved by continuously monitoring the UV absorption of the column effluent. Fluorimetric and electrochemical detection provide improved selectivity and sensitivity for certain vitamins. HPLC allows hundreds of individual separations to be carried out on a given column with high speed, efficiency and reproducibility.

The column

The column efficiency is influenced by the particle size, particle size distribution, packing density and column dimensions. Well packed columns with small, uniformly sized particles provide the highest efficiencies. The majority of HPLC methods for determining vitamins published to date utilize 5 or 10 µm particles of silica or derivatized silica packed into stainless steel tubes having a typical length of 25 cm and a standard i.d. of 4.6 mm. Radially compressed 10 cm × 8 mm i.d. cartridge-type columns (Waters Chromatography Division) manufactured from heavy-wall polyethylene have also found application. The insertion of a short guard column between the injector and analytical column protects the latter against loss of efficiency caused by strongly retained sample components, and from pump or valve wear particles. The guard column is usually packed with the same material as the analytical column, and should not increase band spreading by more than 5–10%. The column packing material is held in the column by fine-porosity frits made from stainless steel or some other material. Rabel (1980) has discussed the care and maintenance of HPLC columns. Membrane filtration of all test extracts is important for the removal of particulate material or high molecular weight compounds which might otherwise enter the guard or analytical column.

The optimum column i.d. depends upon the extent of band broadening that takes place outside the column (i.e. in the injector, connecting tubing and detector flow cell), and electronic broadening within the detector or on transfer to the recording system (Rabel, 1985). In discussing band broadening and electronic broadening, the general term 'dispersion' is used. Improved sensitivity and lower solvent consumption can be achieved by reducing the column i.d., and microbore columns of 2.1 mm i.d. or less are available for use in specifically designed systems having an extremely low extracolumn dispersion. Microbore columns have not gained popularity among food analysts, but the use of 3.2 mm i.d. columns can be recommended as they provide a two-fold increase in sensitivity over standard 4.6 mm i.d. columns, without necessarily requiring low dispersion volume injectors and flow cells.

A short column packed with small particles will produce a similar efficiency as a longer column equally well packed with larger particles, but the short column has the advantage of a decreased separation time. Short columns packed with 3 µm materials are commercially available and have found application in pharmaceutical analysis where speed is important. As is the case with microbore columns, the operation of short '3-micron' columns will result in smaller peak volumes which are susceptible to serious extracolumn dilution unless extracolumn dispersion is stringently minimized (Cooke *et al.*, 1982). For food analysis applications, 5 µm particles are better matched to conventional equipment, hence their popularity.

The mobile phase

Selection of the mobile phase (also referred to as the eluent) is made on the basis of solvent strength and solvent selectivity. The general strategy in optimizing a separation is first to adjust the solvent strength to maintain solute capacity factors (*k* values) in the range 1–10 and then, while holding the solvent strength constant, to alter the selectivity of the mobile phase. Compounds having *k* values in the optimum range of 1–10 can be separated using isocratic elution, i.e. using a mobile phase whose solvent strength remains constant throughout the separation. Compounds having widely different *k* values can often be separated by gradient elution, in which the solvent strength is increased continuously throughout the separation.

Column switching

Column switching is a technique whereby selective fractions of the effluent from one chromatographic column are transferred to the inlet of another column. Automatic on-line column switching is achieved via one or more high-pressure switching valves actuated by time-programmable events from a microprocessor chromatograph. A particularly useful application of column switching is sample clean-up, which, in its simplest form, uses two columns containing identical packing material and one mobile phase. The initial separation takes place on a short column such that unwanted early-eluting components are routed to waste, and the fraction containing the components of interest is diverted on to the analytical column. The valve is then switched back to its initial position so that, while the components of interest are undergoing analysis, the more highly retained components are eluted from the short column directly to waste. A more complicated configuration is on-line two-dimensional HPLC (Majors, 1980a), where fractions from one column are transferred to another column of different chromatographic mode. This configuration is used for better resolution of complex samples, as well as for sample clean-up.

The detector

Three types of in-line HPLC detector have been routinely employed in food analysis to determine vitamins: these are photometric, fluorescence and electrochemical detectors. Each of these three detectors provides a continuous electrical output that is a function of the concentration of solute in the column effluent passing through the flow cell. The photodiode array detector permits simultaneous absorbance detection at several wavelengths and continuous memorizing of spectra during the evolution of a peak. This detector has proved invaluable for the assessment of peak purity and identity in carotenoid analysis. Both amperometry and coulometry have been applied to electrochemical detection. In amperometric detectors only a small proportion (usually < 20%) of the electroactive solute is reduced or oxidized at the surface of a glassy carbon or similar nonporous electrode, while in coulometric detectors the solute is completely reduced or oxidized within the pores of a graphite electrode. A stringent requirement for electrochemical detection is that the solvent delivery system should be virtually pulse-free.

Derivatization

It may be necessary to make a chemical derivative of some vitamins in order to facilitate the use of a more suitable means of detection (usually fluorimetric) and/or a more suitable chromatographic mode. Either pre-column or post-column derivatization may be employed, depending on whether one wishes to chromatograph the derivatized vitamer or the underivatized vitamer. In pre-column derivatization the reaction is carried out before the sample is analysed by HPLC, so it is the derivatized compounds that are actually chromatographed. In post-column derivatization the test solution is injected into the chromatograph, and the separated vitamers in the column effluent are reacted with the derivatizing agent in a reaction coil located between a mixing tee and the detector (Froehlich and Wehry, 1981).

A post-column derivatization system requires a second pump to introduce the derivatizing agent but, once set up, the system provides an automatic and standardized means of preparing the derivatives. There will inevitably be some degree of peak broadening due to the increased distance between the HPLC column and the detector. Another disadvantage is that there is no opportunity to remove or separate excess reagent or impurities within the reagent that might impair the sensitivity of detection. Pre-column derivatization requires manual manipulations, and hence more skill and nonstandardized reaction conditions, unless rigorously controlled. Advantages are the opportunity to clean up the reaction mixture before injection, and the operation of a simpler and more efficient chromatographic system.

Quantification

Two quantification methods are commonly employed in HPLC: external standardization and internal standardization. Both methods compare the size of the peak produced by the component of interest (analyte) with the peak size of a known concentration of a reference compound (standard), and calculate the result on the basis of detector response factors. In principle, either peak height or peak area measurement may be used but, in practice, height may be more precise than area (or vice versa) depending on the chromatographic operating conditions. Factors affecting peak heights and areas are discussed later in this section.

External standardization
In this method the peak size of the analyte in the sample chromatogram is compared with the peak size of an identical compound (external standard) in a separate standard chromatogram. The procedure is as follows.

A series of standard solutions of the pure analyte (A) is prepared spanning the working concentration range in units of $\mu g/ml$. A fixed volume of each standard solution is then injected, and a calibration plot is constructed of peak height (h_A) versus concentration of analyte (C_A). If the system is working properly, the calibration plot should result in a straight line (within a given concentration range), intercepting the origin. The response factor, f, is determined as the slope of the linear part of the plot.

$$f = h_A/C_A$$

The same volume of sample test solution is injected and the peak height (h'_A) is measured. The concentration of analyte (C'_A) in the test solution is calculated from the response factor, thus:

$$C'_A = h'_A/f$$

The concentration of analyte in the original sample, assuming complete extraction, is:

Conc. $(\mu g/g) = [C'_A (\mu g/ml) \times \text{total vol. (ml)}] / [\text{wt. of original sample (g)}]$

If a blank sample can be obtained that is known to be free of the analyte (e.g. a nonfortified sample of a foodstuff that is normally fortified), it can be used in the preparation of the calibration plot as a means of eliminating matrix interferences. That is, several identical sample blanks are separately spiked with graded amounts of the analyte, and a calibration plot is constructed.

The precision of external calibration relies upon constant injection volumes; it is not necessary to know the actual amounts of solute injected. The loop valve injector or an automatic injection device facilitates a more precise control of injection volume than direct on-column syringe injection. As peak measurements are compared from separate chromatograms, the calibration is susceptible to errors arising from changes in the chromatographic conditions (e.g. temperature, flow rate, mobile phase composition). Such errors can be minimized by injecting a standard between every pair of unknowns and recalculating the response factor for every analysis.

Internal standardization
This calibration technique is designed to compensate for losses of analyte during the sample work-up. A known amount of a suitable compound (internal standard) is added to the original sample at the earliest possible point in the extraction stage, the pretreatment is carried out, and the resulting sample test solution is analysed. The quantification is based upon the comparison of the ratio of the internal standard peak size to analyte peak size in the test solution with that ratio in a standard solution containing known amounts of the analyte and internal standard. In this approach, any loss of analyte will be accompanied by the loss of an equivalent amount of internal standard. The procedure is as follows (concentrations in $\mu g/ml$).

A standard solution containing a known concentration (C_A) of the analyte (A) and a known concentration (C_S) of the internal standard (S) is chromatographed. If the peaks for A and S have heights of h_A and h_S, the respective response factors are:

$$f_A = h_A/C_A \quad \text{and} \quad f_S = h_S/C_S$$

$$F = f_A/f_S = (h_A \times C_S)/(C_A \times h_S)$$

where F is the relative response factor. To the sample extract containing an unknown concentration (C'_A) of analyte (A) is added a known concentration (C_S') of the internal standard (S). If the peaks for A and S in the resultant chromatogram have peak heights of h_A' and h_S', respectively, then:

$$C_A' = (1/F) \times [(h_A' \times C_S')/h_S']$$

The concentration of analyte in the original sample, assuming complete extraction, is:

Conc. ($\mu g/g$) $= [C_A'(\mu g/ml) \times$ total vol. (ml)$]/[$wt. of original sample (g)$]$

The ideal internal standard conforms to the following requirements:

1. It must resemble the analyte as closely as possible in its chemical and physical properties (including stability), and behave in a similar manner to the analyte during all steps in the analytical procedure (extraction, derivatization, etc.).
2. It must be commercially available in high purity.
3. It must never occur (before addition) in the original sample.
4. It should not react with any components of the sample or with the HPLC column packing.
5. It must elute near to the analyte, whilst being completely resolved from the analyte and from other neighbouring components.

The main utility of internal standardization is in assays that require extensive sample pretreatment where variable and unpredictable recoveries of analyte may occur. As the calculation involves ratios of peak size, and not the absolute peak size, the injection volumes need not be constant. The concentrations of the analyte and internal standard must, however, remain within the linear range of the detector. Since the procedure involves the measurement of two peaks, the precision of the analysis is decreased by a factor of 1.4 ($\sqrt{2}$).

The inherent errors of the internal standardization procedure will be minimized if the analyte and internal standard yield peaks of similar size, since errors in measuring peaks of about the same size will tend to cancel out. The effect of changing the peak size ratios can be ascertained by chromatographing a series of standard solutions containing different concentrations of analyte and a fixed concentration of internal standard. A calibration plot is then constructed in which the ratios of analyte to internal standard peak height (or peak area) are plotted against the concentration of analyte, i.e. h_A/h_S versus C_A. A linear calibration plot which intercepts the origin is evidence that the relative response factor is constant at all concentration ratios of internal standard and analyte over the concentration range of analyte to be used.

Matrix interferences can be established by comparing the calibration plot with a similar plot prepared with sample blanks.

Peak height or peak area?

In deciding between height and area measurement as a means of obtaining optimum precision, a number of factors relating to system performance must be considered (Bakalyar and Henry, 1976). The key point to remember is that chromatographic development is based on evolution volume, not time. If the mobile phase flow rate should slow down or stop when a solute band is passing through the flow cell of a concentration-sensitive detector, the peak height will be unaltered, but the area will continue to increase as it is accrued on a time basis. Hence area measurement is quite sensitive to deficiencies in the pumping system, and to possible changes in the viscosity and compressibility of solvents that could affect flow rate during gradient elution. Minor changes in mobile phase composition or slight changes in the column temperature will affect retention times, with a consequent effect on peak width and hence peak height. Peak area will remain constant because the height decreases as the width increases. Possible causes of variation in mobile phase composition are evaporation of the organic solvent component and poor reproducibility of gradient programmes. It follows that, if the mobile phase composition and column temperature can be maintained precisely, but the flow rate is variable, peak height measurement yields greater precision than peak area measurement. Conversely, if the mobile phase composition and/or temperature cannot be maintained precisely, but flow rate is constant, peak area measurement is preferred. The use of peak area also relies on the accurate electronic assignment of the beginning and end of the solute peak. The errors will be greater for the narrow peaks at the beginning of a chromatogram, and hence in practice it is sometimes found that peak height measurement is more precise for early-eluting peaks, while peak area is more reproducible for later peaks. Peak height calculations are preferred for trace analysis where peaks are very small or overlap.

2.8.2 Chromatographic modes

Ion exchange chromatography

An ion exchange material comprises a porous support matrix bearing fixed ionogenic groups which, when ionized, function as the ion exchange sites. Depending on their function, ion exchange materials are either anion exchangers or cation exchangers, bearing positively charged and negatively charged functional groups, respectively. The positive charges of anion exchangers result from the protonation of basic groups,

Table 2.2 Characterization of ion exchangers

Type	Functional group	Usable pH range
Strong cation exchanger (SCX)	Sulphonic acid —SO_3^-	Above pH 1
Strong anion exchanger (SAX)	Quaternary amine —NR_3^+	Below pH 11
Weak cation exchanger (WCX)	Carboxylic acid —CO_2^-	Above pH 6
Weak anion exchanger (WAX)	Primary amine —NH_3^+	Below pH 8

while the negative charges of cation exchangers are produced by the protolysis of acidic groups (Table 2.2). The functional groups are located mainly within the extensive pore structure of the matrix. To preserve electrical neutrality, each fixed ion is paired with an exchangeable counterion of opposite charge. The type of counterion specifies the 'form' of the ion exchanger; for example, a strong anion exchanger is usually supplied in the chloride form, i.e. the counterion is Cl^-.

In ion exchange chromatography, the separation of sample ions depends upon the selectivity at the numerous sorption–desorption cycles that take place as the sample material passes through the column. Ions having a strong affinity for the functional groups will be retained on the column, whilst ions that interact only weakly will be easily displaced by competing ions, and eluted early.

Ion exchangers are further classified as strong or weak according to the ionization properties of the basic or acidic functional groups (Table 2.2). The degree of ionization depends upon the pK_a of the functional group and on the pH of the mobile phase, and is directly proportional to the ion exchange capacity. The capacity is maximal when all of the functional groups are ionized. The maximum exchange capacity for strong anion and cation exchangers is maintained over a wide pH range, whereas for weak exchangers the usable pH range is limited (Table 2.2).

Most classical ion exchange resins are polystyrene-divinylbenzene (PS-DVB) copolymers to which the ionogenic functional groups are attached. Such resins exhibit a relatively slow diffusion of solutes within the deep pores containing stagnant mobile phase, and this leads to major band broadening. For this reason, such resins were often operated at elevated temperatures to speed mass transfer through a decrease in mobile phase viscosity. One way of minimizing the diffusion path and improving the efficiency of the separation is to use pellicular particles, which have a nonporous impervious solid core surrounded by a thin coating of active stationary phase. Pellicular packings have been superseded by totally porous microparticulate silica-based packings. Silica-based packings are stable at temperatures up to 80 °C, but strongly acidic (pH < 2) or mildly basic (pH > 7.5) conditions destroy the silicon structure, leading to a drastic increase in column resistance and loss of efficiency. This problem has prompted investigation into new supports for a second generation of microparticulate column packings.

The chief mobile phase parameters that control sample retention and separation selectivity are ionic strength and pH. The role of the buffer component is to maintain the pH at the selected value and to provide the desired solvent strength in terms of the appropriate type of counterion at the right concentration. The ionic strength can be regarded as a measure of the number of counterions present. The sample ions and mobile phase counterions of the same charge compete for the ion exchange sites, and hence an increase in ionic strength will proportionately decrease solute retention, and vice versa. In other words, the solvent strength increases with increasing ionic strength, accompanied by a minimal change in solute selectivity. The ionic strength of the mobile phase can be increased by either increasing the molarity of the buffer solution while holding the pH constant, or adding a nonbuffer salt such as sodium nitrate when it is undesirable to increase the buffer concentration. The primary effect of pH is to control the ionization of weak organic acids and bases in the sample. Increasing the pH leads to an increased ionization of weak acids and decreased ionization of weak bases, and vice versa for a decrease in pH. An increase in ionization in each case leads to increased solute retention.

Water-miscible organic solvents such as acetonitrile, propan-2-ol and ethanol are frequently added as modifiers to the aqueous mobile phase as a means of lowering the viscosity and improving mass transfer kinetics. Typical amounts of added solvent range between 3% and 10% by volume. The effect of the organic modifier upon the ion exchange equilibria is relatively minor, and any significant changes that result from such additions are mainly attributed to hydrophobic mechanisms. In weak anion exchange chromatography an appreciable proportion of an organic acid solute will exist in the nonionized form, and thus behave differently to the ionized form (anion). The resultant peak tailing caused by the mixed-mode chromatography can be eliminated by use of an organic modifier, which also decreases the retention time. In general, using a modifier can dramatically improve a separation, although the effect is unpredictable and has to be determined empirically. It is obviously important to ascertain beforehand that the column packing material is compatible with the proposed organic solvent.

Ion exclusion chromatography

In this technique an ion exchange resin is employed for separating ionic molecules from nonionic or weakly ionic molecules. Ions having the same charge as the functional groups of the support (i.e. co-ions) are repelled by the electrical potential across the exchanger-solution interface (Donnan potential) and excluded from the aqueous phase within the pore volume of the resin beads. Nonionic or weakly ionic molecules are not excluded and, provided they are small enough, may freely diffuse into the matrix, where they can partition between the aqueous phase within the resin beads and the aqueous phase between the resin beads. Therefore, ionized sample solutes pass quickly through the column, whereas nonionic or weakly ionic solutes pass through more slowly. The retention mechanisms of the nonionic solutes include polar attraction between the solute and the resin functional groups (i.e. adsorption), van der Waal's forces between the solute and the hydrocarbon portion of the resin (primarily the benzene rings) and size exclusion. The overall separation is accomplished without any exchange of ions, so the column does not require regeneration after use.

Ion exclusion chromatography using a strong cation exchange resin has been successfully applied to the separation of organic acids, including ascorbic acid. The technique here is to suppress the ionization of the weak organic acid by adding sulphuric acid to the water mobile phase so that the highly ionized sulphate ion is excluded and quickly eluted, while the undissociated organic acid enters the resin pore structure and is retained. The mobile phase pH should be lower than the pK_a of the organic acid to ensure that the acid is undissociated. The volume of aqueous phase within the resin bead must be sufficient to allow partition of the nonionic solutes to take place and, to achieve optimum separation, must be greater than the sample volume. For this reason, PS-DVB types of resin, which are capable of swelling, are used in preference to silica-based exchangers.

Adsorption and polar bonded-phase chromatography

Adsorption or liquid–solid chromatography is a form of normal-phase chromatography in which the surface of microparticulate silica (or other adsorbent) constitutes the polar stationary phase. The silica particles are characterized by their shape (irregular or spherical), size and size distribution, and pore structure (mean pore diameter, specific surface area and specific pore volume). The adsorption sites on the surface are silanol ($\equiv$Si—OH) groups, which are present both as isolated groups and hydrogen-bonded to one another. The mobile phase is a nonpolar solvent (typically hexane) containing a small percentage of a polar organic solvent (e.g. propan-2-ol) which acts as a modifier. The modifier is preferentially adsorbed from the mobile phase by the hydrogen-bonded silanol groups and effectively deactivates these strong adsorption sites. The isolated silanol groups that remain are those responsible for the adsorptive properties of the deactivated silica. The dissolved solute molecules compete with mobile phase molecules for interaction with the remaining weaker adsorption sites, so solute retention can be increased by decreasing the polarity of the mobile phase. Solute retention is very sensitive to changes in temperature, therefore column thermostatting is recommended, especially when peak height measurements are used in quantitative assays.

Separation in normal-phase chromatography is based on the relative polarity of the solutes and their affinity for the stationary phase. Relatively nonpolar solutes prefer the mobile phase and elute first, whilst more polar solutes prefer the stationary phase and elute later. In addition, adsorption chromatography is a powerful means of separating geometrical (*cis–trans*) isomers, the separation mechanism being attributed to a steric fitting of solute molecules with the discrete adsorption sites. Adsorption chromatography is well suited for the analysis of the fat-soluble vitamins, which are of moderate polarity. The analysis of the water-soluble vitamins is precluded because these polar compounds give rise to badly tailing peaks.

Silica columns can tolerate relatively heavy loads of triglyceride and other nonpolar material. Such material is not strongly adsorbed and can easily be washed from the column with 25% diethyl ether in hexane after a series of analyses. Procedures for determining vitamins A and E have been devised in which the total lipid fraction of the food sample is extracted with a nonaqueous solvent, and any polar material that might be present is removed. An aliquot of the nonpolar lipid extract containing the fat-soluble vitamins is then injected into the liquid chromatograph without further purification. Direct injection of the lipid extract is possible because the lipoidal material is dissolved in a nonpolar solvent that is compatible with the predominantly nonpolar mobile phase. Procedures based on this technique are rapid and simple because there is no need to saponify the sample.

An operational disadvantage of adsorption chromatography is the slow equilibration towards water, which is a very strong modifier in deactivating the silica surface. All organic solvents (unless specifically dried) contain an inherent amount of water in the parts per million range that is sufficient to affect solute retention. Gradient elution is best avoided because it is difficult to ensure that equilibration between the silica adsorbent and the changing mobile phase is occurring sufficiently rapidly.

Bonded-phase column packings for use in normal-phase chromatography are available in which the stationary phase is a polar functional group chemically bonded onto the silica surface. Moderately polar nitrile-bonded phases containing propylcyano $[-(CH_2)_3-CN]$ as the functional group generally show less show retention when substituted for silica, but similar selectivity. Amino-bonded phases containing propylamino $[-(CH_2)_3-NH_2]$ functional groups are of high polarity, and the basic nature of the functional groups imparts a quite different selectivity when compared with the acidic surface of silica. A unique type of polar-bonded phase is Partisil PAC (Whatman). This moderately polar silica-based material contains both secondary amine and cyano functional groups in a ratio of 2:1. Polar-bonded phases, unlike unmodified silica, are not sensitive to traces of water and, because they respond rapidly to changes in solvent polarity, can be used in gradient systems. Among the newer generation of packing materials, YMC PVA-Sil (Yamamura Chemicals) is prepared by bonding a polymer coating of vinyl alcohol to silica. As the polymerized PVA completely covers both external and internal surfaces of the silica support, it is protected against aggressive, high pH (up to 9.5) mobile phases.

Reversed-phase chromatography

In this mode a nonpolar stationary phase is used in conjunction with a polar mobile phase. Solute elution is the opposite to that observed in normal-phase chromatography; polar solutes prefer the mobile phase and elute first, while nonpolar solutes prefer the essentially hydrophobic stationary phase and elute later. According to the solvophobic theory of solute retention for reversed-phase chromatography, the very high cohesive density of the mobile phase, arising from the three-dimensional hydrogen-bonding network, causes the less polar solutes to be literally 'squeezed' out of the mobile phase, enabling them to bind with the hydrocarbon ligands of the stationary phase (Horvath and Melander, 1977). The selectivity of the separation results, therefore, almost entirely from specific interactions of the solute with the mobile phase.

Reversed-phase chromatography is ideally suited for the separation of nonpolar to moderately polar compounds. It is also effective in separating the members of an homologous series since hydrophobicity generally increases with the number of carbon atoms in a molecule. Interactions between solute and nonpolar stationary phases involve weaker forces compared with those occurring in normal-phase chromatography, so re-equilibration times following solvent compositional changes are short. This is particularly advantageous during method development and following gradient elution.

Most published vitamin separations involving reversed-phase chromatography have utilized totally porous microparticulate silica-based packings in which the nonpolar stationary phase, usually octadecylsilane (ODS), is chemically bonded to the silica surface through a siloxane bridge. Phenyl-bonded and cyano-bonded phases have also been used in certain applications.

The bonding chemistry invariably used in the manufacture of microparticulate silica-based packings involves the reaction between the surface silanol groups on the silica matrix with an organosilane reagent to form a siloxane ($-Si-O-Si-$) bonded phase, thus:

$$\equiv Si—OH + Cl—\underset{\underset{CH_3}{|}}{\overset{\overset{CH_3}{|}}{Si}}—R \quad \longrightarrow \quad \equiv Si—O—\underset{\underset{CH_3}{|}}{\overset{\overset{CH_3}{|}}{Si}}—R + HCl$$

R is the ligand of interest, usually an octadecyl (C_{18}) hydrocarbon. In a monofunctional silanizing agent, only one of the three remaining substituents is a reactive group, usually either chloro or alkoxy; the other two substituents are usually methyl groups. Silanizing agents containing two and three reactive groups are referred to as di- and trifunctional. A monofunctional agent reacts with the silanol groups on the silica surface to form a monomolecular layer (monomeric phase). The corresponding di-and trifunctional silanizing agents react in a more complicated fashion and, depending on the reaction conditions, can form a monolayer, multilayer or cross-linked polymer of the bonded phase. Monomeric bonded phases exhibit faster diffusion rates in the stationary phase compared with multilayer or cross-linked phases and hence, in theory, provide greater column efficiencies.

Some silica-based reversed-phase packings are 'end-capped', whereby unwanted residual silanol groups trapped by steric hindrance after reaction with the organosilane reagent are initially reacted with a small monofunctional silane reagent. A second reaction using a proprietary reagent ensures that the majority of residual silanols are end-capped to give a more or less completely hydrophobic surface.

The Zorbax StableBond (SB) phases are manufactured using a monofunctional agent whose two nonreactive groups are diisopropyl rather than methyl. The bulky diisopropyl groups sterically protect the siloxane bond and the underlying silica very effectively, allowing the material to be safely used under low pH conditions (down to pH 0.8).

A unique column packing material which can be used with highly aqueous mobile phases is YMC ODS-AQ (Yamamura Chemicals). This material, after ODS-bonding, is treated with a hydrophilic end-capping reagent which provides a surface that is easily wettable. When used with highly aqueous eluents, these phases give longer retention times than conventionally end-capped ODS phases, because the C_{18} ligands are actually lifted off the surface by the action of the eluent penetrating to the hydrophilic end-cap. This creates a greater surface area for reversed-phase interactions compared with conventionally end-capped phases, in which the C_{18} chains lie flat due to repulsion with the polar eluent.

Non-end-capped packings are generally subject to secondary adsorption characteristics created by accessible residual silanol groups, and these reactions can cause undesirable peak tailing of polar solutes.

Capsell Pak C18 SQ 120 (Shiseido) has a silica support whose surface is coated with a thin film of silicone polymer upon which the ODS group is bonded. The polymer coating of this material is very effective in suppressing peak tailing.

Polymer column packings for reversed-phase applications have been developed which are stable over practically the entire pH range. PRP-1 (Hamilton) and PLRP-S (Polymer Laboratories) are examples of PS-DVB copolymers, in which the neutral nonpolar polystyrene surface functions as the stationary phase, and the pH stability range is 1–13. YMC Polymer C18 (Yamamura Chemicals) consists of a C_{18} stationary phase chemically bonded to a highly cross-linked polymethacrylate support. Polymer C18 is more rigid than PS-DVB copolymers, is stable over a pH range of 2–13, and is compatible with a wide range of solvents and buffers. It provides similar selectivity to silica-based ODS materials but, having no residual silanols, does not produce secondary adsorption effects.

Differences in selectivity and chromatographic performance between ODS columns from different manufacturers are attributable to particle characteristics (size, shape and porosity) of the silica substrate and bonded-phase parameters. The latter include the bonded-phase chemistry used in phase preparation, phase coverage and the percentage of accessible residual silanol groups. Carbon loading refers to the amount of ODS ligate bound to the silica surface and is commonly expressed as percentage carbon (weight per weight). The percentage of carbon will depend on the surface area of the silica support, which is governed by the mean pore diameter and by other factors (Cooke and Olsen, 1979). The term carbon loading has little meaning without taking the surface area into account, and a more informative term of surface coverage is carbon content per unit specific surface area, expressed in $\mu mol/m^2$. Take, for example, LiChrosorb RP-18 and Zorbax ODS, which have similar carbon loadings of 17% and 16%, respectively. The surface area of the LiChrosorb material ($150\,m^2/g$) is less than one-half of that of the Zorbax ($350\,m^2/g$). Thus, in terms of carbon content per unit specific area, LiChrosorb RP-18 has a surface coverage that is more than double that of Zorbax ODS. High-loading phases are highly retentive and facilitate a high solute load capacity. Phases with low carbon loadings are less retentive and therefore allow shorter analysis times.

Mobile phases used in reversed-phase chromatography are frequently composed of mixtures of methanol and water or acetonitrile and water. Increasing the proportion of water causes an increased retention of the more hydrophobic solutes relative to the more polar solutes. The surface tension of the eluent plays the major role in governing solute retention, so an increase in temperature, by reducing viscosity, increases column efficiency and shortens retention times.

Several methods for determining fat-soluble vitamins have employed nonaqueous reversed-phase (NARP) chromatography as a means of overcoming the solubility problem encountered with water/organic mobile phases. A typical NARP mobile phase consists of a polar basis (usually acetonitrile), a modifier of lower polarity (e.g. dichloromethane) to act as a solubilizer and to control retention and, occasionally, a small amount of a third solvent with hydrogen-bonding capacity (e.g. methanol) to optimize selectivity. To compensate for the increased affinity of low polarity compounds for the mobile phase, a highly retentive stationary phase is required.

Ionizable compounds in the ionic state elute near the void volumes in reversed-phase chromatography but, if the ionization is suppressed, the undissociated compound will be retained. For a weak acid, ion suppression can be achieved by buffering the mobile phase at 1–2 pH units below the pK_a value, and for a weak base at 1–2 pH units above the pK_b value. Thus weak acids and weak bases can be retained in the pH regions 2–5 and 7–8, respectively. Strong acids and bases would require extreme pH values ($<$ 2 and $>$ 8, respectively) for ion suppression. The ion suppression technique should not be used for silica-based packings that are unstable outside the pH range 2–7.5. Polymeric-based packings have the advantage of being stable throughout the entire practical pH range (pH 1–13) (Lee, 1982). Reversed-phase chromatography with ion suppression has the advantage over conventional ion exchange chromatography in permitting the simultaneous analysis of both ionized and nonionized solutes.

Ion interaction chromatography

Ion interaction chromatography is probably better known as ion pair chromatography, but current theory of retention mechanisms favours the use of the former term. The technique employs the same types of column packing and water/organic mobile phases as those used in conventional reversed-phase HPLC. The pH of the mobile phase is adjusted to encourage ionization of the ionogenic solutes, and retention is controlled by adding to the mobile phase an amphiphilic ion interaction agent bearing an opposite charge to that of the analyte. The ion interaction agent should be univalent, aprotic and soluble in the mobile phase. It should ideally give a low UV-absorbing background, although for special applications a reagent with a strong chromophore can be used to enhance the response of an absorbance detector. The retention behaviour of nonionic solutes is not affected by the presence of the ion interaction agent, so both ionized and nonionized solutes may be resolved in the same chromatographic run. Use of ion interaction chromatography is advantageous for determining water-soluble vitamins because many polar interferences elute in the dead volume, and hydrophobic compounds would be in low concentration in the aqueous extract of the sample.

For the determination of anionic solutes such as ascorbic acid, a variety of organic amines have been used as ion interaction agents, representing primary, secondary, tertiary and quaternary amines. One of the more popular of these is tetrabutylammonium (Bu_4N^+) phosphate, which is commercially available as a prepared 5 mM solution in pH 7.5 buffer (PIC A reagent, Waters Associates). This aprotic quaternary amine interacts with strong and weak acids, and the buffering to pH 7.5 suppresses weak base ions.

For the determination of cationic solutes such as thiamin (a protonated amine), a range of alkyl sulphonates having the formula $CH_3(CH_2)$ nSO_3— (with $n = 4-7$) predominates. Selection of the appropriate reagent is based on solute retention time, which increases with an increase in the length of the alkyl chain. Prepared 5 mM solutions of the sodium salts in pH 3.5 buffer are available from Waters Associates; namely, pentane sulphonic acid (PIC B5), hexane sulphonic acid (PIC B6), heptane sulphonic acid (PIC B7) and octane sulphonic acid (PIC B8). These reagents interact with strong and weak bases, and the buffering to pH 3.5 suppresses weak acid ions.

Most ion interaction chromatographic applications reported for water-soluble vitamin assays up to the present day have utilized 5 or 10 μm silica-based C_{18} bonded-phase packings. Monomeric phases yield better-shaped peaks than do polymeric phases, and high carbon loadings ensure good retention properties (Majors, 1980b). PS-DVB copolymers developed for HPLC have also been utilized for ion interaction chromatography (Iskandarani and Pietrzyk, 1982).

The practice of ion interaction chromatography has been discussed by Gloor and Johnson (1977). Retention and selectivity are optimized mainly by altering the concentration of the ion interaction agent and the pH of the mobile phase. Ionic strength is not a variable for controlling retention and it should be kept as low as possible, commensurate with satisfactory retention characteristics and reproducibility.

Variation of the concentration of ion interaction agent in the mobile phase provides a simple means of controlling solvent strength. An increase in the concentration causes an increase in solute retention but, beyond a certain limit, a further increase in concentration causes a decrease in retention. A possible explanation for this reversal effect is that the increased amount of adsorbed surfactant lowers the interfacial tension between the modified stationary phase and the surrounding aqueous medium to a point at which solute retention is decreased (Stranahan and Deming, 1982). This nonionic theory also accounts for the observed decrease in retention of neutral solutes with increasing concentration of ion interaction agent.

Alterations in the pH of the mobile phase will have a pronounced effect upon separation selectivity for weak acids and weak bases because of the effect of pH upon solute ionization. Maximal retention is obtained where the solute and ion interaction agent are completely ionized. The reagents, being strong acids or salts of strong bases, remain completely dissociated over a wide pH range, so that the pH can be adjusted to an optimal value for the separation. Weak acid solutes ($pK_a > 2$) are usually separated at a pH of 6–7.4, and weak bases at pH 2–5, using a buffer to hold the pH constant. Buffer salts should have poor ion association properties, but good solubilities in the mobile phase. An excessive concentration of buffer salt, or the addition of neutral salt to the mobile phase, results in the surplus ions of such salts competing successfully with analyte ions for association with the adsorbed ion interaction agent, thus causing a decrease in retention. Solute pK_a values are affected by a change in temperature, so significant changes in selectivity can occur with relatively small changes in column temperature. To ensure reproducible separations, it is thus good practice to maintain a constant column temperature with the aid of a column heating oven. Ion interaction chromatography is usually carried out at a few degrees above ambient, although operation at 50–60 °C will improve peak resolution (with a slight decrease in retention) by reducing the viscosity of the mobile phase.

Increasing the proportion of organic modifier increases the solvent strength, resulting in an overall lowering of solute retention. The concentration of organic modifier affects the surface potential (and hence solute retention) by influencing the sorption of the ion interaction agent onto the stationary phase (Bartha, Vigh and Varga-Puchony, 1990).

The general strategy for separating complex mixtures of nonionic and ionic solutes is firstly to adjust the percentage of organic modifier (usually methanol) to obtain optimum retention and separation of nonionic solutes. One then adds a suitable ion interaction reagent in the appropriate buffer to the previously established mobile phase to separate the ionic compounds isocratically. Gradient elution programmes usually involve a decrease in the concentration of ion interaction agent with time as a means of decreasing solute retention.

Ion interaction agents may irreversibly adsorb onto the stationary phase, thereby changing the phase chemistry and reducing the apparent pore volume. Columns used for ion interaction chromatography should therefore be reserved exclusively for this purpose.

Two-dimensional HPLC

In trace component analysis (e.g. determination of naturally occurring vitamins D and K) it is often impossible to obtain an adequate HPLC separation for quantitation using a single column. In this event, the use of two separate HPLC systems may resolve the components. A true two-dimensional combination involves two different chromatographic modes (e.g. normal-phase and reversed-phase) and can be expected to provide better selectivity than the use of two columns operated in the same mode. The first system (semi-preparative HPLC) is designed to isolate a fraction of the sample extract that contains the vitamin analyte and the internal standard. Ideally, the analyte and internal standard should be unresolved from one another so that they can be collected by reference to a single peak in the chromatogram. If the analyte/internal standard peak is masked by co-eluting peaks, the obvious method of collecting the fraction is to refer to the retention time of an analyte standard. However, this method may not be reliable, as in normal-phase HPLC (the separation mode frequently employed) it is difficult to maintain constant retention times during a run. Shearer (1983) overcame this problem in a vitamin K assay procedure by injecting a phylloquinone standard together with a small amount of the sample extract. By comparing the chromatogram thus obtained with a chromatogram produced from the sample extract alone, the phylloquinone peak could be seen superimposed on the background peaks. Thus the elution of vitamin K was related to the UV-absorbing 'fingerprint' given by contaminants in the sample extract.

The fraction of column effluent containing the analyte and internal standard is either collected manually for subsequent re-injection onto the second (analytical) column (off-line operation) or diverted directly onto the second column via a high pressure switching valve (on-line operation). For manual collection, a drop counter system rather than a volume collection system is recommended (Kobayashi, Okano and Takeuchi, 1986). The fraction is collected in a small tapered tube and the solvent is carefully evaporated off under a stream of nitrogen. The residue is then dissolved in a small volume of a suitable solvent for the analytical separation. Because of the sample reconstitution step, the potential problem of mobile phase incompatibility between the two HPLC systems is avoided, and hence any semi-preparative/analytical combination can be used.

On-line operation has the potential for being completely automated, as the switching valve can be actuated by time-programmable events from a microprocessor-based chromatograph. The two mobile phases must be miscible with each other, and the mobile phase from the first column must be of weaker elution strength than that used for the second column. The latter criterion facilitates the concentration of solute onto the head of the second column, this being essential to maintain the effect of the first separation. If a stronger solvent is injected onto the second column, band spreading will occur because the injected solvent will preferentially move the solutes down the column until the concentration of this solvent is diluted sufficiently that solutes begin to be retained (Majors, 1980a).

2.8.3 Applications – concurrent and simultaneous determination of two or three vitamins

HPLC has been applied successfully to the determination of the fat-soluble vitamins and also to several water-soluble vitamins in foodstuffs. The methodology depends to a large extent upon what is being measured. This can be a given vitamin added in fortification, or the total (added plus naturally occurring) content of a given vitamin in a fortified food, or the naturally occurring content of a given vitamin in a nonfortified food. The parent forms of the vitamins, as used in fortification, exhibit different stabilities towards air, light, temperature and pH. Among other factors, the extraction of vitamins from milk and other dairy products is simpler than that from animal and plant tissues.

Owing to the complexities of the food matrix, the simultaneous determination of added vitamins in fortified foods is usually limited to two or three vitamins (Rizzolo and Polesello, 1992). The situation in nonfortified foods is, of course, far more complicated, owing to the difficulties of quantitatively extracting the vitamins from their various bound forms, the need to measure low indigenous concentrations and the requirement to determine the various vitamers of a particular vitamin. There is, therefore, little scope for determining more than one vitamin at a time.

The methods discussed here that determine two or more vitamins concurrently or simultaneously are usually confined to the quantification of supplemental vitamins in fortified foods. Methods for determining individual vitamins are invariably capable of quantifying the major vitamers. These methods are discussed in the chapters on the relevant vitamin.

Fat-soluble vitamins

Selected applications are summarized in Table 2.3.

Vitamins A and D
The unsaponifiable fraction of fortified milk (whole, semi-skimmed and skimmed) can be analysed for vitamins A and D concurrently. An aliquot of the unsaponifiable fraction is first analysed for vitamin A by reversed-phase HPLC. The larger remaining portion of the unsaponifiable fraction is then purified by solid-phase extraction (Reynolds and Judd, 1984) or alumina open-column chromatography (Wickroski and McLean, 1984) and analysed for vitamin D using the same HPLC column as used for the vitamin A.

Table 2.3 HPLC methods used for determining two or three fat-soluble vitamins concurrently or simultaneously in food

Food	Sample preparation	Column	Mobile phase	Vitamins separated	Detection	Reference
Vitamins A and D						
			Reversed-phase chromatography			
Fortified skimmed milk powder	Saponify (hot) in presence of vitamin D_2 or D_3 as internal standard, extract unsaponifiables with petroleum ether/diethyl ether (1:1). Evaporate, dissolve residue in 25 ml MeOH. *For retinol:* remove 5 ml aliquot for HPLC (Solution 1) *For vitamin D:* evaporate remaining 20 ml MeOH solution. Purify by C_{18} solid-phase extraction (Solution 2)	Spherisorb ODS 5 μm 250 × 4.6 mm i.d.	Solutions 1 and 2: MeOH/H_2O (97.5:2.5)	Solution 1: retinol Solution 2: vitamins D_2 and D_3	Solution 1: UV 325 nm Solution 2: UV 265 nm	Reynolds and Judd (1984)
Fortified fluid milk (whole, semi-skimmed, skimmed)	Saponify (ambient), extract unsaponifiables with hexane. Evaporate, dissolve residue in 6 ml hexane. *For retinol:* remove 1 ml aliquot for HPLC (Solution 1) *For vitamin D:* evaporate remaining 5 ml hexane solution. Alumina column chromatography (eluate is Solution 2)	Vydac 201 TP C_{18} 10 μm 250 × 3.2 mm i.d.	Solution 1: MeOH/H_2O (90:10) Solution 2: MeCN/MeOH (90:10)	Solution 1: retinol Solution 2: vitamins D_2 and D_3	Solution 1: UV 325 nm Solution 2: UV 265 nm	Wickroski and McLean (1984)
Vitamins A and E						
			Normal-phase chromatography			
Breakfast cereals fortified with vitamins A and E	Extract sample with a solvent mixture of $CHCl_3$, EtOH and H_2O at 50 °C	μPorasil 10 μm 300 × 4 mm i.d.	Hexane/$CHCl_3$ containing 1% EtOH (85:15)	Retinyl palmitate, α-tocopheryl acetate	UV 280 nm	Widicus and Kirk (1979)
Butter, whole milk powder, infant formula	Saponify (hot), extract unsaponifiables with diethyl ether	LiChrosorb Si-60 5 μm 250 × 4 mm i.d.	Hexane containing 8% 1,4-dioxan	α-Tocopherol, all-*trans*-retinol	UV and fluorescence detectors connected in series UV (retinol) 325 nm Fluorescence (tocopherol) ex. 293 nm em. 326 nm	Kneifel, Ulberth and Winkler-Macheiner (1987)

Sample	Sample preparation	Column	Mobile phase	Compounds	Detection	Reference
Italian cheeses	Saponify (ambient), extract unsaponifiables with diethyl ether	LiChrosorb Si-60 5 μm 250 × 4 mm i.d., column temperature 44 °C	Hexane containing 0.8% 2-PrOH	Total carotenes, α-, β-, γ-, δ-tocopherols, 13-cis-, 9, 13-di-cis-, 9-cis- and all-trans-retinol	Programmable UV/Vis. 450 nm (carotenes) 295 nm (tocopherols) 328 nm (retinols)	Stancher and Zonta (1983)
Italian cheeses	Saponify (hot), extract unsaponifiables with hexane/ethyl acetate (9 + 1)	Ultrasphere Si 5 μm 250 × 4.6 mm i.d.	(A) 1% 2-PrOH in hexane and (B) hexane in a multi-linear gradient elution	Total carotenes, α-, β-, γ-, δ-tocopherols, 13-cis- and all-trans-retinol	Programmable UV/Vis. and fluorescence detectors connected in series Vis. (carotenes) 450 nm Fluorescence (tocopherols): ex. 280 nm em. 325 nm Fluorescence (retinols): ex. 325 nm em. 475 nm	Panfili, Manzi and Pizzoferrato (1994)

Vitamins A, D and E

Reversed-phase chromatography

Sample	Sample preparation	Column	Mobile phase	Compounds	Detection	Reference
Milk, milk powder	Saponify (hot), extract unsaponifiables with hexane	Spheri-5 RP-18 5 μm 220 × 4.6 mm i.d.	MeOH/H$_2$O (99:1) containing 0.1 M lithium perchlorate	Retinol, vitamin D$_3$, α-tocopherol	Amperometric: (oxidative mode), glassy carbon electrode, +1.05 V vs silver–silver chloride reference electrode	Zamarreño et al. (1992)

Abbreviations: see footnote to Table 2.4.

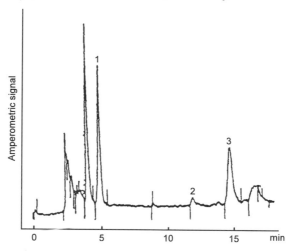

Figure 2.3 Reversed-phase HPLC of vitamins A, D_3 and E in the unsaponifiable fraction of milk. Peaks: (1) retinol; (2) vitamin D_3; (3) α-tocopherol. Conditions as given in Table 2.3. (Reprinted from *Journal of Chromatography*, **623**, 69–74, Zamarreño *et al.*, copyright 1992, with kind permission of Elsevier Science-NL, Sara Burgerhartstraat 25, 1055 KV Amsterdam, The Netherlands.)

Vitamins A and E

For the simultaneous determination of vitamin A and E esters in fortified breakfast cereals, the concentrated lipid fraction was analysed directly by adsorption HPLC using UV measurement at 280 nm (Widicus and Kirk, 1979). Saponification hydrolyses the esters of vitamins A and E to their alcohol forms and allows the tocopherols to be detected fluorimetrically (Kneifel, Ulberth and Winkler-Macheiner, 1987; Panfili, Manzi and Pizzoferrato, 1994).

Vitamins A, D and E

Zamarreño *et al.* (1992) proposed a rapid method for the simultaneous determination of vitamins A, D_3 and E in saponified milk and milk powders using reversed-phase HPLC and amperometric detection. The high sensitivity of the electrochemical detection allows the determination of vitamin D_3 in unenriched fluid milk, which is not possible using UV detection without a preconcentration step. A typical chromatogram showing the indigenous vitamins in fluid bovine milk is depicted in Figure 2.3. The reported on-column detection limits were of the order of 0.07, 4 and 0.2 ng of vitamins A, D_3 and E, respectively.

Water-soluble vitamins

Selected applications are summarized in Table 2.4.

Table 2.4 HPLC methods used for determining two or three water-soluble vitamins concurrently or simultaneously in food

Food	Sample preparation	Column	Mobile phase	Compounds separated	Detection	Reference
Thiamin and riboflavin						
		Reversed-phase chromatography				
Raw and cooked potatoes	Reflux sample with 0.1 N HCl for 30 min, then cool to below 50 °C. Incubate with buffered (pH 4.5) Takadiastase at 45–50 °C for 2 h. Cool, dilute to volume with water and filter *Derivatization*: oxidize thiamin to thiochrome with alkaline $K_3Fe(CN)_6$, filter (0.45 μm)	μBondapak C_{18} 10 μm 300 × 3.9 mm i.d.	Water/MeOH (70:30)	Thiochrome Riboflavin (separate chromatograms)	Fluorescence *Thiochrome* ex. 365 nm em. 435 nm *Riboflavin* ex. 450 nm em. 510 nm	Finglas and Faulks (1984)
Dietetic foods	Digest sample with 0.1 N HCl at 95–100 °C for 30 min, then cool. Adjust pH to 4.5 and incubate with β-amylase and Takadiastase at 37 °C overnight. Dilute to volume with water and filter (0.2 μm) *For riboflavin*: use filtrate directly *For thiamin*: oxidize to thiochrome by treating aliquot of filtrate with alkaline $K_3Fe(CN)_6$. Pass solution through C_{18} Sep-Pak cartridge, wash cartridge with 50 mM sodium acetate then elute thiochrome with MeOH/H_2O (60:40)	μBondapak C_{18} 10 μm 300 × 3.9 mm i.d.	50 mM acetate buffer/(pH 4.5)/MeOH (40:60)	Triochrome Riboflavin (separate chromatograms)	Fluorescence *Thiochrome* ex. 366 nm em. 435 nm *Riboflavin* ex. 422 nm em. 522 nm	Hasselmann et al. (1989)
Soy products	Hydrate dry samples and heat at 90 °C for 30 min. Adjust pH of medium to 2 with 5 N HCl and autoclave at 20 psi for 15 min. Adjust pH of cooled extract to 4.5, centrifuge and filter *For riboflavin*: use filtrate directly *For thiamin*: oxidize to thiochrome by treating aliquot of filtrate with alkaline $K_3Fe(CN)_6$ and neutralizing with conc. H_3PO_4 after 45 s	Ultrasphere C_{18} 5 μm 150 × 4.6 mm i.d.	10 mM acetate buffer (pH 5.5)/MeCN (87:13)	Thiochrome Riboflavin (separate chromatograms)	Fluorescence *Thiochrome* filters *Riboflavin* ex. 436 nm em. 535 nm (filters)	Fernando and Murphy (1990)

Table 2.4 Continued

Food	Sample preparation	Column	Mobile phase	Compounds separated	Detection	Reference
All food types	Autoclave sample with 0.1 N HCl at 121 °C for 30 min, then cool. Adjust pH to 4.5 and incubate with β-amylase and Takadiastase at 37 °C overnight. Add 50% TCA to precipitate soluble proteins, dilute to volume with water and filter *For riboflavin:* analyse filtrate directly *For thiamin:* oxidize to thiochrome by treating aliquot of filtrate with alkaline K₃Fe(CN)₆ and then neutralizing with conc. H₃PO₄ *Clean-up and concentration:* pass the oxidized extract through a preconditioned C₁₈ solid-phase extraction cartridge and wash the cartridge with 50 mM phosphate buffer (pH 7.0). Elute the thiochrome with MeOH/phosphate buffer (80:20)	Novapak C₁₈ 4 µm 150 × 3.9 mm i.d. T = 30 °C	50 mM phosphate buffer (pH 7.0)/ MeOH (70:30)	Thiochrome Riboflavin (separate chromatograms)	Fluorescence *Thiochrome* ex. 366 nm em. 435 nm *Riboflavin* ex. 445 nm em. 522 nm	Ollilainen et al. (1993)
All food types	Autoclave homogenized sample with 0.1 N HCl at 121 °C for 15 min, then cool. Adjust pH to 4.0–4.5 with 2 N sodium acetate and incubate with Claradiastase at 50 °C for 3 h. Add 50% TCA, heat at 90 °C for 15 min, then cool. Adjust pH to 3.5 with 2 N sodium acetate, dilute with water, then filter through paper *Derivatization:* oxidize to thiochrome by treating aliquot of filtrate with alkaline K₃Fe(CN)₆ and neutralize with conc. H₃PO₄ *Clean-up and concentration:* pass the oxidized extract through a preconditioned C₁₈ Sep-Pak cartridge and wash the cartridge sequentially with 5 mM phosphate buffer (pH 7.0) and 5 mM phosphate buffer/MeOH (95:5). Elute the vitamins with 50% aqueous MeOH, dilute to volume and filter (0.45 µm)	Radial-PAK C₁₈ 10 µm 100 × 8 mm T = 30 °C	5 mM phosphate buffer (pH 7.0)/ MeOH (65:35)	Thiochrome Riboflavin (separate chromatograms)	Fluorescence *Thiochrome* ex. 360 nm em. 425 nm *Riboflavin* ex. 440 nm em. 520 nm	Hägg (1994)

Sample	Procedure	Column	Mobile phase	Compounds determined	Detection	Reference
Cereals	Autoclave sample with 0.1 N HCl at 121 °C for 30 min, then cool. Adjust pH to 4.5 with 2 N sodium acetate, dilute with water and filter through paper. *Derivatization*: oxidize thiamin to thiochrome with alkaline K₃Fe(CN)₆ *Clean-up and concentration*: pass the oxidized extract through a preconditioned C₁₈ Sep-Pak cartridge and wash the cartridge with 5 mM ammonium acetate (pH 5.0). Elute the vitamins with MeOH/5 mM ammonium acetate (pH 5.0) and filter (0.45 μm)	μBondapak C₁₈ 10 μm 300 × 3.9 mm i.d.	5 mM acetate buffer (pH 5.0)/MeOH (72:28)	Thiochrome and riboflavin (in same chromatogram)	Fluorescence (wavelength-programmable) *Thiochrome* ex. 370 nm em. 435 nm *Riboflavin* ex. 370 nm em. 520 nm	Sims and Shoemaker (1993)
Peas, beans, liver, skim milk and enriched wheat flour	Autoclave samples with dilute HCl at 121 °C for 15 min, then cool. Adjust pH to 4.0–4.5 and incubate with Takadiastase at 48 °C for 3 h. Deproteinize by adding 50% TCA and heating at 95–100 °C for 15 min. Cool, adjust pH to 3.5, dilute to volume with water and filter. *Derivatization*: oxidize thiamin to thiochrome by treatment with alkaline K₃Fe(CN)₆ followed by conc. H₃PO₄ *Clean-up and concentration*: pass the oxidized extract through a preconditioned C₁₈ Sep-Pak cartridge and wash the cartridge sequentially with 10 mM phosphate buffer (pH 7.0) and 10 mM phosphate buffer/MeOH (95:5). Elute the vitamins with 50% aqueous MeOH and dilute to volume	Radial-PAK C₈ (octyl) 10 μm 100 × 8 mm	10 mM phosphate buffer (pH 7.0)/MeOH (63:37)	Thiochrome and riboflavin (in same chromatogram)	Fluorescence *Thiochrome and riboflavin* ex. 360 nm em. 415 nm (filters) *Riboflavin alone* ex. 450 nm em. 530 nm (filters)	Fellman et al. (1982)
Cereal products	Autoclave ground samples with 0.1 N HCl at 121 °C for 30 min, then cool. Adjust pH to 4.0–4.5 and incubate with Takadiastase at 50 °C for 3 h (minimum). Cool and dilute to volume with water. *Derivatization and clean-up/concentration*: as for Fellman et al. (1982) (see above)	Radial-PAK C₁₈ 10 μm 100 × 8 mm	5 mM phosphate buffer (pH 7.0)/MeOH (65:35)	Thiochrome and riboflavin (in same chromatogram)	Fluorescence (filters)	Reyes and Subryan (1989)

Table 2.4 *Continued*

Food	Sample preparation	Column	Mobile phase	Compounds separated	Detection	Reference
		Ion interaction chromatography				
Cereal products, fortified breakfast cereals	Autoclave sample with 0.1 N HCl at 121 °C for 30 min, cool and centrifuge	μBondapak C$_{18}$ 10 μm 300 × 3.9 mm i.d.	10 mM phosphate buffer (pH 7.0)/MeCN (12.5:87.5) containing 5 mM sodium heptane sulphonate	Thiamin and riboflavin (in same chromatogram)	UV 254 nm	Kamman, Labuza and Warthesen (1980)
Meat and liver	Autoclave homogenized sample with 0.01 N HCl at 121 °C for 30 min, then cool. Add Takadiastase, Claradiastase and papain, adjust to pH 4.5 and incubate at 37 °C for 16–18 h. Filter through paper, adjust to pH 6.5, refilter and dilute with water	Nucleosil C$_{18}$ 3 μm 150 × 4.6 mm i.d. $T = 45\,°C$	10 mM phosphate buffer (pH 3.0)/MeCN (84:16 for meat and 85:15 for liver) containing 5 mM sodium heptane sulphonate	Thiamin and riboflavin (in same chromatogram)	UV 254 nm	Barna and Dworschák (1994)
Cereal products, fresh meat and meat products, fresh fruit and vegetables, eggs, milk, yoghurt	Digest sample with 0.1 N HCl at 95–100 °C for 30 min, cool and dilute to volume. Adjust pH of aliquot to 4.0–4.5 and incubate with Clarase at 45–50 °C for 3 h. Cool and filter. *Clean-up and concentration:* pass filtrate through Sep-Pak C$_{18}$ cartridge, wash cartridge with aqueous 5 mM sodium hexane sulphonate and elute the vitamins with methanolic 5 mM sodium hexane sulphonate	Radial-PAK C$_{18}$ 10 μm 100 × 8 mm	MeOH/water (40:60) containing 5 mM hexane sulphonic acid	Thiamin and riboflavin (in same chromatogram)	Fluorescence *Thiochrome* ex. 360 nm em. 425 nm (filters) after post-column derivatization with alkaline K$_3$Fe(CN)$_6$ *Riboflavin* ex. 360 nm em. 500 nm (filters)	Wimalasiri and Wills (1985) Wills, Wimalasiri and Greenfield (1985)
Enriched cereal-based products	Digest sample with 0.1 N H$_2$SO$_4$ at 95–100 °C for 10 min, then cool. Incubate with buffered Mylase at 56 °C for 1 h. Cool, dilute to volume and filter	μBondapak C$_{18}$ 10 μm 300 × 3.9 mm i.d.	MeOH/water (36:64) containing 5 mM hexane sulphonic acid and 1% acetic acid	Thiamin and riboflavin (in same chromatogram)	Dual fluorescence *Thiochrome* 1st detector: filters, after post-column derivatization with alkaline K$_3$Fe(CN)$_6$ *Riboflavin* 2nd detector: filters	Mauro and Wetzel (1984)

Riboflavin and nicotinic acid

Reversed-phase chromatography

Riboflavin and pyridoxine

Ion interaction chromatography

Nicotinamide and pyridoxine

Sample	Sample preparation	Column	Mobile phase	Analytes	Detection	Reference
Meat (beef, pork, lamb)	Autoclave homogenized sample with 0.1 N HCl at 121 °C for 30 min, then cool. Adjust pH to 4.0–4.5 and incubate with Takadiastase and papain at 42–45 °C for 2.5–3.0 h. Precipitate proteins by adding TCA, heating to 100 °C for 10 min, then cooling, diluting and filtering. Thiamin removed by converting to thiochrome with alkaline $K_3Fe(CN)_6$ and extracting with isobutanol	Alltech C_{10} 10 μm	20 mM phosphate buffer (pH 7.0)/MeOH (70:30)	Nicotinic acid and riboflavin (in same chromatogram) Thiamin (as thiochrome) can be determined separately	*Nicotinic acid* UV 254 nm *Riboflavin* Fluorescence ex. 464 nm em. 540 nm *Thiochrome* Fluorescence ex. 378 nm em. 430 nm Detectors connected in series	Dawson, Unklesbay and Hedrick (1988)
Infant formula (fortified)	Extract sample with water. Deproteinize by pH adjustment to 1.7 and then to 4.6, dilute to volume with water and filter	Spherisorb ODS-1 5 μm 150 × 4.6 mm i.d.	40 mM triethyl ammonium phosphate buffer (pH 3.0) containing 7.5 mM sodium octane sulphonate, 10% MeOH and 5% MeCN	Pyridoxal, pyridoxine, riboflavin	Fluorescence ex. 285 nm em. 546 nm (filter)	Ayi, Yuhas and Deangelis (1986)
Fortified foods (e.g. milk products, powdered meals)	Digest sample with 2 N H_2SO_4 at 95–100 °C for 30 min. Cool, dilute to volume with water and filter	Partisil ODS 10 μm or Ultrasphere ODS 5 μm	2.5 M sodium acetate (80 ml), acetic acid (50 ml) and water (25 ml) containing 1.1 g sodium heptane sulphonate. Mixture diluted to 1 litre with water	Nicotinamide and pyridoxine	*Nicotinamide* UV 260 nm *Pyridoxine* Fluorescence ex. 296 nm em. 396 nm Detectors connected in series	Rees (1989)

Table 2.4 Continued

Food	Sample preparation	Column	Mobile phase	Compounds separated	Detection	Reference
Thiamin, riboflavin and pyridoxine						
Fortified cereal products	Digest ground sample with 0.1 N H_2SO_4 at 95–100 °C for 30 min, then cool. Incubate with buffered Clarase at 55 °C for 1 h, centrifuge and dilute to volume	μBondapak C_{18} 10 μm 250 × 4.6 mm i.d. and 50 × 4.6 mm i.d. short column. Sample clean-up achieved by column switching	MeOH/water/acetic acid (30:69:1) containing 5 mM sodium hexane sulphonate	Pyridoxine, riboflavin and thiamin	Dual fluorescence *Pyridoxine and riboflavin* 1st detector: ex. 288 nm em. 418 nm (cut-off filter) *Thiochrome* 2nd detector: ex. 360 nm em. 460 nm (cut-off filter) after post-column oxidation of thiamin with alkaline $K_3Fe(CN)_6$ to form thiochrome	Wehling and Wetzel (1984)
Infant formulas, and medical foods (fortified)	Precipitate proteins by adding perchloric acid, stirring for 0.5 h then adjusting pH to 3.2. Dilute with mobile phase and refrigerate overnight. Filter through paper, refilter through nylon (0.45 μm)	Nova-PAK C_{18} 300 × 3.9 mm i.d.	Water/MeCN containing sodium hexane sulphonate and NH_4OH, adjusted to pH 3.6 with phosphoric acid	Pyridoxine, riboflavin, and thiamin	Fluorescence (wavelength-programmable) A specially designed flow system allows the same detector to be used for two injection sequences. First sequence: *Pyridoxine:* ex. 295 nm; em. 395 nm *Riboflavin:* ex. 440 nm; em. 565 nm 2nd sequence: *Thiochrome:* ex. 360 nm; em. 435 nm after post-column oxidation of thiamin with alkaline $K_3Fe(CN)_6$ to form thiochrome	Chase et al. (1993)

MeOH, methanol; EtOH, ethanol; 2-PrOH, isopropanol; MeCN, acetonitrile; THF, tetrahydrofuran; TCA, trichloroacetic acid; EDTA, ethylenediaminetetraacetic acid; BHT, butylated hydroxytoluene.

Thiamin and riboflavin

Autoclaving food samples with dilute mineral acid, followed by enzymatic hydrolysis, liberates both thiamin and riboflavin, enabling these vitamins to be determined from one sample hydrolysate.

For the analysis of fortified foods, such as breakfast cereals, thiamin and riboflavin can be determined simultaneously (i.e. in one chromatographic run), and without the need for derivatization, by ion interaction chromatography and photometric detection at 254 nm (Kamman, Labuza and Warthesen, 1980). Most applications, however, use fluorescence detection as a means of obtaining adequate sensitivity for thiamin after conversion to its fluorescent oxidation product, thiochrome, by reaction with alkaline potassium hexacyanoferrate(III). The chromatographic mode employed depends upon whether the derivatization is carried out pre-column or post-column. Pre-column derivatization provides the opportunity for detecting thiochrome and riboflavin simultaneously, using simple reversed-phase chromatography. This technique has been carried out using a single fluorescence detector equipped with suitable excitation and emission filters to obtain adequate responses for each compound (Fellman et al., 1982; Reyes and Subryan, 1989). In order to obtain the optimum wavelength settings to achieve maximum sensitivity for each compound, it would be necessary to install a wavelength-programmable spectrofluorimeter or two fluorescence monitors connected in series. It is usually preferred to determine thiochrome and riboflavin successively (i.e. in two successive chromatographic runs) so that the optimum detection parameters can be selected (Finglas and Faulks, 1984; Hasselmann et al., 1989; Fernando and Murphy, 1990). Although the separation of thiochrome from riboflavin might seem to be unnecessary if they are to be determined successively, it is in fact essential. If these compounds coeluted, the radiation emitted by thiochrome could be reabsorbed by riboflavin by the inner filter effect, causing a diminished fluorescent signal.

With post-column derivatization, thiamin and riboflavin can be chromatographed simultaneously using ion interaction chromatography. This technique, used with two fluorescence detectors, has been applied to the analysis of a wide range of foods (Mauro and Wetzel, 1984; Wills, Wimalasiri and Greenfield, 1985; Wimalasiri and Wills, 1985).

The method of Hasselmann et al. (1989) has been subjected to a collaborative study (Arella et al., 1996). Thiamin and riboflavin were co-extracted by enzyme hydrolysis using a mixture of Takadiastase and β-amylase. Minor modifications to the method were filtration of the solution obtained after the incubation step through a fine filter paper and a slight change of the methanol–0.05 M sodium acetate proportion of the mobile phase (30 : 70 v/v instead of 40 : 60 v/v without adjustment to pH 4.5). With all nine foodstuffs studied, the recovery rate was always greater than 89%, except with chocolate powder, for which it was reduced to *c*.50% for thiamin and 75% for riboflavin. The proposed method, despite this drawback, was confirmed as the official French method for thiamin and riboflavin determination in foodstuffs for nutritional purposes.

Riboflavin and pyridoxine

Ayi, Yuhas and Deangelis (1986) developed a comparatively simple method for the simultaneous determination of supplemental riboflavin and pyridoxine in fortified milk-based and soy-based infant formula products. The method took advantage of the natural fluorescence of these vitamins and was performed using a single fluorescence detector equipped with carefully selected filters. The ion interaction mode of chromatography was capable of isolating pyridoxine from the other B_6 vitamers and from 4-pyridoxic acid. Similarly, riboflavin was well separated from its potential degradation products, lumiflavin and lumichrome.

Nicotinamide and pyridoxine

Rees (1989) used ion interaction chromatography, and absorbance and fluorescence detectors connected in series, for the simultaneous determination of nicotinamide and pyridoxine, respectively, in fortified food products. The extraction process (heating the sample with $2 N H_2SO_4$ for 30 min at 100 °C) caused a 6% conversion of nicotinamide to nicotinic acid, but this was catered for by subjecting the nicotinamide standard to the same treatment as the sample extract. A Partisil ODS column containing residual free silanol groups was used for routine analyses and gave retention times of 8.0 and 5.8 min for nicotinamide and pyridoxine, respectively. If a new product was being analysed, or it was known that the pyridoxine peak was subject to interference from other peaks, the analysis was repeated using a fully end-capped Ultrasphere ODS column. On this column, the order of elution was reversed, nicotinamide and pyridoxine having retention times of 4.2 and 8.6 min, respectively. The detection limit for nicotinamide was 0.1 mg/100 g, and that for pyridoxine was 0.01 mg/100 g. Pyridoxine is usually present in fortified food products at levels 10 times lower than those of nicotinamide, and hence the higher sensitivity of detection for pyridoxine was fortuitous. HPLC results for milk powder, high-protein meal powder and diet meal powder were in good agreement with results obtained by microbiological assay.

Thiamin, riboflavin and pyridoxine

Wehling and Wetzel (1984) accomplished the simultaneous determination of thiamin, riboflavin and pyridoxine in fortified cereal products after extraction of the samples by heating with 0.1 N H$_2$SO$_4$, followed by Clarase treatment to break down the starch. Ion interaction chromatography with isocratic elution separated the three vitamins within 20 min, but more strongly retained components took up to 50 min to elute. The 50 min chromatographic analysis time was halved by the use of on-line sample clean-up, utilizing a short (5 cm) C$_{18}$ column in addition to the C$_{18}$ analytical (25 cm) column, and a column switching arrangement. Aliquots of the test extract were injected on to the short column, with the effluent being routed directly to waste. Solvent flow through the analytical column and detector was maintained independently by a second pump. At 0.6 min after injection, the fraction of effluent containing the vitamins was diverted on to the analytical column, and at 3.5 min the valve was switched back to its original position. During the remaining time, as the vitamins were undergoing analysis, the more strongly retained components were eluted from the short column directly to waste. Fluorescence detection was used to measure the native fluorescence of pyridoxine and riboflavin. Thiamin, in the form of thiochrome, was detected by a second fluorescence monitor after post-column derivatization, the strong alkaline conditions suppressing the fluorescence of pyridoxine and riboflavin. The concentration of potassium hexacyanoferrate(III) added for post-column thiochrome formation was decreased significantly from levels used in other procedures in order to minimize the quenching effect of excess reagent during detection. Detection limits were 2 µg/g for pyridoxine and 1 µg/g for both thiamin and riboflavin.

2.9 CAPILLARY ELECTROPHORESIS

Electrophoresis is the migration of charged particles in solution in an electric field gradient. Conventional zone electrophoresis on solid supports and gels has not been easily adapted to on-line sample application, detection, quantification or automated operation, and to meet these needs capillary electrophoresis has been developed. For this purpose, most capillaries are made of fused silica that has been coated with an external layer of polyimide to make it very flexible and less susceptible to fracture. The window for the on-line absorbance or fluorescence detection cell is made by removing a small section of the polyimide coating. Besides being well suited for automation, the use of open tubular capillaries offers several important advantages over solid supports and gels. The electro-osmotic flow causes all solutes to elute at one end of the capillary, allowing on-line monitoring of sample components using HPLC-like detectors. It is important to keep the temperature under control as the electroosmotic flow and the mobility of sample ions are dependent on the viscosity of the solution. An enhanced heat dissipation from small-diameter capillaries permits the use of unusually high voltages, which lead to extremely efficient separations with a dramatic decrease in analysis time. Also contributing to the high efficiency is a flat plug-like flow profile, so the zone broadening is much less than that obtained with laminar flow. The main disadvantage of using capillary electrophoresis for quantitative analysis may be the high concentration limits of detection that arise from the short optical pathlengths and small introduction volumes inherent to the technique. This drawback is partly compensated for by the fact that the sharp peaks produced make integration easier (Lindeberg, 1996).

Two modes of capillary electrophoresis have found application in vitamin analysis. Capillary zone electrophoresis is limited to the separation of charged solutes according to their mobility in free solution. Micellar electrokinetic capillary chromatography facilitates the separation of neutral solutes as well as charged solutes according to their partition coefficients between solution and micelle.

2.9.1 Capillary zone electrophoresis

The apparatus in its simplest form consists of a buffer-filled length of capillary tubing suspended between two reservoirs filled with buffer. Samples are introduced at one end and, under the influence of an applied electric field, migrate toward the other end of the capillary at different rates according to their electrophoretic mobility. An in-line detector monitors the eluted solutes and produces an electropherogram which is analogous to a chromatogram. The basic theory of capillary zone electrophoresis and system parameters have been described by Jorgenson and Lukacs (1983).

The separation mechanisms that take place in capillary zone electrophoresis depend on whether the internal surface of the fused silica column is coated or uncoated. When the capillary is uncoated, solute migration is dominated by the process of electroosmosis, which causes the flow of the buffer solution itself. Electroosmosis occurs because the uncoated silica surface, under normal aqueous conditions with small binary electrolytes, has an excess of anionic charge resulting from ionization of the silanol groups. Counterions to these anions are in the stagnant double layer adjacent to the capillary walls. This cationic nature extends into the diffuse layer, which is mobile. When an electric field is applied, the cationic counterions in the diffuse layer, accompanied by their spheres of hydration, migrate towards the cathode, causing a concomitant flow of solvent through the capillary (Ewing, Wallingford and Olefirowicz, 1989). Electroosmotic flow is much stronger than the electrophoretic mobility of an ion, therefore all solutes are carried towards the cathode, regardless of charge. Cations with the highest charge/mass ratio will be eluted first followed by cations with reduced ratios. Neutral solutes move with the speed of the electroosmotic flow and follow the cations. Finally, the anions elute, those with the highest charge/mass ratio appearing last. An increase in pH causes more ionization of silanol groups and hence an increased electroosmotic flow rate. Other variables that affect flow rate are viscosity, ionic strength, the voltage and the dielectric constant of the buffer (Olechno *et al.*, 1990).

When a coated fused-silica capillary is used, the electroosmotic flow is strongly reduced and the neutral solutes exhibit a negligible mobility under an electric field. In this case the separation is only affected by the differences in solute charge (Chiari *et al.*, 1993).

2.9.2 Micellar electrokinetic capillary chromatography

When an ionic surfactant is added to the buffer solution above a certain critical concentration, the surfactant monomers form spherical aggregates called micelles. The monomers are arranged with the charged groups facing the surface of the micelle and the hydrophobic tails forming the interior. Micellar electrokinetic capillary chromatography (MECC) is most commonly performed with anionic surfactants, especially sodium dodecyl sulphate, which forms micelles with a large net negative charge, giving them a high electrophoretic mobility towards the anode. However, as the electroosmotic flow of the aqueous phase is predominant, the micelles are transported towards the cathode, but at a slower rate than the bulk aqueous phase. The separation is based on the partitioning of the neutral solute molecules between the aqueous mobile phase and the hydrophobic interior of the slower moving micelle (Ewing, Wallingford and Olefirowicz, 1989). This separation mechanism is much the same as observed in traditional liquid–liquid partition chromatography, with the micellar phase functioning as a 'pseudo-stationary' phase. MECC will separate ionic solutes as well as neutral solutes, the retention mechanisms being a combination of charge/mass ratios, hydrophobicity and charge interactions at the surface of the micelles (Olechno *et al.*, 1990). The choices of anionic and cationic surfactants further increases the diversity of MECC.

2.9.3 Applications

Water-soluble vitamins

Both capillary zone electrophoresis and MECC have been applied to the determination of ascorbic acid in fruit juices and plant tissues (Chapter 15).

Fujiwara, Iwase and Honda (1988) reported the simultaneous determination of seven water-soluble vitamins within 22 minutes by MECC

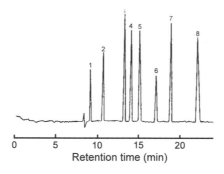

Figure 2.4 Electropherogram of water-soluble vitamins present in a commercial vitamin injection product. Peaks: (1) nicotinamide; (2) pyridoxine hydrochloride; (3) cyanocobalamin; (4) ascorbic acid; (5) riboflavin-5'-phosphate (sodium salt); (6) nicotinic acid; (7) ethyl *p*-aminobenzoate (internal standard); (8) thiamin hydrochloride. Fused-silica capillary tube, 80 cm × 100 μm i.d.; carrier, 0.02 M phosphate solution (pH 9.0) containing 0.05 M sodium dodecyl sulphate; current applied, 75 μA; detection wavelength, 254 nm (Reprinted from *Journal of Chromatography*, **447**, 133–40, Fujiwara *et al.*, copyright 1988, with kind permission of Elsevier Science-NL, Sara Burgerhartstraat 25, 1055 KV Amsterdam, The Netherlands.)

using 0.05 M sodium dodecyl sulphate as the surfactant at a buffered pH of 9.0. Figure 2.4 is an electropherogram showing the separation of thiamin hydrochloride, riboflavin phosphate (sodium salt), nicotinamide, nicotinic acid, pyridoxine hydrochloride, cyanocobalamin and L-ascorbic acid. Similar conditions were used in another study (Nishi *et al.*, 1989) to separate thiamin, riboflavin, riboflavin phosphate, nicotinic acid, nicotinamide, vitamin B_6 vitamers (pyridoxine, pyridoxal, pyridoxamine, pyridoxal phosphate and pyridoxamine phosphate) and cyanocobalamin (vitamin B_{12}).

The analysis of B_6 vitamers has been reported using MECC and laser-excited fluorescence (Burton, Sepaniak and Maskarinec, 1986) and electrochemical (Yik *et al.*, 1991) detection.

2.10 CONTINUOUS-FLOW ANALYSIS

Continuous-flow analysis has been applied to the determination of certain vitamins using either segmented-flow or continuous-flow methods.

2.10.1 Segmented-flow methods

The Technicon Auto Analyzer II modular system, which was introduced commercially in 1957 for use in clinical analysis, has been adapted to automate a number of official (Association of Official Analytical Chemists, AOAC) fluorimetric and colorimetric methods for determining vitamins in foods. The chemical reactions take place in a continuously flowing stream segmented by air bubbles to limit sample dispersion and to promote mixing of sample with diluent or reagent by so-called bolus flow. The mixing is enhanced by arrangement of the tubing as a horizontally orientated helical coil. For some applications, a dialyser is incorporated into the system. A peristaltic proportioning pump moves sample (or standard) and reagents through the system to the photometric or fluorimetric detector. The concentration of reaction product corresponding to a state of chemical equilibrium is recorded as a flat-topped profile, the height of which can be compared with that of a standard to give the analyte concentration (Snyder *et al.*, 1976). Segmented-flow methods are usually operated at sampling rates of 30–60 per hour.

Extraction of the vitamins from the food matrix must be performed manually, and so the modified official methods can only be truly described as semi-automated procedures. Comparison of results obtained by semi-automated and corresponding manual procedures show overall good agreement. As well as reducing the labour requirement, the automation significantly improves the precision of reactions in which timing and reagent volume are critical.

2.10.2 Flow-injection analysis

This is a more recent innovation, in which there is no air segmentation and it is not necessary for a state of chemical equilibrium to be reached. In flow-injection analysis, the sample is introduced into the carrier stream as a discrete plug, and the presence of a sample–carrier interface allows diffusion-controlled dispersion of the sample as it is swept through narrow-bore tubing to create a concentration gradient. The flow-through detector monitors the change in concentration of the reaction product, which is displayed as a well-defined peak. Quantification of analyte concentration can be achieved by comparing the peak height of the sample with that of a standard. In practice, the area of the peak is measured by integration, since the peak width will not vary significantly. Flow-injection systems permit faster sampling rates and consume less reagent than comparable segmented-flow methods. The technique depends on precise sample injection, reproducible timing and controlled sample dispersion. These features preclude the requirement for reaching the state of chemical equilibrium, since the residence time of the sample in the analytical system is constant. Optimization of a flow-injection system for a given application is achieved by empirical selection of manifold design (tubing bore size and length), sample and reagent volume, and flow rate (Betteridge, 1978; Ranger, 1981). Osborne and

Tyson (1988) have reviewed the principles, instrumentation and techniques of flow-injection analysis, and considered the scope of its actual and potential applications in food and beverage analysis.

2.10.3 Applications

Fat-soluble vitamins

Segmented-flow analysis has been used to automate a fluorimetric method for determining vitamin A in milk, including the saponification step (Thompson and Madère, 1978). Bourgeois, George and Cronenberger (1984) also used segmented-flow analysis to automate a fluorimetric technique for determining α-tocopherol in the unsaponifiable fraction of foods and feeds.

Water-soluble vitamins

Segmented-flow analysis has been used to automate the chemical stages of the AOAC fluorimetric methods for thiamin, riboflavin and vitamin C, and the AOAC colorimetric method for niacin (Roy and Conetta, 1976; Roy, 1979). The methodology of these applications is discussed in the relevant chapters on these vitamins. Dunbar and Stevenson (1979) reported the simultaneous determination of thiamin and riboflavin in infant formulas using segmented-flow analysis and a common extraction procedure involving both acid and enzymatic hydrolysis.

Russell and Vanderslice (1992) applied flow-injection analysis to the AOAC fluorimetric method for riboflavin. Several flow-injection analysis methods have been reported for determining vitamin C in foods (Chapter 15).

2.11 MICROBIOLOGICAL METHODS FOR B-GROUP VITAMINS

2.11.1 General principles

The general principles of microbiological assays are discussed here. Specific techniques for individual B-group vitamins are discussed in the relevant chapters. Accepted official (AOAC) microbiological methods for nutrition labelling are listed in Table 2.5.

Turbidimetric microbiological assays

These methods, as applied to the determination of the B-group vitamins, are based on the absolute requirement of certain microorganisms for the vitamin; that is, the organisms can multiply only when the vitamin is

Table 2.5 Accepted official microbiological methods for nutrition labelling (from Sullivan and Carpenter, 1993)

Vitamin	AOAC No.	Assay organism (ATCC No.)[a]	Matrices
Riboflavin (vitamin B_2 activity)	960.46	*L. casei* subsp. *rhamnosus* (7469)	All
Niacin	960.46	*L. plantarum* (8014)	All
Niacin	985.34	*L. plantarum* (8014)	Milk-based infant formula
Vitamin B_6	961.15	*S. carlsbergensis* (9080)	All
Vitamin B_6	985.32	*S. carlsbergensis* (9080)	Milk-based infant formula
Pantothenic acid (free form only)	960.46	*L. plantarum* (8014)	All
Folic acid (free from only)	960.46	*Enterococcus hirae* (8043)[b]	All
Vitamin B_{12} activity	960.46	*L. delbrueckii,* 313 strain (7830)	All
Vitamin B_{12} activity	986.23	*L. delbrueckii,* 313 strain (7830)	Milk-based infant formula

[a] American Type Culture Collection.
[b] *L. casei* subsp. *rhamnosus* (7469) commonly preferred, but not collaboratively studied.

present in the surrounding medium. Aliquots of a standard solution of the vitamin to be determined, or aliquots of the sample extract containing the vitamin, are added to an initially translucent basal nutrient medium, complete in all respects except for the vitamin in question. Following inoculation with the assay organism, the organism multiplies in proportion to the vitamin content of the standard or sample, and the extent of the growth is ascertained by measuring the turbidity produced. Over a defined concentration range, the measured response will be directly proportional to the amount of vitamin present and, within this range, the sample solution and standard vitamin solution can be compared accurately. The response is highly dependent on whether bound forms of the vitamin are released during the extraction stage of the assay.

Methods based on the measurement of metabolic carbon dioxide

The radiometric microbiological assay (RMA) is based upon the measurement of radioactive $^{14}CO_2$ generated from the metabolism of a ^{14}C-labelled nutrient by the test organism, and has been applied to the determination of niacin (Guilarte and Pravlik, 1983), vitamin B_6 (Guilarte, 1983), biotin (Guilarte, 1985), folate (Chen, Hill and McIntyre, 1983) and vitamin B_{12} (Chen and McIntyre, 1979) in foods, and also to pantothenic acid in human blood and milk (Guilarte, 1989). In these applications, the radioactivity is measured automatically by means of a commercially available gas flow system incorporating an ionization chamber (Guilarte, 1991a). The RMA combines the biological specificity of measuring a vitamin-dependent microbiological metabolic reaction with the sensitivity and accuracy of radioactive decay measurement. Extraction methods suitable for the analysis of thiamin, niacin, pantothenic acid and biotin with the RMA have been discussed by Guilarte (1991b). Sample preparation is simplified due to the fact that coloured, turbid or precipitated debris do not interfere with the $^{14}CO_2$ output or detection; furthermore, the scrupulous cleaning of glassware is unnecessary.

A nonradiometric technique for measuring metabolic CO_2 employs an infra-red CO_2 analyser, which measures automatically the infra-red radiation absorbed by the CO_2 band at 4.2 µm (Goli and Vanderslice, 1989).

2.11.2 Procedures for turbidimetric microbiological assays

In the standard procedure, the appropriate basal medium, free of the vitamin to be determined, is prepared at twice its final concentration (double strength). Multiple aliquots of a standard solution of the pure vitamin and of suitably prepared extracts of the test food are added to a series of uniform test tubes in amounts suitable to produce gradations in growth between no growth and maximum growth. The contents of all tubes are diluted with water to the same volume, and an equal volume of the translucent basal medium is added. The tubes are sterilized, cooled to a uniform temperature, and then inoculated with an actively growing culture of the test organism. The assay tubes are incubated for 22–24 h at a constant temperature near the optimum for the test organism until growth has reached the maximum permitted by the limiting vitamin present. The growth response to standard and test extract is then determined by measuring the turbidity produced. The data obtained from the standards are used to construct a standard curve from which the vitamin concentrations of the various sample aliquots are derived. The use of multiple aliquots allows a validity check to be carried out: the vitamin concentration found should be directly proportional to the volume of aliquot taken. The amount of vitamin present in the original sample is then calculated at the different test levels, and the results are averaged to obtain the final result.

Assay organisms

Lactic acid bacteria approach the ideal for microbiological assay work and are the most widely used assay organisms. Their nutritional requirements are specific and complex, they grow readily in synthetic and semi-synthetic media, and they are nonpathogenic. They are not prone to mutation and have maintained their characteristics unimpaired after many years of subculture in the laboratory. Their growth can be easily followed by turbidity measurement. The rod forms (genus, *Lactobacillus*) are microaerophilic and the coccus forms (genera, *Enterococcus, Leuconostoc*) are facultative aerobes (Snell, 1950). This ability to grow well in limited amounts of air means that test tubes of liquid medium can conveniently be used for assay purposes.

Lactic acid bacteria do not respond to all forms of vitamin B$_6$, and certain species of yeasts are used instead for the determination of this vitamin. Because yeasts grow aerobically, such assays have the inconvenience of requiring constant and uniform shaking during incubation.

Protozoa have more highly developed ingestive and digestive systems than do bacteria and yeasts, and therefore exhibit a more mammalian-like response to the various naturally occurring forms of the vitamins. However, the test growth period for protozoa is longer than that for lactic acid bacteria (3–5 days versus 24–48 h), conditions of growth are more demanding, and growth response is more difficult to measure.

The assay organisms are particular strains obtained from various culture collections such as the American Type Culture Collection (ATCC). Cultures are despatched in freeze-dried form sealed under vacuum in glass ampoules, which must be opened under strictly aseptic conditions. Growth may be slower than usual immediately after resuscitation, and the culture must be activated by making at least three successive daily transfers in liquid culture medium before preparing agar stab or slope cultures.

Assay organisms commonly used for determining B-group vitamins are listed in Table 2.6.

Assay media

Media for the microbiological assay of the B-group vitamins are commercially available in a dehydrated form from Difco Laboratories, Michigan, USA. They are reconstituted for use simply by suspending the required weighed amount in distilled water, heating to effect solution, and autoclaving. The use of dehydrated media allows a fresh batch of medium to be made up in the amount required with the minimum of time and effort. Alternatively, the media can be prepared in the laboratory from individual ingredients (Association of Vitamin Chemists, Inc., 1966; Pearson, 1967).

General assay procedure

The standard assay procedure using lactic acid bacteria can be broken down into a number of steps:

Table 2.6 Commonly used assay organisms and basal media for determining B-group vitamins

Vitamin	Organism (ATCC No.)	Basal medium (Difco Code No.)
Thiamin	L. viridescens (12706)	Bacto-Thiamin Assay Medium LV (0808)
Riboflavin	L. casei subsp. rhamnosus (7469)	Bacto-Riboflavin Assay Medium (0325)
Riboflavin	E. faecalis (10100)	According to Barton-Wright (1967)
Niacin (as nicotinic acid)	L. plantarum (8014)	Bacto-Niacin Assay Medium (0322)
B_6	S. carlsbergensis (9080)	Bacto-Pyridoxine Y Medium (0951)
B_6	K. apiculata (9774)	According to Barton-Wright (1971)
Pantothenic acid	L. plantarum (8014)	Bacto-Pantothenate Assay Medium (0604)
Biotin	L. plantarum (8014)	Bacto-Biotin Assay Medium (0419)
Total folate	L. casei subsp. rhamnosus (7469)	Bacto-Folic Acid Casei Medium (0822)
B_{12}	L. delbrueckii subsp. lactis (4797)	Bacto-Vitamin B_{12} Assay Medium (0360)
B_{12}	O. malhamensis (11532)	According to Ford (1953)

1.   Maintenance of stock cultures
2.   Preparation of the inoculum culture
3.   Preparation of the basal medium
4.   Extraction of the vitamin from the test material
5.   Setting up the assay
6.   Quantification.

Maintenance of stock cultures
It is obviously essential to maintain a pure culture of each assay organism and to preserve its viability and sensitivity. It is possible to sustain these requirements over a period of several years by employing a regular schedule of subculturing into a nutrient medium of suitable composition. The frequency of subculturing depends upon the composition of the medium and upon the particular assay organism or group of organisms used. Stock cultures of lactic acid bacteria are maintained as stab cultures in a semi-synthetic agar-based maintenance medium, which contains all of the nutritional factors essential for the organism's normal growth and metabolism. Such media invariably contain yeast extract and glucose,

plus other nutrients. A buffer salt is included to prevent the pH from rapidly dropping to levels which would inhibit growth, although growth will proceed until a pH of at least 4 is reached (Barton-Wright, 1952). The medium is formulated to give an initial pH (typically 6.8) that is somewhat higher than the optimum pH of 5.5–6.5 for most lactic acid bacteria (Snell, 1948). This is to allow for the production of acid that occurs when the medium is sterilized by autoclaving.

At regular intervals, three fresh stab cultures are prepared in maintenance medium from a stock culture. The usual procedure is to distribute 10 ml quantities of medium into lipless test tubes, 16–20 mm in diameter, and autoclave the plugged tubes for 20 min at 15 lb pressure (121 °C). After cooling, stab inoculations are made into the solidified agar. The tubes are incubated under the appropriate conditions for each vitamin (typically overnight at 37 °C) that produce distinctly visible growth along the line of the stab. Since the growth rate of an organism is a function of temperature, the incubation must be precisely controlled to within ±0.5 °C of the selected temperature. The stab cultures thus prepared are stored in the refrigerator under aseptic conditions. One of these cultures is reserved as the stock culture until the next time for transfer, when it will be used to prepare three more agar stabs. The other two are held available during the interval between transfer for the preparation of inoculum cultures required for assay.

Preparation of the inoculum culture
Stab cultures of lactic acid bacteria that have been stored in the refrigerator must be activated by making at least three successive daily stab transfers before preparing the inoculum. The inoculum is prepared the day before the day of the assay by subculturing the activated agar stab into a tube containing sterile inoculum medium. A new inoculum should not be made from a previously prepared inoculum, because of the possible danger of contamination; this practice may also promote changes in the nutritional requirements of the assay organism.

Several different methods of preparing the inoculum have been proposed. The most satisfactory methods take into account the capacity of the assay organism for storing sufficient amounts of vitamins (particularly pantothenic acid, biotin, folate and vitamin B_{12}) to produce considerable growth in the assay blank tubes. Such methods are designed to reduce cellular vitamin accumulation to a minimum, and also to eliminate the risk of carry-over of vitamin and other components of the nutrient medium that might interfere in the subsequent assay. It is important that the inoculum be suitably diluted. If insufficient dilution is made, the blanks may be too high, and slight differences in the drop size of the inoculum may introduce variations in the amount of growth in the assay tubes which are unrelated to the amount of vitamin present in the material being assayed. On the other hand, if the inoculum is too dilute, growth may not be complete by the end of the assay period. It is also desirable to inoculate the assay tubes from inoculum cultures which are in the logarithmic (acceleration) phase of growth.

In methods described by Snell (1950), Barton-Wright (1952) and Strohecker and Henning (1966), the inoculum culture is grown in a liquid basal medium to which has been added the vitamin to be assayed in an amount that barely supports the growth of the assay organism. This depletion technique limits the accumulation of the vitamin in the cell and thus maintains the sensitivity of the organism towards the vitamin. A 10 ml quantity of such a medium can be prepared by taking 5 ml of double-strength basal medium and adding 2.5 ml of standard vitamin solution plus 2.5 ml of distilled water, and then autoclaving. The inoculum broth thus prepared is inoculated with agar stab culture and incubated overnight at 37 °C. The incubation time must not exceed 22–24 h for the lactic acid bacteria or yeasts, or the organism will become attenuated (weakened) and poor growth responses will be encountered. The cells of the resultant culture are washed with isotonic saline (0.9% NaCl solution) to reduce carry-over of vitamin into the assay tubes during the subsequent inoculation. This operation is performed aseptically by centrifuging the tube contents and resuspending the cells in a suitable volume of saline for direct use as an inoculum, or after an appropriate dilution has been made. In some methods, two or even three centrifugal washings are employed for certain vitamins.

Difco supply special dehydrated inoculum media, whose formulations are the same as those of the corresponding maintenance media, except that the agar component is omitted (Difco Laboratories, 1985). The cells grown in such media are harvested by centrifugation and washed with sterile isotonic saline or, in accordance with the United States Pharmacopeia (USP) procedure for determining vitamun B_{12}, with single-strength basal assay medium.

Bell (1974) found that procedures using centrifugal washing gave an inoculum in the lag phase of growth, thus necessitating unduly long assay incubation periods. Problems of airborne bacterial contamination were also encountered. Bell prepared inocula for the determination of riboflavin, niacin, pantothenic acid, biotin, folate and vitamin B_{12} using Bacto-Micro Inoculum Broth (Difco Code 0320) as the inoculum medium. The common procedure for all these determinations was to subculture the assay organism from the most recent agar stab into 5 ml of sterile inoculum broth (Difco) and incubate overnight at 37 °C. The following morning, one drop of this subculture was added to 5 ml of single-strength basal medium (basal medium diluted with an equal volume of water) containing a controlled amount of the vitamin being assayed. These amounts were 25 ng nicotinic acid/ml, 20 ng pantothenic acid/ml,

0.05 ng biotin/ml, 2 ng riboflavin/ml, 1.0 ng folic acid/ml and 0.04 ng cyanocobalamin/ml (vitamin B_{12}). Both the inoculum broth and the basal medium were at 37 °C during transfer of the organism. After a further 6 h incubation at 37 °C, two drops of the resultant culture were transferred to 10 ml of single-strength basal medium (without added vitamin). This final suspension was used as the inoculum and contained cells in the logarithmic growth phase.

The inoculum for the determination of thiamin using *L. viridescens* was prepared in a similar manner to that described above, except that the inoculum medium was Bacto-APT Broth (Difco Code 0655) and the incubation temperature was 30 °C.

To prepare the inoculum for the determination of vitamin B_6, a loop of *S. carlsbergensis* from the agar slope was added to 5 ml of single-strength medium containing a glass bead, and this subculture was incubated for 20 h at 27 °C with constant shaking. One millilitre of the suspension obtained was pipetted into a tube containing 5 ml of sterile single-strength basal medium and mixed to give the inoculum culture.

Preparation of the basal medium

The basal medium contains all the nutritional factors necessary for the normal growth and metabolism of the assay organism, except the vitamin to be determined. It is formulated from highly purified natural products, synthetic vitamins and other reagent-grade compounds. An assay medium used for lactic acid bacteria must contain a fermentable carbohydrate, a variable assortment of essential amino acids, various vitamins and mineral salts, certain purine and pyrimidine bases, and an appropriate buffer system. Glucose is universally used as a source of carbon and energy; the amino acids are provided in the form of acid-hydrolysed casein plus tryptophan or a mixture of the specific amino acids; and the buffer salt is usually sodium acetate, which also has a stimulatory effect on growth. Dehydrated media for use with assays utilizing lactic acid bacteria and *S. carlsbergensis* are available from Difco Laboratories. The Difco code numbers of basal media and references to publications which give other media compositions are listed in Table 2.6. In traditional assay procedures, the assay media are prepared at double strength, either from commercial dehydrated formulations (if available) or from individual ingredients.

In addition to factors essential for bacterial growth, other substances which stimulate growth must be considered. Ideally, the medium should contain sufficient amounts of all growth factors and stimulatory substances so that the effects of these nutrients added with the hydrolysed food extract being assayed will be eliminated. In practice, the ideal is seldom achieved and the adequacy of a basal medium depends upon the experimental conditions (Snell, 1948). A basal medium which, when supplied with the missing vitamin, contains all of the nutrients *essential* for growth, but is lacking in one or more substances which markedly stimulate growth, may give satisfactory assays if the period of incubation is long enough to eliminate the effects of the nonessential growth stimulants. Such a medium will also be satisfactory for assaying samples which are rich in the vitamin to be determined, as the interfering material will be diluted out. Conversely, such a medium will tend to give erroneous results if used with a short incubation period, and with samples of low vitamin potency.

Fatty acids are notorious growth stimulants for a number of lactic acid bacteria (Williams, Broquist and Snell, 1947) but it is not customary to add these acids to the basal medium; rather they are removed from the food sample after the extraction step.

Extraction of the vitamin from the test material
The vitamins are extracted from the food matrix in a form that can be utilized by the particular assay organism being used. This generally involves autoclaving the food sample in the presence of acid or, for acid-labile vitamins, digesting the sample with suitable enzymes. After precipitating the proteins at their isoelectric point (*c*.pH4), the pH of the extract is adjusted to that of the basal medium (typically pH 6.8). This step is necessary to ensure that the pH of the medium is not altered by the addition of different amounts of the extract. The extract is then diluted to bring the concentration of the vitamin to be assayed within the range of the standard curve. Hopefully, the dilution factor is sufficiently high to dilute out any interfering substances that would cause drift and invalidate the assay. The minimum dilutions of foods necessary to avoid the inhibitory effects of food preservatives and neutralization salts have been calculated by Voigt, Eitenmiller and Ware (1979). Finally, the extracts are filtered to remove the precipitated protein and lipoidal material, and to obtain a clear solution for assay.

Setting up the assay
At all stages during the analytical procedure, the solutions must be protected from daylight.

Aliquots of the working standard vitamin solution are added in increasing volumes up to 5 ml to a duplicate series of tubes for the construction of a standard curve; duplicate blanks containing no vitamin are included. Similar volumes of the neutralized test extracts are added to a single series of tubes. A range of concentration levels of each test extract is assayed in the expectation that at least three will fall on the standard curve. Fresh glass-distilled water is added to all tubes to bring the volume in each tube to 5.0 ml, after which 5.0 ml of double-strength basal medium is added so that the total volume is 10 ml. The addition of basal medium to the extract and water gives better mixing of the two solutions than adding the extract and water to the medium. The entire rack of filled tubes is covered with a sheet of aluminium foil. Sterilization at 15 lb pressure (121 °C), according to Pearson (1967), is based on lower levels of glucose used in earlier medium formulations, and can lead to caramelization and darkening of the solutions. Barton-Wright (1952) recommended autoclaving under the milder conditions of 10 lb pressure for 10 min as a means of avoiding this problem. The tubes are removed immediately from the autoclave when atmospheric pressure is reached, and the tubes are cooled to below the subsequent incubation temperature. It is imperative that all tubes are cooled to the same temperature, because turbidity is a measure of the rate of growth rather than the extent of growth, and small differences in the temperature between tubes at the start of the incubation influence the growth rate. Temperature equilibration of the whole rack of tubes is best achieved using a water bath or convection incubator.

All tubes are inoculated with one drop of freshly prepared inoculum culture using a sterile pipette or a 10 ml syringe fitted with a 20-gauge needle. If the plunger is removed from the syringe barrel after starting the inoculation through the needle, uniform drops will continue to fall from the needle. It is not necessary to flame the tubes during inoculation. The inclusion of a control run, which is not inoculated but is otherwise treated identically, is advisable to check the sterility of the basal medium. For turbidimetric assays using lactic acid bacteria or yeasts, the tubes are incubated for 22–24 h at a constant temperature near the optimum for these organisms. Suitable temperatures are 37 °C for *L. casei*, *L. plantarum* and *L. delbrueckii*; 30 °C for *L. viridescens*; and 28 °C (with constant shaking) for *S. carlsbergensis* and *Kloeckera apiculata*.

The setting-up procedure described by Bell (1974) involves fewer manipulations than the conventional procedure, owing to the volumes of standard or sample assay solutions being reduced from millilitre aliquots to microlitre aliquots ranging from 0 to 250 µl. The addition of such small volumes to 10 ml of assay medium has two advantages: no significant volume change occurs so that single-strength medium can be used; and the need for adjustment of the pH value of samples extracted with acids is greatly reduced. For example, no pH adjustment is required when acids less than 0.1 N are used, and for strongly acid hydrolysates the solution need only be adjusted to pH 1–2.

Quantification
At the end of the incubation period, the cells are uniformly suspended by shaking the tubes, and time is allowed for the air bubbles to disperse before measurement. The turbidities of all tubes are measured in a nephelometer using a neutral filter, colorimeter with a filter in the region of 640 nm (Association of Vitamin Chemists, Inc., 1966) or

spectrophotometer at 540–660 nm wavelength (Difco Laboratories, 1985). The turbidity may be expressed as an absorption (extinction), as a transmittance (in % T), as a difference $100 - T$ (in %) or simply as a galvanometer reading. The arithmetic means of the replicates are calculated, and the means for the standard solutions are plotted on semilogarithmic graph paper with the turbidity values as ordinates (linear scale) and concentrations in nanograms per millilitre as abscissae (logarithmic scale). The calibration curve is drawn through these points. The vitamin concentrations of the sample tubes are read off from the calibration curve and the concentration values for the original samples are calculated from simple dilution factors. Concentration values for a given sample calculated from at least three dilutions should check within the limits of error of the assay, which is usually considered to be ± 10–15% (Association of Vitamin Chemists, Inc., 1966); that is, they should not differ by more than 15% from their common mean. If this condition is not fulfilled, the determination must be repeated.

2.12 BIOSPECIFIC METHODS FOR SOME OF THE B-GROUP VITAMINS

2.12.1 General aspects

Introduction

Biospecific methods of analysis for selected vitamins of the B group can be broadly classified as immunoassays and protein-binding assays. Immunoassays are based on the specific interaction of an antibody with its antigen, and are represented by the radioimmunoassay and the enzyme-linked immunosorbent assay (ELISA). Protein-binding assays utilize natural vitamin-binding proteins and are represented by the radioassay and the nonisotopic competitive protein-binding assay. Biospecific assays can be performed on complex biological matrices, so they require minimal sample clean-up. The analytical stages of immunoassays and protein-binding assays can be automated using equipment that is commercially available, but the methods can only be described as semiautomated, as it is necessary to liberate the vitamins from their bound forms using manual extraction procedures.

The immunological reaction

If a test animal such as a rabbit is given repeated small injections of an immunogenic antigen, antibodies against the antigen are produced by lymphoid tissues following recognition of the antigenic determinant. Proteins of molecular weight greater than 5000 can usually be both antigens and immunogens. Smaller compounds, though antigenic, must be coupled to a large protein carrier such as albumin to be immunogenic. When coupled, the small compound is called a hapten, and the carrier–hapten complex is called a conjugate. Usually, the immunogen is mixed with an adjuvant, which when injected serves to both enhance and prolong the immune response (Hawker, 1973).

The following terms are encountered in immunoassays.

- **Antibody** A binding protein (immunoglobulin) which is synthesized by the immune system of an animal in response to the injection of an immunogenic antigen.
- **Antigen** A substance capable of binding to a specific antibody.
- **Immunogen** A substance that, when injected into a suitable animal, elicits an immune response.
- **Antiserum** The serum of the test animal containing polyclonal antibodies.
- **Polyclonal antibodies** Antibodies that are present in the antiserum of an immunized animal and which are derived from several clones of lymphocyte. They are reactive for several antigenic sites.
- **Monoclonal antibodies** Antibodies derived from a single clone of lymphocytes produced in cell culture by hybridoma cells, which are formed by the fusion of lymphocytes with myeloma cells (cancerous lymphocytes) from an immunized animal donor. The antibody molecules, being chemically identical, exhibit identical binding properties.
- **Cross-reactivity** Ability of substances, other than the antigen, to bind to the antibody, and the ability of substances other than the antibody to bind the antigen. Cross-reactants may be substances that carry on their surface a molecular configuration similar to the antigenic determinants on the antigen being measured.
- **Antigenic determinant** Structural feature of an antigen which defines the recognition pattern of an antibody.
- **Affinity** The energy with which the combining sites of an antibody bind its specific antigen. It is analogous to the association constant (K_A) in physical chemistry and has the dimensions of moles/litre.
- **Avidity** There are several populations of antibodies with different affinities in a polyclonal antiserum, the mean affinity being referred to as its avidity. The high-affinity antibodies dictate the sensitivity of an immunoassay.

The production of monoclonal antibodies (Galfré and Milstein, 1981) is more costly, labour intensive and time-consuming than the production of polyclonal antiserum. However, the provision of potentially unlimited amounts of a homogeneous reagent is a major advantage in the development of commercial assay kits. The higher specificity compared with polyclonal antibodies is another advantage. On the demerit side, mono-clonal antibodies rarely exhibit such a high affinity for the antigen as do polyclonal antibodies, and this can result in a less sensitive assay. The affinity is less important for the sensitivity of an excess reagent assay than it is for a competitive assay.

2.12.2 Radioassay

Radioassays or, more precisely, radiometric competitive protein-binding assays have been applied to the determination of biotin, folate and vitamin B_{12} in biological materials (see relevant chapters). The assays are based on the radioisotope-dilution principle, whereby the unknown quantity of the vitamin in the test material, after first being liberated from bound materials, is used to dilute the radioactivity of an added measured quantity of tracer (radioactively labelled vitamin). The analysis usually involves an initial heating step to denature indigenous binding proteins. The radioassay procedure is carried out as follows.

Into a centrifuge tube are placed measured volumes of a suitable buffer solution, the test extract or unlabelled vitamin standard, and the tracer. The labelled vitamin is available commercially in powdered form or in solution, and can be standardized against the unlabelled vitamin stan-dard by the isotopic dilution method of Lau *et al.* (1965). The standard-ization technique allows the actual quantity of labelled vitamin to be calculated for any percentage change in the binding capacity of the protein for the labelled vitamin. A soluble natural vitamin-binding pro-tein is then added in a predetermined quantity that has a maximal capacity to bind only some of the labelled vitamin present. Typical bind-ing capacities are 80–90% for biotin assays (Hood, 1979), 50–60% for folate assays (Waxman, Schreiber and Herbert, 1971) and 60–80% for vitamin B_{12} assays (Lau *et al.*, 1965). The binding protein has a high affinity and specificity for the vitamin in question, but it does not dis-criminate between labelled and unlabelled vitamin.

The tubes are stored at ambient temperature in the dark for a prescribed period. During this time, unlabelled and labelled vitamin will compete stoichiometrically for the limited number of binding sites on the protein molecule. The amount of labelled vitamin that is subsequently bound is inversely related to the amount of indigenous vitamin present. Activated charcoal coated with haemoglobin, albumin or dextran is added and the tube contents are mixed thoroughly. The charcoal coating acts as a mol-ecular sieve, allowing the unbound vitamin to pass through and be adsorbed onto the charcoal, but excluding the protein-bound vitamin. The unbound vitamin is separated from bound vitamin by centrifugation.

The specific radioactivity in the supernatant fluid (bound fraction) or in the pellet (unbound fraction) is measured in counts per minute (cpm) in a liquid scintillation counter for β-emitters such as ^{3}H or ^{14}C isotopes or in a gamma counter for γ-emitters such as ^{125}I, ^{75}Se or ^{57}Co isotopes. Included in the assay procedure is a control tube which contains only the tracer and coated charcoal (plus buffer solution to make up the volume). The control represents the amount of radioactivity that is not bound to the charcoal and is due to radioactive degradation products of the tracer. The cpm for the control is subtracted from those for the unknown and the standards in order to obtain net counts. All cpm data must be corrected for counting efficiency if significant differences in efficiency are observed between samples and standards due to quenching effects.

Quantification is achieved by assaying a range of unlabelled vitamin standards of known concentration. Let B represent the amount of bound tracer (net cpm) corresponding to each concentration of standard, and B_M represent the amount of bound tracer (net cpm) corresponding to a zero amount of standard (i.e. the maximal binding capacity of the fixed amount of protein). A linear calibration curve is obtained on logit-log paper (1 × 3 cycle log–log paper) by plotting the percentage of tracer bound at each concentration of standard (B/B_M × 100) as the logit function (ordinate) versus the log concentration (abscissa) of standard in nanograms per millilitre. Alternatively, the reciprocal of the percentage of tracer bound versus concentration can be plotted as a straight line on nonlogarithmic graph paper (Dakshinamurti *et al.*, 1974). The concentration of vitamin in the assay solution can be obtained from the standard curve by interpolation of the percentage of tracer binding found or by calculation from the regression equation of the standard curve (Hood, 1977).

2.12.3 Radioimmunoassay

This is a variation of the radioassay and is based on the competition for a fixed, but limited, number of antibody binding sites by antigen (the vitamin analyte) and a trace amount of radiolabelled antigen added to the sample extract. Thus, the presence of larger amounts of unlabelled analyte results in less radioactivity being bound to the antibody. The free and antibody-bound fractions are separated by adsorption or precipitation, followed by centrifugation, and the radioactivity in the supernatant or precipitate is measured. A comparison of the ratio of the bound to free labelled analyte with that obtained from a series of standards permits quantification of unknown samples.

A radioimmunoassay for pantothenic acid in foods is described in Chapter 11.

2.12.4 Enzyme-linked immunosorbent assay (ELISA)

An ELISA is an enzyme-linked immunoassay in which one of the reactants is immobilized by physical adsorption onto the surface of a solid phase. In its simplest form, as used in food analysis applications, the solid phase is provided by the plastic surface of a 96-well microtitration plate. The ELISA can be performed manually, with the aid of push-button dispensers, or it can be totally automated, complete with computer for calculation of standard curves, statistical analysis of data and data storage.

There are many variants of the ELISA, but in a discussion of basic principles, they fall into two main types, namely competitive and non-competitive (reagent excess) immunoassays.

In the direct competitive ELISA, the analyte vitamin molecules and added enzyme–vitamin conjugate compete for a limited number of binding sites on the immobilized antibody. The proportion of added enzyme–vitamin conjugate present in either the free or bound phases after equilibrium has been reached depends on the amount of analyte initially present. The phases are separated by emptying the well contents and washing the plate. The amount of bound enzyme is then determined by addition of substrate and spectrophotometric measurement of the coloured product. A variation of this format is the indirect competitive ELISA, in which the analyte and immobilized analyte compete for a limited number of binding sites on the enzyme-labelled antibody. The characteristic feature of competitive ELISAs is that the higher the optical density, the lower is the amount of analyte present.

The generally preferred ELISA format for vitamin assays in food analysis is a two-site noncompetitive assay used in the indirect mode. This format employs two antibodies: a primary anti-vitamin antibody raised against a hapten–protein conjugate, and an enzyme-labelled, species-specific second antibody, which binds specifically to the primary antibody. The scheme for performing such an ELISA is depicted in Figure 2.5. A protein conjugate of the vitamin is immobilized to the well surface of the microtitration plate, the attached protein being different to that used for the immunogen. The protein adsorbs passively and strongly to the plastic and, once coated, plates can usually be stored for several months. To perform the assay, the sample or standard is added to the well, followed by a limited amount of primary antibody. After incubation, the antibody becomes distributed between immobilized vitamin and free vitamin according to the amount of analyte initially present. After phase separation, achieved by well emptying and washing, the second antibody is added in excess, and the plate is incubated for a second time. Excess unbound material is removed and substrate is added. Optical densities are measured after a suitable time, and unknown samples are quantified by reference to the behaviour of vitamin standards.

In contrast to the competitive ELISA, the noncompetitive assay uses an excess of antibody, so that the optical densities increase with increasing

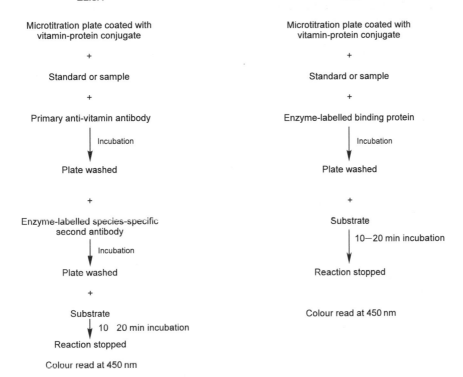

Figure 2.5 Comparison of methodologies for the two-site noncompetitive ELISA (indirect mode) and the competitive PBA (indirect mode). (Reprinted from *Journal of Micronutrient Analysis*, **7**, 261–70, Lee *et al.*, copyright 1990, with kind permission from Elsevier Science Ltd, The Boulevard, Langford Lane, Kidlington, OX5 1GB, UK.)

amount of analyte. Although the competitive assay produces its greatest signal (optical density) for low concentrations of analyte, the noncompetitive assay gives lower detection limits. It is also more specific, since the two antibodies recognize separate antigenic determinants on the analyte. Two other advantages of the noncompetitive assay are that the affinity of the primary antibody is less important than in the competitive assay, since excess antibody is used, and the accuracy of pipetting the primary antibody is less critical, since it is no longer a limiting factor (Gould, 1988). Enzyme-labelled second antibodies are widely available commercially, active against different species, and labelled with a variety of enzymes. In summary, noncompetitive assays are more reliable and more rugged than competitive assays, with the added advantages of improved sensitivity and specificity.

An ELISA for pantothenic acid in food is described in Chapter 11.

2.12.5 Nonisotopic protein-binding assay

A nonisotopic competitive protein-binding assay (PBA) in the indirect mode has been developed for food analysis applications using the 96-well microtitration plate as the solid phase. Individual methods have been reported for the determination of biotin (Finglas, Faulks and Morgan, 1986), folate (Finglas, Faulks and Morgan, 1988) and vitamin B_{12} (Alcock, Finglas and Morgan, 1992) using avidin, folate-binding protein and R-protein as the respective vitamin-specific binding proteins. The principle of the assay is based on the competition between immobilized vitamin and free vitamin (analyte) in the assay solution for a limited number of binding sites on the enzyme-linked vitamin-binding protein. The amount of protein bound to the well surface is inversely proportional to the concentration of free vitamin in the assay solution and is determined, after plate washing, by measuring the enzyme activity. The scheme for performing such an assay is compared with an ELISA format in Figure 2.5.

REFERENCES

Ahuja, S. (1976) Derivatization in gas chromatography. *J. Pharm. Sci.*, **65**, 163–82.

Alcock, S.C., Finglas, P.M. and Morgan, M.R.A. (1992) Production and purification of an R-protein–enzyme conjugate for use in a microtitration plate protein-binding assay for vitamin B_{12} in fortified food. *Food Chem.*, **45**, 199–203.

Arella, F., Lahély, S., Bourguignon, J.B. and Hasselmann, C. (1996) Liquid chromatographic determination of vitamins B_1 and B_2 in foods. A collaborative study. *Food Chem.*, **56**, 81–6.

Association of Vitamin Chemists, Inc. (1966) *Methods of Vitamin Assay*, 3rd edn, Interscience Publishers, New York, pp. 1–20.

Ayi, B.K., Yuhas, D.A. and Deangelis, N.J. (1986) Simultaneous determination of vitamins B_2 (riboflavin) and B_6 (pyridoxine) in infant formula products by reverse phase liquid chromatography. *J. Ass. Off. Analyt. Chem.*, **69**, 56–9.

Bakalyar, S.R. and Henry, R.A. (1976) Variables affecting precision and accuracy in high-performance liquid chromatography. *J. Chromat.*, **126**, 327–45.

Ball, G.F.M. (1988) *Fat-Soluble Vitamin Assays in Food Analysis*. Elsevier Applied Science, London.

Barna, E. and Dworschák, E. (1994) Determination of thiamine (vitamin B_1) and riboflavin (vitamin B_2) in meat and liver by high-performance liquid chromatography. *J. Chromat., A.*, **668**, 359–63.

Bartha, A., Vigh, G. and Varga-Puchony, Z. (1990) Basis of the rational selection of the hydrophobicity of the ion-pairing reagent in reversed-phase ion-pair high-performance liquid chromatography. *J. Chromat.*, **499**, 423–34.

Barton-Wright, E.C. (1952) *The Microbiological Assay of the Vitamin B-Complex and Amino Acids*, Sir Isaac Pitman & Sons, Ltd, London.

Barton-Wright, E.C. (1967) The microbiological assay of certain vitamins in compound feeding stuffs. *J. Ass. Publ. Analysts*, **5**, 8–23.

Barton-Wright, E.W. (1971) The microbiological assay of the vitamin B_6 complex (pyridoxine, pyridoxal and pyridoxamine) with *Kloeckera brevis*. *Analyst, Lond.*, **96**, 314–18.

Bell, J.G. (1971) The microbiological assay of B-complex vitamins. In *Proceedings of the University of Nottingham Residential Seminar on Vitamins* (ed. M. Stein), Churchill Livingstone, Edinburgh and London, pp. 165–87.

Bell, J.G. (1974) Microbiological assay of vitamins of the B group in foodstuffs. *Lab. Pract.*, **23**, 235–42, 252.

Betteridge, D. (1978) Flow injection analysis. *Analyt. Chem.*, **50**, 832A–3A, 835A–6A, 839A–40A, 842A–4A, 846A.

Bourgeois, C.F., George, P.R. and Cronenberger, L.A. (1984) Automated determination of α-tocopherol in food and feed. Part 2. Continuous flow technique. *J. Ass. Off. Analyt. Chem.*, **67**, 631–4.

Bui, M.H. (1987) Sample preparation and liquid chromatographic determination of vitamin D in food products. *J. Ass. Off. Analyt. Chem.*, **70**, 802–5.

Burton, D.E., Sepaniak, M.J. and Maskarinec, M.P. (1986) Analysis of B₆ vitamers by micellar electrokinetic capillary chromatography with laser-excited fluorescence detection. *J. Chromatogr. Sci.*, **24**, 347–51.

Chase, G.W., Landen, W.O. Jr, Soliman, A.-G.M. and Eitenmiller, R.R. (1993) Method modification for liquid chromatographic determination of thiamine, riboflavin, and pyridoxine in medical foods. *J. AOAC Int.*, **76**, 1276–80.

Chen, M. and McIntyre, P.A. (1979) Measurement of the trace amounts of vitamin B₁₂ present in various foods by a new radiometric microbiologic technique. In *Trace Organic Analysis: a New Frontier in Analytical Chemistry* (eds H.S. Hertz and S.N. Chesler), National Bureau of Standards, Washington, DC, pp. 257–65.

Chen, M.F., Hill, J.W. and McIntyre, P.A. (1983) The folacin contents of foods as measured by a radiometric microbiologic method. *J. Nutr.*, **113**, 2192–6.

Chiari, M., Nesi, M., Carrea, G. and Righetti, P.G. (1993) Determination of total vitamin C in fruits by capillary zone electrophoresis. *J. Chromat.*, **645**, 197–200.

Cooke, N.H.C. and Olsen, K. (1979) Chemically bonded alkyl reversed-phase columns. *Am. Lab.*, **11**, 45, 46, 48, 50, 52–5, 58–60.

Cooke, N.H.C., Archer, B.G., Olsen, K. and Berick, A. (1982) Comparison of three- and five-micrometer column packings for reversed-phase liquid chromatography. *Analyt. Chem.*, **54**, 2277–83.

Dakshinamurti, K., Landman, A.D., Ramamurti, L. and Constable, R.J. (1974) Isotope dilution assay for biotin. *Analyt. Biochem.*, **61**, 225–31.

Davídek, J., Velíšek, J., Černá, J. and Davídek, T. (1985) Gas chromatographic determination of pantothenic acid in foodstuffs. *J. Micronutr. Anal.*, **1**, 39–46.

Dawson, K.R., Unklesbay, N.F. and Hedrick, H.B. (1988) HPLC determination of riboflavin, niacin, and thiamin in beef, pork, and lamb after alternate heat-processing methods. *J. Agric. Food Chem.*, **36**, 1176–9.

Difco Laboratories (1985) Media for the microbiological assay of vitamins. In *Difco Manual*, 10th edn, Difco Laboratories, Detroit, MI, pp. 1055–1114.

Dunbar, W.E. and Stevenson, K.E. (1979) Automated fluorometric determination of thiamine and riboflavin in infant formulas. *J. Ass. Off. Analyt. Chem.*, **62**, 642–7.

Echols, R.E., Miller, R.H. and Foster, W. (1986) Analysis of thiamine in milk by gas chromatography and the nitrogen–phosphorus detector. *J. Dairy Sci.*, **69**, 1246–9.

Echols, R.E., Miller, R.H. and Thompson, L. (1985) Evaluation of internal standards and extraction solvents in the gas chromatographic determination of thiamine. *J. Chromat.*, **347**, 89–97.

Echols, R.E., Miller, R.H., Winzer, W. *et al.* (1983) Gas chromatographic determination of thiamine in meats, vegetables and cereals with a nitrogen–phosphorus detector. *J. Chromat.*, **262**, 257–63.

Elkins, E.R. and Dudek, J.A. (1985) Sampling for vitamin analysis. In *Methods of Vitamin Assay*, 4th edn (eds J. Augustin, B.P. Klein, D. Becker and P.B. Venugopal), John Wiley & Sons, New York, pp. 135–51.

Ewing, A.G., Wallingford, R.A. and Olefirowicz, T.M. (1989) Capillary electrophoresis. *Analyt. Chem.*, **61**, 292A–294A, 296A, 298A, 300A–303A.

Fellman, J.K., Artz, W.E., Tassinari, P.D. *et al.* (1982) Simultaneous determination of thiamin and riboflavin in selected foods by high-performance liquid chromatography. *J. Food Sci.*, **47**, 2048–50, 2067.

Fernando, S.M. and Murphy, P.A. (1990) HPLC determination of thiamin and riboflavin in soybeans and tofu. *J. Agric. Food Chem.*, **38**, 163–7.

Finglas, P.M. and Faulks, R.M. (1984) The HPLC analysis of thiamin and riboflavin in potatoes. *Food Chem.*, **15**, 37–44.

Finglas, P.M., Faulks, R.M. and Morgan, M.R.A. (1986) The analysis of biotin in liver using a protein-binding assay. *J. Micronutr. Anal.*, **2**, 247–57.

Finglas, P.M., Faulks, R.M. and Morgan, M.R.A. (1988) The development and characterisation of a protein-binding assay for the determination of folate – potential use in food analysis. *J. Micronutr. Anal.*, **4**, 295–308.

Ford, J.E. (1953) The microbiological assay of 'vitamin B_{12}'. The specificity of the requirement of *Ochromonas malhamensis* for cyanocobalamin. *Br. J. Nutr.*, **7**, 299–306.

Froehlich, P. and Wehry, E.L. (1981) Fluorescence detection in liquid and gas chromatography. Techniques, examples, and prospects. In *Modern Fluorescence Spectroscopy*, Vol. 3 (ed. E.L. Wehry), Plenum Press, New York, pp. 35–94.

Fujiwara, S., Iwase, S. and Honda, S. (1988) Analysis of water-soluble vitamins by micellar electrokinetic capillary chromatography. *J. Chromat.*, **447**, 133–40.

Galfré, G. and Milstein, C. (1981) Preparation of monoclonal antibodies: strategies and procedures. *Methods Enzymol.*, **73B**, 3–46.

Garfield, F.M. (1989) Sampling in the analytical scheme. *J. Ass. Off. Analyt. Chem.*, **72**, 405–11.

Gloor, R. and Johnson, E.L. (1977) Practical aspects of reverse phase ion pair chromatography. *J. Chromatogr. Sci.*, **15**, 413–23.

Goli, D.M. and Vanderslice, J.T. (1989) Microbiological assays of folacin using a CO_2 analyzer system. *J. Micronutr. Anal.*, **6**, 19–33.

Gould, B.J. (1988) The use of enzymes in ultrasensitive immunoassays. In *Immunoassays for Veterinary and Food Analysis – 1* (eds B.A. Morris, M.N. Clifford and R. Jackman), Elsevier Applied Science, London and New York, pp. 53–65.

Guilarte, T.R. (1983) Radiometric microbiological assay of vitamin B_6: assay simplification and sensitivity study. *J. Ass. Off. Analyt. Chem.*, **66**, 58–61.

Guilarte, T.R. (1985) Analysis of biotin levels in selected foods using a radiometric-microbiological method. *Nutr. Rep. Int.*, **32**, 837–45.

Guilarte, T.R. (1989) A radiometric microbiological assay for pantothenic acid in biological fluids. *Analyt. Biochem.*, **178**, 63–6.

Guilarte, T.R. (1991a) Radiometric microbiological assay of B vitamins. Part 1: assay procedure. *J. Nutr. Biochem.*, **2**, 334–8.

Guilarte, T.R. (1991b) Radiometric microbiological assay of B vitamins. Part 2: extraction methods. *J. Nutr. Biochem.*, **2**, 399–402.

Guilarte, T.R. and Pravlik, K. (1983) Radiometric-microbiologic assay of niacin using *Kloeckera brevis*: analysis of human blood and food. *J. Nutr.*, **113**, 2597–94.

Hägg, M. (1994) Effect of various commercially available enzymes in the liquid chromatographic determination with external standardization of thiamine and riboflavin in foods. *J. AOAC Int.*, **77**, 681–6.

Hamaker, H.C. (1986) A statistician's approach to repeatability and reproducibility. *J. Ass. Off. Analyt. Chem.*, **69**, 417–28.

Hasselmann, C., Franck, D., Grimm, P. *et al.* (1989) High-performance liquid chromatographic analysis of thiamin and riboflavin in dietetic foods. *J. Micronutr. Anal.*, **5**, 269–79.

Hawker, C.D. (1973) Radioimmunoassay and related methods. *Analyt. Chem.*, **45**, 878A–882A, 884A, 886A, 888A, 890A.

Hawthorne, S.B. (1990) Analytical-scale supercritical fluid extraction. *Analyt. Chem.*, **62**, 633A–636A, 638A–642A.

Healy, M., Jenkins, T. and Poliakoff, M. (1989) SFC: a hyphenated future. *Lab. News*, 7 August, 6–8.

Hood, R.L. (1977) The use of linear regression analysis in the isotope dilution assay of biotin. *Analyt. Biochem.*, **79**, 635–8.

Hood, R.L. (1979) Isotopic dilution assay for biotin: use of [^{14}C]biotin. *Methods Enzymol.*, **62D**, 279–83.

Horvath, C. and Melander, W. (1977) Liquid chromatography with hydrocarbonaceous bonded phases; theory and practice of reversed phase chromatography. *J. Chromatogr. Sci.*, **15**, 393–404.

Horwitz, W. (1990) Nomenclature for sampling in analytical chemistry (Recommendations 1990). *Pure Appl. Chem.*, **62**, 1193–208.

Ibáñez, E., Herraiz, M. and Reglero, G. (1993) Use of micropacked columns for quantitative SFC. *J. High Resolution Chromat.*, **16**, 615–8.

Ibáñez, E., Alvarez, P.J.M., Reglero, G. and Herraiz, M. (1993) Large particle micropacked columns in supercritical fluid chromatography. *J. Microcol. Sep.*, **5**, 371–81.

Ibáñez, E., Tabera, J., Reglero, G. and Herraiz, M. (1995) Optimization of separation of fat-soluble vitamins by supercritical fluid chromatography using serial micropacked columns. *J. Agric. Food Chem.*, **43**, 2667–71.

Iskandarani, Z. and Pietrzyk, D.J. (1982) Ion interaction chromatography of organic anions on a poly(styrene-divinylbenzene) adsorbent in the presence of tetraalkylammonium salts. *Analyt. Chem.*, **54**, 1065–71.

Jorgenson, J.W. and Lukacs, K.D. (1983) Capillary zone electrophoresis. *Science*, **222**, 266–72.

Kamman, J.F., Labuza, T.P. and Warthesen, J.J. (1980) Thiamin and riboflavin analysis by high performance liquid chromatography. *J. Food Sci.*, **45**, 1497–9, 1504.

Kmostak, S. and Kurtz, D.A. (1993) Rapid determination of supplemental vitamin E acetate in feed premixes by capillary gas chromatography, *J. AOAC Int.*, **76**, 735–41.

Kneifel, W., Ulberth, F. and Winkler-Macheiner, U. (1987) HPLC methods for the simultaneous determination of retinol and tocopherol in butter and whole-milk powder. *Deutsche Lebensm. Rundschau*, **83**, 137–9 (in German).

Kobayashi, T., Okano, T. and Takeuchi, A. (1986) The determination of vitamin D in foods and feeds using high-performance liquid chromatography. *J. Micronutr. Anal.*, **2**, 1–24.

Kratochvil, B. and Taylor, J.K. (1981) Sampling for chemical analysis. *Analyt. Chem.*, **53**, 924A–6A, 928A, 930A, 932A, 934A, 936A, 938A.

Landen, W.O., Jr (1980) Application of gel permeation chromatography and nonaqueous reverse phase chromatography to high pressure liquid chromatographic determination of retinyl palmitate in fortified breakfast cereals. *J. Ass. Off. Analyt. Chem.*, **63**, 131–6.

Landen, W.O., Jr (1982) Application of gel permeation chromatography and nonaqueous reversed-phase chromatography to high performance liquid chromatographic determination of retinyl and α-tocopheryl acetate in infant formulas. *J. Ass. Off. Analyt. Chem.*, **65**, 810–16.

Landen, W.O., Jr (1985) Liquid chromatographic determination of vitamins D_2 and D_3 in fortified milk and infant formulas. *J. Ass. Off. Analyt. Chem.*, **68**, 183–7.

Landen, W.O., Jr and Eitenmiller, R.R. (1979) Application of gel permeation chromatography and nonaqueous reverse phase chromatography to high pressure liquid chromatographic determination of retinyl palmitate and β-carotene in oil and margarine. *J. Ass. Off. Analyt. Chem.*, **62**, 283–9.

Lau, K.-S., Gottlieb, C., Wasserman, L.R. and Herbert, V. (1965) Measurement of serum vitamin B_{12} level using radioisotope dilution and coated charcoal. *Blood*, **26**, 202–14.

Lee, D.P. (1982) Reversed-phase HPLC from pH 1 to 13. *J. Chromatogr. Sci.*, **20**, 203–8.

Lee, H.A., Mills, E.N.C., Finglas, P.M. and Morgan, M.R.A. (1990) Rapid biospecific methods of vitamin analysis. *J. Micronutr. Anal.*, **7**, 261–70.

Lim, K.L., Young, R.W., Palmer, J.K. and Driskell, J.A. (1982) Quantitative separation of B_6 vitamers in selected foods by a gas–liquid chromatographic system equipped with an electron-capture detector. *J. Chromat.*, **250**, 86–9.

Lindeberg, J. (1996) Capillary electrophoresis in food analysis. *Food Chem.*, **55**, 73–94.

Majors, R.E. (1980a) Multidimensional high performance liquid chromatography. *J. Chromatogr. Sci.*, **18**, 571–9.

Majors, R.E. (1980b) Practical operation of bonded-phase columns in high-performance liquid chromatography. In *High-performance Liquid Chromatography, Advances and Perspectives*, Vol. 1 (ed. C. Horvath), Academic Press, New York, pp. 75–111.

Majors, R.E. (1986) Sample preparation for HPLC and gas chromatography using solid-phase extraction. *Liquid Chromat.–Gas Chromat Mag.*, **4**(10), 972, 980, 982, 984.

Majors, R.E. (1991) Supercritical fluid extraction – an introduction. *Liquid Chromat.–Gas Chromat. Int.*, **4**(3), 11–12, 14, 16–17.

Marks, C. (1988) Determination of free tocopherols in deodorizer distillate by capillary gas chromatography. *J. Am. Oil Chem. Soc.*, **65**, 1936–9.

Marsili, R. and Callahan, D. (1993) Comparison of a liquid solvent extraction technique and supercritical fluid extraction for the determination of α- and β-carotene in vegetables. *J. Chromatogr. Sci.*, **31**, 422–8.

Mauro, D.J. and Wetzel, D.L. (1984) Simultaneous determination of thiamine and riboflavin in enriched cereal based products by high-performance liquid chromatography using selective detection. *J. Chromat.*, **299**, 281–7.

Nishi, H., Tsumagari, N., Kakimoto, T. and Terabe, S. (1989) Separation of water-soluble vitamins by micellar electrokinetic chromatography. *J. Chromat.*, **465**, 331–43.

Olechno, J.D., Tso, J.M.Y., Thayer, J. and Wainright, A. (1990) Capillary electrophoresis: a multifaceted technique for analytical chemistry. Part 1. Separations. *Am. Lab.*, **22**, 51, 52, 54, 55–9.

Ollilainen, V., Vahteristo, L., Uusi-Rauva, A. *et al.* (1993) The HPLC determination of total thiamin (vitamin B_1) in foods. *J. Food Comp. Anal.*, **516**, 152–65.

Osborne, B.G. and Tyson, J.F. (1988) Review: flow injection analysis – a new technique for food and beverage analysis. *Int. J. Food Sci. Technol.*, **23**, 541–54.

Panfili, G., Manzi, P. and Pizzoferrato, L. (1994) High-performance liquid chromatographic method for the simultaneous determination of tocopherols, carotenes, and retinol and its geometric isomers in Italian cheeses. *Analyst, Lond.*, **119**, 1161–5.

Pearson, W.N. (1967) Principles of microbiological assay. In *The Vitamins. Chemistry, Physiology, Pathology, Methods*, 2nd edn, Vol. VII (eds P. György and W.N. Pearson), Academic Press, New York, pp. 1–26.

Pomeranz, Y. and Meloan, C.E. (1978) *Food Analysis: Theory and Practice*, AVI Pub. Co. Inc., Westport, CT., pp. 11–20.

Rabel, F.M. (1980) Use and maintenance of microparticle high performance liquid chromatography columns. *J. Chromatogr. Sci.*, **18**, 394–408.

Rabel, F.M. (1985) Instrumentation for small-bore liquid chromatography. *J. Chromatogr. Sci.*, **23**, 247–52.

Ranger, C.B. (1981) Flow injection analysis. Principles, techniques, application, design. *Analyt. Chem.*, **53**, 20A–22A, 24A, 26A, 28A, 30A, 32A.

Rees, D.I. (1989) Determination of nicotinamide and pyridoxine in fortified food products by HPLC. *J. Micronutr. Anal.*, **5**, 53–61.

Reyes, E.S.P. and Subryan, L. (1989) An improved method of simultaneous HPLC assay of riboflavin and thiamin in selected cereal products. *J. Food Comp. Anal.*, **2**, 41–7.

Reynolds, S.L. (1985) The use of HPLC in the determination of fat-soluble vitamins in a variety of milk-based food products. *Proc. Inst. Food Sci. Technol. (UK)*, **18**, 43–50.

Reynolds, S.L. and Judd, H.J. (1984) Rapid procedure for the determination of vitamins A and D in fortified skimmed milk powder using high-performance liquid chromatography. *Analyst, Lond.*, **109**, 489–92.

Rizzolo, A. and Polesello, S. (1992) Review. Chromatographic determination of vitamins in foods. *J. Chromat.*, **624**, 103–52.

Roy, R.B. (1979) Application of Technicon AutoAnalyzer II to the analysis of water-soluble vitamins in foodstuffs. In *Topics in Automated Chemical Analysis*, Vol. I, Technicon Ind. Systems, Tarrytown, New York, pp. 138–62.

Roy, R.B. and Conetta, A. (1976) Automated analysis of water-soluble vitamins in food. *Food Technol.*, **30**(10), 94, 95, 98, 100, 103, 104.

Russell, L.F. and Vanderslice, J.T. (1992) Comments on the standard fluorometric determination of riboflavin in foods and biological tissues. *Food Chem.*, **43**, 79–82.

Saito, M., Yamauchi, Y., Inomata, K. and Kottkamp, W. (1989) Enrichment of tocopherols in wheat germ by directly coupled supercritical fluid extraction with semipreparative supercritical fluid chromatography. *J. Chromatogr. Sci.*, **27**, 79–85.

Schneiderman, M.A., Sharma, A.K., Mahanama, K.R.R. and Locke, D.C. (1988) Determination of vitamin K_1 in powdered infant formulas, using supercritical fluid extraction and liquid chromatography with electrochemical detection. *J. Ass. Off. Analyt. Chem.*, **71**, 815–7.

Shearer, M.J. (1983) High-performance liquid chromatography of K vitamins and their antagonists. *Adv. Chromat.*, **21**, 243–301.

Sims, A. and Shoemaker, D. (1993) Simultaneous liquid chromatographic determination of thiamine and riboflavin in selected foods. *J. AOAC Int.*, **76**, 1156–60.

Slover, H.T., Thompson, R.H. Jr, Davis, C.S. and Merola, G.V. (1985) Lipids in margarines and margarine-like foods. *J. Am. Oil Chem. Soc.*, **62**, 775–86.

Snell, E.E. (1948) Use of microorganisms for assay of vitamins. *Physiol. Rev.*, **28**, 255–82.

Snell, E.E. (1950) Microbiological methods in vitamin research. In *Vitamin Methods*, Vol. I (ed. P. György), Academic Press, New York, pp. 327–505.

Snyder, J.M., Taylor, S.L. and King, J.W. (1993) Analysis of tocopherols by capillary supercritical fluid chromatography and mass spectrometry. *J. Am. Oil Chem. Soc.*, **70**, 349–54.

Snyder, L., Levine, J., Stoy, R. and Conetta, A. (1976) Automated chemical analysis: update on continuous-flow approach. *Analyt. Chem.*, **48**, 942A–944A, 946A, 948A, 950A, 952A, 954A, 956A.

Spanos, G.A., Chen, H. and Schwartz, S.J. (1993) Supercritical CO_2 extraction of β-carotene from sweet potatoes. *J. Food Sci.*, **58**, 817–20.

Stancher, B. and Zonta, F. (1983) HPLC of fat-soluble vitamins in cheese. New method for determining the total biological activity of vitamins A and E. *Riv. Ital. Sostanze Grasse*, **60**, 371–5 (in Italian).

Stranahan, J.J. and Deming, S.N. (1982) Thermodynamic model for reversed-phase ion-pair liquid chromatography. *Analyt. Chem.*, **54**, 2251–6.

Strohecker, R. and Henning, H.M. (1966) *Vitamin Assay. Tested Methods*, Verlag Chemie, Weinheim.

Sullivan, D.M. and Carpenter, D.E. (eds) (1993) *Methods of Analysis for Nutrition Labeling*, AOAC International, Arlington, VA.

Tanaka, A., Iijima, M., Kikuchi, Y. *et al.* (1989) Gas chromatographic determination of nicotinamide in meats and meat products as 3-cyanopyridine. *J. Chromat.*, **466**, 307–17.

Tarli, P., Benocci, S. and Neri, P. (1971) Gas-chromatographic determination of pantothenates and panthenol in pharmaceutical preparations by pantoyl lactone. *Analyt. Biochem.*, **42**, 8–13.

Taylor, J.K. (1986) Role of collaborative and cooperative studies in evaluation of analytical methods. *J. Ass. Off. Analyt. Chem.*, **69**, 398–400.

Thompson, J.N. and Madère, R. (1978) Automated fluorometric determination of vitamin A in milk. *J. Ass. Off. Analyt. Chem.*, **61**, 1370–3.

Ulberth, F. (1991) Simultaneous determination of vitamin E isomers and cholesterol by GLC. *J. High Resolut. Chromat.*, **14**, 343–4.

Velíšek, J. and Davídek, J. (1986) Gas–liquid chromatography of vitamins in foods: the water-soluble vitamins. *J. Micronutr. Anal.*, **2**, 25–42.

Velíšek, J., Davídek, J., Mňuková, J. and Pištěk, T. (1986) Gas chromatographic determination of thiamin in foods. *J. Micronutr. Anal.*, **2**, 73–80.

Vivilecchia, R.V., Lightbody, B.G., Thimot, N.Z. and Quinn, H.M. (1977) The use of microparticulates in gel permeation chromatography. *J. Chromatogr. Sci.*, **15**, 424–33.

Voigt, M.N., Eitenmiller, R.R. and Ware, G.O. (1979) Vitamin analysis by microbial and protozoan organisms: response to food preservatives and neutralization salts. *J. Food Sci.*, **44**, 723–8, 737.

Walsh, J.H., Wyse, B.W. and Hansen, R.G. (1979) A comparison of microbiological and radioimmunoassay methods for the determination of pantothenic acid in foods. *J. Food Biochem.*, **3**, 175–89.

Waxman, S., Schreiber, C. and Herbert, V. (1971) Radioisotopic assay for measurement of serum folate levels. *Blood*, **38**, 219–28.

Wehling, R.L. and Wetzel, D.L. (1984) Simultaneous determination of pyridoxine, riboflavin, and thiamin in fortified cereal products by high-performance liquid chromatography. *J. Agric. Food Chem.*, **32**, 1326–31.

White, C.M., Gere, D.R., Boyer, D. *et al.* (1988) Analysis of pharmaceuticals and other solutes of biochemical importance by supercritical fluid chromatography. *J. High Resolution Chromat. Chromatogr. Commun.*, **11**, 94–8.

Wickroski, A.F. and McLean, L.A. (1984) Improved reverse phase liquid chromatographic determination of vitamins A and D in fortified milk. *J. Ass. Off. Analyt. Chem.*, **67**, 62–5.

Widicus, W.A. and Kirk, J.R. (1979) High performance liquid chromatographic determination of vitamins A and E in cereal products. *J. Ass. Off. Analyt. Chem.*, **62**, 637–41.

Wiedemer, R.T., McKinley, S.L. and Rendl, T.W. (1986) Advantages of wide-bore capillary columns. *Int. Lab.*, **16**, May, 68, 70, 72, 74, 76, 77.

Williams, W.L., Broquist, H.P. and Snell, E.E. (1947) Oleic acid and related compounds as growth factors for lactic acid bacteria. *J. Biol. Chem.*, **170**, 619–30.

Wills, R.B.H., Wimalasiri, P. and Greenfield, H. (1985) Comparative determination of thiamin and riboflavin in foods by high-performance liquid chromatography and fluorometric methods. *J. Micronutr. Anal.*, **1**, 23–9.

Wimalasiri, P. and Wills, R.B.H. (1985) Simultaneous analysis of thiamin and riboflavin in foods by high-performance liquid chromatography. *J. Chromat.*, **318**, 412–16.

Woodrow, I.L., Torrie, K.M. and Henderson, G.A. (1969) A rapid method for the determination of riboflavin in dried milk products. *J. Inst. Can. Technol. Aliment.*, **2**, 120–2.

Yik, Y.F., Lee, H.K., Li, S.F.Y. and Khoo, S.B. (1991) Micellar electrokinetic capillary chromatography of vitamin B_6 with electrochemical detection. *J. Chromat.*, **585**, 139–44.

Zamarreño, M.M.D., Pérez, A.S., Pérez, C.G. and Méndez, J.H. (1992) High-performance liquid chromatography with electrochemical detection for the simultaneous determination of vitamin A, D_3 and E in milk. *J. Chromat.*, **623**, 69–74.

3

Vitamin A and the provitamin A carotenoids

3.1 INTRODUCTION

In 1915 McCollum and Davis isolated from animal fats and fish oils a 'fat-soluble A' that was essential to rats for growth and also cured eye disorders. In 1921 Bloch reported that a diet containing full milk and cod-liver oil cured xerophthalmia in infants and concluded that the eye affliction was due to the absence of the fat-soluble A in the diet. In the meantime it was discovered that green vegetables also possess fat-soluble A activity and in 1930 Moore provided evidence that carotene was converted to vitamin A in the body. The biochemical function of vitamin A in vision was established by Wald in 1935.

Vitamin A-active compounds are represented by retinoids (preformed vitamin A) and provitamin A carotenoids. The retinoids, as defined by Sporn, Roberts and Goodman (1984), comprise retinol, retinaldehyde and retinoic acid, together with their naturally occurring and synthetic analogues. Carotenoids are yellow, orange, red or violet pigments that are responsible for the colour of many vegetables and fruits. Certain almost colourless carotenoids also exist, such as phytofluene, which fluoresces intensely under ultraviolet (UV) irradiation. In nature, the carotenoids are synthesized exclusively by higher plants and photosynthetic microorganisms, in which they play fundamental roles in metabolism. Although animals are unable to synthesize carotenoids *de novo*, they can assimilate them through their diet.

3.2 CHEMICAL STRUCTURE AND NOMENCLATURE

Vitamin A

The structures of retinoids found in foods and fish-liver oils are shown in Figure 3.1. The parent vitamin A compound, retinol, has the empirical formula $C_{20}H_{30}O$ and a molecular weight (MW) of 286.44. It is systematically named 9,13-dimethyl-7-(1,1,5-trimethyl-6-cyclohexen-5-yl)-7,9,11, 13-nonatetraen-15-ol (Harris, 1967) and comprises a β-ionone (cyclohexenyl) ring attached at the carbon-6 position to a side chain composed of four isoprene units. The four double bonds in the polyene side chain give rise to *cis–trans* (Z-E) isomerism. Theory predicts the existence of a possible 16 isomers of retinol, but most of these exhibit steric hindrance, and some are too labile to exist (Schwieter and Isler, 1967). The predominant isomer, all-*trans*-retinol, possesses maximal (100%) vitamin A activity and is frequently accompanied in foodstuffs by smaller amounts of 13-*cis*-retinol. 9-*Cis*- and 9,13-di-*cis*-retinol occur in small amounts in fish-liver oils. 3-Dehydroretinol (vitamin A_2) represents the major form of vitamin A in the liver and flesh of freshwater fish.

Retinyl acetate $(C_{22}H_{32}O_2;$ MW $= 328.5)$ and retinyl palmitate $(C_{36}H_{60}O_2;$ MW $= 524)$ are used commercially in synthetic form to supplement the vitamin A content of foodstuffs. Current industrial processes for synthesizing retinyl acetate start from acetone and proceed through the key intermediate, β-ionone (O'Leary, 1993).

Provitamin A carotenoids

Carotenoids are classified chemically as carotenes, which are hydrocarbons, and xanthophylls, which have an oxygen group either on the ring or in the chain. These oxygen-containing groups include hydroxyl, carbonyl, carboxylic acid, ester, epoxide, glycoside and ether (Simpson

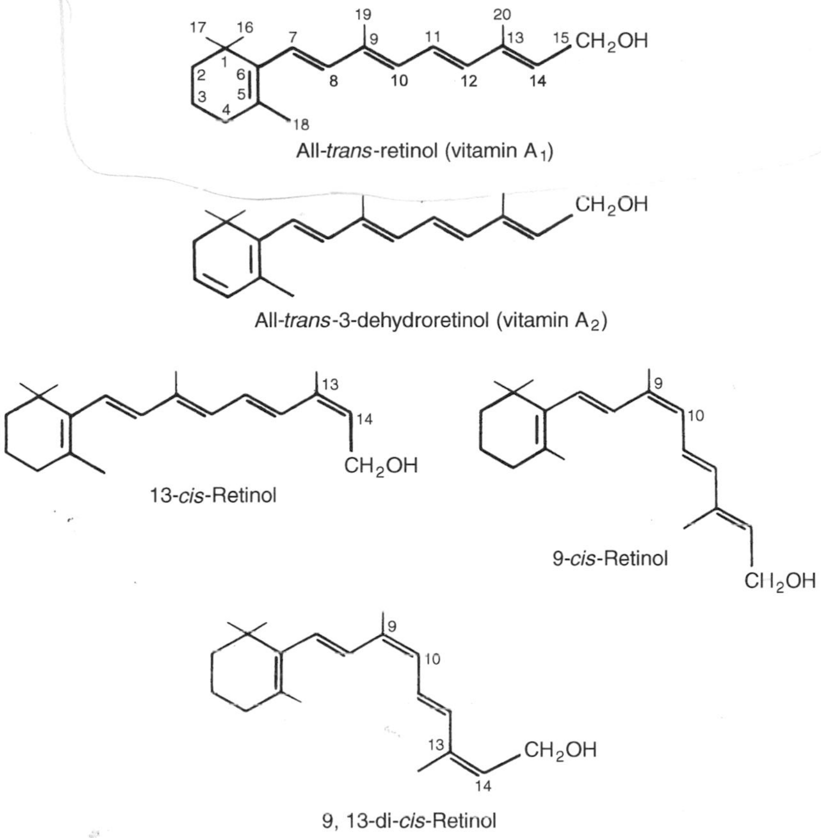

All-*trans*-retinol (vitamin A₁)

All-*trans*-3-dehydroretinol (vitamin A₂)

13-*cis*-Retinol

9-*cis*-Retinol

9, 13-di-*cis*-Retinol

Figure 3.1 Chemical structures of retinoids found in foods and fish-liver oils.

and Chichester, 1981). Most naturally occurring carotenoids contain 40 carbon atoms, corresponding to 8 isoprene units. In some instances, C_{40}-carotenoids undergo partial oxidative cleavage in the plant tissues to give shortened molecules known as apocarotenoids, which bear either aldehyde or carboxyl groups at the points of attack (Zechmeister, 1962).

From a nutritional viewpoint, the carotenoids are classified as provitamins and inactive carotenoids. The structures of the principal provitamin A carotenoids that occur in foods are shown in Figure 3.2. The nutritionally most important carotenoid, β-carotene ($C_{40}H_{56}$; MW = 536), is composed of two molecules of retinol joined tail to tail, thus the compound possesses maximal provitamin A activity. The structures of all other provitamin A carotenoids incorporate only one molecule of retinol, hence theoretically contribute 50% of the biological value of β-carotene. Over 400 naturally occurring carotenoids have been crystallized and fully

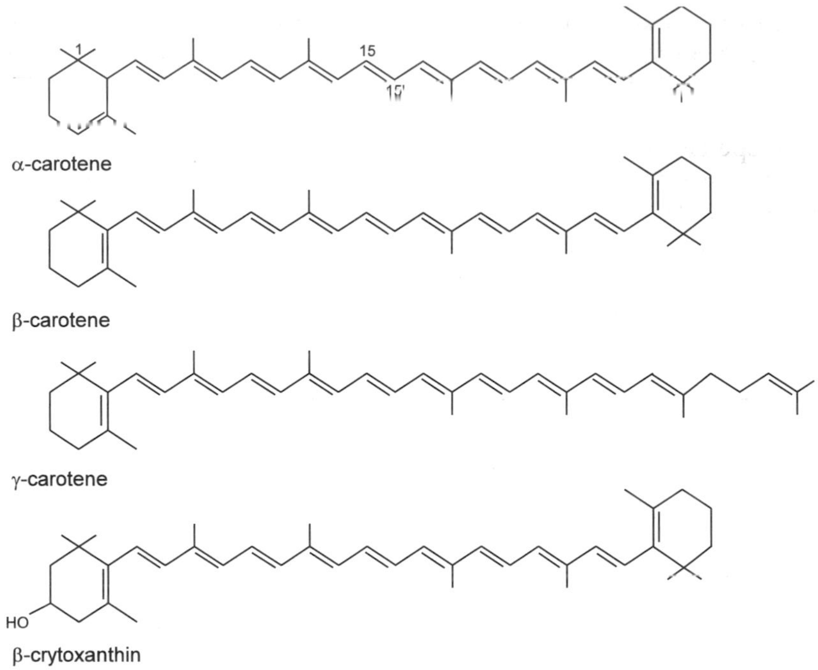

α-carotene

β-carotene

γ-carotene

β-crytoxanthin

Figure 3.2 Chemical structures of the principal provitamin A carotenoids that occur in foods.

characterized; of these, about 50 possess provitamin A activity in varying degrees (Olson, 1988).

In most vegetables and in many fruits, β-carotene constitutes more than 85% of the total provitamin A activity. Notable exceptions are carrots and oranges, which contain both β-carotene and α-carotene in the ratio of 2:1 (Bureau and Bushway, 1986). β-Cryptoxanthin is the predominant provitamin A carotenoid in orange juice (Quackenbush and Smallidge, 1986) and in some varieties of sweet corn (Lee, McCoon and LeBowitz, 1981). In many fruits and vegetables the concentrations of provitamin A carotenoids are low relative to the concentrations of inactive carotenoids. For example, lutein is the most abundant carotenoid in green leafy vegetables (Khachik, Beecher and Whittaker, 1986), lycopene predominates in tomatoes (Tan, 1988), and capsanthin is the major pigment in red peppers (Mínguez-Mosquera and Hornero-Méndez, 1993). Other inactive carotenoids found in fruits and vegetables include violaxanthin, neoxanthin, zeaxanthin, zeta-carotene, phytoene and phytofluene (Khachik *et al.*, 1991). Milk products, egg yolk, shellfish and crustacea also contain carotenoids, which are derived from the animal's diet.

In plant and animal tissues the carotenoids are usually found associated with lipid fractions in noncovalent association with membranes and lipo-proteins, and they accumulate, together with chlorophylls, in the chloro-plasts of green leaves (Goodwin and Britton, 1988). They also occur as very fine dispersions in aqueous systems, such as orange juice. Carotenoids exist primarily in the all-*trans* configuration, but small amounts of *cis*-isomeric forms have been found in fresh fruits and vegetables. Traditional food processing and preservation methods, especially canning, induce *cis–trans* isomerization. The main *cis* isomers of β-carotene that have been found in fresh and processed fruits and vegetables are 13-*cis* and 9-*cis* (O'Neil and Schwartz, 1992). A third isomer, 15-*cis*-β-carotene, has also been reported to occur in several fruits and vegetables (O'Neil and Schwartz, 1992). The main portion of the xanthophylls exists as mono or bis esters of saturated long-chain fatty acids, such as myristic, lauric and palmitic acids (Khachik and Beecher, 1988; Philip and Chen, 1988).

Commercial processes for synthesizing β-carotene start either from vitamin A or from a C_{14} aldehyde intermediate produced during the synthesis of vitamin A. β-Carotene is also isolated commercially from natural sources, usually an alga, to satisfy a small market for a natural source of this provitamin (O'Leary, 1993). In addition to β-carotene, two other provitamin A carotenoids are produced commercially; β-apo-8'-carotenaldehyde ($C_{30}H_{40}O$; MW − 416.6) and the ethyl ester of β-apo-8'-carotenoic acid ($C_{32}H_{44}O_2$; MW = 460.7). These two compounds are more commonly known as apocarotenal and apocarotenoic ester, respectively.

3.3 BIOPOTENCY

Any conversion of all-*trans* forms of retinol and carotenoid precursors to *cis* forms would cause a reduction in the net vitamin A activity of the food product. By both growth and liver storage bioassays, 13-*cis*-retinol pos-sesses 75% of the activity of the all-*trans* isomer; 9-*cis*- and 9,13-di-*cis*-retinol possess, respectively, 21% and 24% relative activities (Ames, 1966). 13-*Cis*-β-carotene and 9-*cis*-β-carotene exhibit, respectively, 53% and 38% of the provitamin A activity of all-*trans*-β-carotene (Rodriguez-Amaya 1989). Dietary retinaldehyde possesses about 90% of the biologi-cal activity of all-*trans*-retinol and 3-dehydroretinol is about 40% as active (Sivell *et al.*, 1984).

3.4 DEFICIENCY SYNDROMES

Animals

Animals maintained on a diet containing retinoic acid, but no retinol, grow well and appear outwardly healthy (except for a deficiency in visual pigment), but they are incapable of reproduction. The effect of vitamin A deficiency upon reproduction is manifested in the male rat by the cessation of spermatogenesis and in the female rat by resorption of the foetus midway through pregnancy.

Humans

An early sign of vitamin A deficiency is night blindness, which is caused by an insufficient amount of visual purple in the retina. Night blindness refers to the lengthening of the time required for the eyesight to adapt from light conditions to dark conditions.

Another sign of vitamin A deficiency is where the epithelial cells of the skin and mucous membranes lining the respiratory, gastrointestinal and urinogenital tracts cease to differentiate, and lose their secretory function. The undifferentiated cells are flattened and multiply at an increased rate, so that the cells pile up on one another and the surface becomes keratinized. This condition promotes dry skin and loss of hair sheen; the symptoms of loss of appetite may be due to keratinization of the taste buds. The lack of protective mucus in the affected mucosae leads to an increased susceptibility to infections. Xerophthalmia refers to keratinization of the conjunctiva (the mucous membrane covering the eye) which later spreads to the cornea, causing ulceration. The ultimate condition is keratomalacia (softening of the cornea) which, if not treated, leads to permanent blindness.

The clinical effects of vitamin A deficiency in adults are usually seen only in people whose diet has been deficient for a long time in both dairy produce and vegetables. Induced vitamin A deficiency experiments in human volunteers resulted in night blindness and some follicular keratosis (blockage of the sebaceous glands by horny plugs), but there was no xerophthalmia (Passmore and Eastwood, 1986). Xerophthalmia mainly affects very young children and is regarded as the most serious vitamin deficiency disease in the world today.

3.5 BIOCHEMICAL AND PHYSIOLOGICAL FUNCTIONS

The vital role of vitamin A in maintaining life, growth and general health is commonly referred to as its systemic role, and can be fulfilled by either retinol or retinoic acid. The other essential functions of vitamin A, namely vision and reproduction, cannot be supported by retinoic acid and specifically require either retinol or retinaldehyde. These differences can be explained by the reversible conversion of retinol to retinaldehyde and the irreversible conversion of retinaldehyde to retinoic acid (Figure 3.3).

The function of vitamin A in vision is well defined, and is based on the binding of 11-*cis*-retinaldehyde with the protein opsin to form visual

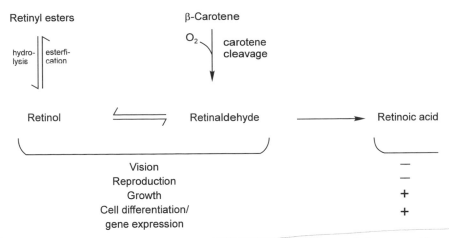

Figure 3.3 Biochemical relationships among dietary and cellular retinoids. (Reproduced with permission from Ross, 1993, © *J. Nutr.*, **123**, 346–50, American Institute of Nutrition.)

purple (rhodopsin) in the retina (Pitt, 1985). No unifying mechanism has been proposed that can satisfactorily explain all of the other functions of vitamin A.

Vitamin A is clearly involved in the differentiation of epithelial cells, as shown by the replacement of mucus-secreting cells by cells producing keratin in vitamin A deficiency. A leading hypothesis for this involvement is that vitamin A induces the synthesis of specific glycoproteins through the direct control of gene expression in a manner analogous to that of the steroid hormones. In this hypothesis, retinol is taken up by target cells and oxidized in part to retinoic acid. Both retinoids enter the nucleus in combination with specific cellular binding proteins, and stimulate and inhibit transcription at specific sites on the DNA to produce messenger RNAs that are coded for different proteins (Olson, 1986, 1988).

Vitamin A is involved in growth, and the first sign of deficiency is a reduction in food intake through loss of appetite. Another function of the vitamin is in maintaining the normal activity of the immune system. Ziegler (1989, 1991) has reviewed epidemiological studies which have correlated the intake of carotenoid-rich fruits and vegetables with protection from some forms of cancer. These observational studies, however, do not show cause and effect. In a balanced review of the potential role of β-carotene and other carotenoids in disease prevention, Mayne (1996) concluded that major public health benefits could be achieved by increasing the consumption of carotenoid-rich fruits and vegetables. However, intervention trials of supplemental β-carotene indicate that supplements are of little or no value in preventing the major cancers occurring in well-nourished populations, and may actually increase, rather than reduce, lung cancer incidence in smokers.

3.6 PHYSICOCHEMICAL PROPERTIES OF VITAMIN A AND CAROTENOIDS

Appearance and solubility

Retinol and retinyl acetate are yellow crystalline powders; retinyl palmitate is a pale yellow oil or crystalline mass. β-Carotene is a reddish-brown to deep violet crystalline powder; β-apo-8'-carotenaldehyde is a deep violet crystalline powder; β-apo-8'-carotenoic acid ethyl ester is a rust red crystalline powder.

Vitamin A is insoluble in water; soluble in alcohol; and readily soluble in diethyl ether, petroleum ether, chloroform, acetone, and fats and oils. β-Carotene is insoluble in water; very sparingly soluble in alcohol, fats and oils; sparingly soluble in ether and acetone; and slightly soluble in chloroform and benzene.

Spectroscopic properties

Absorption

The molar absorptivity (ε) of all-*trans*-retinol at the absorption maximum (λ_{max}) of 325 nm in isopropanol is 52 300 (Boldingh *et al.*, 1951), which corresponds to an absorptivity ($A_{1\,cm}^{1\%}$) of *c.*1830. Retinol and its esters exhibit practically equal molar absorptivities when dissolved in a given solvent.

β-Carotene, by virtue of its deep orange colour, exhibits very strong absorption in the visible region of the spectrum. Reported $A_{1\,cm}^{1\%}$ values for β-carotene are 2592 at 450 nm in hexane and 2560 at 452 nm in ethanol (Nierenberg, Peng and Alberts 1988).

Fluorescence

Retinol and retinyl esters exhibit strong native fluorescence. The fluorescence excitation spectra of vitamin A compounds correspond to their absorption spectra with wavelength maxima in the 324–328 nm region; emission takes place between 470 nm and 490 nm (λ_{max} 470 nm) (Olson, 1991). The fluorescence response of 13-*cis*-retinol is less than that of all-*trans*-retinol, the relative fluorescence depending upon the solvent (Egberg, Heroff and Potter, 1977; Lawn, Harris and Johnson, 1983).

Stability

Retinol is readily oxidized by molecular oxygen to yield the 5,6-epoxide and 5,8-furanoxide among the oxidation products. Oxidation of vitamin

A results in an almost complete loss of biological activity. The acetate and palmitate esters of retinol are somewhat more stable towards oxidation than the alcohol. Vitamin A is stable towards alkali, but is extremely sensitive towards acids, which can cause rearrangements of the double bonds and *cis–trans* isomerization (Schwieter and Isler, 1967). Heat also causes steroisomerization. The stabilized vitamin A preparations supplied to the food industry are not prone to isomerization, and remain in the all-*trans* form after storage for several years (Parrish, 1977).

Solutions of all-*trans*-retinol or retinyl palmitate in hexane undergo slow isomerization to the lower potency *cis* isomers when exposed to white light; the photoisomerization rate is greatly increased in the presence of chlorinated solvents. No significant isomerization occurred within 23 h under gold fluorescent light (wavelengths > 500 nm) in chloroform, dichloromethane, or hexane solutions (Landers and Olson, 1986). Irradiation also causes double bond rearrangement to form inactive retro compounds (Thompson, 1982).

The carotenoids are highly stable within their natural plant cell environment, but, once isolated, they are sensitive to light, heat and acids, which promote *cis–trans* isomerization (Zechmeister, 1962). The xanthophylls are particularly susceptible to these agents and are also destroyed in alkaline environments. Plant tissues contain lipoxygenases which catalyse the oxidation of carotenoids.

3.7 ANALYSIS

3.7.1 Scope of analytical techniques

Of the preformed vitamin A commonly found in foods, only all-*trans*-retinol and smaller amounts of 13-*cis*-retinol, both in esterified form, are usually present in significant quantities. For the analysis of vitamin A-fortified foods, HPLC can be applied to determine either the total retinol content or the added retinyl ester (acetate or palmitate), depending on the extraction technique employed. The vitamin A activity of plant foods is usually based on the HPLC determination of the three most ubiquitous provitamins, namely α- and β-carotene and β-cryptoxanthin. It is important to isolate the provitamins from inactive carotenoids such as lycopene. The most satisfactory HPLC methods are those that are capable of separating all-*trans*-β-carotene from its *cis* isomers.

3.7.2 Expression of dietary values

The need to establish an international standard for use in vitamin A assays was recognized before the discovery of the various isomeric forms of carotene and before vitamin A had been purified and crystallized. The original International Standard for vitamin A, introduced in 1931, was a stock of supposedly pure crystalline β-carotene. It was soon discovered that this material contained α-carotene as well as β-carotene and so it was replaced in 1934 with a stock of pure crystalline β-carotene. One International Unit (IU) of vitamin A was defined as that amount which had the same vitamin A activity in rats as 0.6 μg of the International Standard Preparation of β-carotene. In 1949, shortly after pure crystalline vitamin A became commercially available, a second International Standard was established based on retinyl acetate. One IU was defined as the amount of activity contained in 0.344 μg of all-*trans*-retinyl acetate, which is equivalent to 0.300 μg of all-*trans*-retinol (Rodriguez and Irwin, 1972).

In 1960, an expert committee of the World Health Organisation (WHO) decided to abandon the use of international standards for vitamin A, because of the commercial availability of high purity crystalline retinyl acetate and also β-carotene. A subsequent committee of the Food and Agricultural Organisation (FAO) and the WHO proposed in 1965 that the vitamin A value of diets no longer be expressed in international units. Instead, the value would be designated in terms of retinol equivalents expressed in micrograms of retinol. The retinol equivalent allows the contributions from retinoids to be combined with those from provitamin A carotenoids to yield a single numerical value for the vitamin A content of a food, and is defined as the amount of retinol plus the equivalent amount of retinol that can be obtained from the provitamin A carotenoids. The term 'retinol equivalent' is purely a dietary concept for estimating the vitamin A activity in foods, and is not an equivalency in the usual chemical sense. The term, therefore, applies specifically to retinol and the provitamin A carotenoids; retinoic acid, for example, cannot be expressed as a retinol equivalent (Bieri and McKenna, 1981).

In defining the equivalency of β-carotene and other provitamins in terms of retinol, several assumptions have to be made. On the basis of the enzymatic cleavage of one molecule of β-carotene into two molecules of retinaldehyde, 1 μg of β-carotene would be roughly equivalent to 1 μg of retinol. However, β-carotene is absorbed less efficiently than is retinol, and not all of the β-carotene absorbed is converted into retinol. In the rat, under conditions of suboptimal intake, it is well established that 1 μg of retinol has the same biological activity as 2 μg of pure all-*trans*-β-carotene in oil (FAO/WHO, 1967). Evidence for a similar relationship in humans was provided by data from the Sheffield study in which small oral doses of synthetic all-*trans*-retinyl acetate and of synthetic all-*trans*-β-carotene were used to cure visual defects. These signs of vitamin A deficiency were induced by feeding diets deficient in vitamin A and carotene for periods ranging from 6 to 25 months (Sauberlich *et al.*, 1974). In the light of the Sheffield study and of the known relationship in the rat, it is assumed that about 50% of the β-carotene absorbed from the diet can be converted into retinol. This 50% yield is borne out by data from rat growth assays, which showed that major provitamin A carotenoids, such as α-carotene and β-cryptoxanthin, exhibit about 50% of the activity of β-carotene (Olson, 1988).

The β-carotene naturally present in food is not absorbed as efficiently as is pure β-carotene administered in oil. Studies of carotene absorption in humans have revealed a wide variation, from 1% to 88%, from a variety of yellow and green vegetables. The FAO/WHO Committee decided in 1965 that 33% availability from diets was the best approximation that could be made for practical purposes. The Committee recommended that, in the absence of more specific data for foods, the availability of β-carotene be taken as one-third and that the efficiency of conversion in the body be accepted as one-half of the available β-carotene: hence the utilization efficiency in the human is taken as one-sixth. Thus, in the human, 1 μg of β-carotene in the diet is taken to have the same biological activity as 0.167 μg of retinol (FAO/WHO, 1967). For the other provitamins that theoretically yield only one-half as much retinol as does β-carotene, the utilization efficiency is one-twelfth.

To evaluate diets in terms of retinol equivalents (RE), the following formula may be applied:

$$RE = \mu g \text{ retinol} + \frac{\mu g \text{ β-carotene}}{6} + \frac{\mu g \text{ other provitamin A carotenoids}}{12}$$

For comparison with values in the older literature, the IU values can be converted into retinol equivalents as follows.

To convert IU into RE on the basis of retinol,

$$1\,RE = 1\,\mu g \text{ retinol}$$
$$1\,IU = 0.3\,\mu g \text{ retinol}$$

Therefore, 1 RE = 1/0.3 = 3.33 IU vitamin A activity from retinol.

Since the IU was based on studies that did not take into account the poor absorption and availability of carotenoids in foods, the equivalency of retinol and β-carotene in the IU system differs from that of the RE system. Thus:

$$\text{In the RE system, } 1\,\mu g \text{ retinol} \equiv 6\,\mu g \text{ β-carotene}$$
$$\text{In the RE system, } 1\,\mu g \text{ retinol} \equiv 2\,\mu g \text{ β-carotene}$$

To convert IU into RE on the basis of β-carotene, one must first multiply the IU value by a factor of 3 (6/2) to make the equivalency the same as that of the RE system, and then muliply by 3.33.

Table 3.1 Units for expressing vitamin A values (Reprinted from Olson, 1991, *Handbook of Vitamins*, 2nd edn (ed. L. J. Machlin) p. 12 by courtesy of Marcel Dekker Inc.)

Compound[a]	μg/IU	IU/μg	μg/RE	RE/μg
Retinol	0.300	3.33	1.000	1.000
Retinyl acetate	0.344	2.91	–	–
Retinyl palmitate	0.55	1.82	–	–
β-Carotene	0.6 (1.8)[b]	0.56	6	0.167
Mixed provitamin carotenoids	1.2 (3.6)	0.28	12	0.083

IU, International Units; RE, retinol equivalents.
[a] All-*trans* isomers.
[b] The value of 1.8 μg/IU agrees with the convention of considering 6 μg of β-carotene as 1 μg RE (Olson, 1984).

Therefore, 1 RE = 3 × 3.33 = 10.0 IU vitamin A activity from β-carotene.

RE = number of IU from retinol/3.33 + number of IU from β-carotene/10

The units for expressing vitamin A values are summarized in Table 3.1.

The estimate for the utilization efficiency of β-carotene given by the FAO/WHO in 1965 (i.e. 6 g β-carotene corresponds to 1 μg retinol) is considered to be applicable for an intake of 1000–4000 Ng β-carotene per meal. At lower intakes, within the range of daily requirements or below, carotenoids are absorbed much more efficiently and the conversion factor is therefore higher. Conversely, at intakes in excess of 4000 μg, the conversion factor becomes progressively lower as the absorption system becomes saturated (Brubacher and Weiser, 1985). Possible alternative conversion factors have been suggested by the FAO/WHO (1988) (Table 3.2).

Table 3.2 Possible alternative factors for determination of biological activity of β-carotene in foods at various levels[a]

Intake of β-carotene (μg) per meal	Amount of β-carotene (μg) equivalent to 1 μg of retinol	Factor by which the weight of β-carotene is multiplied to give an equivalent weight of retinol
1000	4	0.25
1000–4000	6	0.167[b]
More than 4000[c]	10	0.10

[a] The bioavailability of β-carotene in oil at each intake is up to two times higher than in food.
[b] This figure is in common use. As per definition, 1 μg of β-carotene is considered equal to 0.167 μg retinol equivalents.
[c] Because the absorption efficiency of carotenoids declines markedly with increased intake, factors in this high range are intake-dependent.

In nutrition surveys, the β-carotene content of foods should be expressed in weight units rather than in retinol equivalents and the appropriate conversion factor should be used.

For food labelling purposes, which require the actual amounts of vitamin A in the food rather than the nutritional value, data obtained by physicochemical assay are expressed on a weight basis. In animal-derived foods (e.g. liver, milk, chicken meat, whole eggs), where the vitamin exists predominantly as preformed vitamin A, the units are either micrograms of retinol or retinol equivalents. In plant-derived foods (e.g. fruits, vegetables and cereals), where the provitamins predominate, the units are β-carotene equivalents in micrograms of β-carotene. By definition, one β-carotene equivalent is equal to 1 μg of all-*trans*-β-carotene or 2 μg of other, largely all-*trans*, provitamin A carotenoids in the foods. The relationship is based on the 50% activity of other provitamins relative to that of β-carotene in the rat growth assay (Food Labelling Regulations, 1984).

It is apparent that the vitamin A value for many foods will be a rough approximation. In foods of animal origin the analysis for retinol will be a reasonably accurate estimation of the retinol equivalents. In plant foods and composite diets the calculation of retinol equivalents may contain a considerable error, owing to the variability of absorption of carotenoids. The situation is further complicated by the effects of cooking and degree of comminution upon the bioavailability of carotenoids.

3.7.3 Bioassays

Three bioassays for determining vitamin A activity have been commonly used to measure the response of animals towards foods and diets (Bliss and Roels, 1967). The growth response assay and the vaginal smear assay are curative methods, which involve feeding the test sample and a standard preparation of the vitamin to separate groups of rats, which have previously been fed a vitamin A-deficient maintenance diet. The extent to which the deficiency sign (lack of growth or vaginal cornification) has been cured is then measured. The third bioassay, the liver-storage assay, depends on the direct physicochemical determination of vitamin A *in vivo* after administration of a known dose.

Growth response assay

The growth response assay is the most widely used bioassay method for vitamin A. Young rats are fed on a vitamin A-deficient diet for a 3–4-week period until their body stores are depleted of vitamin A and they cease to grow. The rats are then divided into four treatment groups for a 4-week test period. Two treatment groups are fed weekly supplements of a vitamin A standard (all-*trans* retinyl acetate in cottonseed oil) at two dosage levels, and the other two treatments are fed similar levels of the unknown sample under test. The gain in weight during the 4-week treatment period is a linear function of the logarithm of the dose, and the potency of the test sample is calculated by standard methods. An assay requires 40 rats (10 rats for each treatment group) plus extra rats as a negative control to ensure that no stray sources of vitamin A have been introduced during the experiment (Bliss and Roels, 1967; Green, 1970).

Vaginal smear assay

An early sign of vitamin A deficiency in the female rat is interruption of the normal oestrous cycle with the persistence of cornified cells in vaginal smears. Ovariectomized rats are used in the vaginal smear assay to increase the specificity of response. The assay procedure entails placing ovariectomized rats on a vitamin A-deficient maintenance diet and determining the advent of depletion by examining vaginal smears under the microscope for the predominance of squamous epithelial cells. The rats are assigned randomly to seven groups, each of 10 or more individually housed animals, and continued on the vitamin A-deficient diet. Of the seven groups, three groups receive the reference standard, three the unknown, and one serves as a negative control. The response of each rat is the number of days from the first day of treatment until the day of recovery, which is indicated by a predominance of leucocytes in the vaginal smears. The response is a linear function of the logarithm of the dose over a range of 25 to about 150 IU.

Liver-storage assay

The liver-storage assay is based on the accumulation of vitamin A by the depleted rat and storage in its liver. The assay has the advantage over the growth response assay in that the depletion period is one-third as long, and the treatment period lasts only 2 or 3 days instead of 4 weeks. It is, however, about 1000 times less sensitive than the growth response assay. The assay procedure entails placing male weanling rats on a vitamin A-deficient diet for 9–18 days, by which time the liver stores are depleted of vitamin A. The rats are then fed for 2 or 3 days with two dosage levels of the standard and of the unknown, after which the livers are removed and analysed for vitamin A. The vitamin A content of the liver is directly proportional to the ingested dose over a range of 500–10,000 IU. The assay requires four equal groups of 10 rats plus a fifth group as a negative control. The liver-storage assay is more specific than the growth response assay in that retinoic acid, which is highly active in promoting growth, is not stored in the liver (Sporn and Roberts, 1984).

3.7.4 Extraction techniques

Vitamin A

Saponification
Saponification is the universally accepted method of extracting vitamin A from almost any type of food commodity or composite food sample. Exceptions are fish roe and hens' eggs, which contain 90% and 10%, respectively, of their vitamin A activity in the form of retinaldehyde (Sivell *et al.*, 1984). Retinaldehyde is destroyed by alkali treatment, and hence its determination requires special treatment (Parrish, 1977).

Retinol, being a slightly polar compound, is efficiently extracted from the diluted saponification digest with a 50/50 mixture of diethyl ether and light petroleum ether (boiling range 40–60 °C). The soaps are soluble in diethyl ether, so the ether extracts have to be washed with successive portions of distilled water until the washes are free from alkali (colourless on addition of phenolphthalein). The washings must be performed by inverting the separating funnel gently several times to avoid the formation of stable emulsions, which are produced when soaps, water and hydrophobic solvent are shaken in the absence of ethanol. The use of hexane is advantageous in that soaps are not extracted, but large amounts of soaps confer hydrophobic properties to the ethanol water mixture. Therefore, the minimum number of extractions needed to achieve a quantitative recovery of retinol is affected by the amount of fat in the sample (Thompson, 1986). It is also important, when using hexane as the extracting solvent, to maintain the ethanol concentration in the extraction system to below 40% (Thompson *et al.*, 1982).

Zahar and Smith (1990) reported a rapid saponification method for the extraction of vitamin A from milk and other fluid dairy products, which avoids the need for several extractions and washings using separating funnels. Into a series of 50 ml stoppered centrifuge tubes is placed 2 ml of sample, 5 ml of absolute ethanol containing 1% (w/v) pyrogallol, and 2 ml of 50% (w/v) aqueous KOH. The tubes are stoppered, agitated carefully and placed in a water bath at 80 °C for 20 min with periodic agitation. After saponification, the tubes are cooled with running water and then placed in an ice-water bath before adding 20 ml of diethyl ether/ petroleum ether (50/50) containing 0.01% (w/v) butylated hydroxytoluene (BHT) as antioxidant. The tubes are again stoppered and vortex-mixed vigorously for 1 min, then allowed to stand for 2 min, and again vortexed for 1 min. To each tube is added 15 ml of ice-cold water, and the tubes are inverted at least 10 times. After centrifugation, 10 ml of the upper organic layer is accurately removed by pipette into a tube, and the solvent is evaporated to dryness in a stream of nitrogen or under vacuum at 40 °C using a rotary evaporator. The residue is dissolved in 1.0 ml of methanol (for milk samples) ready for analysis by HPLC.

Direct solvent extraction

Thompson, Hatina and Maxwell (1980) reported a simple and rapid extraction technique, which could be applied to the analysis of fortified fluid milks in which the vitamin A ester (palmitate or acetate) is added in the form of an oily premix, and thoroughly dispersed in the bulk product. A 2.0 ml aliquot of the milk sample is mixed with 5 ml of absolute ethanol in a 15 ml stoppered centrifuge tube, and allowed to stand in the dark for 5 min. The milk constituents in this mixture are suspended in 71% aqueous ethanol, which denatures the proteins and fractures the fat globules. Hexane (5.0 ml) is added and the tube contents are vortexed for 30 s, then allowed to stand for 2 min. The mixing and standing procedure is repeated twice. Distilled water (3 ml) is added to induce the aqueous and organic phases to separate, and the tube is inverted several times. After centrifugation, the upper phase is composed of a hexane solution of the milk lipids containing the vitamin A, and the lower phase is composed of aqueous ethanol in which are dissolved salts, denatured proteins and polar lipids. The interface is a solid disc containing a mixture of upper and lower phases plus insoluble protein. The upper layer is removed using a Pasteur pipette, and an aliquot of this solution is analysed by HPLC.

Provitamin A carotenoids

Plant material should be extracted as rapidly as possible after it is obtained so as to avoid oxidative degradation of the carotenoids. If the material cannot be extracted immediately, it should be stored in a freezer or freeze-dried. The plant material should be undamaged, as rapid enzymatic degradation of carotenoids occurs immediately when the leaves are cut. Fresh and frozen fruits and vegetables should be blanched to inactivate enzymes before blending to a purée consistency, adding a known volume of water if necessary. Sun-dried fruits such as apricots, peaches and prunes should be ground in a food chopper, and a representative sample allowed to rehydrate by steeping in a known volume of water for several hours before blending. Fruits canned in thick syrup should be washed two or three times with water before blending.

Saponification

Plant materials such as papaya, which contain β-cryptoxanthin ester among other xanthophyll esters, require saponification to hydrolyse these esters and simplify the analysis. Saponification is also advisable for the analysis of green leafy vegetables, as the alkali treatment breaks down chlorophylls which could otherwise sensitize photoisomerization of the carotenoids (Goodwin and Britton, 1988). Although carotenes are not degraded to any significant extent during conventional 'hot' saponification, as used for extracting vitamin A, such treatment causes a significant loss of xanthophylls (Khachik, Beecher and Whittaker, 1986). Kimura, Rodriguez-Amaya and Godoy (1990) tested six widely used saponification procedures and recommended one in which the carotenoids are dissolved in petroleum ether, an equal volume of 10% methanolic KOH is added, and the mixture is left standing overnight (about 16 h) in the dark at room temperature. This treatment was shown to retain β-carotene and β-apo-8′-carotenaldehyde, while completely hydrolysing β-cryptoxanthin ester.

Direct solvent extraction
Nonhydrolytic extraction methods cannot provide assurance that the total β-cryptoxanthin is quantified, as a proportion of this xanthophyll may occur in the esterified form. Furthermore, such methods are inadvisable for the analysis of green leafy vegetables, since the coextracted chlorophylls sensitize photoisomerization of the carotenoids, and appreciable amounts of *cis* isomers can be produced during even a brief exposure of a chlorophyll-containing extract to light (Goodwin and Britton, 1988).

For the determination of carotenoids in fruits and nonleafy vegetables, which contain a large percentage of water, direct solvent extraction using a suitable water-miscible organic solvent is appropriate. The extracting solvent employed is important in determining the susceptibility of carotenoids towards isomerization. Little or no isomerization takes place when tetrahydrofuran (THF), methanol or acetonitrile are used, but when chlorinated solvents (chloroform, dichloromethane) are used, there is significant formation of 9-*cis* and 13-*cis* isomers (Pesek, Warthesen and Taoukis, 1990). THF is suitable, because it not only readily solubilizes carotenoids and chlorophylls, but it also prevents the formation of emulsions by denaturing the associated proteins (Khachik, Beecher and Whittaker, 1986). THF is known to promote peroxide formation, so it must be stabilized with an antioxidant such as BHT. The extraction may be carried out in the presence of anhydrous sodium sulphate as a drying agent. The addition of magnesium carbonate to the extraction system serves to neutralize traces of organic acids that can cause destruction and structural transformation of carotenoids.

In an extraction procedure described by Khachik and Beecher (1987), homogenized vegetables are blended with anhydrous sodium sulphate (200% of the weight of the test portion of vegetable), magnesium carbonate (10% of the weight of the test portion) and THF. The extract is filtered under vacuum, and the solid materials are re-extracted with THF until the resulting filtrate is colourless. Most of the solvent is removed on a rotary evaporator at 30 °C, and the concentrated filtrate is partitioned between petroleum ether and water to remove the majority of contaminating non-terpenoid lipids. The water layer is washed with petroleum ether several times, and the resulting organic layers are combined, dried over anhydrous sodium sulphate, and evaporated to dryness. The residue is taken up in a small volume of the HPLC solvent for analysis.

Marsili and Callahan (1993) compared an ethanol/pentane solvent extraction procedure with a supercritical CO_2 extraction procedure for the HPLC determination of α- and β-carotene in vegetables. A combination of static and dynamic modes of extraction with ethanol modifier at 338 atm and 40 °C was necessary in order to achieve optimum recovery with the SFE procedure. The extracted material was recovered by depressurization of the CO_2 across a solid-phase trap and rinsed from the trap into a 2 ml vial with HPLC-grade hexane for injection. β-Carotene results obtained using the SFE procedure averaged 23% higher than results using the solvent extraction process.

3.7.5 High-performance liquid chromatography

Vitamin A

HPLC methods for determining vitamin A, and in some cases β-carotene as well, are summarized in Table 3.3.

Vitamin A can be adequately and simply determined by HPLC using absorbance detection. Fluorescence detection, although possible, offers no real advantage over absorbance detection for routine food analysis applications, and the linear response range is more limited. In addition, the fluorescence response of 13-*cis*-retinol under reversed-phase conditions was estimated to be only 33% that of all-*trans*-retinol compared with a relative absorptivity of 92% (Lawn, Harris and Johnson, 1983).

Quantification
Several different quantification procedures for vitamin A have been described in the literature, some using retinol directly as a standard and some using retinyl acetate, which is converted to retinol by saponification. The latter approach is generally preferred because crystalline all-*trans*-retinyl acetate is commercially available in high purity and is free from *cis* isomers. Commercial sources of retinol are oily preparations and are at best only about 70% pure. There are two ways of preparing a retinol standard from retinyl acetate.

1. A relatively large amount (typically 25 mg) of retinyl acetate is saponified and extracted, and the residue is dissolved in isopropanol to give a stock solution of retinol, which can be stored for 2–3 months in a refrigerator. This stock solution is diluted with isopropanol to give a suitable working standard solution, whose concentration is determined spectrophotometrically ($A_{1\,cm}^{1\%} = 1830$ at λ_{max} 325 nm) immediately before use as an external standard in the HPLC procedure.

2. An accurately prepared standard solution of retinyl acetate (i.e. a solution of known concentration) is taken through the saponification and extraction procedure along with each batch of samples, and the resultant retinol solution is used as an external standard without spectrophotometric standardization. This technique, which is recommended by COST 91 (Brubacher, Müller-Mulot and Southgate, 1985), compensates for losses of vitamin A incurred during the saponification and subsequent manipulations (i.e. the calculated vitamin A value is recovery-corrected).

Normal-phase systems

Adsorption chromatography using a silica column is capable of separating several *cis–trans* isomers of retinol and dehydroretinol (Stancher and Zonta, 1984). If a foodstuff is supplemented with retinyl acetate, adsorption chromatography can be used to distinguish between supplemental vitamin A and indigenous vitamin A (mainly retinyl palmitate), as shown by Woollard and Woollard (1981) in the analysis of fortified whole milk powder. Woollard and Indyk (1989) extended the technique to identify the minor indigenous esters of vitamin A in milk and milk products. The adsorption of carotenes on silica is very weak, and they are eluted as an unresolved group near the solvent front. However, the added β-carotene in margarine (Thompson, Hatina and Maxwell, 1980) and cheese (Stancher and Zonta, 1982) can be quantified alongside the vitamin A because of the known absence of other carotenes.

For the determination of vitamin A esters, a nonhydrolytic extraction procedure must obviously be employed. This entails extracting the entire lipid fraction with a nonaqueous solvent, and removing the polar material. An aliquot of the lipid extract is then injected onto the silica column with no further purification. As previously stated, silica columns can tolerate relatively heavy loads of triglycerides, which are not strongly adsorbed.

Saponification of the sample simplifies the analysis by converting the naturally occurring and supplemental esters of vitamin A to retinol. The extract containing the unsaponifiable material can be dissolved in a solvent which is similar in composition to the mobile phase. For the analysis of vitamin A-rich foods, it is possible to inject an aliquot of the unconcentrated unsaponifiable extract directly onto the silica column, as demonstrated by Egberg, Heroff and Potter (1977).

Reversed-phase systems

The removal of triglycerides from the food sample by saponification or other means presents the opportunity to utilize reversed-phase chromatography. The unsaponifiable matter is conventionally extracted into a solvent (e.g. diethyl ether/petroleum ether (50/50) or hexane) that is

Table 3.3 HPLC methods used for the determination of vitamin A compounds in food

Food	Sample preparation	Column	Mobile phase	Compounds separated	Detection	Reference
Normal-phase chromatography						
Margarine	Dissolve sample in heptane containing 500 mg BHT and 200 mg α-tocopherol/1. Pass emulsion through glass column containing 10 g Na_2SO_4 and 20 g NaCl	LiChrosorb Si-60 5 μm 290 × 4.6 mm i.d.	Heptane/diisopropyl ether (95:5)	Retinyl palmitate or retinyl acetate	UV 325 nm	Aitzetmüller, Pilz and Tasche (1979)
Fortified fluid milk (whole, semi-skimmed, skimmed)	Mix 2 ml milk and 5 ml absolute ethanol in a centrifuge tube, let stand 5 min. Vortex-mix with 5 ml hexane, let stand 2 min. Repeat mixing and standing procedure twice. Add 3 ml water, mix by inversion, centrifuge	LiChrosorb Si-60 5 μm 250 × 3.2 mm i.d.	Hexane/diethyl ether (98:2)	All-*trans*-retinyl palmitate	UV 325 nm	Thompson, Hatina and Maxwell (1980)
Margarine	Dissolve sample in hexane, shake with 60% aqueous EtOH, centrifuge	LiChrosorb Si-60 5 μm 250 × 3.2 mm i.d.	Hexane/diethyl ether (98:2)	Carotene, all-*trans*-retinyl palmitate	Vis. 453 nm (carotene) UV 325 nm (retinyl palmitate)	Thompson, Hatina and Maxwell (1980)
Milk, infant formulas	Saponify (ambient), extract unsaponifables with hexane/diethyl ether (85:15)	Apex Silica 3 μm 150 × 4.5 mm i.d.	1-5% 2-PrOH in heptane	13-*cis*- and all-*trans*-retinol	UV 340 nm (filter)	Thompson and Duval (1989)
Cheese	Saponify (ambient), extract unsaponifables with hexane	LiChrosorb Si-60 5 μm 250 × 4 mm i.d.	Hexane/methyl ethyl ketone (90:10)	Carotene, all-*trans*-retinol	Vis. 450 nm (carotene) UV 340 nm (retinol)	Stancher and Zonta (1982)
Milk	Dilute 400 mg milk with water/MeOH/EtOH (55:9:36) and saponify in a culture tube. Extract unsaponifables with heptane/isopropyl ether (3:1), centrifuge. Repeat extraction	Perkin-Elmer HS-5-Silica 125 × 4 mm i.d.	Heptane containing 2-PrOH (60 ml/l)	Retinol	Fluorescence: ex. 344 nm em. 472 nm	Jensen (1994)

Food	Sample preparation	Column	Mobile phase	Analytes	Detection	Reference
Various foods	Saponify (ambient), dilute solution with water and absolute ethanol to yield a volumetric ratio of water to EtOH of 1:1. Pipette a 20 ml aliquot onto a Kieselguhr cartridge and elute with petroleum ether. Evaporate eluate and dissolve residue in isooctane	Spherisorb SW silica gel 3 μm 100 × 2 mm i.d. (narrow-bore)	0.3% 1-octanol in hexane	13-*cis*- and all-*trans*-retinol	UV 325 nm	Brinkmann *et al.* (1995)

Reversed-phase chromatography

Food	Sample preparation	Column	Mobile phase	Analytes	Detection	Reference
All food types	Saponify (hot), extract unsaponifiables with diethyl ether/petroleum ether (1 + 1)	μBondapak C$_{18}$ 10 μm 300 × 3.9 mm i.d.	MeOH/H$_2$O (90:10)	Retinol	UV 325 nm	Brubacher, Müller-Mulot and Southgate (1985)
Unfortified fluid dairy products	Saponify (hot) 2 ml sample in a centrifuge tube, extract unsaponifiables once with diethyl ether/petroleum ether (1:1), centrifuge	Nova-Pak C$_{18}$ 150 × 3.9 mm i.d.	MeOH/H$_2$O (95:5)	Retinol	UV 325 nm	Zahar and Smith (1990)
Fortified fluid milk (whole, skimmed), infant formulas, margarine	Saponify (hot) 1 ml sample of milk or formula or 50 mg margarine in a centrifuge tube, extract unsaponifiables five times with hexane	LiChrosorb RP-18 10 μm 250 × 3.2 mm i.d.	MeOH/H$_2$O (90:10) (retinol), MeOH/H$_2$O (99:1) (β-carotene)	Retinol and β-carotene (separate chromatograms)	UV 325 nm (retinol) Vis. 453 nm (β-carotene)	Thompson and Maxwell (1977)
Selected foods of animal origin and processed foods	Saponify (hot), extract unsaponifiables with hexane	μBondapak C$_{18}$ 10 μm 300 × 3.9 mm i.d.	MeCN/MeOH/ethyl acetate (88:10:2)	Retinol, α- and β-carotenes	UV 313 nm (retinol) Vis. 436 nm (carotenes)	Tee and Lim (1992)
Breakfast cereals, margarine, butter	Saponify (hot), precipitate soaps with acetic acid in MeCN, dilute with water	Vydac 201 TP C$_{18}$ 10 μm 250 × 3.2 mm i.d.	MeCN/H$_2$O (65:35)	13-*cis*- and all-*trans*-retinol	UV 328 nm	Egberg, Heroff and Potter (1977)

Abbreviations: see footnote to Table 2.4.

incompatible with a semi-aqueous mobile phase. It then becomes necessary to evaporate the unsaponifiable extract to dryness and to redissolve the residue in a small volume of methanol (if methanol is the organic component of the mobile phase). For the analysis of breakfast cereals, margarine and butter, Egberg, Heroff and Potter (1977) avoided the time-consuming extraction of the unsaponifiable matter and the evaporation step by acidifying the unsaponifiable matter with acetic acid in acetonitrile to precipitate the soaps. An aliquot of the filtered extract could then be injected, after dilution with water, onto an ODS column eluted with a compatible mobile phase (65% acetonitrile in water).

Reversed-phase chromatography using semi-aqueous mobile phases can separate all-*trans*-retinol from 13-*cis*-retinol, albeit rather poorly. Further separation of the minor *cis* isomers is not achieved, but this is of little concern for most purposes. In practice, peak measurements are made to encompass both the all-*trans* and 13-*cis* isomers, ignoring the reduced biopotency of the latter. Alternatively, the peak areas of the two isomers can be measured separately and either added together or multiplied by their respective relative potencies of 100% and 75% and then added.

The carotenes are retained under reversed-phase conditions, allowing β-carotene and retinol to be determined in the same unsaponifiable extract using an appropriate mobile phase and detection wavelength (Thompson and Maxwell, 1977).

Reversed-phase chromatography can be performed without hydrolysing the vitamin A esters by using high-pressure gel permeation chromatography to remove the glycerides from crude lipid extracts. This technique was employed in conjunction with NARP HPLC for the determination of retinyl palmitate and β-carotene in margarine (Landen and Eitenmiller, 1979) and retinyl palmitate in breakfast cereals (Landen, 1980). NARP chromatography is not capable of separating the *cis* and *trans* isomers of vitamin A (Lawn, Harris and Johnson, 1983).

Provitamin A carotenoids

The vitamin A activity of plant foods is usually based on the determination of the three most ubiquitous provitamins, namely α- and β-carotene and β-cryptoxanthin. The most satisfactory methods are those which can separate all-*trans*-β-carotene from its *cis* isomers.

Normal-phase systems

Silica stationary phases are unsuitable for the analysis of plant materials as the carotenes are very weakly adsorbed, while the xanthophylls are highly retained. A further disadvantage of silica columns is their inability to separate α-carotene from β-carotene, even when gradient elution is attempted (Fiksdahl, Mortensen and Liaaen-Jensen, 1978).

Calcium hydroxide (lime) is an excellent adsorbent for resolving caro-tene stereoisomers, and easily resolves α- and β-carotenes. Lime columns are not commercially available and therefore it is difficult to standardize separations between laboratories. Chandler and Schwartz (1987), using a lime column and a mobile phase of hexane containing 0.3% acetone, separated six carotene isomers from canned carrots. These were, in order of elution, two *cis*-α-carotenes, all-*trans*-α-carotene, 13-*cis*-β-caro-tene, all-*trans*-β-carotene and 9-*cis*-β-carotene. O'Neil, Schwartz and Catignani (1991) reported the isocratic separation of all-*trans*-β-carotene and its 9-, 13- and 15-*cis* isomers.

Microparticulate alumina especially prepared for HPLC is available commercially and permits the determination of all-*trans*-β-carotene with-out its lower potency *cis* isomers interfering. Using a mobile phase of isooctane containing 0.5% stabilized THF, the *cis* isomers of β-carotene are eluted before the all-*trans* isomer to form a single composite peak in the chromatogram (Bushway, 1985). γ-Carotene and β-cryptoxanthin are retained, but α-carotene co-elutes with the *cis* isomers of β-carotene and therefore cannot be accurately quantified.

The interactions between carotenes and nitrile-bonded phases are very weak, thus the limitations described for silica apply (Rüedi, 1985). Amino-bonded phases eluted with isooctane containing 0.5% stabilized THF offer a rapid means of determining all-*trans*-β-carotene without its *cis* isomers interfering. However, this system is inadequate if α-carotene is present, because α- and β-carotenes co-elute (Bushway, 1985).

Reversed-phase systems
Reversed-phase chromatography is preferred to normal-phase chroma-tography for the determination of provitamin A carotenoids in foods, because the carotenes are retained and the separation of α- and β-caro-tene is easily achieved. Owing to the weak hydrophobic forces on which the separation mechanism depends, there is little risk of on-column degradation of carotenoids. The xanthophylls are eluted well before the carotenes, the latter requiring strong mobile phases containing little or no water to displace them. The more recent published methods (Table 3.4) employ NARP chromatography, which overcomes the poor solubility of carotenes in conventional semi-aqueous mobile phases.

The separation of all-*trans*-β-carotene from its principal 9-*cis* and 13-*cis* isomers has been achieved using silica-based polymeric ODS bonded-phase column packings such as Vydac TP (Bushway, 1985, 1986; Quack-enbush and Smallidge, 1986; Quackenbush, 1987; Khachik, Beecher and Lusby, 1989; Saleh and Tan, 1991) and Spheri-5 ODS (Lesellier *et al.*, 1989; Granado *et al.*, 1991, 1992). Under the isocratic NARP conditions employed, 15-*cis*-β-carotene co-elutes with the 13-*cis*-isomer (O'Neil, Schwartz and Catignani, 1991; Granado *et al.*, 1991). Vydac TP and

138

Table 3.4 HPLC methods used for the determination of provitamin A carotenoids in food

Food	Sample preparation	Column	Mobile phase	Compounds separated	Detection	Reference
Reversed-phase chromatography						
Fruits, vegetables	Extract sample with THF + Na$_2$SO$_4$ + MgCO$_3$	Partisil ODS 5 μm 250 × 4.6 mm i.d.	MeCN/THF/H$_2$O (85:12.5:2.5)	α- and β-Carotenes, β-cryptoxanthin	Vis. 470 nm	Bureau and Bushway (1986)
Fruit, vegetables	Extract sample with THF	Vydac 218 TP54 C$_{18}$ 5 μm 250 × 4.6 mm i.d.	MeOH/MeCN/THF (56:40:4)	α-Carotene, all-*trans*-β-carotene (separated from *cis*-β-carotenes)	Vis. 470 nm	Bushway (1986)
Raw and cooked vegetables	Extract sample with THF + Na$_2$SO$_4$ + MgCO$_3$	Spheri-5 ODS 5 μm 220 × 4.6 mm i.d.	MeCN/CH$_2$Cl$_2$/MeOH (70:20:10)	α-Carotene, all-*trans*-β-carotene (separated from *cis*-β-carotenes)	Vis. 450 nm	Granado *et al.* (1992)
Vegetables	Extract sample with THF + Na$_2$SO$_4$ +MgCO$_3$ + nonapreno-β-carotene (internal standard). Partition into petroleum ether and water	Spheri-5 RP-18 5 μm 220 × 4.6 mm i.d.	MeCN/MeOH/CH$_2$Cl$_2$ (55:22:23)	α- and β-Carotenes	Vis. 470 nm	Khachik and Beecher (1987)
Vegetables	Saponify (hot), extract unsaponifiables with diisopropyl ether	Hypersil-ODS 3 μm 250 × 4.6 mm i.d.	MeCN/MeOH/CHCl$_3$/H$_2$O (250 + 200 + 90 + 11)	β-Carotene	Vis. 445 nm	Speek, Temalilwa and Schrijver (1986)
Orange juice	Saponify (hot), extract unsaponifiables with hexane. Magnesia column chromatography	Vydac 201 TP54 C$_{18}$ 5 μm 250 × 4.6 mm i.d.	MeOH/CHCl$_3$ (94:6)	β-Cryptoxanthin, α-carotene, all-*trans*-β-carotene (separated from *cis*-β-carotenes)	Vis. 475 nm	Quackenbush and Smallidge (1986) (Figure 3.4)
Olive oil	Saponify (ambient), extract unsaponifiables with diethyl ether	Supelcosil LC-18 5 μm 150 × 4.6 mm i.d.	MeCN/2-PrOH/1,2-dichloroethane, (92.5:5.0:2.5)	α- and β-Carotenes	Vis. 458 nm	Stancher, Zonta and Bogoni (1987)

Sample	Procedure	Column	Mobile phase	Compounds	Detection	Reference
Malaysian vegetables and fruits	Saponify (hot), extract unsaponifiables with hexane	μBondapak C_{18} 10 μm 300 × 3.9 mm i.d.	MeCN/MeOH/ethyl acetate (88:10:2)	β-Cryptoxanthin, γ-, α- and β-carotenes	Vis. 436 nm	Tee and Lim (1991)
Vegetables	Supercritical fluid extraction	Vydac 201 TP54 C_{18} 5 μm 250 × 4.6 mm i.d.	MeOH/MeCN/CH_2Cl_2/hexane (65:27:4:4)	α- and β-Carotenes	Vis. 450 nm	Marsili and Callahan (1993)
Fruits, vegetables	Extract sample with THF/MeOH (1:1) +$MgCO_3$ + internal standard (β-apo-8'-carotenal or, for green vegetables, echinenone). Filter. Partition into petroleum ether containing 0.1% BHT Saponify (ambient), extract with petroleum ether (saponification applied to fruit samples and peppers only)	Vydac 201 TP54 C_{18} 5 μm (metal frits replaced with Teflon frits) 250 × 4.6 mm i.d.	MeCN/MeOH/CH_2Cl_2 (75:20:15) containing 0.1% BHT and 0.05% triethylamine. (The MeOH contains 0.05 M ammonium acetate)	β-Cryptoxanthin, α-carotene, all-*trans*-β-carotene (separated from *cis*-β-carotenes)	Vis. 450 nm	Hart and Scott (1995)
Chinese vegetables	Extract sample with hexane/acetone/EtOH/toluene (10:7:6:7) +$MgCO_3$ + internal standard (β-apo-8'-carotenal). Filter. Saponify (ambient), extract unsaponifiables with hexane	Ultramex C-18 5 μm 250 × 4.6 mm i.d.	MeCN/MeOH/ethyl acetate (75:15:10)	β-apo-8'-carotenal (internal standard), β-cryptoxanthin, α- and β-carotenes	Vis. 450 nm	Chen *et al.* (1993)
Carrots	Extract sample with hexane/acetone/absolute EtOH/toluene (10:7:6:7) + 40% methanolic KOH. Leave for 16h in the dark for saponification. Extract unsaponifiables with hexane	Vydac 201 TP54 C_{18} 5 μm 250 × 4.6 mm i.d.	MeOH/CH_2Cl_2 (99:1)	α- and β-Carotenes, (separated from their 15-*cis* isomers and from 13-*cis*-β-carotene)	Vis. 450 nm	Chen and Chen (1994)

Table 3.4 *Continued*

Food	Sample preparation	Column	Mobile phase	Compounds separated	Detection	Reference
Dark green vegetables (cowpea leaves, spring cabbage, Italian spinach)	Blanch leaves in boiling water for 3 min (cabbage and cowpea leaves) or 1 min (spinach) Homogenize by blending with water containing 0.5% ascorbic acid. Extract an aliquot of the resultant mixture with acetone/ petroleum ether (3:2) containing 0.5% BHT by shaking for 10 min. Re-extract until colourless (3 extractions) Saponify the combined solvent extracts by adding a saturated solution of KOH in EtOH and standing for 15 min at ambient temperature. Wash solvent layer with 10% NaCl solution followed by three portions of water to remove the acetone. Dry over sodium sulphate, evaporate to near dryness and dissolve the residue in MeOH/CH$_2$Cl$_2$ 9:1 containing 0.5% BHT	Vydac TP-201 C$_{18}$ 5 μm 250 × 4.6 mm i.d.	MeOH/CH$_2$Cl$_2$/H$_2$O (80:15.2:4.8)	All-*trans*-, 9-*cis*-, 13-*cis*-β-carotene, all-*trans*-α-carotene	Vis. 450 nm	Nyambaka and Ryley (1996)

Abbreviations: see footnote to Table 2.4.

Table 3.5 Properties of polymeric ODS bonded phase column packings

Property	Vydac 218 TP[a]	Vydac 201 TP[a]	Spheri-5 ODS[b]
Surface configuration	Polymeric	Polymeric	Polymeric
End-capped	Yes	No	Yes
Percentage carbon loading (w/w)	9	9	14
Mean pore diameter (Å)	300	300	80
Specific surface area (m^2/g)	80	80	200

[a] Separations Group
[b] Brownlee Laboratories

Spheri-5 ODS column packings have very different particle characteristics and surface coverages, as shown in Table 3.5. It appears that the

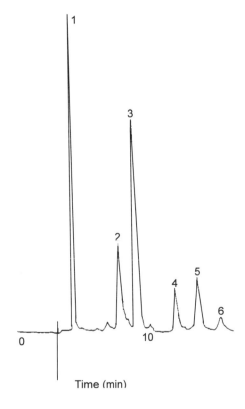

Figure 3.4 HPLC analysis of carotenoids in the unsaponifiable matter from frozen orange juice concentrate after magnesia column clean-up. Peaks: (1) Sudan I (internal standard); (2) zeinoxanthin; (3) β-cryptoxanthin; (4) α-carotene; (5) β-carotene; (6) *cis*-β-carotene. Column, Vydac 201 TP54 C$_{18}$; mobile phase, methanol/chloroform (94:6); VIS detection, 475 nm. (From Quackenbush and Small-ridge, 1986. (Reprinted from *The Journal of AOAC International*, 1986, vol. 69, No. 5, p. 771. Copyright, 1986, by AOAC International, Inc.)

polymeric surface configuration, which is common to Vydac TP and Spheri-5 ODS, is responsible for the separation of the *cis–trans* isomers of β-carotene, end-capping being unimportant. Indeed Lesellier *et al* (1989) demonstrated that a polymeric (as opposed to a monomeric) ODS phase is necessary for the isocratic separation of all-*trans*-α-carotene and all-*trans*-β-carotene from their respective *cis* isomers using NARP chromatography under usual conditions. The separation of the major carotenoids in an extracted sample of frozen orange juice concentrate is shown in Figure 3.4.

Owing to reported losses of carotenoids by rapid decomposition through oxidation in the presence of a stainless steel column frit (Nierenberg and Lester, 1986), it has been recommended to purchase HPLC columns fitted with inert metal-free frits (MacCrehan, 1990; Scott, 1992).

3.8 BIOAVAILABILITY

3.8.1 Physiological aspects

Overview of vitamin A metabolism

Essential to vitamin A metabolism is the role of lipoproteins and specific retinoid-binding proteins in transporting the vitamin through an aqueous environment, and preventing damage to membranes by excessive amounts of free retinol. The major proteins and lipoproteins involved in vitamin A metabolism are listed in Table 3.6.

Dietary preformed vitamin A and carotenoids are incorporated into mixed micelles in the intestinal lumen and absorbed. Within the enterocytes, provitamin A carotenoids are converted to vitamin A and retinyl esters are released into the bloodstream via the lymphatic system in the form of chylomicrons. The chylomicrons undergo lipolysis and the resultant chylomicron remnants are taken up by the liver for storage in stellate cells. Upon demand, vitamin A is mobilized from the liver in the form of a complex between retinol and retinol-binding protein (RBP). In the plasma, this complex (*holo*-RBP) forms a larger complex with a protein called transthyretin. Vitamin A is taken up by target cells and interacts with a specific cellular retinol-binding protein (CRBP). Plasma concentrations of vitamin A are maintained relatively constant through a homeostatically controlled interchange of vitamin A among plasma, liver and extrahepatic tissues.

Intestinal absorption and metabolism

The absorption of β-carotene and metabolic events within the enterocyte are shown diagrammatically in Figure 3.5.

Table 3.6 Principal lipoproteins and retinoid-binding proteins involved in vitamin A metabolism in humans (Reproduced with permission from Ross, 1993, © *J. Nutr.*, **123**, 346–350, American Institute of Nutrition.)

Binding protein[a]	Retinoid[b]	Principal organs or tissues
Intercellular transport		
Chylomicron	Retinyl esters	Intestine → liver
Retinol-binding protein (RBP)	Retinol	Liver → extrahepatic organs
Intracellular transport		
Cellular retinol-binding protein (CRBP)	Retinol	Liver, kidney, intestine, eye, spleen
Cellular retinol-binding protein type II (CRBP-II)	Retinol	Small intestine
Cellular retinoic acid-binding protein (CRABP)	Retinoic acid	Seminal vesicle, vas deferens, skin, eye

[a] Excluding proteins involved in vision.
[b] All-*trans* isomers.

Ingested preformed vitamin A and provitamin A carotenoids are liberated from their association with membranes and lipoproteins by the action of pepsin in the stomach and of proteolytic enzymes in the small intestine. In the stomach the free carotenoids and retinyl esters congregate in fatty globules, which then pass into the duodenum. In the presence of bile salts, the globules are broken up into smaller globules, which renders them more easily digestible by a variety of pancreatic lipases and

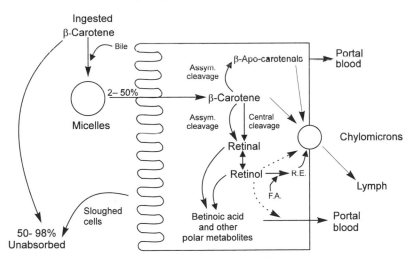

Figure 3.5 Intestinal absorption of β-carotene. R.E., retinyl esters; F.A., fatty acids. (Reproduced with permission from Erdman, Bierer and Gugger, 1993, © *Ann. N.Y. Acad. Sci.*, **691**, 76–85, Academy of Sciences New York.)

results in the formation of mixed micelles. Extensive hydrolysis of retinyl esters also takes place within the duodenum, catalysed mainly by a relatively nonspecific carboxylic ester hydrolase that can act on a wide variety of esters as substrates. This enzyme has been given a variety of names in the literature, the most common being pancreatic nonspecific lipase and cholesterol esterase (Goodman and Blaner, 1984).

The retinol and carotenoids within the mixed micelles are absorbed along with other fat-soluble vitamins and lipids into the mucosal cells (enterocytes) of the upper small intestine. *In vivo* and *in vitro* experiments using perfused intestines of unanaesthetized rats and everted intestinal sacs showed apparent saturation kinetics with physiological concentrations of retinol and no dependence on metabolic energy (Hollander and Muralidhara, 1977). These findings suggest that retinol from natural food sources is absorbed by facilitated diffusion. At higher concentrations of retinol, a linear relationship was found between the concentration and absorption rate, indicating a process of simple diffusion. The uptake of β-carotene was also shown to occur by simple diffusion at perfusate concentrations of 0.5–11 μM (Hollander and Ruble, 1978).

Within the enterocytes retinol is esterified with long chain fatty acids, preferentially with palmitic acid. Two microsomal enzymes are involved in the esterification, namely acyl coenzyme A:retinol acyltransferase (ARAT), and lecithin:retinol acyltransferase (LRAT), their roles being apparently dependent on the amount of vitamin A absorbed. Retinol complexed to cellular retinol-binding protein type II (CRBP-II) is esterified by LRAT during absorption of a normal load of vitamin A. When large amounts are absorbed CRBP-II becomes saturated and uncomplexed membrane-bound ARAT takes over. The excess retinyl esters might be temporarily stored in cytoplasmic lipid droplets for subsequent release in chylomicrons after hydrolysis in the droplets and re-esterification by LRAT. It seems that CRBP-II functions both to direct retinol to the microsomes for esterification by LRAT and to prevent retinol from participating in the ARAT reaction (Blomhoff *et al.*, 1991).

Also inside the enterocytes β-carotene and other provitamin A carotenoids are cleaved centrally by a cytosolic enzyme termed β-carotenoid-15, 15′-dioxygenase to yield two moles (or one) of retinaldehyde. The possibility of asymmetric (random) cleavage of β-carotene in mammals has been debated as an additional mechanism for producing vitamin A (Glover, 1960; Ganguly and Sastry, 1985; Olson, 1989). Such a mechanism would yield only one mole of retinaldehyde per mole of β-carotene consumed. Evidence that asymmetric cleavage may have biological significance is based partly on the appearance of β-àpo-carotenals after *in vivo* dosing with β-carotene in rats and chickens (Sharma *et al.*, 1977) and on the *in vitro* conversion of β-carotene into β-apo-carotenals in homogenates prepared from human, monkey, ferret and rat intestine (Wang *et al.*, 1991). Thus far, no mammalian carotenoid dioxygenase with asymmetric bond specificity has been identified and direct evidence is required to support the random cleavage hypothesis. In the absence of any other characterized process, central cleavage is considered to be the predominant mechanism.

Most of the retinaldehyde formed from carotenoids is reversibly reduced to retinol which is then esterified. Some of the retinaldehyde is oxidized irreversibly to retinoic acid. Retinyl esters derived from both carotenoids and preformed vitamin A are incorporated into chylomicrons which are then released into the lymph. The chylomicrons also contain appreciable amounts of intact carotenoids, which are ultimately stored in adipose tissues and in various organs of the body. The enzymes responsible for reducing retinaldehyde to retinol and esterifying the retinol are complexed with CRBP-II, which orchestrates these reactions within the enterocyte (Kakkad and Ong, 1988; Ong, Kakkad and Macdonald, 1987).

Hepatic uptake and storage

On entering the bloodstream the chylomicrons are attacked by lipoprotein lipase to produce chylomicron remnants which are taken up almost entirely by the liver. Uptake by the hepatocytes takes place by receptor-mediated endocytosis and the retinyl esters are rapidly hydrolysed to retinol at the plasma membrane or in early endosomes. Some of the carotenoids taken up with the chylomicron remnants are probably converted to retinoids. The retinol formed in the hepatocyte can be released into the bloodstream in response to the body's immediate needs but the majority is transferred to stellate cells, where it is esterified and stored in lipid globules. The storage capacity is high: only when the stellate cells can accept no more retinol does hypervitaminosis A occur.

The vitamin A stored in stellate cells can be readily mobilized for use in a highly regulated process. Thus an individual's plasma vitamin A levels remain quite constant over a wide range of dietary intakes and liver stores. Only when liver reserves of vitamin A are nearly depleted do plasma concentrations of retinol decrease significantly. Upon demand, the retinyl esters are hydrolysed to retinol, which combines with RBP in a 1:1 mole ratio to form *holo*-RBP. This complex is secreted into the bloodstream where it becomes associated with a larger protein called transthyretin, which also binds thyroid hormone. Within RBP, retinol resides in a hydrophobic cleft which protects it from oxidation during transport. The formation of the larger transthyretin–*holo*-RBP complex may minimize the loss of *holo*-RBP in the urine during its passage through the kidney.

Tissue uptake and metabolism

Cell-surface receptors on target tissues recognize the protein portion of *holo*-RBP, but not retinol itself. After interaction with the receptor, *holo*-RBP, in all likelihood, is internalized within the cell by receptor-mediated endocytosis. Other less specific mechanisms may contribute to uptake. Thereafter, retinol is released into the cytosol and the RBP is modified in conformation and released from the cell. This modified RBP no longer binds retinol, no longer interacts with transthyretin, and ultimately is catabolized, primarily by the kidney (Olson, 1994).

Within target cells, retinol combines with CRBP in the cytosol. Some retinol can also be oxidized through retinaldehyde to retinoic acid, which combines with cellular retinoic acid-binding protein (CRABP). Two main functions of CRBP appear to be: to limit the concentration of free retinol in cells, thereby protecting the ligand from nonspecific oxidation and the cell's membranes from damage by excessive retinol; and to specify the metabolism of vitamin A by directing the retinol to the appropriate enzyme reaction sites (Ross, 1993). CRBP may also be involved in facilitating the specific interaction of retinol with retinol-binding sites in the cell nucleus (Goodman and Blaner, 1984).

Excretion

Approximately 5–20% of ingested vitamin A and a larger percentage of carotenoids are not absorbed from the intestinal tract and are eliminated from the body in the faeces. Some 10–40% of the absorbed vitamin A is oxidized and/or conjugated in the liver and then is secreted into the bile. Although some of these biliary metabolites, such as retinoyl β-glucuronide, are reabsorbed to some extent and returned to the liver in an enterohepatic cycle, most of the biliary metabolites are eliminated in the faeces (Olson, 1988).

3.8.2 Dietary sources and their bioavailability

Dietary sources

All sources of vitamin A are derived ultimately from provitamin A carotenoids, which are present in higher plants, certain animal sources, and in lower forms of animal and plant life. Humans obtain preformed vitamin A exclusively from animal sources. The retinoids are present in animal tissues and in milk as a consequence of the enzymatic conversion of ingested provitamin A carotenoids in the intestinal wall of the animal.

In general, much of the dietary vitamin A intake is derived from red, yellow and green fruits and vegetables containing mainly α-, β- and

Table 3.7 Distribution of provitamin A carotenoids in selected vegetables and fruits (Heinonen *et al.*, 1989)

Item	μg/100 g				
	α-Carotene	β-Carotene	γ-Carotene	β-Cryptoxanthin	RE[a]
Potato					
new	–	3.2	–	–	0.5
old	tr	7.7	–	nd	1.3
Carrot	530	7600	nd	–	1300
Broccoli	tr	1000	nd	24	170
Cauliflower	–	11	–	–	1.9
White cabbage	tr	66	nd	–	11
Lettuce	tr	980	–	–	160
Spinach	–	3300	–	–	550
Tomato	–	660	170	nd	120
Pepper					
red	nd	2900	nd	nd	480
yellow	92	150	–	–	33
green	tr	240	nd	–	40
Apple	–	39	–	nd	6.5
Orange	19	38	nd	nd	7.9
Banana	12	14	nd	nd	3.4
Peach	tr	86	nd	51	19

[a] Retinol equivalents = 0.167 (μg β-carotene) + 0.083 (μg α-carotene + μg γ-carotene + μg β-cryptoxanthin).
–, Not detected at a detection limit of 0.5 μg/100 g; nd, not determined; tr, trace.

γ-carotenes (Table 3.7). Carrots and green leafy vegetables are rich sources of provitamin A. In some countries of Africa, red palm oil contributes up to 75% of the total dietary vitamin A, and it is also an important food source of vitamin A in South America and Southeast Asia. This oil is extracted from the fleshy mesocarp of the palm nut; the oil extracted from the palm kernel is without value as a source of provitamin A (Underwood, 1984). Provitamin carotenoids may also be obtained from animal food products such as egg yolk, milk and butter, and a small amount from animal tissues where they are deposited mostly in adipose fat. Negligible sources of provitamin A carotenoids include cereals (except yellow maize), vegetable oils, potatoes (but not sweet potatoes), jams and syrups.

The carotenoid content of vegetables and fruits varies according to the variety and the maturity. For example, the carotene content of carrots grown in the United Kingdom varies from 4300 to 11 000 μg/100 g (Holland, Unwin and Buss, 1991) and lettuce varieties with darker green leaves will have a higher carotenoid content than paler varieties. In ripening fruit, the decrease in chlorophylls is often accompanied by an increase in carotenoid content.

Table 3.8 Vitamin A activity in meat, liver, milk, eggs and fish

Item	µg RE/100 gᵃ	Reference
Ground beef	9	(1)
Liver		(1)
pig	58 000	
ox	20 000	
Milk, standardized		(2)
1.9% fat	19	
3.9% fat	37	
Egg, whole	190	(3)
Baltic herring		(4)
fillets	8	
smoked	40	
Sardines in oil	33	(4)
Tuna in oil	6	(4)
Cod	2	(4)

ᵃ Retinol equivalent = µg all-*trans* retinol +(µg 13-*cis* retinol × 0.75)+(µg retinaldehyde × 0.90) + (µg β-carotene × 0.167)
(1) Ollilainen *et al.* (1988); (2) Ollilainen *et al.* (1989a); (3) Sivell, Wenlock and Jackson (1982); (4) Ollilainen *et al.* (1989b).

The distribution of preformed vitamin A in some common foods and fish-liver oils is given in Table 3.8. The liver of meat animals is a rich source of vitamin A as this organ is the main storage site of the vitamin. Fish liver oils, particularly halibut liver oil, are especially rich in the vitamin, but these are diluted for use in pharmaceutical products as they are too potent to be consumed as foods. Whole milk, butter, cheese and eggs are important dietary sources. The edible portion of fatty fish (e.g. herring, mackerel, pilchards, sardines and tuna) contains moderate amounts of vitamin A, but white fish, apart from the haddock, contain only trace amounts. In most of the above foods, the vitamin A occurs mainly in the form of retinol esterified with long chain fatty acids, particularly palmitic acid. However, in eggs the unesterified alcohol represents the major form of the vitamin, with retinaldehyde and retinyl esters constituting lesser amounts (Parrish, 1977). *Cis* isomers of vitamin A occur in foods to varying extents, with fish-liver oils and eggs containing as much as 35 and 20%, respectively, of their total retinol in this form (Sivell *et al.*, 1984).

Foods are supplemented with vitamin A in the form of standardized preparations of fatty acyl esters, i.e. retinyl acetate, propionate or palmitate. Nowadays, most food manufacturers use retinyl palmitate in the fortification process, as this compound has been shown to exhibit the greatest stability among the retinyl esters toward oxidation (Widicus and Kirk, 1979). The preparations are available commercially either as dilutions in high quality vegetable oils containing added vitamin E as an antioxidant, or as dry stabilized beadlets in which the vitamin A is dispersed in a solid matrix of gelatine and sucrose or gum acacia and sucrose. The oily preparations are used to supplement fat-based foods such as margarines and shortenings, while the dry preparations are used in dried food products such as milk powder, infant foods and dietetic foods (Kläui, Hausheer and Huschka, 1970).

Synthetic preparations of β-carotene, apocarotenal and apocarotenoic ester are used primarily as colouring agents for food to impart, standardize or enhance natural colour. Suspensions of micronized crystals of carotenoids in high quality vegetable oils containing added vitamin E are used in colouring fat-based foods such as margarines, butter, shortenings, cheese and French dressings. Water-dispersible forms of carotenoids in the form of dry gelatine-protected beadlets have been developed for the colouring of water-based foods such as cake mixes, puddings and dried and canned soups. Carotenoids added to juices and carbonated beverages in the form of liquid emulsions show adequate stability due to the antioxidant property of ascorbic acid (Kläui, Hausheer and Huschke, 1970).

Bioavailability

Several distinct differences exist between preformed vitamin A and provitamin A carotenoids (provitamins) with regard to their digestion and absorption. When foods containing normal physiological amounts of these compounds are ingested, vitamin A is absorbed with an efficiency of 70–90% compared with 20–50% for the provitamins. The retinyl esters are utilized as efficiently as retinol itself if compared on the basis of their retinol content. The absorption efficiency of vitamin A remains high as the amount ingested increases beyond physiological levels, whereas that of the provitamins falls markedly with increased ingestion to less than 10%.

Among the various nutritional factors that influence the bioavailability of carotenoids is the poor digestibility of fibrous plant cells. Crystalline β-carotene in antioxidant-stabilized commercial form is absorbed with the highest efficiency (c.50%) whereas the absorption of carotenoids from raw carrot can be as low as 1% (Erdman, Poor and Dietz, 1988). In humans, the 29 mg of natural β-carotene contained in a 272 g serving of cooked carrots produced a plasma carotenoid response that was only 21% of that produced by a 30 mg capsule of pure β-carotene after a single ingestion (Brown *et al.*, 1989); the corresponding response after chronic intake was 18% (Micozzi *et al.*, 1992). A number of investigators have reported that carotene absorption, especially from carrots, is directly proportional to particle size (Rodriguez and Irwin, 1972), which suggests that the degree of cellular rupture is implicated in rendering carotenes available. Other factors which may contribute to the lower bioavailability of carotenoids relative to vitamin A is their passive mode of absorption, and their relatively slow rate of cleavage to vitamin A in the intestine.

Among various nutrients which influence vitamin A bioavailability, fat and protein are the most important. The presence of adequate amounts of dietary fat is essential in forming micelles, and the absorption of retinol and carotenoids is markedly reduced when diets contain very little (<5 g per day) fat (National Research Council, 1989). Conversely, adult human subjects placed on a high-fat diet showed significant increases in plasma β-carotene levels as compared with those placed on a low-fat diet (Dimitrov *et al.*, 1988). A study conducted in a Ruandan village area in Africa showed that supplementation of the carotenoid-sufficient but low-fat diet with 18 g per day of olive oil increased the absorption of vegetable carotenoids from 5% to 25% in boys showing clear signs of vitamin A deficiency (Bitôt spots) (Roels, Trout and Dujacquier, 1958). In a later study carried out in India, a daily supplement of 5 g of groundnut oil was enough to enhance carotene absorption in children (Jayarajan, Reddy and Mohanram, 1980).

The presence of peroxidized fat and oxidizing agents in the food depresses the absorption and utilization of retinol and carotenoids. Non-peroxidized fat contains vitamin E, which, being an antioxidant, has a protective effect on these compounds at both intestinal and cellular levels.

Direct effects of protein-calorie malnutrition include a depression in both intestinal carotenoid cleavage activity, and synthesis and release of plasma RBP (Olson, 1988). Deficiencies of iron and zinc also adversely affect vitamin A transport, storage and utilization.

Results from studies with chicks (Erdman, Fahey and White, 1986) and rats (Khokhar and Kapoor, 1990) suggest that various types of dietary fibre reduce the bioavailability of β-carotene. In a human study, Rock and Swendseid (1992) reported that the addition of 12 g of pectin (a type of dietary fibre) to a controlled meal reduced plasma β-carotene concentrations by more than 50%. Typical dietary levels of fibre in the United States should not significantly affect carotenoid bioavailability (Erdman, Poor and Dietz, 1988).

3.8.3 Effects of food processing, storage and cooking

In foods, the indigenous retinyl esters and carotenoids are dissolved in the lipid matrix, where they are protected from the oxidizing action of atmospheric oxygen by vitamin E and other antioxidants that might be present. On depletion of the antioxidants, the retinyl esters and carotenoids become vulnerable to oxidation. Under these circumstances, unsaturated fats also begin to oxidize and turn rancid, with the production of highly active peroxides. The peroxides produced during oxidative rancidity attack, and finally destroy, the vitamin A and carotenoids. Thus factors that accelerate oxidative rancidity, such as access of air, heat, light, traces of certain metals (notably copper and, to a lesser extent, iron), and storage time, will also result in the destruction of the vitamin A compounds and carotenoids (Kläui, Hausheer and Huschke, 1970).

Commercial preparations of vitamin A used in fortifying low-fat foods such as beverages and cereal products are stabilized liquid or dehydrated emulsions and therefore would be expected to be well absorbed.

Mild cooking dramatically increases the bioavailability of the carotene content of vegetables by liberating the carotenes from their association with membranes and lipoproteins. However, over-cooking can cause loss of vitamin A activity due to the action of heat and oxygen to form *cis* isomers and epoxy-carotenoids. Optimal retention of carotene is obtained by steaming vegetables or cooking with minimal water until the vegetables are cooked but still crisp (Erdman, Poor and Dietz, 1988).

The thermal processing of vegetables by canning lowers the vitamin A values by 15–35% (Sweeney and Marsh, 1971; Ogunlesi and Lee, 1979). Canning of sweet potatoes and carrots results in the conversion of 25% of all-*trans*-β-carotene to 13- or 9-*cis* isomers (Chandler and Schwartz, 1987). Dehydrated foods are highly susceptible to the loss of vitamin A and provitamin A during storage because of their propensity to undergo oxidation in a medium of low water activity (Tannenbaum, Young and Archer, 1985). The blanching of vegetables before dehydration exerts a protective effect (Cain, 1975).

In margarines, the various naturally occurring vitamin A isomers and provitamin A carotenoids that may have been present in the original crude oils are removed during the refining process. Thus the only vitamin A that is present in the final margarine is the all-*trans*-retinyl palmitate or acetate that is added during production (Aitzetmüller, Pilz and Tasche, 1979).

Vitamin A is more stable in whole milk products than in low-fat or skimmed milk products, presumably because of natural antioxidants present in milk fat (Coulter and Thomas, 1968). Vitamin A added to low-fat milk is destroyed more rapidly by light than is indigenous vitamin A. A certain portion of the native vitamin A is lost rapidly on exposure to light, with the remaining portion being resistant to further destruction (de Man, 1981). In ultra-high-temperature (UHT) milk, sterilization by direct steam injection and evaporative cooling removes virtually all of the oxygen from the milk; indirect heating and evaporative cooling leaves about one-third of the initial oxygen content; and indirect heat followed by cooling in a heat exchanger causes no change in oxygen content. None of these three treatments caused any significant loss of vitamin A in UHT milk either during processing, or during subsequent storage in aluminium foil-lined cartons at room temperature for up to 90 days (Ford *et al.*, 1969). The stability of vitamin A in direct UHT-treated milk, and also in pasteurized milk, was confirmed by Le Maguer and Jackson (1900). However, Woollard and Fairweather (1985) reported that both natural (retinyl palmitate) and supplemental (retinyl acetate) vitamin A in UHT milk sterilized by indirect heating underwent rapid degradation. In two trials the total loss of vitamin A was 21 and 25% after three weeks' storage; the loss increased to 33 and 45% after 15 weeks' storage, after which there was no further loss up to 40 weeks. The cause of the vitamin degradation was attributed to residual oxygen in the milk.

The loss of vitamin A in milk is confined to the all-*trans* isomers of the indigenous retinyl palmitate and the supplemental retinyl acetate. *Cis* isomers, which are essentially absent in untreated milk before UHT treatment, but which appear after processing, maintain their concentrations until the onset of proteolysis and gelation (Woollard and Indyk, 1986). Vitamin A isomerization and degradation in dried-milk powder occur in a similar fashion to liquid milk (Woollard and Indyk, 1986).

3.9 DIETARY INTAKE

In nationwide food consumption surveys published in 1986 and 1987 the average vitamin A intakes for adult men and women in the United States were 1419 and 1170 RE, respectively. Less than one-third of total vitamin A activity in the diets came from carotenoids. Table 3.9 shows the percentage vitamin A contribution of the major food sources in the American diet.

The RDA for vitamin A is 1000 RE for men and 800 RE for women. An additional 500 RE per day is recommended during the first six months of lactation and a 400 RE increment for the second six months (National Research Council, 1989).

Effects of high intakes

An excessive intake of vitamin A produces symptoms of toxicity (Miller and Hayes, 1982; Omaye, 1984; Bendich and Langseth, 1989). Normally, toxicity results from the indiscriminate use of pharmaceutical supplements, and not from the consumption of usual diets. The only naturally occurring products that contain sufficient vitamin A to induce toxicity in humans are the livers of animals at the top of long food chains, such as large marine fish and carnivores (e.g. bear and dog).

In acute hypervitaminosis A, the excess retinol circulating in the bloodstream is not subject to the normal regulation of RBP binding, and the unbound retinol disrupts the integrity of the cell membranes. Ingestion of

Table 3.9 Major American food sources of vitamin A (from Olson, 1994)

Food source	Percentage contribution[a]
Milk and dairy products	31
Vegetables and fruits	22
Meat and meat products	16
Fats, oils and fortified margarine	10
Fortified grain	9
Eggs	8
Miscellaneous	4

[a] Expressed in retinol equivalents.

acute high doses of vitamin A (2–5 million IU) produces symptoms of severe headache, dizziness, abdominal pain, nausea and diarrhoea.

Chronic hypervitaminosis A in adults resulting from daily intakes of 100 000 IU of vitamin A for several months produces symptoms of anorexia, headache, blurred vision, muscle soreness after exercise, hair loss, bleeding lips, cracking and peeling skin, and nose bleed. These symptoms are relieved promptly when vitamin A dosing is discontinued. Prolonged high vitamin A consumption may eventually result in cirrhosis of the liver.

Both excess and deficiency of vitamin A in pregnant animals can produce teratogenic effects (malformations) in the foetus. In humans, congenital malformations associated with maternal over-use of high doses of vitamin A have been reported but no cause- and-effect relationship has been established (Bendich and Langseth, 1989).

Assessment of nutritional status

Serum levels of vitamin A are tightly maintained and may show normal values in an individual with depleted liver stores of the vitamin. Therefore, serum retinol concentration alone is not a reliable indicator of vitamin A status. The relative dose–response (RDR) test is based on the principle that when an individual's liver stores of vitamin A are low, RBP accumulates in the hepatocytes, ready to be released as soon as vitamin A is available. An oral challenge with vitamin A provokes a prominent rise in circulating RBP-associated retinol (*holo*-RBP) in those individuals with low hepatic vitamin reserves. As currently applied, an initial blood sample is taken from the fasted subject and a large oral dose of retinyl acetate or palmitate in oil is administered immediately afterwards. A second blood sample is collected 5 h after dosing and the RDR is calculated as follows:

$$RDR = [(A_5 - A_0)/A_5] \times 100\%$$

where A_0 and A_5 = concentration of retinol in the initial and 5 h blood samples, respectively.

RDR values higher than 50% are characteristic of acute vitamin A deficiency, values between 20 and 50% indicate marginal status, and values lower than 20% indicate adequate status.

Krasinski *et al.* (1990) demonstrated a slower clearance of postprandial retinyl esters from the circulation in elderly subjects compared with younger adults and children, implying that the 5 h interval between blood sampling is inappropriate in elderly subjects. This was confirmed by Bulux *et al.* (1992) who reported that retinol concentration was delayed to between the sixth and seventh hour after dosing. Thus in elderly subjects (over 60 years of age) the test procedure requires additional blood samples to avoid missing the diagnostic response.

Subjects with malabsorption problems can be dosed by intramuscular injection.

3.10 SYNOPSIS

In foods derived from animal products, only all-*trans*-retinol and smaller amounts of 13-*cis*-retinol, both in esterified form, are usually present in significant quantities. The major provitamin A carotenoids found in fruits and vegetables are α- and β-carotene and β-cryptoxanthin. Nowadays the determination of preformed vitamin A and provitamin A carotenoids is almost exclusively the province of HPLC following a suitable extraction procedure. HPLC conditions used for carotenoid analysis must permit separation of the major provitamins from inactive carotenoids. Selective stationary phases allow the separation of all-*trans*-β-carotene from its less active *cis* isomers.

With regard to natural (unsupplemented) foods, vitamin A is absorbed in the small intestine at an efficiency of 70–90% compared with 20–50% for the provitamins. The reduced bioavailability of carotenoids compared with vitamin A is due at least partly to the poor digestibility of plant tissues. Thus mildly cooked carrots have a higher bioavailability than raw carrots.

REFERENCES

Aitzetmüller, K., Pilz, J. and Tasche, R. (1979) Fast determination of vitamin A palmitate in margarines by HPLC. *Fette Seifen Anstrichmittel*, **81**, 40–3.

Ames, S.R. (1966) Methods for evaluating vitamin A isomers. *J. Ass. Off. Analyt. Chem.*, **49**, 1071–8.

Bendich, A. and Langseth, L. (1989) Safety of vitamin A. *Am. J. Clin. Nutr.*, **49**, 358–71.

Bieri, J.G. and McKenna, M.C. (1981) Expressing dietary values for fat-soluble vitamins: changes in concepts and terminology. *Am. J. Clin. Nutr.*, **34**, 289–95.

Bliss, C.I. and Roels, O.A. (1967) Bioassay of vitamin A potency. In *The Vitamins. Chemistry, Physiology, Pathology, Methods*, 2nd edn, Vol. VI (eds P. György and W.N. Pearson), Academic Press, New York, pp. 197–210.

Blomhoff, R., Green, M.H., Green, J.B. *et al.* (1991) Vitamin A metabolism: new perspectives on absorption, transport, and storage. *Physiol. Rev.*, **71**, 951–90.

Boldingh, J., Cama, H.R., Collins, F.D. *et al.* (1951) Pure all-*trans* vitamin A acetate and the assessment of vitamin A potency by spectrophotometry. *Nature*, **168**, 598.

Brinkmann, E., Dehne, L., Oei, H.B. *et al.* (1995) Separation of geometrical retinol isomers in food samples by using narrow-bore high-performance liquid chromatography. *J. Chromat. A*, **693**, 271–9.

Brown, E.D., Micozzi, M.S., Craft, N.E. *et al.* (1989) Plasma carotenoids in normal men after a single ingestion of vegetables or purified β-carotene. *Am. J. Clin. Nutr.*, **49**, 1258–65.

Brubacher, G.B. and Weiser, H. (1985) The vitamin A activity of β-carotene. *Int. J. Vitam. Nutr. Res.*, **55**, 5–15.

Brubacher, G., Müller-Mulot, W. and Southgate, D.A.T. (1985) *Methods for the Determination of Vitamins in Foods. Recommended by COST 91*, Elsevier Applied Science, London.

Bulux, J., Carranza, E., Castañeda, C. *et al.* (1992) Studies on the application of the relative-dose-response test for assessing vitamin A status in older adults. *Am. J. Clin. Nutr.*, **56**, 543–7.

Bureau, J.L. and Bushway, R.J. (1986) HPLC determination of carotenoids in fruits and vegetables in the United States. *J. Food Sci.*, **51**, 128–30.

Bushway, R.J. (1985) Separation of carotenoids in fruits and vegetables by high performance liquid chromatography. *J. Liquid Chromat.*, **8**, 1527–47.

Bushway, R.J. (1986) Determination of α- and β-carotene in some raw fruits and vegetables by high-performance liquid chromatography. *J. Agric. Food Chem.*, **34**, 409–12.

Cain, R.F. (1975) Factors influencing the nutritional quality and fortification of fruits and vegetables. In *Technology of Fortification of Foods*, Food and Nutrition Board, NRC, Nat. Acad. Sci., Washington, DC.

Chandler, L.A. and Schwartz, S.J. (1987) HPLC separation of *cis–trans* carotene isomers in fresh and processed fruits and vegetables. *J. Food Sci.*, **52**, 669–72.

Chen, T.M. and Chen, B.H. (1994) Optimization of mobile phases for HPLC of *cis–trans* carotene isomers. *Chromatographia*, **39**, 346–54.

Chen, B.H., Chuang, J.R., Lin, J.H. and Chiu, C.P. (1993) Quantification of provitamin A compounds in Chinese vegetables by high-performance liquid chromatography. *J. Food Protection*, **56**, 51–4.

Coulter, S.T. and Thomas, E.L. (1968) Enrichment and fortification of dairy products and margarine. *J. Agric. Food Chem.*, **16**, 158–62.

de Man, J.M. (1981) Light-induced destruction of vitamin A in milk. *J. Dairy Sci.*, **64**, 2031–2.

Dimitrov, N.V., Meyer, C., Ullrey, D.E. *et al.* (1988) Bioavailability of β-carotene in humans. *Am. J. Clin. Nutr.*, **48**, 298–304.

Egberg, D.C., Heroff, J.C. and Potter, R.H. (1977) Determination of all-*trans* and 13-*cis* vitamin A in food products by high-pressure liquid chromatography. *J. Agric. Food Chem.*, **25**, 1127–32.

Erdman, J.W. Jr, Bierer, T.L. and Gugger, E.T. (1993) Absorption and transport of carotenoids. *Ann. N.Y. Acad. Sci.*, **691**, 76–85.

Erdman, J.W. Jr, Fahey, G.C. Jr and White, C.B. (1986) Effects of purified dietary fiber sources on β-carotene utilization by the chick. *J. Nutr.*, **116**, 2415–23.

Erdman, J.W. Jr, Poor, C.L. and Dietz, J.M. (1988) Factors affecting the bioavailability of vitamin A, carotenoids, and vitamin E. *Food Technol.*, **41**(10) 214–16, 219, 221.

FAO/WHO (1967) Requirements of vitamin A, thiamine, riboflavine and niacin, WHO *Technical Report Series No. 362*, World Health Organization, Geneva.

FAO/WHO (1988) Requirements of vitamin A, iron, folate and vitamin B_{12}. *FAO Food and Nutrition Series No. 23*, Food and Agriculture Organization of the United Nations, Rome.

Fiksdahl, A., Mortensen, J.T. and Liaaen-Jensen, S. (1978) High-pressure liquid chromatography of carotenoids. *J. Chromat.*, **157**, 111–7.

Food Labelling Regulations (1984) *Statutory Instrument 1984, No. 1305, as amended*, H.M. Stationery Office, London.

Ford, J.E., Porter, J.W.G., Thompson, S.Y. *et al.* (1969) Effects of ultra-high temperature (UHT) processing and of subsequent storage on the vitamin content of milk. *J. Dairy Res.*, **36**, 447–54.

Ganguly, J. and Sastry, P.S. (1985) Mechanism of conversion of β-carotene into vitamin A – central cleavage versus random cleavage. *Wld Rev. Nutr. Diet.*, **45**, 198–220.

Glover, J. (1960) The conversion of β-carotene into vitamin A. *Vitam. Horm.*, **38**, 371–86.

Goodman, De W.S. and Blaner, W.S. (1984) Biosynthesis, absorption, and hepatic metabolism of retinol. In *The Retinoids*, Vol. 2 (eds M.S. Sporn, A.B. Roberts and De W.S. Goodman), Academic Press, Inc., New York, pp. 1–39.

Goodwin, T.W. and Britton, G. (1988) Distribution and analysis of carotenoids. In *Plant Pigments* (ed. T.W. Goodwin), Academic Press, London, pp. 61–132.

Granado, F., Olmedilla, B., Blanco, I. and Rojas-Hidalgo, E. (1991) An improved HPLC method for the separation of fourteen carotenoids, including 15-/13- and 9-*cis*-β-carotene isomers, phytoene and phytofluene. *J. Liquid Chromat.*, **14**, 2457–75.

Granado, F., Olmedilla, B., Blanco, I. and Rojas-Hidalgo, E. (1992) Carotenoid composition in raw and cooked Spanish vegetables. *J. Agric. Food Chem.*, **40**, 2135–40.

Green, J. (1970) Distribution of fat-soluble vitamins and their standardization and assay by biological methods. In *Fat-soluble Vitamins* (ed. R.A. Morton), Pergamon Press, Oxford, pp. 71–97

Harris, R.S. (1967) Vitamin A and carotene. 1. Nomenclature and formulas. In *The Vitamins. Chemistry, Physiology, Pathology, Methods*, 2nd edn, Vol. I (eds W.H. Sebrell, Jr and R.S. Harris), Academic Press, New York, pp. 3–5.

Hart, D.J. and Scott, K.J. (1995) Development and evaluation of an HPLC method for the analysis of carotenoids in foods, and the measurement of the carotenoid content of vegetables and fruits commonly consumed in the UK. *Food Chem.*, **54**, 101–111.

Heinonen, M.I., Ollilainen, V., Linkola, E.K. *et al.* (1989) Carotenoids in Finnish foods: vegetables, fruits and berries. *J. Agric. Food Chem.*, **37**, 655–9.

Holland, B., Unwin, I.D. and Buss, D.H. (1991) *Fifth supplement to McCance and Widdowson's The Composition of Foods: Vegetables, Herbs and Spices*, Royal Society of Chemistry, H.M. Stationery Office, London.

Hollander, D. and Muralidhara, K.S. (1977) Vitamin A_1 intestinal absorption in vivo: influence of luminal factors on transport. *Am. J. Physiol.*, **232**, E471–7.

Hollander, D. and Ruble, P.E. Jr (1978) β-Carotene intestinal absorption: bile, fatty acid, pH, and flow rate effects on transport. *Am. J. Physiol.*, **235**, E686–91.

Jayarajan, P., Reddy, V. and Mohanram, M. (1980) Effect of dietary fat on absorption of β-carotene from green leafy vegetables in children. *Indian J. Med. Res.*, **71**, 53–6.

Jensen, S.K. (1994) Retinol determination in milk by HPLC and fluorescence detection. *J. Dairy Res.*, **61**, 233–40.

Kakkad, B.P. and Ong, D.E. (1988) Reduction of retinaldehyde bound to cellular retinol-binding protein (type II) by microsomes from rat small intestine. *J. Biol. Chem.*, **263**, 12916–9.

Khachik, F. and Beecher, G.R. (1987) Application of a C-45-β-carotene as an internal standard for the quantification of carotenoids in yellow/orange vegetables by liquid chromatography. *J. Agric. Food Chem.*, **35**, 732–8.

Khachik, F. and Beecher, G.R. (1988) Separation and identification of carotenoids and carotenol fatty acid esters in some squash products by liquid chromatography. I. Quantification of carotenoids and related esters by HPLC. *J. Agric. Food Chem.*, **36**, 929–37.

Khachik, F., Beecher, G.R. and Lusby, W.R. (1989) Separation, identification, and quantification of the major carotenoids in extracts of apricots, peaches, cantaloupe, and pink grapefruit by liquid chromatography. *J. Agric. Food Chem.*, **37**, 1465–73.

Khachik, F., Beecher, G.R. and Whittaker, N.F. (1986) Separation, identification, and quantification of the major carotenoid and chlorophyll constituents in extracts of several green vegetables by liquid chromatography. *J. Agric. Food Chem.*, **34**, 603–16.

Khachik, F., Beecher, G.R., Goli, M.B. and Lusby, W.R. (1991) Separation, identification, and quantification of carotenoids in fruits, vegetables and human plasma by high performance liquid chromatography. *Pure Appl. Chem.*, **63**, 71–80.

Khokhar, S. and Kapoor, A.C. (1990) Effect of dietary fibres on bioavailability of vitamin A and thiamine. *Plant Foods Human Nutr.*, **40**, 259–65.

Kimura, M., Rodriguez-Amaya, D.B. and Godoy, H.T. (1990) Assessment of the saponification step in the quantitative determination of carotenoids and provitamins A. *Food Chem.*, **35**, 187–95.

Kläui, H.M., Hausheer, W. and Huschke, G. (1970) Technological aspects of the use of fat-soluble vitamins and carotenoids and of the development of stabilized marketable forms. In *Fat-soluble Vitamins* (ed. R.A. Morton), Pergamon Press, New York, pp. 113–59.

Krasinski, S.D., Cohn, J.S., Schaefer, E.J. and Russell, R.M. (1990) Postprandial plasma retinyl ester response is greater in older subjects compared with younger subjects. *J. Clin. Invest.*, **85**, 883–92.

Landen, W.O. Jr (1980) Application of gel permeation chromatography and nonaqueous reverse phase chromatography to high pressure liquid chromatographic determination of retinyl palmitate in fortified breakfast cereals. *J. Ass. Off. Analyt. Chem.*, **63**, 131–6.

Landen, W.O. Jr and Eitenmiller, R.R. (1979) Application of gel permeation chromatography and nonaqueous reverse phase chromatography to high pressure liquid chromatographic determination of retinyl palmitate and β-carotene in oil and margarine. *J. Ass. Off. Analyt. Chem.*, **62**, 283–9.

Landers, G.M. and Olson, J.A. (1986) Absence of isomerization of retinyl palmitate, retinol, and retinal in chlorinated and nonchlorinated solvents under gold light. *J. Ass. Off. Analyt. Chem.* **69**, 50–5.

Lawn, R.E., Harris, J.R. and Johnson, S.F. (1983) Some aspects of the use of high-performance liquid chromatography for the determination of vitamin A in animal feeding stuffs. *J. Sci. Food Agric.*, **34**, 1039–46.

Lee, C.Y., McCoon, P.E. and LeBowitz, J.M. (1981) Vitamin A value of sweet corn. *J. Agric. Food Chem.*, **29**, 1294–5.

Le Maguer, I. and Jackson, H. (1983) Stability of vitamin A in pasteurized and ultra-high temperature processed milk. *J. Dairy Sci.*, **66**, 2452–8.

Leseuilet, L., Marty, C., Beroot, C. and Tchapla, A. (1989) Optimization of the isocratic non-aqueous reverse phase (NARP) HPLC separation of *trans/cis* α- and β-carotenes. *J. High Resolution Chromat.*, **12**, 447–54.

MacCrehan, W.A. (1990) Determination of retinol, α-tocopherol, and β-carotene in serum by liquid chromatography. *Methods Enzymol.*, **189**, 172–81.

Marsili, R. and Callahan, D. (1993) Comparison of a liquid solvent extraction technique and supercritical fluid extraction for the determination of α- and β-carotene in vegetables. *J. Chromatogr. Sci.*, **31**, 422–8.

Mayne, S.T. (1996) Beta-carotene, carotenoids, and disease prevention in humans. *FASEB, J.*, **10**, 690–701.

Micozzi, M.S., Brown, E.D., Edwards, B.K. *et al.* (1992) Plasma carotenoid response to chronic intake of selected foods and β-carotene supplements in men. *Am. J. Clin. Nutr.*, **55**, 1120–5.

Miller, D.R. and Hayes, K.C. (1982) Vitamin excess and toxicity. In *Nutritional Toxicology*, Vol. I (ed. J.N. Hathcock), Academic Press, New York, pp. 81–133.

Mínguez-Mosquera, M.I. and Hornero-Méndez, D. (1993) Separation and quantification of the carotenoid pigments in red peppers (*Capsicum annuum* L.), paprika, and oleoresin by reversed-phase HPLC. *J. Agric. Food Chem.*, **41**, 1616–20.

National Research Council (1989) Fat-soluble vitamins. In *Recommended Dietary Allowances*, 10th edn, National Academy Press, Washington, DC, pp. 78–114.

Nierenberg, D.W. and Lester, D.C. (1986) High pressure liquid chromatographic assays for retinol and beta-carotene: potential problems and solutions. *J. Nutr. Growth Cancer*, **3**, 215–25.

Nierenberg, D.W., Peng, Y.-M. and Alberts, D.S. (1988) Methods for determination of retinoids, α-tocopherols, and carotenoids in human serum, plasma, and other tissues. In *Nutrition and Cancer Prevention. Investigating the Role of Micronutrients* (eds T.E. Moon and M.S. Micozzi), Marcel Dekker, Inc., New York, pp. 181–209.

Nyambaka, H. and Ryley, J. (1996) An isocratic reversed-phase HPLC separation of the steroisomers of the provitamin A carotenoids (α- and β-carotene) in dark green vegetables. *Food Chem.*, **55**, 63–72.

Ogunlesi, A.T. and Lee, C.Y. (1979) Effects of thermal processing on the stereo-isomerization of major carotenoids and vitamin A value of carrots. *Food Chem.*, **4**, 311–18.

O'Leary, M.J. (1993) Industrial production. In *The Technology of Vitamins in Food* (ed. P. Berry Ottaway), Blackie Academic & Professional, Glasgow, pp. 63–89.

Ollilainen, V., Heinonen, M., Linkola, E. *et al.* (1988) Carotenoids and retinoids in Finnish foods: meat and meat products. *J. Food Comp. Anal.*, **1**, 178–88.

Ollilainen, V., Heinonen, M., Linkola, E. *et al.* (1989a) Carotenoids and retinoids in Finnish foods: dairy products and eggs. *J. Dairy Sci.*, **72**, 2257–65.

Ollilainen, V., Heinonen, M., Linkola, E. *et al.* (1989b) Retinoids and carotenoids in Finnish foods: fish and fish products. *J. Food Comp. Anal.*, **2**, 93–103.

Olson, J.A. (1986) Physiological and metabolic basis of major signs of vitamin A deficiency. In *Vitamin A Deficiency and its Control* (ed. J.C. Bauernfeind), Academic Press, Inc., New York, pp. 19–57.

Olson, J.A. (1988) Vitamin A, retinoids, and carotenoids. In *Modern Nutrition in Health and Disease*, 7th edn (eds M.E. Shils and V.R. Young), Lea & Febiger, Philadelphia, pp. 292–312.

Olson, J.A. (1989) Provitamin A function of carotenoids: the conversion of β-carotene into vitamin A. *J. Nutr.*, **119**, 105–8.

Olson, J.A. (1991) Vitamin A. In *Handbook of Vitamins*, 2nd edn (ed. L.J. Machlin), Marcel Dekker, Inc., New York, pp. 1–57.

Olson, J.A. (1994) Vitamin A, retinoids, and carotenoids. In *Modern Nutrition in Health and Disease*, 8th edn, Vol. 1 (eds M.E. Shils, J.A. Olson and M. Shike), Lea & Febiger, Philadelphia, pp. 287–307.

Omaye, S.T. (1984) Safety of megavitamin therapy. In *Nutritional and Toxicological Aspects of Food Safety* (ed. M. Friedman), Plenum Press, New York, pp. 169–203.

O'Neil, C.A. and Schwartz, S.J. (1992) Chromatographic analysis of *cis/trans* carotenoid isomers. *J. Chromat.*, **624**, 235–52.

O'Neil, C.A., Schwartz, S.J. and Catignani, G.L. (1991) Comparison of liquid chromatographic methods for determination of *cis-trans* isomers of β-carotene. *J. Ass. Off. Analyt. Chem.*, **74**, 36–42.

Ong, D.E., Kakkad, B. and MacDonald, P.N. (1987) Acyl-CoA-independent esterification of retinol bound to cellular retinol-binding protein (type II) by microsomes from rat intestine. *J. Biol. Chem.*, **262**, 2729–36.

Parrish, D.B. (1977) Determination of vitamin A in foods – a review. *CRC Crit. Rev. Food Sci. Nutr.*, **9**, 375–94.

Passmore, R. and Eastwood, M.A. (1986) *Davidson and Passmore Human Nutrition and Dietetics*, 8th edn, Churchill Livingstone, New York.

Pesek, C.A., Warthesen, J.J. and Taoukis, P.S. (1990) A kinetic model for equilibration of isomeric β-carotenes. *J. Agric. Food Chem.*, **38**, 41–5.

Philip, T. and Chen, T.-S. (1988) Development of a method for the quantitative estimation of provitamin A carotenoids in some fruits. *J. Food Sci.*, **53**, 1703–1706.

Pitt, G.A.J. (1985) Vitamin A. In *Fat-Soluble Vitamins. Their Biochemistry and Applications* (ed. A.T. Diplock), Heinemann, London, pp. 1–75.

Quackenbush, F.W. (1987) Reverse phase HPLC separation of *cis-* and *trans*-carotenoids and its application to β-carotenes in food materials. *J. Liquid Chromat.*, **10**, 643–53.

Quackenbush, F.W. and Smallidge, R.L. (1986) Nonaqueous reverse phase liquid chromatographic system for separation and quantitation of provitamins A. *J. Ass. Off. Analyt. Chem.*, **69**, 767–72.

Rock, C.L. and Swendseid, M.E. (1992) Plasma β-carotene response in humans after meals supplemented with dietary pectin. *Am. J. Clin. Nutr.*, **55**, 96–9.

Rodriguez, M.S. and Irwin, M.I. (1972) A conspectus of research on vitamin A requirement of man. *J. Nutr.*, **102**, 909–68.

Rodriguez-Amaya, D.B. (1989) Critical review of provitamin A determination in plant foods. *J. Micronutr. Anal.*, **5**, 191–225.

Roels, O.A., Trout, M. and Dujacquier, R. (1958) Carotene balances on boys in Ruanda where vitamin A deficiency is prevalent. *J. Nutr.*, **65**, 115–27.

Ross, A.C. (1993) Overview of retinoid metabolism. *J. Nutr.*, **123**, 346–50.

Rüedi, P. (1985) HPLC – a powerful tool in carotenoid research. *Pure Appl. Chem.*, **57**, 793–800.

Saleh, M.H. and Tan, B. (1991) Separation and identification of *cis/trans* carotenoid isomers. *J. Agric. Food Chem.*, **39**, 1438–43.

Sauberlich, H.E., Hodges, R.E., Wallace, D.L. *et al.* (1974) Vitamin A metabolism and requirements in the human studied with the use of labeled retinol. *Vit. Horm.*, **32**, 251–75.

Scott, K.J. (1992) Observations of some of the problems associated with the analysis of carotenoids in foods by HPLC. *Food Chem.*, **45**, 357–64.

Schwieter, U. and Isler, O. (1967) Vitamins A and carotene. II. Chemistry. In *The Vitamins. Chemistry, Physiology, Pathology, Methods*, 2nd edn, Vol. I (eds W.H. Sebrell, Jr and R.S. Harris), Academic Press, New York, pp. 5–99.

Sharma, R.V., Mathur, S.N., Dmitrovskii, A.A. *et al.* (1977) Studies on the metabolism of β-carotene and apo β-carotenoids in rats and chickens. *Biochim. Biophys. Acta*, **486**, 183–94.

Simpson, K.L. and Chichester, C.O. (1981) Metabolism and nutritional significance of carotenoids. *Ann. Rev. Nutr.*, **1**, 351–74.

Sivell, L.M., Wenlock, R.W. and Jackson, P.A. (1982) Determination of vitamin D and retinoid activity in eggs by HPLC. *Human Nutr. Appl. Nutr.*, **36A**, 430–7.

Sivell, L.M., Bull, N.L., Buss, D.H. *et al.* (1984) Vitamin A activity in foods of animal origin. *J. Sci. Food Agric.*, **35**, 931–9.

Speek, A.J., Temalilwa, C.R. and Schrijver, J. (1986) Determination of β-carotene content and vitamin A activity of vegetables by high-performance liquid chromatography and spectrophotometry. *Food Chem.*, **19**, 65–74.

Sporn, M.B. and Roberts, A.B. (1984) Biological methods for analysis and assay of retinoids – relationships between structure and activity. In *The Retinoids*, Vol. 1 (eds M.B. Sporn, A.B. Roberts and De W.S. Goodman), Academic Press, Inc., New York, pp. 235–79.

Sporn, M.B., Roberts, A.B. and Goodman, D.S. (1984) Introduction. In *The Retinoids*, Vol. 1 (eds M.B. Sporn, A.B. Roberts and De W.S. Goodman), Academic Press, Inc. (London) Ltd, pp. 1–5.

Stancher, B. and Zonta, F. (1982) High performance liquid chromatographic determination of carotene and vitamin A and its geometric isomers in foods. Applications to cheese analysis. *J. Chromat.*, **238**, 217–25.

Stancher, B. and Zonta, F. (1984) High-performance liquid chromatography of the unsaponifiable from samples of marine and freshwater fish: fractionation and identification of retinol (vitamin A$_1$) and dehydroretinol (vitamin A$_2$) isomers. *J. Chromat.*, **287**, 353–64.

Stancher, B., Zonta, F. and Bogoni, P. (1987) Determination of olive oil carotenoids by HPLC. *J. Micronutr. Anal.*, **3**, 97–106.

Sweeney, J.P. and Marsh, A.C. (1971) Effect of processing on provitamin A in vegetables. *J. Am. Diet. Assoc.*, **59**, 238–43.

Tan, B. (1988) Analytical and preparative chromatography of tomato paste carotenoids. *J. Food Sci.*, **53**, 954–9.

Tannenbaum, S.R., Young, V.R. and Archer, M.C. (1985) Vitamins and minerals. In *Food Chemistry*, 2nd edn (ed. O.R. Fennema), Marcel Dekker, Inc., New York, pp. 477–544.

Tee, E.-S. and Lim, C.-L. (1991) Carotenoid composition and content of Malaysian vegetables and fruits by the AOAC and HPLC methods. *Food Chem.*, **41**, 309–39.

Tee, E.-S. and Lim, C.-L. (1992) Re-analysis of vitamin A values of selected Malaysian foods of animal origin by the AOAC and HPLC methods. *Food Chem.*, **45**, 289–96.

Thompson, J.N. (1982) Trace analysis of vitamins by liquid chromatography. In *Trace Analysis*, Vol. II (ed. J.F. Lawrence), Academic Press, New York, pp. 1–67.

Thompson, J.N. (1986) Problems of official methods and new techniques for analysis of foods and feeds for vitamin A. *J. Ass. Off. Analyt. Chem.*, **69**, 727–38.

Thompson, J.N. and Duval, S. (1989) Determination of vitamin A in milk and infant formula by HPLC. *J. Micronutr. Anal.*, **6**, 147–59.

Thompson, J.N. and Maxwell, W.B. (1977) Reverse phase high pressure liquid chromatography of vitamin A in margarine, infant formula, and fortified milk. *J. Ass. Off. Analyt. Chem.*, **60**, 766–71.

Thompson, J.N., Hatina, G. and Maxwell, W.B. (1980) High performance liquid chromatographic determination of vitamin A in margarine, milk, partially skimmed milk, and skimmed milk. *J. Ass. Off. Analyt. Chem.*, **63**, 894–8.

Thompson, J.N., Hatina, G., Maxwell, W.B. and Duval, S. (1982) High performance liquid chromatographic determination of vitamin D in fortified milks, margarine and infant formulas. *J. Ass. Off. Analyt. Chem.*, **65**, 624–31.

Underwood, B.A. (1984) Vitamin A in animal and human nutrition. In *The Retinoids*, Vol. 1 (eds M.B. Sporn, A.B. Roberts and D.S. Goodman), Academic Press, Inc., New York, pp. 281–392.

Wang, X.-D., Tang, G.-W., Fox, J.G. *et al.* (1991) Enzymatic conversion of β-carotene into β-apo-carotenals and retinoids by human, monkey, ferret, and rat tissues. *Arch. Biochem. Biophys.*, **285**, 8–16.

Widicus, W.A. and Kirk, J.R. (1979) High performance liquid chromatographic determination of vitamins A and E in cereal products. *J. Ass. Off. Analyt. Chem.*, **62**, 637–41.

Woollard, D.C. and Fairweather, J.P. (1985) The storage stability of vitamin A in fortified ultra-high temperature processed milk. *J. Micronutr. Anal.*, **1**, 13–21.

Woollard, D.C. and Indyk, H. (1986) The HPLC analysis of vitamin A isomers in dairy products and their significance in biopotency estimations. *J. Micronutr. Anal.*, **2**, 125–46.

Woollard, D.C. and Indyk, H. (1989) The distribution of retinyl esters in milks and milk products. *J. Micronutr. Anal.*, **5**, 35–52.

Woollard, D.C. and Woollard, G.A. (1981) Determination of vitamin A in fortified milk powders using high performance liquid chromatography. *NZ J. Dairy Sci. Technol.*, **16**, 99–112.

Zahar, M. and Smith, D.E. (1990) Vitamin A quantification in fluid dairy products: rapid method for vitamin A extraction for high performance liquid chromatography. *J. Dairy Sci.*, **73**, 3402–7.

Zechmeister, L. (1962) *Cis-trans Isomeric Carotenoids, Vitamin A and Arylpolyenes*, Springer-Verlag, Vienna.

Ziegler, R.G. (1989) A review of epidemiologic evidence that carotenoids reduce the risk of cancer. *J. Nutr.*, **119**, 116–22.

Ziegler, R.G. (1991) Vegetables, fruits, and carotenoids and the risk of cancer. *Am. J. Clin Nutr.*, **53**, 251S–9S.

4

Vitamin D

4.1 INTRODUCTION

Sir Edward Mellanby in 1921 was the first to show clearly that rickets is a nutritional disease, and that cod-liver oil contains a factor that prevents it. In 1922 McCollum and co-workers treated cod-liver oil by bubbling oxygen through it. This destroyed the antixerophthalmic properties but left the antirachitic properties, thus indicating the presence of two factors: factor A (or vitamin A) and the antirachitic factor, which they later termed 'fat-soluble' vitamin D. Zucker and co-workers in 1922 found that vitamin D was present in the unsaponifiable fraction of cod-liver oil, and suggested that it was closely related to cholesterol.

The beneficial effect of sunlight in curing rickets in children had been recognized in the early nineteenth century. In 1924, Steenbock and Black discovered that rat rations exposed to UV radiation had the same beneficial effects as when rachitic rats were irradiated. Soon afterward, Hess and Weinstock independently reported the increased vitamin D activity of foods exposed to UV radiation. In 1925, several independent workers demonstrated that it is certain sterols in foods that are activated.

Vitamin D is a precursor of a metabolite which functions as a steroid hormone, and therefore is more correctly classed as a prohormone than as a vitamin. However, many human populations depend upon dietary sources of vitamin D because of insufficient biosynthesis of the vitamin due to inadequate skin exposure to sunlight. For this reason, vitamin D is included among the fat-soluble vitamins, despite the fact that physiologically it is a prohormone.

4.2 CHEMICAL STRUCTURE AND NOMENCLATURE

The term vitamin D, as applied in this text, refers to the secosteroids ergocalciferol (vitamin D_2) and cholecalciferol (vitamin D_3), which are irradiation products of the steroids ergosterol and 7-dehydrocholesterol, respectively. Ergosterol is the provitamin D_2 in plants, fungi and invertebrates and 7-dehydrocholesterol is the provitamin D_3 in vertebrates. Irradiation of the parent steroid results in breakage of the B ring at the 9,10-carbon bond, resulting in the conjugated triene system of double bonds. The numbering system of the carbon atoms of the vitamin D molecule is identical to that of the parent steroid (Figure 4.1). Vitamin D_2 ($C_{28}H_{44}O$; MW = 396.63) and vitamin D_3 ($C_{27}H_{44}O$; MW = 384.2) differ structurally only in the C-17 side chain, the former having an additional C-22 to C-23 double bond and a C-24 methyl group (Figure 4.1). The breakage of the B ring frees the A ring from the rigid C and D rings, giving the vitamin D molecule a high degree of conformational mobility. As a result of this mobility, the A ring undergoes rapid interconversion between two chair conformations so that the substituents alternate rapidly and continually between axial and equatorial positions. This is likely to present special problems for vitamin D receptors that are not encountered by the receptors for other steroid hormones (Collins and Norman, 1991).

Vitamin D can be produced on an industrial scale by a total chemical synthesis without involving photochemical irradiation (Collins and Norman, 1991).

4.3 BIOGENESIS

In vertebrates, vitamin D_3 can be produced by the action of sunlight on the provitamin D_3 sterol, 7-dehydrocholesterol, in the skin. The provitamin is synthesized in the sebaceous glands, then secreted onto the skin

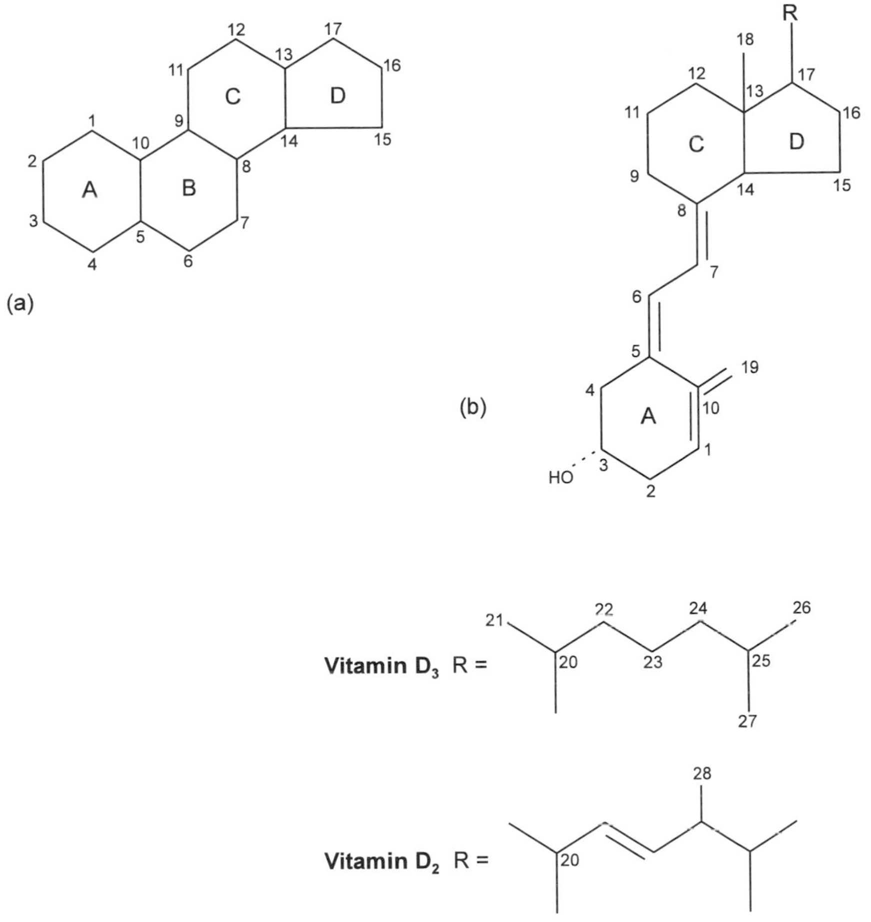

(a)

(b)

Vitamin D₃ R =

Vitamin D₂ R =

Figure 4.1 Structural relationship of (a) the parent steroid nucleus to (b) vitamin D.

surface, and re-absorbed into the various layers of the epidermis. The UV radiation can readily penetrate the skin to the level of the epidermis, and the vitamin D_3 thus formed eventually reaches the bloodstream and is transported to the liver bound to a specific plasma transport protein known as the vitamin D-binding protein (DBP).

The biogenesis of vitamin D_3 is a two-stage nonenzymatic reaction, which involves the opening of the sterol B ring and double bond rearrangement (Figure 4.2). 7-Dehydrocholesterol is irradiated to form the intermediate previtamin D_3 compound which, aided by the warmth of the body, is isomerized to vitamin D_3 over a period of about 36 h (DeLuca, 1988). The wavelength range for the biogenic process by solar UV radiation lies between 285 and 315 nm, with maximum activity at

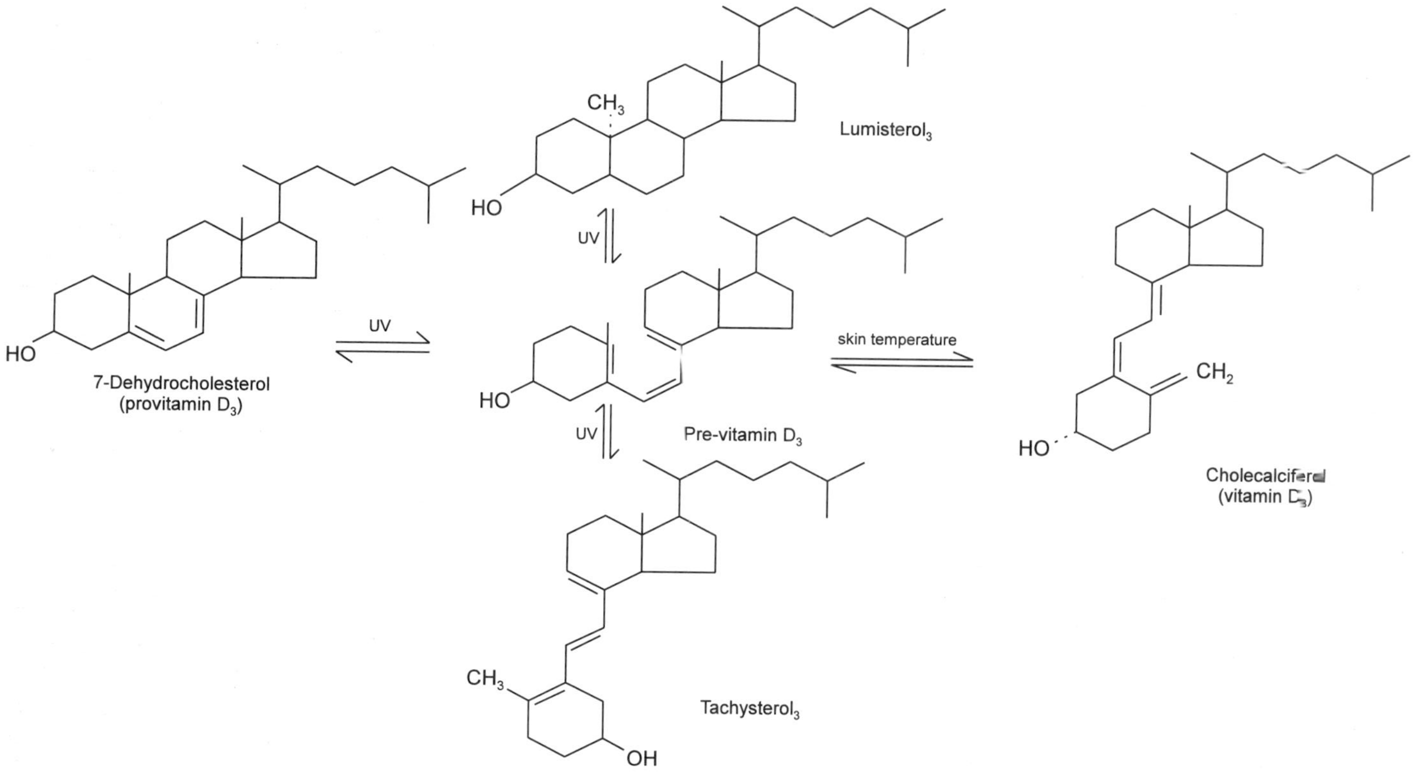

Figure 4.2 The biogenesis of vitamin D_3. (From DeLuca, 1978.)

305 nm (Lawson, 1985). Ordinary fluorescent tubes are of no benefit as the UV radiation is absorbed by the glass. Only about 15% of the amount of 7-dehydrocholesterol present is converted to previtamin D_3, and this occurs within the first 30 min of exposure to solar wavelengths. Further exposure does not increase the level of previtamin D_3 but results in the formation of biologically inactive lumisterol and tachysterol (Holmes and Kummerow, 1983).

It has been estimated that approximately 10–15 min of summer sun exposure of hands and face will produce 10 µg of vitamin D_3 in light-skinned individuals, which is sufficient to meet the daily requirement (DeLuca, 1988). Pigmented skin is less efficient in the production of vitamin D; approximately six times the amount of simulated solar ir-radiation was required in individuals with dark skin to raise serum vitamin D_3 levels to the same extent as in light-skinned individuals (Clemens *et al.*, 1982). The age-related decline in skin thickness contributes to the decline in serum 25-hydroxyvitamin D observed in persons aged above 69 years (Need *et al.*, 1993).

4.4 BIOPOTENCY

The biological activity of vitamin D_3 is mediated through its metabolite $1\alpha,25$-dihydroxyvitamin D_3. Another metabolite, 25-hydroxyvitamin D_3, which is found in significant quantities in animal-derived foods, induces all the responses of vitamin D_3, but its biological activity is 2 to 5 times higher (DeLuca, 1988).

The potency of vitamin D_2 is (for humans) considered to be equivalent to that of D_3 and vitamin D_2 has been freely substituted for D_3 as a dietary substitute (Parrish, 1979). It has been known for some time that birds and New World monkeys respond efficiently only to vitamin D_3; thus they are apparently able to discriminate against the vitamin D_2 series of compounds. Horst, Napoli and Littledike (1982) demonstrated for the first time that the pig and rat also discriminate in their metabolism of the two forms of vitamin D. Discrimination by the pig, a species which evolved with adequate exposure to sunlight, is predictably in favour of a vitamin D_3 metabolite. The rat, on the other hand, discriminates in favour of a vitamin D_2 metabolite, perhaps because it evolved as a nocturnal animal consuming grains (which contain predominantly vitamin D_2) as a significant part of its diet.

Previtamin D exhibits *c.* one-half of the antirachitic activity of vit-amin D when administered to experimental animals. Reported activities are 34%, 40% and 56% for previtamin D_2 in the rat, and 35% for previtamin D_3 in the chick relative to vitamins D_2 and D_3, respectively (Keverling Buisman *et al.*, 1968). Previtamin D itself is considered to be inactive, and the biological activity shown by this isomer is attributable to its *in vivo* conversion to vitamin D (Mulder, de Vries and Borsje, 1971).

Animal tissues contain a small proportion of vitamin D esterified with both saturated and unsaturated fatty acids. All the esters have biological activity equivalent on a molar basis to that shown by cholecalciferol (Lawson, 1985).

4.5 DEFICIENCY SYNDROMES

The primary cause of clinical vitamin D deficiency is lack of skin exposure to sunlight and the secondary cause is insufficient dietary intake of the vitamin.

Vitamin D deficiency in children causes rickets, in which the bones do not develop properly through inadequate deposition of calcium and phosphorus, and become deformed. The equivalent disease in adults is called osteomalacia, in which there is a reabsorption of calcium and phosphate from bone already laid down, causing the bones to become very brittle.

4.6 BIOCHEMICAL AND PHYSIOLOGICAL FUNCTIONS

It was not known until as recently as 1971 that vitamin D must be metabolized first to its 25-hydroxy and then to its $1\alpha,25$-dihydroxy derivative before it is able to function biologically at physiological concentrations. The enzymatic hydroxylation reactions are the same for both vitamins D_2 and D_3.

The $1\alpha,25$-dihydroxy metabolite of vitamin $D[1,25(OH)_2D]$ acts as a hormone in the regulation of calcium and phosphorus metabolism. The maintenance of normal serum calcium and phosphorus concentrations prevents hypocalcaemic tetany, and provides for normal mineralization of bone. Vitamin D also functions in some manner to improve muscle strength and tone (DeLuca, 1978).

The calcium and phosphorus homeostatic mechanisms, showing only the fully established hormonal actions, are represented diagrammatically in Figure 4.3. In response to a lowering of the blood calcium level, the parathyroid glands are stimulated to secrete parathyroid hormone (PTH) which in turn stimulates the production of $1,25(OH)_2D$ in the kidney. In addition to acting as a trophic hormone for $1,25(OH)_2D$, PTH promotes the reabsorption of calcium by the kidney and the increased urinary excretion of phosphate. PTH and $1,25(OH)_2D$ act together in mobilizing calcium, accompanied by phosphate, from the bone fluid compartment into the bloodstream. The $1,25(OH)_2D$ by itself also promotes the intestinal absorption of calcium and phosphate by two independent mechanisms. The net result of these hormonal actions is the restoration of the normal serum calcium level, which removes the stimulus upon the

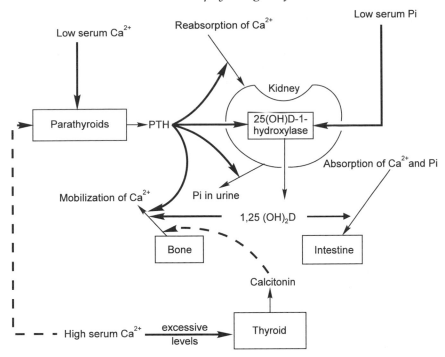

Figure 4.3 Diagrammatic representation of the calcium and phosphorus homeo-static mechanisms. The heavy solid lines represent stimulatory effects and the dotted lines represent inhibitory effects. Pi, inorganic phosphate.

parathyroids, thus suppressing the secretion of PTH and shutting down the renal production of $1,25(OH)_2D$. If the serum calcium should rise above a certain level, the thyroid gland is stimulated to secrete calcitonin, which inhibits the mobilization of calcium from the bone fluid compartment (DeLuca, 1978).

The restoration of the normal serum calcium level in response to low serum calcium is not accompanied by a rise in serum phosphate, because PTH independently causes a phosphate diuresis. Phosphate regulation takes place independently of PTH. A lowering of the serum phosphate levels stimulates the production of $1,25(OH)_2D$ without PTH being secreted. In the absence of PTH the mobilization of calcium from bone will be retarded, but the intestinal absorption of calcium and phosphate will be stimulated by $1,25(OH)_2D$. The kidney will reabsorb phosphate because there is no PTH to cause a phosphate diuresis. The net effect is an elevation of serum phosphate unaccompanied by an elevation of serum calcium.

$1,25(OH)_2D$ regulates its own synthesis as the renal 1-hydroxylase activity is inhibited by $1,25(OH)_2D$ in a negative feed-back system.

The $1,25(OH)_2D$ metabolite increases the intestinal absorption of calcium by inducing the synthesis of a specific calcium-binding protein (CaBP), which plays a role in the active transport of calcium (accompanied by phosphate) across the intestinal mucosa. In its molecular mode of action $1,25(OH)_2D$ acts like a typical steroid hormone. It is produced by an endocrine gland (the kidney) in response to a physiological stimulus (low serum calcium) and physiological amounts of the metabolite are transported to the target tissues (the intestinal mucosal cells). Within the mucosal cell the $1,25(OH)_2D$ binds to a highly specific cytoplasmic receptor protein and the complex thus formed is transported across the nuclear membrane into the nucleus. The undissociated complex attaches to a specific binding site on the chromatin and stimulates RNA polymerase to transcribe the genetic code for CaBP to single-stranded messenger RNA (mRNA) by making, through base pairing, a complementary copy of a segment of the informational strand of DNA. The mRNA carries the genetic code from the nucleus to the site of protein synthesis, the ribosomes, in the cytoplasm. Protein synthesis is effected by the process of translation whereby transfer RNAs (tRNA) carry the individual amino acids in the correct sequence to recognition sites (codons) on the mRNA template.

4.7 PHYSICOCHEMICAL PROPERTIES

Appearance and solubility

Vitamins D_2 and D_3 are white to yellowish crystalline powders that are insoluble in water; soluble in 95% ethanol, acetone, fats and oils; and readily soluble in benzene, chloroform and ether.

Spectroscopic properties

Vitamins D_2 and D_3 exhibit identical UV absorption spectra with a λ_{max} at 265 nm (λ_{min} at 228 nm) and an ε value of 18 000 in ethanol or hexane. The ε value is less than that predicted for a conjugated *cis*-triene structure because the degree of conjugation is reduced by the C-19 methylene group being above the plane of the other two double bonds (Lawson, 1985).

Vitamins D_2 and D_3 do not exhibit native fluorescence (Nair, 1966).

Stability

Grady and Thakker (1980) studied the effects of temperature (25 °C and 40 °C) and relative humidity (RH) (dry air at 45% RH and very moist air at 85% RH) on the stability of vitamin D_2 and D_3 powders in the absence of light. After 21 days, vitamin D_2 samples stored in dry air at 25 °C had only

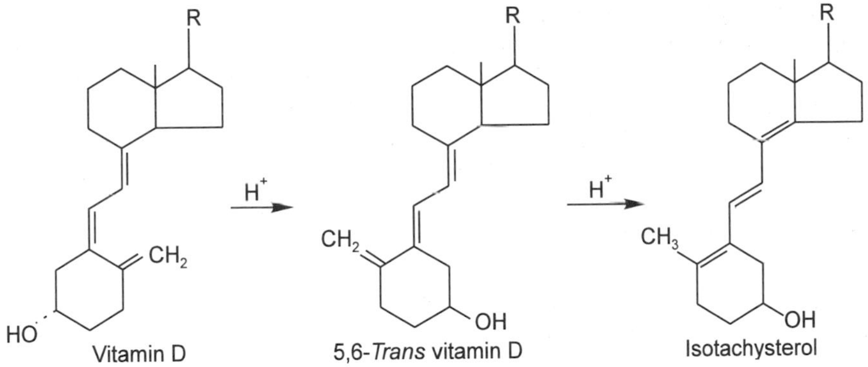

Figure 4.4 Isomerization of vitamin D under acid conditions. (From DeLuca, 1978.)

about 66% of the vitamin remaining, whilst samples stored under very moist air had > 95% remaining. Vitamin D_3 was more stable, and had > 99% remaining under both sets of conditions. Both compounds were less stable at the higher temperature. After seven days, samples of vitamin D_2 stored at 40 °C and 85% RH had 87% remaining; those stored at 40 °C and 45% RH had only 29% remaining. Conversely, samples of vitamin D_3 stored for seven days at 40 °C and 85% RH had 15% of the vitamin remaining; those stored at 40 °C and 45% RH had about 65% remaining.

In oily solutions at 0 °C the biological activity of vitamin D drops by about 50% within three to five years, whilst in emulsions this drop takes place within three weeks (Krampitz, 1980). Vitamin D is also unstable towards acids. Under conditions of even mild acidity, the vitamin D molecule isomerizes to form the 5,6-*trans* isomer and the isotachysterol isomer (DeLuca, 1978) (Figure 4.4). Vitamin D is stable towards alkali (Kutsky, 1973).

In solution, vitamin D exhibits reversible thermal isomerization to previtamin D, forming an equilibrium mixture. In the solid state, vitamin D does not isomerize. Equations and calculations have been published to determine the ratio of previtamin D to vitamin D as a function of the temperature and the reaction time (Keverling Buisman *et al.*, 1968). The previtamin D/vitamin D ratios and the equilibrium times attained at different temperatures are given in Table 4.1. At 100 °C equilibrium is reached in less than 30 min, and 28% of the mixture consists of previtamin D. At 0 °C no more than 4% previtamin D is found, and conversion is protracted over many months. When equilibrated at 20 °C, the ratio of previtamin D to vitamin D is 7 : 93. The isomerization rates of vitamins D_2 and D_3 are virtually equal (Hanewald, Mulder and Keuning, 1968) and are not affected by solvent, light or catalysis (Mulder, de Vries and Borsje, 1971). Because of the uncertainty of the previtamin D to vitamin D ratio, it

Table 4.1 The previtamin D/vitamin D equilibrium (from Mulder, de Vries and Borsje, 1971)

Temperature (°C)	% Previtamin D	% Vitamin D	Equilibrium time[a]
−20	2	98	16 years
10	4	96	350 days
20	7	93	30 days
30	9	91	10 days
40	11	89	3.5 days
50	13	87	1.3 days
60	16	84	0.5 days
80	22	78	0.1 days
100	28	72	30 min
120	35	65	7 min

[a] The time necessary to reach equilibrium, starting with pure vitamin D or pure previtamin D.

is important in physicochemical assays to determine the potential vitamin D, i.e. the combined value of previtamin D and vitamin D.

Apart from thermal isomerism, solutions of vitamin D in organic solvents are very stable, provided that oxygen, light and acids are excluded. Relatively dilute solutions of vitamin D_3 (10 µg/ml) could be refluxed for at least 1 h with a number of different solvents, and the solvent removed by rotary evaporation, or in a stream of nitrogen, without loss of UV absorbance. Solvents studied (with their boiling points) have been diethyl ether (34.6 °C), dichloromethane (40.5 °C), chloroform (61.5 °C), methanol (64.7 °C), ethanol (78.5 °C) and isooctane (99.3 °C). The deleterious effect of water has been demonstrated by the observation that 10–20 µg/ml concentrations of vitamin D_3 dissolved in water, containing 50% or less of methanol or ethanol, exhibited loss of UV absorbance as a function of time. In contrast, similar concentrations of vitamin D_3 dissolved in 80–100% alcohol exhibited long-term maintenance of UV absorbance (Chen *et al.*, 1965).

The stability of vitamin D in fats and oils corresponds to the stability of the fat itself, as described previously for vitamin A. Vitamin D is, however, more stable than vitamin A under comparable conditions.

Once freed from the protection of the food matrix, vitamin D is susceptible to decomposition by oxygen, light, acidity and water, as demonstrated by loss of UV absorbance at 265 nm. Conditions which promote destruction of vitamin D include exposure of thin films to air (especially with heat), acidic conditions, and dispersion of an alcoholic solution of the vitamin into an aqueous phase in the presence of dissolved oxygen. The decomposition products, which may be different in each case, are separable from vitamin D by some form of chromatography (Chen *et al.*, 1965).

4.8 ANALYSIS

4.8.1 Scope of analytical techniques

Most of the published HPLC methods for determining vitamin D in foods are concerned with estimating the vitamin D content in fortified products such as milk in various forms, infant formulas and margarine. In fortified foods, the amount of naturally occurring vitamin D (if any) is usually relatively very small, and it is only necessary to determine the vitamin D that is added. Even so, fortification levels are very low (e.g. 7.5–12.5 µg/ 100 g in milk powder) (Woollard, 1987), and the determination of vitamin D is by no means a simple task.

A vitamin D bioassay will account for the activity of previtamin D as well as vitamin D and its various active metabolites. A valid estimate of the vitamin D value of a food should therefore represent 'potential vitamin D', i.e. the sum of the vitamin D and previtamin D contents.

When determining naturally occurring vitamin D in animal products for nutritional evaluation purposes, 25-hydroxyvitamin D_3 [25(OH)D_3] should be included, as this metabolite contributes significantly to the total biological activity, particularly in milk. It is present in dairy products, eggs and meat tissues in sufficient concentration to permit its determination by HPLC using an absorbance detector. In bovine milk the concentration of this metabolite is < 1 ng/ml (Koshy and VanDerSlik, 1979) and hence it is usually determined by a competitive protein-binding assay after fractionation of the extracted sample by HPLC (Hollis and Frank, 1986).

Reports of water-soluble vitamin D sulphate being present in large quantities in milk (Asano *et al.*, 1981; Le Boulch, Cancela and Miravet, 1982) have not been confirmed in later studies (Okano *et al.*, 1986). The water-soluble vitamin D activity in milk can now be explained by the presence of protein-bound 25(OH)D (van den Berg, 1997). In any event, synthetic vitamin D sulphate has been shown to possess negligible biological activity (Reeve, DeLuca and Schnoes, 1981).

The vitamin D content of foods is frequently expressed in IU (1 IU = 25 ng of crystalline vitamin D_3). There is as yet no consensus on which conversion factor should be used for expressing the content of 25(OH)D in IU. For calculation of the 'total' vitamin D activity present in foods a factor of 5 [1 IU = 5 ng 25(OH)D] has been used (van den Berg, 1997).

4.8.2 Bioassay

The calcification of bone is a highly specific biological response to vitamin D and forms the basis of the classical assay methods. The response is extremely sensitive; 0.1 µg of vitamin D per day is more than adequate for a white rat.

Rat line test

The rat line test is the most widely used biological method for determining the vitamin D content of foods, and is the AOAC method for determining vitamin D in unfortified milk (AOAC, 1990). The test is a curative method in which a single dose of vitamin D promotes bone recalcification in the rachitic rat. The extent to which the rickets has been cured is assessed by measuring the extent of the new calcification, which increases as a linear function of the dose of vitamin D. The procedure involves feeding young rats on a rachitogenic (vitamin D-deficient) maintenance diet during a 3-week depletion period until severe rickets develops. Separate groups of rachitic rats are then fed diets that have been supplemented either with graded amounts of a standard preparation of vitamin D_3 or with the unknown test sample over a 7–10-day curative period. Food samples must be saponified before assay and the unsaponifiable matter taken up in the same oil as used to dilute the vitamin D_3 standard. It is essential to ensure that the standard solution and the sample solution are equilibrated to the same previtamin D/vitamin D ratio. This can be achieved by heating the standard and sample solutions simultaneously at 80 °C for 2.5 h, and then quickly cooling (Keverling Buisman *et al.*, 1968).

At the end of the curative period the animals are sacrificed and the proximal half of the tibia or the distal half of the radius or ulna is removed. The bone is cleaned of adhering tissue and a longitudinal median section is made so as to expose a plane surface. The sectioned bones are placed in a solution of silver nitrate whereupon the calcium phosphate of the newly calcified area is converted to silver phosphate. On exposure to light, the silver phosphate is reduced photochemically, and the resultant deposit of colloidal silver in the calcified area is visible as a distinct black stain. The degree of calcification of each sectioned bone is scored by matching it with a diagram or photograph of a series of bones at various stages of healing. The potency of the test sample is calculated by statistical analysis of the scores obtained for standards and unknowns (Kodicek and Lawson, 1967).

4.8.3 Extraction techniques

Saponification
Saponification is obligatory for the determination of vitamin D in fatty foods because of the need to remove the vast excess of triglycerides present. Saponification at elevated temperatures results in the thermal isomerization of vitamin D to previtamin D, and the consequent need to determine the potential vitamin D content. Thompson, Maxwell and L'Abbe (1977) reported that saponification of milk by refluxing for

30 min at 83 °C in the presence of pyrogallol resulted in a 10–20% loss of added vitamin D due to thermal isomerization. Several workers have avoided this problem by employing 'cold' saponification, i.e. overnight digestion with ethanolic KOH and pyrogallol at ambient temperature with slow constant stirring. Whatever the saponification temperature, it is necessary to perform the reaction in an inert atmosphere of oxygen-free nitrogen (Indyk and Woollard, 1984).

A 50/50 mixture of diethyl ether/petroleum ether is suitable for extracting vitamin D from the unsaponifiable material (Jackson, Shelton and Frier, 1982; Reynolds and Judd, 1984; Sertl and Molitor, 1985) and allows vitamins A and D to be co-extracted. For the determination of vitamin D in fortified milks, margarine and infant formulas, Thompson *et al.* (1982) extracted the unsaponifiable matter three times with hexane in the presence of a 6:4 ratio of water to ethanol. The combined hexane layers were then washed with 55% aqueous ethanol, after the initial 5% aqueous KOH and water washes, to remove material (including 25-hydroxyvitamin D) that was more polar than vitamin D. This extraction process was based on partition studies which showed that insignificant amounts of vitamin D were extracted from hexane by aqueous ethanol when the ratio of ethanol to water was less than 6:4.

Direct solvent extraction
For the subsequent determination of vitamin D in fortified nonfat dried milk by HPLC, Cohen and Wakeford (1980) extracted the lipid fraction into dichloromethane containing sodium phosphate tribasic solution and BHT. In another HPLC method, the lipid fraction of fortified low-fat milk was extracted by homogenization in isopropanol/dichloromethane, with magnesium sulphate added to remove water, and BHT (Landen, 1985). In a simplified method for screening vitamin D levels in fortified skimmed milk, the milk sample was mixed with water, ethanol and ammonium hydroxide then extracted four times with diethyl ether/hexane. The dried residue obtained from the combined organic phase could be analysed by HPLC without the need for purification (O'Keefe and Murphy, 1988).

4.8.4 High-performance liquid chromatography

HPLC methods for determining vitamin D are presented in Table 4.2

Preliminary clean-up procedures
The removal of sterols, vitamin E compounds, carotenoids and other interfering material from the unsaponifiable fraction of food samples has been achieved using one or more of the following techniques: co-precipitation of sterols with digitonin, precipitation of sterols from a

Table 4.2 HPLC methods used for the determination of vitamin D in food

Food	Sample preparation	Semi-preparative HPLC	Quantitative HPLC	Compounds separated	Reference
			Normal-phase		
Infant formulas	Saponify (ambient), extract unsaponifiables with hexane. Convert vitamin D to isotachysterol with acidified butanol	Supelcosil LC-18-DB 5 μm 250 × 4.6 mm i.d. MeCN/MeOH (90:10) UV 301 nm	Spherisorb silica 5 μm 250 × 2 mm i.d. Hexane/ethyl acetate/MeOH (97:2.5:0.05) UV 301 nm	Isotachysterols D_2 or D_3 (same retention time)	Agarwal (1989)
			Reversed-phase		
Fortified fluid milk (whole, low-fat, skimmed)	Saponify (ambient), extract unsaponifiables with petroleum ether. Co-precipitate sterols with digitonin. Extract with petroleum ether. Alumina column chromatography		Vydac TP 201 C_{18} 10 μm 250 × 3.2 mm i.d. MeCN/MeOH (90:10) UV 265 nm	Vitamins D_2 and D_3	Muniz, Wehr and Wehr (1982)
Fortified fluid milk (skimmed), whole milk powder, milk powder with soybean, chocolate milk powder, diet food	Digest starchy samples with Takadiastase before saponification. Saponify (hot), extract unsaponifiables with petroleum ether. Silica solid-phase extraction (high-fat samples only)		Hypersil ODS 5 μm 120 × 4 mm i.d. (two columns connected in series) 0.5% H_2O in MeOH UV 265 nm	Vitamin D_2 or D_3	Bui 1-87
Fortified whole milk powder	Add vitamin D_2 or D_3 to sample as internal standard. Saponify (ambient), extract unsaponifiables with petroleum ether/diethyl ether (90:10). Precipitate sterols from a methanolic solution		Radial-PAK cartridge containing either Resolve C_{18} or Nova-PAK C_{18} 5 μm (two cartridges connected in series) MeOH/THF/H_2O (93:2:5) UV 254 and 280 nm (dual)	Vitamins D_2 and D_3	Indyk and Woollard (1985)
Infant formulas	Add vitamin D_2 or D_3 to sample as internal standard. Saponify (ambient), extract unsaponifiables with petroleum ether/diethyl ether (90:10). Silica solid-phase extraction		Two Radial-PAK cartridges as above MeOH/THF/H_2O (92:2:6) UV 254 and 280 nm (dual)	Vitamins D_2 and D_3	Indyk and Woollard (1985)

Sample	Sample preparation	HPLC (normal phase)	HPLC (reverse phase)	Vitamins	Reference
Milk (unfortified)	Add vitamin D_2 to sample as internal standard. Saponify (ambient), extract unsaponifiables with hexane/diethyl ether (90:10). Silica solid-phase extraction	Radial-PAK cartridge containing Resolve silica 5 μm Hexane/2-PrOH (99:1) UV 265 nm	Radial-PAK cartridge containing Resolve C_{18} 5 μm, column temperature 30 °C MeOH/THF/H_2O (93:2:5) UV 265 nm	Vitamins D_2 and D_3	Kurmann and Indyk (1994)
Milk	Saponify (ambient), extract unsaponifiables with petroleum ether/diethyl ether (90:10). Silica solid-phase extraction		Vydac 201 TP54 C_{18} 5 μm 250 × 4.6 mm i.d. MeCN/MeOH (90:10) UV 254 nm	Vitamins D_2 and D_3	Renken and Warthesen (1993)
Infant formulas	Add vitamin D_2 to sample as internal standard. Saponify (hot), extract unsaponifiables with hexane. Silica solid-phase extraction		Vydac 201 TP54 C_{18} 5 μm 250 × 4.6 mm i.d., column temperature 27 °C MeCN/MeOH (91:9) UV 265 nm	Vitamins D_2 and D_3	Sliva et al. (1992)
Margarine, fats and oils	Add vitamin D_2 to sample as internal standard. Saponify (hot), extract unsaponifiables with diethyl ether	LiChrosorb Si-60 7 μm 250 × 4.6 mm i.d. Hexane/2-PrOH/THF (98:1:1) UV 264 nm	ChromSphere C_{18} 8 μm 100 × 3 mm i.d. MeCN/CHCl$_3$/MeOH (91:5:3) UV 265 nm	Vitamins D_2 or D_3	Rychener and Walter (1985)
Fortified fluid milk (whole)	Saponify (ambient), extract unsaponifiables with hexane	Supelcosil LC-Si 5 μm 150 × 4.6 mm i.d. Cyclohexane/hexane (50:50) containing 0.5% 2-PrOH UV 254 nm	Radial-PAK cartridge containing Resolve C_{18} 5 μm or Spherisorb ODS 10 μm MeCN/MeOH (90:10) UV 254 nm	Vitamins D_2 or D_3	Thompson et al. (1982)
Margarine, infant formulas	As above, but with alumina column chromatography before semi-preparative HPLC	As above, but concentration of 2-PrOH changed to 0.25%	As above	Vitamins D_2 and D_3	Thompson et al. (1982)
Margarine, vegetable oils, fortified milk	Add vitamin D_2 to sample as internal standard. Saponify (hot), extract unsaponifiables with hexane	Polygosil 60 5 μm 300 × 8 mm i.d., column temperature 30 °C Isooctane/CHCl$_3$/THF/isobutane (94:3:2:1) UV 254 nm	Vydac 201 TP54 C_{18} 5 μm 250 × 4.6 mm i.d., column temperature 30 °C MeCN/MeOH/CHCl$_3$ (82:12:6) UV 265 nm	Vitamins D_2 and D_3	Johnsson et al. (1989)
Infant formulas	Add vitamin D_2 to sample as internal standard. Saponify (hot), extract unsaponifiables with petroleum ether/diethyl ether (90:10)	Polygosil 60 5 μm 250 × 8 mm i.d. Isooctane/isobutanol (99:1) UV 265 nm	Hypersil ODS 5 μm 250 × 4.6 mm i.d. (two columns connected in series) 100% MeOH UV 265 nm	Vitamins D_2 and D_3	Konings (1994)

Table 4.2 *Continued*

Food	Sample preparation	Semi-preparative HPLC	Quantitative HPLC	Compounds separated	Reference
Margarine	Add vitamin D_2 or D_3 to sample as internal standard. Saponify (ambient), extract unsaponifiables with hexane. Clean-up on an alumina-digitonin/Celite column	Nucleosil 50-5 5 µm 250 × 4 mm i.d. Hexane containing 0.5% 2-PrOH UV 265 nm	Spherisorb ODS-2 3 µm 250 × 4 mm i.d. MeCN/MeOH/CHCl₃ (91:3:6) UV 265 nm	Vitamins D_2 and D_3	Ho̶mberg (1993)
Nutritionally complete liquid-formula diet	Saponify (hot), extract unsaponifiables with diethyl ether. Add vitamin D_2 as internal standard to the extracted ether solution	Nucleosil 50-5 5 µm 250 × 4.6 mm i.d. Hexane containing 0.5% 2-PrOH UV 265 nm	Hitachi Gel 3056 reversed-phase column 5 µm 250 × 4.6 mm i.d. MeCN/MeOH/50% perchloric acid (970 + 30 + 1.2) containing 0.057 M sodium perchlorate Dual cell electrochemical detector (redox mode): +0.65 V (oxidation) −0.20 V (reduction)	Vitamins D_2 and D_3	Ha̶gawa (1992)
Raw meat and liver, milk and milk products	Add vitamin D_2 and $25(OH)D_2$ to homogenized sample as internal standards. Saponify (ambient), extract unsaponifiables with diethyl ether/petroleum ether (1:1)	µPorasil silica 10 µm 300 × 3.9 mm i.d. Gradient elution with hexane/2-PrOH to obtain a vitamin D fraction and a $25(OH)D$ fraction UV 265 nm *Vitamin D fraction:* Vydac 201 TP54 C_{18} 5 µm 250 × 4.6 mm i.d. MeOH/H₂O (93:7) UV 265 nm *25(OH)D fraction:* Vydac 201 TP54 C_{18} 5 µm MeOH/H₂O (83:17) UV 265 nm	*Vitamin D_3:* Zorbax ODS + Vydac 201 TP54 C_{18} (connected in series) MeOH/H₂O (96:4) UV 255 nm *$25(OH)D_3$:* Spherisorb S5NH₂ 5 µm 250 × 4.6 mm i.d. + µPorasil 10 µm (connected in series) Hexare/2-PrOH (97:3) UV 265 nm	Vitamins D_2 and D_3 $25(OH)D_2$ and $25(OH)D_3$	Matt̶l *et al.* (1995)

Sample	Preparation	HPLC conditions		Analytes	Reference
Meat and fat from livestock fed normal and excessive quantities of vitamin D	Saponification (hot), extract unsaponifiables with hexane/ CH$_2$Cl$_2$ (85:15). Alumina column chromatography to obtain a vitamin D fraction and a 25(OH)D fraction	*Vitamin D fraction:* Apex silica 3 μm 150 × 4.5 mm i.d. Cyclohexane/hexane (1:1) containing 0.25% 2-PrOH UV 254 nm *25(OH)D fraction:* Radial-PAK cartridge containing Resolve C$_{18}$ 5 μm Dry 100% MeOH UV 254 nm	*Vitamin D$_3$:* Radial-PAK cartridge containing Resolve C$_{18}$ 5 μm Dry 100% MeOH UV 254 nm *25(OH)D$_3$ by normal-phase HPLC:* Apex silica 3 μm Heptane/2-PrOH (96:4) UV 254 nm	Vitamins D$_2$ and D$_3$ 25(OH)D$_3$	Thompson and Plouffe (1993)

Abbreviations: see footnote to Table 2.4.

methanolic solution, adsorption chromatography on open columns of alumina, solid-phase extraction and semi-preparative HPLC.

In a typical solid-phase extraction technique, the unsaponifiable resi-due is dissolved in hexane and loaded onto a preconditioned silica cartridge. The vitamin D is retained and nonpolar material passes through. The cartridge is then flushed with a solvent that is sufficiently polar to remove further interfering material without displacing the vit-amin D. The vitamin D is finally eluted with a slightly more polar solvent, leaving behind adsorbed material that is more polar than vitamin D. This technique was used by Bui (1987) in the analysis of high-fat fortified milk products and diet foods. The bulk of the sterols was removed with 3 ml of hexane/ethyl acetate (85 + 15), after which the vitamin D was eluted with 5 ml of hexane/ethyl acetate (80 + 20). Sliva *et al.* (1992) dissolved the unsaponifiable residue in dichloromethane/isopropanol (99.8 + 0.2), rather than in hexane, and applied this solution to a silica cartridge. After flushing with 2 ml of dichloromethane/isopropanol (99.9 + 0.2), vitamins D_2 and D_3 were eluted with 7 ml of dichloromethane/isopro-panol (99.8 + 0.2).

In an alternative solid-phase extraction technique, Reynolds and Judd (1984) dissolved the unsaponifiable residue obtained from fortified skimmed milk powder in 2 ml of ethanol, added 1 ml of water, and applied this solution to a C_{18} reversed-phase cartridge. The cartridge was flushed with 15 ml of methanol/THF/water (1 + 1 + 2) to remove material that was more polar than vitamin D. The vitamin D was eluted with 5 ml of methanol, leaving the nonpolar material retained on the sorbent.

Normal-phase systems
Normal-phase HPLC, using either silica or polar-bonded stationary phases, isocratically separates vitamin D_2 or D_3 from their respective previtamin forms and inactive isomers (de Vries *et al.*, 1979). Vitamin D ($D_2 + D_3$), 25-hydroxyvitamin D_2 and 25-hydroxyvitamin D_3 can be separated from one another and from other hydroxylated metabolites (Jones and DeLuca, 1975), but vitamins D_2 and D_3 cannot be resolved from one another. For nutritional evaluation purposes, it is not essential to distinguish between naturally occurring vitamins D_2 and D_3, because the two vitamers exhibit similar biopotency in humans, and their thermal isomerization rates are virtually equal. In the analysis of fortified products, any naturally occurring vitamin D is usually relatively small and, since fortification is carried out with either vitamin D_2 or D_3 (now-adays, usually the latter), it is only necessary to determine one of the D vitamers.

The thermal isomerization of vitamin D that occurs during saponifi-cation of the sample at refluxing temperatures poses a problem in the estimation of potential vitamin D, as the previtamin D peak is invariably masked by co-eluting matrix peaks. Also, in some methods, the previtamin is separated and discarded during chromatographic clean-up. Fortunately, the reversibility of the isomerization reaction is very slow, and therefore the percentage of the previtamin will remain virtually unchanged during the subsequent stages of the analytical procedure. This equilibrium allows the potential vitamin D to be calculated from measurements of the vitamin D peak alone, and the same principle applies to the hydroxylated metabolites of vitamin D (Vanhaelen-Fastré and Vanhaelen, 1981). This can be done by saponifying a standard vitamin D solution in parallel with the sample and using the resultant solution as an external standard in the quantification (Kobayashi, Okano and Takeuchi, 1986).

Agarwal (1989) circumvented the isomerization problem by converting previtamin D and vitamin D to a common derivative, isotachysterol, by treatment of the unsaponifiable residue with acidified butanol. The isotachysterol was detected at its λ_{max} of 301 nm, which provided greater sensitivity and selectivity compared with the detection wavelength of 265 nm usually used for vitamin D.

Bekhof and van den Bedem (1988) developed an on-line multidimensional HPLC technique using a nitrile bonded-phase semi-preparative column and an amino bonded-phase analytical column. The respective mobile phase compositions of 0.15% 1-pentanol in hexane and 0.35% 1-pentanol in hexane were adjusted, if necessary, to achieve a capacity factor for vitamin D_3 of 6–7 on each column. The method described was found to be suitable for the determination of vitamin D in the unsaponifiable fraction obtained from fortified evaporated milk, and skimmed and whole milk powder. Interference from vitamin E compounds made it unreliable for the analysis of vegetable oil-based infant formulas.

Reversed-phase systems
Reversed-phase chromatography, as in the normal-phase mode, isocratically separates vitamin D_2 or D_3 from their corresponding previtamin and inactive isomers (de Vries *et al.*, 1979), and hence estimates of the potential vitamin D must be made from measurement of the vitamin D peak alone. Unlike normal-phase chromatography, the reversed-phase mode, using non-end-capped ODS stationary phases, can separate vitamin D_2 from D_3, permitting the measurement of the two adjacent peaks. The 25-hydroxylated metabolites of vitamins D_2 and D_3 can be separated from one another using a Vydac 201 TP column (Mattila *et al.*, 1993). The separation of vitamin D_2 from vitamin D_3, and 25(OH)D_2 from 25(OH)D_3, allows the D_2 form of the vitamin or its 25-hydroxylated metabolite to be used as an internal standard for quantifying the corresponding D_3 form. Internal standardization is desirable, in view of the

multi-step extraction/purification procedure, to compensate for any losses of vitamin D that may be incurred. When using vitamin D_3 as an internal standard, the previtamin D/vitamin D ratio for vitamins D_2 and D_3 will be the same at any given temperature, because the isomerization rates of vitamins D_2 and D_3 are virtually equal. Therefore, the quantification will compensate for the formation of previtamin D and give a result for the potential vitamin D content of the sample.

Normal-phase/reversed-phase HPLC is the ideal combination for semi-preparative and analytical separations in two-dimensional HPLC, as vitamins D_2 and D_3 co-elute during the semi-preparative stage, allowing a narrow retention window to be collected for analysis using internal standardization (Rychener and Walter, 1985). By this means, Johnsson *et al.* (1989) obtained a vitamin D_3 detection limit of $0.1 \mu g/kg$ for milk and milk products. The RSD for replicates of fortified milk samples was 0.53% ($n = 7$).

Indyk and Woollard (1985a) reported that removal of the cholesterol from the unsaponifiable fraction of fortified whole-milk powder by methanolic precipitation and filtration was an adequate clean-up procedure, making semi-preparative HPLC unnecessary. This simplified procedure was made possible by the use of two analytical columns connected in series, which adequately separated vitamins D_2 and D_3 from one another and from vitamins A and E. The analysis of infant formulas (Indyk and Woollard, 1985b) required clean-up by silica solid-phase extraction to remove the minor tocopherols and tocotrienols, which constituted potential sources of interference. Tandem column chromatography was also used by Bui (1987) in a simplified procedure for determining vitamin D in various fortified milks and milk products. The method developed by Bristol-Myers Squibb (Sliva *et al.*, 1992) for the determination of vitamin D in infant formulas and enteral products has been studied collaboratively (Sliva and Sanders, 1996) and adopted as first action by AOAC International.

Mattila *et al.* (1995) described a two-dimensional HPLC procedure for determining vitamin D_3 and $25(OH)D_3$ in meat and milk products. Samples were saponified in the presence of vitamin D_2 and $25(OH)D_2$ as internal standards, and the extracted unsaponifiable matter was subjected to normal-phase semi-preparative HPLC to obtain a fraction containing $25(OH)D_2 + 25(OH)D_3$ and a fraction containing $D_2 + D_3$. The collected fractions were evaporated and purified by reversed-phase HPLC. Fractions were again collected, after which vitamin D_3 was analysed by reversed-phase HPLC and $25(OH)D_3$ by normal-phase HPLC.

Hasegawa (1992) applied electrochemical detection in the redox mode to the determination of vitamin D in medical nutritional products. The on-column detection limit was estimated to be about 200 pg.

4.9 BIOAVAILABILITY

4.9.1 Physiological aspects

Absorption and transport

The vitamin D activity in the human diet is contributed mainly by free vitamin D and 25(OH)D. Ingested vitamin D is solubilized within mixed micelles in the duodenum and absorbed in the jejunum along with other lipids (Schachter, Finkelstein and Kowarski, 1964). Esters of vitamin D, if present, are hydrolysed during solubilization in the mixed micelles (van den Berg, 1997). Vitamin D is incorporated into chylomicrons within the enterocytes and the chylomicrons convey the vitamin in the mesenteric lymph to the systemic circulation. When chylomicrons mix with blood plasma, a significant fraction of the vitamin D is transferred from the chylomicrons to plasma protein, most likely the vitamin D-binding protein (DPB) (Dueland *et al.*, 1982). After lipolysis of the chylomicrons, the vitamin D remaining on the chylomicron remnants, and also the vitamin D bound to protein, is initially taken up by the liver (Dueland *et al.*, 1983a).

Various studies in rats (Maislos, Silver and Fainaru, 1981; Sitrin *et al.*, 1982; Dueland *et al.*, 1983b) and humans (Stamp, 1974) have shown that 25(OH)D is absorbed at a faster rate than vitamin D, indicating that these compounds are absorbed by different mechanisms. The difference in absorption rate has been attributed to 25-hydroxyvitamin D being directly absorbed primarily into portal blood and to a lesser extent into lymph (Maislos, Silver and Fainaru, 1981; Sitrin *et al.*, 1982). However, the data produced by Dueland *et al.* (1983b) suggested that both compounds are absorbed mainly in the lymph, vitamin D being carried primarily with chylomicrons and 25(OH)D being mainly transported with protein. This protein is probably DBP, which has a greater affinity for 25(OH)D than for vitamin D (Haddad and Walgate, 1976). Other evidence to support the protein-bound transport of 25(OH)D in the lymph was the demonstration by Compston *et al.* (1981) that very little 25(OH)D is carried in the chylomicrons after intestinal absorption in humans.

Oral doses of vitamin D_3 have frequently been unsuccessful in correcting vitamin D deficiency in patients with chronic cholestatic liver diseases such as primary biliary cirrhosis. On the other hand, oral treatment with 25(OH)D has resulted in reversal of osteomalacia in such patients (Reed *et al.*, 1980). This may partly reflect the superior absorption of 25(OH)D compared with vitamin D in patients with cholestasis (Sitrin and Bengoa, 1987). The lack of effect of vitamin D in these cases could be at least partly explained by its dependence on bile for its absorption, bile salts being required for the formation of chylomicrons. Absorption of 25(OH)D is less dependent on bile salts.

Functional metabolism

In the liver the vitamin D of both cutaneous and dietary origin is hydroxylated to 25(OH)D or is catabolized to inactive compounds for excretion
in the bile. The rate of 25-hydroxylation is related to substrate supply.
Almost all of the 25(OH)D produced in the liver is released without delay
into the bloodstream, where it becomes attached to DBP. The blood
constitutes the largest single pool of 25(OH)D, since extrahepatic tissues
take up only small amounts. During vitamin D deprivation the 25(OH)D
body pool is maintained through the prolonged release of vitamin D from
its skin reservoir and from storage sites in muscle and adipose tissue.

Large oral doses of vitamin D overcome any feedback regulation of 25-
hydroxylation that may exist and plasma 25(OH)D levels can rise to more
than 400 ng/ml, leading to vitamin D toxicity. In contrast, extensive UV
irradiation of the skin does not cause hypervitaminosis D and raises
25(OH)D levels in plasma to no more than 80 ng/ml (Fraser, 1983).

25(OH)D is transported to the kidney, where it undergoes its second
hydroxylation to $1,25(OH)_2D$. The kidney is not the exclusive site of 1α-
hydroxylation, as the placenta has the capacity to provide additional
$1,25(OH)_2D$ to fulfil the need during pregnancy (DeLuca, 1988). The
1α-hydroxylation is tightly regulated in the manner of a typical endocrine
system. The $1,25(OH)_2D$ is transported on the DBP to the target tissues of
intestine, bone and kidney, where it carries out its hormonal functions. In
the kidney 25(OH)D is converted also into $24,25(OH)_2D$ [at one time
incorrectly identified as $21,25(OH)_2D$] which circulates in the bloodstream in concentrations that increase with increasing amounts of vitamin D ingested. It is now evident that $24,25(OH)_2D$ possesses no
biological function and represents an alternative means of removing
25(OH)D (DeLuca, 1988).

Biliary excretion and enterohepatic circulation

The major excretory route of the final products of vitamin D catabolism in
humans is the bile. Clements, Chalmers and Fraser (1984) administered
[^{3}H]vitamin D_3 orally to human volunteers to determine the fate of
ingested vitamin; in addition, [^{14}C]vitamin D_3 was given intravenously
to simulate vitamin D formed in response to sunlight in the skin. The
vitamin D given intravenously was bound to a protein to simulate
physiological conditions. A significant proportion of labelled vitamin D,
administered orally or intravenously, was excreted in bile as highly
polar, biologically inactive metabolites, many of which were conjugated
with glucuronic acid. Less than 4% of labelled compounds in the bile
were present as 25(OH)D or its glucuronide conjugate. Conjugation of the
metabolites facilitates biliary excretion, cyclic molecules with a molecular
weight below 300 being poorly excreted in bile. A rat study of the

metabolism of $1,25(OH)_2D_3$ (Onisko *et al.*, 1980) showed only small amounts (1–8% of the metabolites in bile, dependent on dose given) of unchanged $1,25(OH)_2D$. A large fraction of the biologically inactive metabolites were side chain cleavage products.

Early evidence for the existence of an enterohepatic cycle for vitamin D metabolites in humans was provided by Arnaud *et al.* (1975). Following a single intravenous injection of $[^3H]25(OH)D_3$, one-third of the radioactivity was secreted into the lumen of the duodenum, probably in the bile. Based on measurement of faecal radioactivity, at least 85% of the tritiated material secreted into the duodenum was estimated to have been reabsorbed. A rat study conducted by Gascon-Barré (1982) confirmed the existence of an enterohepatic cycle for vitamin D metabolites. Rats were dosed with $[^3H]25(OH)D_3$ by the intravenous or intraduodenal route and the biliary material was collected and given intraduodenally to another group of rats. The compounds derived from $[^3H]25(OH)D_3$ were efficiently re-excreted in newly secreted bile but plasma analysis indicated that these compounds did not reach the systemic circulation in significant quantities. This latter observation and the virtual absence of active vitamin D metabolites in the bile (Clements, Chalmers and Fraser, 1984) suggest that the enterohepatic circulation does not constitute a functional conservation mechanism analogous to the recycling of bile salts. On the other hand, it should not be excluded that the constant cycling of vitamin D catabolites from intestine to liver to intestine might prevent additional liver extraction of active vitamin D metabolites from the systemic circulation, indirectly contributing, in that manner, to the conservation of circulating vitamin D metabolites (Gascon-Barré, 1986).

4.9.2 Dietary sources and their bioavailability

Dietary sources

The proportion of vitamin D obtained from the diet is normally very small compared with that synthesized in skin in response to sunlight. Cereals, vegetables and fruit contain no vitamin D, whilst meat, poultry and white fish contribute insignificant amounts. The richest natural sources of vitamin D_3 are fish-liver oils, especially halibut-liver oil, but they are regarded as pharmaceuticals rather than foods. Fatty fish, such as herring, sardines, pilchards and tuna, are rich natural food sources; smaller amounts of the vitamin are found in mammalian liver, eggs and dairy products (Table 4.3). Eggs from hens receiving a vitamin D supplement will have a considerably higher vitamin D content than eggs from unsupplemented hens. The concentration of vitamin D_3 in milk shows a seasonal variation, which is related to the amount of sunlight available to convert 7-dehydrocholesterol in the animal's skin to vitamin D_3. Vitamin

Table 4.3 Vitamin D content of certain foods

Food	µg Vitamin D/100 g	Reference
Milk, pasteurized		(1)
summer	0.13	
winter	0.03	
Butter	0.8	(1)
Margarine, fortified, UK	8.0	(1)
Cheese, cheddar	0.25	(2)
Egg, whole, battery	1.2	(3)
Mammalian liver	0.75	(2)
Herring	22	(1)
Pilchard, canned	8	(1)
Tuna, canned	6	(1)

(1) Lawson (1985); (2) Passmore and Eastwood (1986); (3) Sivell, Wenlock and Jackson (1982)

D_2 also occurs in milk, but in smaller concentrations than vitamin D_3. Unlike vitamin D_3, vitamin D_2 is derived by UV irradiation of ergosterol in sun-dried green forage (hay); ergosterol cannot be converted by the animal into vitamin D_2 (Cremin and Power, 1985). Except for eggs and fatty fish, a serving of food containing only natural sources of vitamin D would probably supply less than 1 µg of vitamin D (Parrish, 1979).

The vitamin D activity in animal products is contributed by both vitamin D and its metabolite, 25(OH)D. Typical values of the hydroxylated metabolite (µg/100 g) are bovine muscle, 0.2–0.3; bovine liver, 0.3–0.5; bovine kidney, 0.5–1.0 (Koshy and VanDerSlik, 1977) and chicken egg yolk, 1.0 (Mattila *et al.*, 1993). Being more potent than vitamin D, 25(OH)D is of nutritional significance. Reeve, Jorgensen and DeLuca (1982) calculated that, in milk, 25(OH)D accounts for 75% of the total vitamin D activity as estimated by the calcium transport assay.

In the United Kingdom, the Margarine Regulations (1967) state that 'every ounce of margarine shall contain not less than 80 IU and not more than 100 IU of vitamin D', equating to limits of between 7 and 9 µg/100 g. Other foodstuffs commonly enriched with vitamin D include skimmed milk powder, evaporated milk, milk-based beverages, breakfast cereals, dietetic products of all kinds, baby foods and soup powders. Vitamin D_2 is either added as an oily solution, or in combination with a vitamin A formulation (Kläui, Hausheer and Huschke, 1970).

Bioavailability

In the only known study of vitamin D bioavailability from natural sources (van den Berg, 1997), the average relative bioavailability of vitamin D_2 from meat sources was estimated to be about 60% as compared with a vitamin supplement.

Vitamin D given with milk has been reported to be 3–10 times more potent than that given with oil, the stimulatory factor being attributable to the lactalbumin fraction (Holmes and Kummerow, 1983). It is not known whether enhanced absorption or some other factor is responsible for the greater apparent biological activity of vitamin D given in milk.

In the rat, chronic ethanol ingestion promotes the biliary loss of 25(OH)D and this loss may be a contributing factor in the impaired vitamin D status in alcoholics (Gascon-Barré and Joly, 1981).

Human subjects receiving a high-fibre diet exhibited a reduced plasma half-life of $25(OH)D_3$, indicating a more rapid elimination of the metabolite from the body (Batchelor and Compston, 1983).

4.9.3 Effects of food processing, storage and cooking

Food processing, cooking and storage of foods do not generally affect the activity of vitamin D. The vitamin will withstand smoking of fish, pasteurization and sterilization of milk, and spray-drying of eggs, although it is generally considered to be destroyed in oxidizing fats.

Indyk, Littlejohn and Woollard (1996) used HPLC to evaluate the stability of supplemental vitamin D_3 in spray-dried milk. Measured losses through the pasteurization, high-pressure evaporation and drying processes were demonstrated to be statistically insignificant ($P > 0.05$). It has been common practice to add an 'overage' of vitamin D to the supplemental amount to allow for up to 30% destruction of the vitamin during the drying process, but this is a questionable risk factor to regulatory compliance.

4.10 DIETARY INTAKE

The recommended daily allowance of vitamin D for humans exposed to inadequate sunlight is 10 µg. To maintain adequate plasma 25(OH)D levels without any input from skin irradiation would require ingestion of 12.5 µg or more of vitamin D per day in the form of dietary supplements (Fraser, 1983). A serving of food containing only natural sources of vitamin D and no egg or fish would probably supply less than 1 µg of the vitamin. Under conditions of ample solar exposure, the vitamin D requirement is met entirely by biogenesis in the skin.

The serum 25(OH)D concentration reflects vitamin D stores in humans and its measurement by protein-binding assay is an index of nutritional status (Burnand *et al.*, 1992).

Effects of high intakes

Vitamin D, in common with vitamin A, is toxic when ingested in large amounts. In adults, daily intakes of vitamin D in excess of 50 000 IU produce toxicity symptoms of anorexia, dehydration, muscle weakness, headache, nausea, vomiting, polyuria (excessive urine production) and polydipsia (frequent drinking because of extreme thirst). Serum calcium levels are increased as a result of bone demineralization, and calcium is deposited in the kidneys, causing hypertension, renal failure, cardiac insufficiency and anaemia. As in vitamin A toxicity, hypervitaminosis D results from oversupplementation with pharmaceutical products, and not from the consumption of usual diets. Hypervitaminosis D does not result from unlimited exposure to sunshine. Skin tanning or the aggregation of the pigment melanin in the skin creates a filter for UV radiation, and prevents conversion of the 7-dehydrocholesterol to vitamin D_3. The toxicity of excessive amounts of ingested vitamin D is attributable to the pharmacological effects of 25(OH)D in high concentration; circulating concentrations of $1,25(OH)_2D$ are not greatly increased in vitamin D-intoxicated patients (Miller and Hayes, 1982; Omaye, 1984).

4.11 SYNOPSIS

The formation of cholecalciferol (vitamin D_3) by biogenesis in the skin provides a large part, if not all, of the vitamin D requirement in healthy people, given adequate exposure to sunlight. Vitamin D of dietary origin is transported in chylomicrons to the liver where it is converted to 25(OH)D, the major circulating form of the vitamin. This metabolite is transported to the kidney where it is further hydroxylated to $1,25(OH)_2D$, the physiologically active form of the vitamin.

Information on the bioavailability of vitamin D in foods is scarce, but an average relative bioavailability of 60% from meat sources has been reported.

The vitamin D bioassay is based on the ability of the vitamin to promote bone recalcification in rachitic rats and is the official (AOAC) method for determining vitamin D in unfortified milk.

Many HPLC methods have been published for determining the vitamin D content in fortified foods. Although it is only necessary to determine the vitamin D that has been added, fortification levels are very low and the method often entails both semi-preparative and quantitative HPLC of the saponified sample.

When determining naturally occurring vitamin D for nutritional evaluation purposes, 25(OH)D should be included, as this metabolite contributes significantly to the total biological activity. 25(OH)D is present in dairy products, eggs and meat tissues in sufficient concentration to permit its determination by HPLC using an absorbance detector. In bovine milk the concentration of this metabolite is $< 1\,\mathrm{ng/ml}$ and hence it is usually determined by a competitive protein-binding assay after fractionation of the extracted sample by HPLC.

REFERENCES

Agarwal, V.K. (1989) Liquid chromatographic determination of vitamin D in infant formula. *J. Ass. Off. Analyt. Chem.*, **72**, 1007–9.

AOAC (1990) Vitamin D in milk, vitamin preparations, and feed concentrates. Rat bioassay. Final action. In *AOAC Official Methods of Analysis*, 15th edn (ed. K. Helrich), Association of Official Analytical Chemists, Inc., Arlington, VA, 936.14.

Arnaud, S.B., Goldsmith, R.S., Lambert, P.W. and Go, V.L.W. (1975) 25-Hydroxyvitamin D_3: evidence of an enterohepatic circulation in man. *Proc. Soc. Expt. Biol. Med.*, **149**, 570–2.

Asano, T., Hasegawa, T., Suzuki, K. *et al.* (1981) Determination of vitamin D_3-sulfate in milk by high-performance liquid chromatography. *Nutr. Rep. Int.*, **24**, 451–6.

Batchelor, A.J. and Compston, J.E. (1983) Reduced plasma half-life of radio-labelled 25-hydroxyvitamin D_3 in subjects receiving a high-fibre diet. *Br. J. Nutr.*, **49**, 213–6.

Bekhof, J.J. and van den Bedem, J.W. (1988) Study on the determination of vitamin D in fortified milk, milk powder and infant formula by HPLC using a column switching technique. *Neth. Milk Dairy J.*, **42**, 423–35.

Bui, M.H. (1987) Sample preparation and liquid chromatographic determination of vitamin D in food products. *J. Ass. Off. Analyt. Chem.*, **70**, 802–5.

Burnand, B., Sloutskis, D., Gianoli, F. *et al.* (1992) Serum 25-hydroxyvitamin D: distribution and determinants in the Swiss population. *Am. J. Clin. Nutr.*, **56**, 537–42.

Chen, P.S. Jr, Terepka, R., Lane, K. and Marsh, A. (1965) Studies of the stability and extractability of vitamin D. *Analyt. Biochem.*, **10**, 421–34.

Clemens, T.L., Henderson, S.L., Adams, J.S. and Holick, M.F. (1982) Increased skin pigment reduces the capacity of skin to synthesize vitamin D_3. *Lancet*, **I**, 74–6.

Clements, M.R., Chalmers, T.M. and Fraser, D.R. (1984) Enterohepatic circulation of vitamin D: a reappraisal of the hypothesis. *Lancet*, **I** (No. 8391), 1376–9.

Cohen, H. and Wakeford, B. (1980) High pressure liquid chromatographic determination of vitamin D_3 in instant nonfat dried milk. *J. Ass. Off. Analyt. Chem.*, **63**, 1163–7.

Collins, E.D. and Norman, A.W. (1991) Vitamin D. In *Handbook of Vitamins*, 2nd edn (ed. L.J. Machlin), Marcel Dekker, Inc., New York, pp. 59–98.

Compston, J.E., Merrett, A.L., Hammett, F.G. and Magill, P. (1981) Comparison of the appearance of radiolabelled vitamin D_3 and 25-hydroxy-vitamin D_3 in the chylomicron fraction of plasma after oral administration in man. *Clin. Sci.*, **60**, 241–3.

Cremin, F.M. and Power, P. (1985) Vitamins in bovine and human milks. In *Developments in Dairy Chemistry*, Vol. III (ed. P.F. Fox), Elsevier Applied Science Publishers, London, pp. 337–98.

DeLuca, H.F. (1978) Vitamin D. In *Handbook of Vitamin Research*, Vol. II (ed. H.F. DeLuca), Plenum Press, New York, pp. 69–132.

DeLuca, H.F. (1988) Vitamin D and its metabolites. In *Modern Nutrition in Health and Disease* (eds M.E. Shils and V.R. Young), Lea & Febiger, Philadelphia, pp. 313–27.

de Vries, E.J., Zeeman, J., Esser, R.J.E. *et al.* (1979) Analysis of fat-soluble vitamins. XXI. High pressure liquid chromatographic assay methods for vitamin D in vitamin D concentrates. *J. Ass. Off. Analyt. Chem.*, **62**, 129–35.

Ducland, S., Pedersen, J.I., Helgerud, P. and Drevon, C.A. (1982) Transport of vitamin D_3 from rat intestine. *J. Biol. Chem.*, **257**, 146–50.

Dueland, S., Helgerud, P., Pedersen, J.I. *et al.* (1983a) Plasma clearance, transfer, and distribution of vitamin D_3 from intestinal lymph. *Am. J. Physiol.*, **245**, E326–31.

Dueland, S., Pedersen, J.I., Helgerud, P. and Drevon, C.A. (1983b) Absorption, distribution, and transport of vitamin D_3 and 25-hydroxyvitamin D_3 in the rat. *Am. J. Physiol.*, **245**, E463–7.

Fraser, D.R. (1983) The physiological economy of vitamin D. *Lancet*, **I** (No. 8331), 969–72.

Gascon-Barré, M. (1982) Biliary excretion of [^{3}H]-25-hydroxyvitamin D_3 in the vitamin D-depleted rat. *Am. J. Physiol.*, **242**, G522–32.

Gascon-Barré, M. (1986) Is there any physiological significance to the entero-hepatic circulation of vitamin D sterols? *J. Am. College Nutr.*, **5**, 317–24.

Gascon-Barré, M. and Joly, J.-G. (1981) The biliary excretion of [^{3}H]-25-hydroxy-vitamin D_3 following chronic ethanol administration in the rat. *Life Sci.*, **28**, 279–86.

Grady, L.T. and Thakker, K.D. (1980) Stability of solid drugs: degradation of ergocalciferol (vitamin D_2) and cholecalciferol (vitamin D_3) at high humidities and elevated temperatures. *J. Pharm. Sci.*, **69**, 1099–102.

Haddad, J.G. Jr and Walgate, J. (1976) 25-Hydroxyvitamin D transport in human plasma. Isolation and partial characterization of calcifidol-binding protein. *J. Biol. Chem.*, **251**, 4803–9.

Hanewald, K.H., Mulder, F.J. and Keuning, K.J. (1968) Thin-layer chromato-graphic assay of vitamin D in high-potency preparations. *J. Pharm. Sci.*, **57**, 1308–12.

Hasegawa, H. (1992) Vitamin D determination using high-performance liquid chromatography with internal standard-redox mode electrochemical detection and its application to medical nutritional products. *J. Chromat.*, **605**, 215–20.

Hollis, B.W. and Frank, N.E. (1986) Quantitation of vitamin D_2, vitamin D_3, 25-hydroxyvitamin D_2, and 25-hydroxyvitamin D_3 in human milk. *Meth. Enzy-mol.*, **123**, Part H, 167–76.

Holmes, R.P. and Kummerow, F.A. (1983) The relationship of adequate and excessive intake of vitamin D to health and disease. *J. Am. College Nutr.*, **2**, 173–99.

Homberg, E. (1993) Vitamin D determination in margarine. *Fat Sci. Technol.*, **95**, 181–5 (in German).

Horst, R.L., Napoli, J.L. and Littledike, E.T. (1982) Discrimination in the metab-olism of orally dosed ergocalciferol and cholecalciferol by the pig, rat and chick. *Biochem. J.*, **204**, 185–9.

Indyk, H. and Woollard, D.C. (1984) The determination of vitamin D in milk powders by high performance liquid chromatography. *NZ J. Dairy Sci. Technol.*, **19**, 19–30.

Indyk, H. and Woollard, D.C. (1985a) The determination of vitamin D in supple-mented milk powders by HPLC. II. Incorporation of internal standard. *NZ J. Dairy Sci. Technol.*, **20**, 19–28.

Indyk, H. and Woollard, D.C. (1985b) The determination of vitamin D in fortified milk powders and infant formulas by HPLC. *J. Micronutr. Anal.*, **1**, 121–41.

Indyk, H., Littlejohn, V. and Woollard, D.C. (1996) Stability of vitamin D_3 during spray-drying of milk. *Food Chem.*, **57**, 283–6.

Jackson, P.A., Shelton, C.J. and Frier, P.J. (1982) High performance liquid chro-matographic determination of vitamin D_3 in foods with particular reference to eggs. *Analyst, Lond.*, **107**, 1363–9.

Johnsson, H., Halén, B., Hessel, H. *et al.* (1989) Determination of vitamin D_3 in margarines, oils and other supplemented food products using HPLC. *Int. J. Vitam. Nutr. Res.*, **59**, 262–8.

Jones, G. and DeLuca, H.F. (1975) High-pressure liquid chromatography: separation of the metabolites of vitamins D_2 and D_3 on small-particle silica columns. *J. Lipid Res.*, **16**, 448–53.

Keverling Buisman, J.A., Hanewald, K.H., Mulder, F.J. *et al.* (1968) Evaluation of the effect of isomerization on the chemical and biological assay of vitamin D. *J. Pharm. Sci.*, **57**, 1326–9.

Kläui, H.M., Hausheer, W. and Huschke, G. (1970) Technological aspects of the use of fat-soluble vitamins and carotenoids and of the development of stabilized marketable forms. In *Fat-soluble Vitamins* (ed. R.A. Morton), Pergamon Press, New York, pp. 113–59.

Kobayashi, T., Okano, T. and Takeuchi, A. (1986) The determination of vitamin D in foods and feeds using high-performance liquid chromatography. *J. Micronutr. Anal.*, **2**, 1–24.

Kodicek, E. and Lawson, D.E.M. (1967) Vitamin D. In *The Vitamins. Chemistry, Physiology, Pathology, Methods*, 2nd edn, Vol. VI (eds P. György and W.N. Pearson), Academic Press, New York, pp. 211–44.

Konings, E.J.M. (1994) Estimation of vitamin D in baby foods with liquid chromatography. *Neth. Milk Dairy J.*, **48**, 31–9.

Koshy, K.T. and VanDerSlik, A.L. (1977) High-performance liquid chromatographic method for the determination of 25-hydroxycholecalciferol in the bovine liver, kidney, and muscle. *J. Agric. Food Chem.*, **25**, 1246–9.

Koshy, K.T. and VanDerSlik, A.L. (1979) 25-Hydroxycholecalciferol in cow milk as determined by high-performance liquid chromatography. *J. Agric. Food Chem.*, **27**, 650–2.

Krampitz, G. (1980) *Vitamin D in Animal Nutrition*, F. Hofmann-La Roche and Co. Ltd, Basle, Switzerland.

Kurmann, A. and Indyk, H. (1994) The endogenous vitamin D content of bovine milk: influence of season. *Food Chem.*, **50**, 75–81.

Kutsky, R.J. (1973) *Handbook of Vitamins and Hormones*, Van Nostrand Reinhold, New York.

Landen, W.O. Jr (1985) Liquid chromatographic determination of vitamins D_2 and D_3 in fortified milk and infant formulas. *J. Ass. Off. Analyt. Chem.*, **68**, 183–7.

Lawson, E. (1985) Vitamin D. In *Fat-Soluble Vitamins. Their Biochemistry and Applications* (ed. A.T. Diplock), Heinemann, London, pp. 76–153.

Le Boulch, N., Cancela, L. and Miravet, L. (1982) Cholecalciferol sulfate identification in human milk by HPLC. *Steroids*, **39**, 391–8.

Maislos, M., Silver, J. and Fainaru, M. (1981) Intestinal absorption of vitamin D sterols: differential absorption into lymph and portal blood in the rat. *Gastro enterology*, **80**, 1528–34.

Margarine Regulations (1967) *Statutory Instrument, No. 1867*, as amended, H.M. Stationery Office, London.

Mattila, P., Piironen, V., Uusi-Rauva, E. and Koivistoinen, P. (1993) Determination of 25-hydroxycholecalciferol content in egg yolk by HPLC. *J. Food Comp. Anal.*, **6**, 250–5.

Mattila, P.H., Piironen, V.I., Uusi-Rauva, E.J. and Koivistoinen, P.E. (1995) Contents of cholecalciferol, ergocalciferol, and their 25-hydroxylated metabolites in milk products and raw meat and liver as determined by HPLC. *J. Agric Food Chem.*, **43**, 2394–9.

Miller, D.R. and Hayes, K.C. (1982) Vitamin excess and toxicity. In *Nutritional Toxicology*, Vol. I (ed. J.N. Hathcock), Academic Press, New York, pp. 81–133.

Mulder, F.J., de Vries, E.J. and Borsje, B. (1971) Chemical analysis of vitamin D in concentrates and its problems. 12. Analysis of fat-soluble vitamins. *J. Ass. Off. Analyt. Chem.*, **54**, 1168–74.

Muniz, J.F., Wehr, C.T. and Wehr, H.M. (1982) Reverse phase liquid chromatographic determination of vitamins D₂ and D₃ in milk. *J. Ass. Off. Analyt. Chem.*, **65**, 791–7.

Nair, P.P. (1966) Quantitative methods for the study of vitamin D. In *Advances in Lipid Research*, Vol. IV (eds R. Paoletti and D. Kritchevsky), Academic Press, New York, pp. 227–56.

Need, A.G., Morris, H.A., Horowitz, M. and Nordin, B.E.C. (1993) Effects of skin thickness, age, body fat, and sunlight on serum 25-hydroxyvitamin D. *Am. J. Clin. Nutr.*, **58**, 882–5.

Okano, T., Kuroda, E., Nakao, H. *et al.* (1986) Lack of evidence for existence of vitamin D and 25-hydroxyvitamin D sulfates in human breast and cow's milk. *J. Nutr. Sci. Vitaminol.*, **32**, 449–62.

O'Keefe, S.F. and Murphy, P.A. (1988) Rapid determination of vitamin D in fortified skim milk. *J. Chromat.*, **445**, 305–9.

Omaye, S.T. (1984) Safety of megavitamin therapy. In *Nutritional and Toxicological Aspects of Food Safety* (ed. M. Friedman), Plenum Press, New York, pp. 169–203.

Onisko, B.L., Esvelt, R.P., Schnoes, H.K. and DeLuca, H.F. (1980) Metabolites of 1α,25-dihydroxyvitamin D₃ in rat bile. *Biochemistry*, **19**, 4124–30.

Parrish, D.B. (1979) Determination of vitamin D in foods: a review. *CRC Crit. Rev. Food Sci. Nutr.*, **12**, 29–57.

Passmore, R. and Eastwood, M.A. (1986) *Davidson and Passmore Human Nutrition and Dietetics*, 8th edn, Churchill Livingstone, New York.

Reed, J.S., Meredith, S.C., Nemchausky, B.A. *et al.* (1980) Bone disease in primary biliary cirrhosis: reversal of osteomalacia with oral 25-hydroxyvitamin D. *Gastroenterology*, **78**, 512–7.

Reeve, L.E., Deluca, H.F. and Schnoes, H.K. (1981) Synthesis and biological activity of vitamin D₃-sulfate. *J. Biol. Chem.*, **256**, 823–6.

Reeve, L.E., Jorgensen, N.A. and DeLuca, H.F. (1982) Vitamin D compounds in cows' milk. *J. Nutr.* **112**, 667–72.

Renken, S.A. and Warthesen, J.J. (1993) Vitamin D stability in milk. *J. Food Sci.*, **58**, 552–6, 566.

Reynolds, S.L. and Judd, H.J. (1984) Rapid procedure for the determination of vitamins A and D in fortified skimmed milk powder using high-performance liquid chromatography. *Analyst, Lond.*, **109**, 489–92.

Rychener, M. and Walter, P. (1985) A simplified and improved determination of vitamin D in fat, oil and margarine by HPLC. *Mitt. Gebiete Lebensm. Hyg.*, **76**, 112–24 (in German).

Schachter, D., Finkelstein, J.D. and Kowarski, S. (1964) Metabolism of vitamin D. I. Preparation of radioactive vitamin D and its intestinal absorption in the rat. *J. Clin. Invest.*, **43**, 787–96.

Sertl, D.C. and Molitor, B.E. (1985) Liquid chromatographic determination of vitamin D in milk and infant formula. *J. Ass. Off. Analyt. Chem.*, **68**, 177–82.

Sitrin, M.D. and Bengoa, J.M. (1987) Intestinal absorption of cholecalciferol and 25-hydroxycholecalciferol in chronic cholestatic liver disease. *Am. J. Clin. Nutr.*, **46**, 1011–5.

Sitrin, M.D., Pollack, K.L., Bolt, M.J.G. and Rosenberg, I.H. (1982) Comparison of vitamin D and 25-hydroxyvitamin D absorption in the rat. *Am. J. Physiol.*, **242**, G326–32.

Sivell, L.M., Wenlock, R.W. and Jackson, P.A. (1982) Determination of vitamin D and retinoid activity in eggs by HPLC. *Human Nutr. Appl. Nutr.*, **36A**, 430–7.

Sliva, M.G. and Sanders, J.K. (1996) Vitamin D in infant formula and enteral products by liquid chromatography: collaborative study. *J. AOAC Int.*, **79**, 73–80.

Sliva, M.G., Green, A.E., Sanders, J.K. *et al.* (1992) Reversed-phase liquid chroma-tographic determination of vitamin D in infant formulas and enteral nu-tritionals. *J. Ass. Off. Analyt. Chem.*, **75**, 566–71.

Stamp, T.C.B. (1974) Intestinal absorption of 25-hydroxycholecalciferol. *Lancet*, **II** (No. 7873) 121–3.

Thompson, J.N. and Plouffe, L. (1993) Determination of cholecalciferol in meat and fat from livestock fed normal and excessive quantities of vitamin D. *Food Chem.*, **46**, 313–18.

Thompson, J.N., Maxwell, W.B. and L'Abbe, M. (1977) High pressure liquid chromatographic determination of vitamin D in fortified milk. *J. Ass. Off. Analyt. Chem.*, **60**, 998–1002.

Thompson, J.N., Hatina, G., Maxwell, W.B. and Duval, S. (1982) High perform-ance liquid chromatographic determination of vitamin D in fortified milks, margarine and infant formulas. *J. Ass. Off. Analyt. Chem.*, **65**, 624–31.

van den Berg, H. (1997) Bioavailability of vitamin D. *Eur. J. Clin. Nutr.*, **51**, Suppl. 1, S76–9.

Vanhaelen-Fastré, R. and Vanhaelen, M. (1981) Separation and determination of the D vitamins by HPLC. In *Steroid Analysis by HPLC. Recent Applications* (ed. M.P. Kautsky), Marcel Dekker, Inc., New York, pp. 173–251.

Woollard, D.C. (1987) Quality control of the fat-soluble vitamins in the New Zealand dairy industry. *Food Technol. Aust.*, **39**, 250–3.

5

Vitamin E

5.1 INTRODUCTION

In 1922, H.M. Evans and Katherine S. Bishop reported from the University of California that pregnant rats fed on formulated diets containing all of the then known nutritional factors did not reach term. Death and resorption of the foetuses were prevented by supplementation of the formulated diet with fresh lettuce, thus implying the existence of a hitherto unknown nutritional factor. By 1924, these initial studies had been independently verified by Sure, who recommended that the unknown factor be officially designated as vitamin E. The active fat-soluble substance was isolated from wheat germ by Evans in 1936. On recognition that more than one compound possessed vitamin E activity, the vitamin was given the generic name, tocopherol, which is derived from the Greek language meaning 'to bring forth in childbirth'. By 1944, it was found that a multiplicity of deficiency syndromes occurs in animals deprived of vitamin E. The structure of the vitamin was elucidated by Fernholz in 1937 and synthesis was accomplished by Karrer in the following year.

Vitamin E is represented by a family of structurally related compounds (vitamers) which show pronounced quantitative differences in biological activity. Eight of these vitamers are known to occur in nature, having been isolated from vegetable oils and other plant materials. In addition, at least three others have been synthesized in the laboratory, and a total of 14 vitamers is theoretically possible.

5.2 CHEMICAL STRUCTURE AND NOMENCLATURE

The eight naturally occurring vitamers of vitamin E comprise four tocopherols and four corresponding tocotrienols, which are designated as alpha- (α), beta- (β), gamma- (γ) and delta- (δ) according to the number and position of substituent methyl groups on the chromanol ring (Table 5.1). The tocopherols are methyl-substituted derivatives of tocol, which has a saturated side chain and is systematically named 2-methyl-2-(4′,8′,12′-trimethyltridecyl)chroman-6-ol. The tocotrienols are identical to the tocopherols but with three double bonds in the side chain; they are systematically named 2-methyl-2-(4′,8′,12′-trimethyltrideca-3′,7′,11′-trienyl)chroman-6-ol.

The tocol molecule exhibits optical isomerism attributable to the three asymmetric carbon atoms at positions 2, 4′ and 8′, thus the tocopherols can exist as one of eight possible diastereoisomers, i.e. four enantiomeric pairs (racemates). The tocotrienols possess only one centre of asymmetry at position 2, in addition to sites of geometrical isomerism at positions 3′ and 7′ in the unsaturated side chain. The stereochemical structures of tocol and its unsaturated analogue are depicted in Figure 5.1. The *RS* system of asymmetric configuration is used to specify the chirality of vitamin E compounds in accordance with the present IUPAC rules (IUPAC-IUB, 1974). In this system of nomenclature, a methyl group attached to an asymmetric carbon atom

Table 5.1 Designation of tocopherols and tocotrienols (Pennock, Hemming and Kerr, 1964)

Substitution	Tocopherol (T)	Tocotrienol (T3)
5,7,8-Trimethyl	α-T	α-T3 (formerly ζ-(zeta) tocopherol)
5,8-Dimethyl	β-T	β-T3 (formerly ϵ-(epsilon) tocopherol)
7,8-Dimethyl	γ-T	γ-T3 (formerly η-(eta) tocopherol)
8-Methyl	δ-T	δ-T3

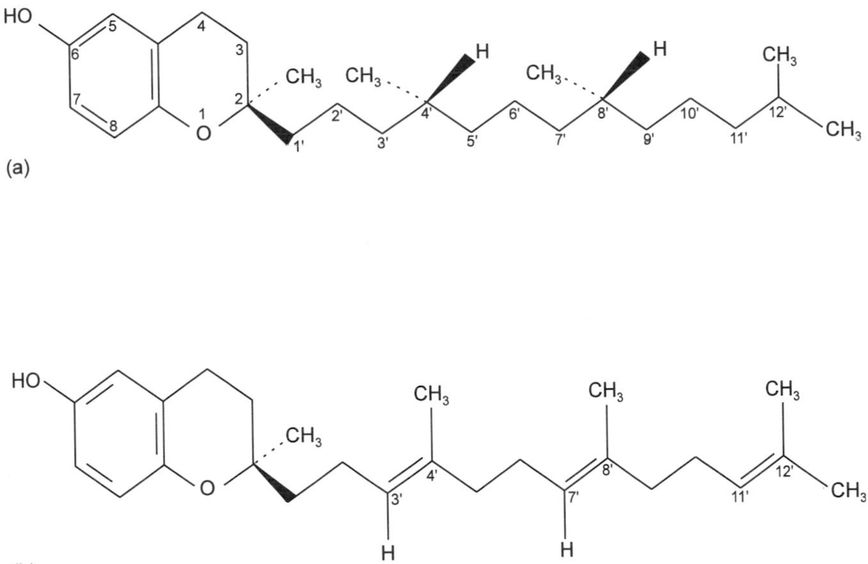

(a)

(b)

Figure 5.1 Stereochemical structures of tocol and tocotrienol. (a) *RRR*-tocol; (b) 2*R*, 3′-*trans*, 7′-*trans*-tocotrienol.

by a dotted line indicates an *R*-configuration, and by a solid wedged line an *S*-configuration. If the groups are attached by normal solid lines, the stereochemical designation is unspecified. In nature, the tocopherols exist exclusively as their 2*R*, 4′*R*, 8′*R* stereoisomers (*RRR*-forms), while the tocotrienols exist exclusively in the 2*R*, 3′-*trans*, 7′-*trans* configuration (Kasparek, 1980). The best known and most biologically active vitamer is *RRR*-α-tocopherol ($C_{29}H_{50}O_2$; MW = 430.72), more commonly known as *d*-α tocopherol.

The term 'vitamin E' should be used as the generic descriptor for all tocol and tocotrienol derivatives that exhibit qualitatively the biological activity of α-tocopherol. The term 'tocopherol' correctly refers to the methyl-substituted derivatives of tocol and is not synonymous with the term 'vitamin E'. The tocopherols and tocotrienols may be referred to collectively as tocochromanols.

The nomenclature for some α-tocopherols is given in Table 5.2. *RRR*-α-tocopherol is obtained on a commercial scale by extraction from vegetable oils. Since it is not isolated without chemical processing, it cannot legally be called 'natural', but it can be described as 'derived from natural sources'. Totally synthetic α-tocopherol is produced by the condensation of trimethylhydroquinone with synthetic phytol or isophytol. This method of synthesis results in all-racemic 2*RS*, 4′*RS*, 8′*RS*-α-tocopherol

Vitamin E

Table 5.2 Nomenclature for some α-tocopherols (IUPAC-IUB, 1974)

Configuration	Designated name	Trivial name	Description
2R, 4′R, 8′R	RRR-α-tocopherol	d-α-tocopherol	The only isomer of α-tocopherol found in nature
2S, 4′R, 8′R	2-epi-α-tocopherol	l-α-tocopherol	C-2 epimer of RRR form
2R, 4′R, 8′R and 2S, 4′R, 8′R mixture in equal proportions	2-ambo-α-tocopherol		Semisynthetic (produced from natural phytol)
2RS, 4′RS, 8′RS (mixture of four enantiomeric pairs)	all-rac-α-tocopherol	dl-α-tocopherol	Totally synthetic (produced from synthetic phytol or isophytol)

(*all-rac-α*-tocopherol), which is a mixture of all eight possible diastereo-isomers in virtually equal proportions (Weiser and Vecchi, 1981; Scott *et al.*, 1982). The four enantiomeric pairs are *RRR/SSS, RRS/SSR, RSS/SRR,* and *RSR/SRS.*

The principal commercially available forms of vitamin E used in the food, feed and pharmaceutical industries are the acetate esters of *RRR-α*-tocopherol and *all-rac-α*-tocopherol. Another commercial preparation is the hydrogen succinate of *RRR-α*-tocopherol. In commercial circles, *all-rac-α*-tocopheryl acetate ($C_{31}H_{52}O_3$; MW = 472.76) is commonly referred to by the trivial name of *dl-α*-tocopheryl acetate. Rather than discarding the *dl* system of nomenclature completely, the IUPAC-IUB Commission on Biochemical Nomenclature (1974) has proposed that the *d* and *dl* prefixes (lower-case italics) used to describe *RRR-α*-tocopheryl acetate and *all-rac-α*-tocopheryl acetate, respectively, should be slightly altered to [d] and [dl] (lower-case romans in brackets).

Another compound, produced by the condensation of trimethylhydro-quinone with natural phytol, is 2-*ambo-α*-tocopherol, which is an equi-molar mixture of the *RRR* and *SRR* epimers (Weiser and Vecchi, 1981). *SRR-α*-tocopherol is designated as 2-*epi-α*-tocopherol, but is more commonly known as *l-α*-tocopherol. Samples of 2-*ambo-α*-tocopherol obtained from laboratory-scale synthesis are available for research purposes, but this compound is not a commercial source of vitamin E.

The *RS* system of biological nomenclature is essential for describing structural formulae with stereochemical accuracy, but it is not necessary for communications among nutritionists and clinicians, particularly since analytical methods used commonly do not distinguish the stereoisomers. In this text, therefore, α-, β-, γ- and δ-tocopherols will be referred to without their stereochemical designation.

5.3 BIOPOTENCY

The currently recognized biological activities, expressed in IU/mg, of commercial forms of vitamin E are given in Table 5.3. The biological equivalency of 1.36 for *RRR*-α-tocopheryl acetate was determined experimentally by Harris and Ludwig (1949) using prevention of foetal resorption during gestation in rats as the criterion of physiological response. This value, 1.36, is based on 15 bioassays involving 629 rats, run over a two-year period with confidence limits of 1.22–1.50. The vitamin E International Standard used in these studies was actually 2-*ambo*-α-tocopheryl acetate and not *all-rac*-α-tocopheryl acetate, the isomeric composition of the standard not being correctly identified at that time (Ames, 1979). Weiser and Vecchi (1981) compared *all-rac*-α-tocopheryl acetate with 2-*ambo*-α-tocopheryl acetate and found no statistically significant difference in biopotency (a potency ratio of 1 to 1.10). Direct comparisons of *RRR*-α-tocopheryl acetate with authentic *all-rac*-α-tocopheryl acetate using the resorption–gestation assay yielded potency ratios of exactly 1.36 (Leth and Søndergaard, 1977) and 1.48 (Weiser and Vecchi, 1981). The validity of the 1.36 potency ratio between *RRR*-α-tocopheryl acetate and *all-rac*-α-tocopheryl acetate has been re-evaluated extensively using other biological parameters besides the rat resorption–gestation assay. Marusich, Ackerman and Bauernfeind (1967) examined published experimental data obtained from resorption–gestation and erythrocyte haemolysis in the rat; liver storage, plasma levels, muscular dystrophy and encephalomalacia in the chick; and creatinuria in the rabbit. The overall average biological equivalency for *RRR*-α-tocopheryl acetate was 1.41 (range 1.21–1.66), which is well within the limits originally obtained by Harris and Ludwig (1949). The combined experimental work supports the accepted biological activity of 1.36 IU/mg for *RRR*-α-tocopheryl acetate.

The potency values of 1.10, 1.49 and 1.21 for the other commercial forms of vitamin E listed in Table 5.3 have been derived from the

Table 5.3 Biological activities of commercial forms of vitamin E

Form	IU/mg
All-rac-α-tocopheryl acetate (*dl*-α-tocopheryl acetate)	1.00
RRR-α-tocopheryl acetate (*d*-α-tocopheryl acetate)	1.36[a]
All-rac-α-tocopherol	1.10[b]
RRR-α-tocopherol	1.49[b]
RRR-α-tocopheryl succinate	1.21[b]

[a] Determined experimentally using the rat resorption–gestation assay (Harris and Ludwig, 1949).
[b] Calculated from the value of the corresponding acetate by multiplication with the ratio of the molecular weights.

Table 5.4 Biological activities of three commercial forms of vitamin E as determined using three different rat bioassays (Leth and Søndergaard, 1983)

Form	Biological activity relative to all-rac-α-tocopheryl acetate		
	Resorption–gestation	Erythrocyte haemolysis	Liver storage
All-rac-α-tocopherol	0.79[a]	0.85	1.03
RRR-α-tocopherol	1.06[a]	1.13	1.32
RRR-α-tocopheryl succinate	0.90[b]	–	–

[a] Doses given in vegetable oil.
[b] Doses given in absolute alcohol.

corresponding acetate by multiplication with the ratio of the molecular weight. The assumption that the esters of α-tocopherol have, on a molar basis, the same biological activity as the nonesterified α-tocopherol (Brubacher and Wiss, 1972) has been generally accepted for many years. However, Leth and Søndergaard (1983), using the resorption–gestation assay, obtained biological activities of 0.79, 1.06 and 0.90 for *all-rac*-α-tocopherol, *RRR*-α-tocopherol and *RRR*-α-tocopheryl succinate, respectively, relative to *all-rac*-α-tocopheryl acetate; these activities are considerably lower than the corresponding expected values of 1.10, 1.49 and 1.21. Biological activities obtained using two additional rat bioassays were also reported (Table 5.4) and show a steady increase from resorption–gestation via erythrocyte haemolysis to liver storage bioassays. This trend corresponds rather well to the complexity of the physiological processes involved in the three different types of bioassay. Presumably, the biopotency is smallest using the resorption–gestation assay because little vitamin E is left when the subcellular level is reached. On the basis of the resorption–gestation assay data, Leth and Søndergaard (1983) recommended the use of a potency ratio of 0.80 for *all-rac*-α-tocopherol and 1.0 for *RRR*-α-tocopherol relative to *all-rac*-α-tocopheryl acetate, instead of the respective ratios of 1.10 and 1.49.

Table 5.5 Relative percentage biopotencies of vitamin E compounds by the rat resorption–gestation assay

Compound[a]	Relative biopotency (%)	Reference
α-Tocopherol	100	Joffe and Harris (1943)
β-Tocopherol	40	Joffe and Harris (1943)
γ-Tocopherol	8	Joffe and Harris (1943)
δ-Tocopherol	1	Stern *et al.* (1947)
α-Tocotrienol	21	Bunyan *et al.* (1961)
β-Tocotrienol	4	Bunyan *et al.* (1961)

[a] *RRR* forms.

The biopotencies of the various tocopherols and tocotrienols with respect to different methods of evaluation have been discussed by Century and Horwitt (1965). Table 5.5 lists the relative biopotencies evaluated by the resorption–gestation assay.

The naturally occurring *RRR* form of α-tocopherol has been shown to be 1.3 times more potent than synthetic *all-rac-α*-tocopherol in the rat foetal resorption assay (Weiser and Vecchi, 1982). Each of the eight stereoisomers present in *all-rac-α*-tocopherol possess different potencies, the *RRR* isomer being the most potent. The chirality of the C-2 carbon atom has the greatest influence upon potency: the *SSS* isomer is about one-third as potent as the *RRR* isomer.

5.4 DEFICIENCY SYNDROMES IN ANIMALS AND HUMANS

Whilst clinical vitamin E deficiency is difficult to demonstrate in humans, a deficiency in animals results in a variety of pathological conditions that affect the muscular, cardiovascular, reproductive and central nervous systems as well as the liver, kidney and erythrocytes (red blood cells). There is a marked difference between animal species in their susceptibility to different deficiency disorders. A complex biochemical interrelationship exists between vitamin E and the trace element selenium. Unsaturated fat, sulphur-containing amino acids and synthetic fat-soluble antioxidants are also implicated in some disorders. Consequently, in order to experimentally induce a particular deficiency syndrome in a given species, it is usually necessary to adjust the balance of these nutrients in the diet. The most extensively studied deficiency syndromes are listed in Table 5.6.

Animals

Foetal resorption
In female rats deprived of vitamin E all reproductive events are normal up to implantation of the fertilized ova. Several days later, however, the developing foetus shows abnormalities followed by intra-uterine death, rapid autolysis and resorption. A defect in the foetal blood vessels may be the primary event leading to death of the foetus (Nelson, 1980). This disease can be prevented by administering an adequate dose of vitamin E as late as the tenth day of pregnancy. The synthetic antioxidant N, N^1-diphenyl-*p*-phenylenediamine (DPPD) is at least as effective as α-tocopherol in preventing foetal resorption, but ethoxyquin, which readily prevents encephalomalacia in chicks, is inactive (Draper *et al.*, 1964). Selenium compounds have no effect on foetal resorption in rats.

Table 5.6 Some vitamin E deficiency diseases (from Scott, 1980)

Disease	Experimental animal	Tissue affected	Severity dependent on dietary PUFA	Prevented by			
				Vitamin E	Se	Synthetic anti-oxidants	Cystine amino acids
Foetal resorption	Female rat	Vascular system of foetus	Yes	Yes	No	Yes	No
Erythrocyte haemolysis	Rat, chick, human (premature infant)	Erythrocytes	Yes	Yes	No	Yes	No
Encephalo-malacia	Chick	Cerebellum	Yes	Yes	No	Yes	No
Exudative diathesis	Chick	Vascular system	No	Yes	Yes	No	No
Liver necrosis	Rat, pig	Liver	No	Yes	Yes	No	No
Testicular atrophy	Male rat	Testes	No	Yes	No	No	No
Necrotizing myopathy	Rabbit	Skeletal muscle	No	Yes	No	?	No
	Chick	Skeletal muscle	–[a]	Yes	No	No	Yes

[a] A low level (0.5%) of linoleic acid was necessary to produce dystrophy; higher levels did not increase the amount of vitamin E required for prevention.

Erythrocyte haemolysis

Erythrocyte plasma membranes are particularly vulnerable to lipid per-oxidation because of their direct exposure to molecular oxygen and the presence of haemoproteins which are catalysts of peroxidation. Erythro-cytes isolated from blood samples of vitamin E-depleted rats exhibit spontaneous haemolysis when added to dilute solutions of dialuric acid, whereas erythrocytes of rats receiving vitamin E are resistant to this haemolysis. This early manifestation of vitamin E deficiency can be prevented by certain synthetic antioxidants administered to the animal or added to the cell suspension *in vitro* as well as by vitamin E. Selenium compounds have no effect on erythrocyte haemolysis.

Encephalomalacia

This nutritional disorder occurs in growing chicks fed vitamin E-deficient diets containing adequate amounts of selenium for the prevention of exudative diathesis and sufficient methionine or cystine for the preven-tion of necrotizing myopathy. Encephalomalacia is manifested by lesions of the cerebellum, the part of the brain concerned with coordination of movement. The cerebellum is softened, swollen and oedematous with minute haemorrhages on the surface and greenish-yellow necrotic areas. The necrosis may be the result of thrombosis in the capillaries. Once established, the lesions are irreversible. The main symptoms are ataxia of gait and stance, backward or downward retraction of the head, tremors, spasms of the limb muscles, and eventually prostration, stupor and death within a few hours. The incidence and severity of the disease are markedly increased with increasing levels of linoleic acid in the diet. Low concentrations of synthetic antioxidants such as DPPD and ethoxyquin in the diet readily prevent encephalomalacia, but selenium has no effect.

Exudative diathesis

This is a vascular disease of chicks which develops as a result of feeding diets that are low in both vitamin E and selenium. The disease can be induced for experimental purposes by feeding diets based on Torula yeast, which is low in both micronutrients and contains substantial amounts of unsaturated fatty acids. The most obvious manifestation is a massive accumulation of a greenish fluid under the skin of the breast and abdomen. Internally, the oedema extends to the muscles and many organs, including the heart and lungs. The oedema is the result of a leakage of plasma from the capillaries caused by an increased permeability of the capillary walls. The disease can be prevented by administration of either vitamin E or selenium, provided that the selenium deficiency is not too severe (Thompson and Scott, 1969). A severe deficiency of selenium causes degeneration of the exocrine component of the pancreas and consequent impairment of dietary lipid absorption, which will affect the absorption of vitamin E (Thompson and Scott, 1970). In this event, extremely high doses of vitamin E are required to prevent exudative diathesis. Some synthetic antioxidants, including DPPD and ethoxyquin, are also effective, but only at concentrations distinctly greater than those required to prevent encephalomalacia.

Liver necrosis

Necrotic liver degeneration develops in weanling rats after commencement of a diet based on Torula yeast, which is deficient in both vitamin E and selenium and low in sulphur-containing amino acids. Necrosis is preceded by degeneration of the sinusoidal cellular plasma membrane and lipid peroxidation has been detected late in the progress of the disease. The onset of necrosis is delayed by cystine, which appears to have a sparing action on the amount of vitamin E or selenium required to prevent the disease.

Testicular atrophy

In male rats depleted of vitamin E from early life there is no testicular injury until the onset of sexual maturity, when a progressive degeneration of the germinal epithelium of the seminiferous tubules occurs and the testes atrophy. The resultant sterility does not respond to vitamin E and is truly permanent.

Necrotizing myopathy

This disease is manifested as a progressive muscular weakness which affects the skeletal muscles of many vertebrate species. It was originally called nutritional muscular dystrophy, but this term suggests an aetiological relationship between the myopathy of vitamin E deficiency and human muscular dystrophy. Although many of the pathological lesions are similar in these two diseases, human muscular dystrophy is genetically determined and does not respond to vitamin E treatment.

Necrotizing myopathy is characterized histologically by marked variation in the cross-sectional diameter of the muscle fibres, segmental fragmentation with interstitial oedema and necrosis and, in the later stages, extensive replacement of muscle tissue by connective tissue. The disease can be detected in its early stages by an increased excretion of creatine in the urine (creatinuria), which is the result of a loss of creatine from the affected muscles. Creatine excretion is often expressed as the creatine : creatinine ratio, the excretion of creatinine being relatively constant on a body weight basis.

Necrotizing myopathy in rabbits, guinea pigs, rats and monkeys responds primarily to vitamin E. Selenium is not capable of completely replacing vitamin E in these species, although it does reduce the vitamin E requirement. The myopathy, as studied in the chick, does not respond to dietary synthetic antioxidants at levels several times those needed to prevent encephalomalacia. The disease is induced in the chick when the dietary vitamin E is accompanied by a deficiency in the sulphur-containing amino acids, methionine and cystine. Approximately 0.5% of dietary linoleic acid (but not linolenic acid) is necessary to produce myopathy. Concentrations above 0.5% linoleic acid do not increase the amount of vitamin E required for prevention. The chick appears to be unique in that the myopathy can be prevented in the absence of vitamin E by supplementing the diet with cystine or methionine. Cystine is about twice as effective as methionine on an equal sulphur basis (Scott, 1970).

Humans

Aside from instances of severe malnutrition or fat malabsorption, a vitamin E deficiency is rare in humans. This is due to the occurrence of the vitamin in a wide variety of foods, its widespread storage distribution throughout the body tissues, and the consequent extended period required for depletion.

No recognizable deficiency syndrome has been demonstrated in adult humans. However, various symptoms have been reported in infants; these include haemolytic anaemia, oedema, colic and failure to thrive (Smith *et al.*, 1971). In children with vitamin E inadequacy a low blood tocopherol level is accompanied by creatinuria (Horwitt, 1965). Evidence of a vitamin E deficiency-related myopathy was obtained in an adult male patient with pancreatic insufficiency, whose symptoms of pronounced muscle weakness and creatinuria were markedly improved by administering α-tocopherol three times per day. When tocopherol administration was halted, the symptoms returned and could again be reversed by giving this vitamin (Vester and Williams, 1963).

It is well documented that a diet rich in polyunsaturated fat, but which does not contain a correspondingly high amount of vitamin E, induces deficiency signs in animals. In a long-term human study (the Elgin project), adult males received a diet in which about half of the fat content was composed of vitamin E-stripped lard. After 30 months this fraction of the fat content was replaced by stripped corn oil and 9 months later the amount of stripped corn oil was doubled. No manifestations of anaemia were observed and it was not until the 72nd month that a well controlled study of erythrocyte survival was performed. The data obtained showed that the erythrocytes of the depleted subjects were being destroyed at a rate about 8–10% faster than in the subjects in the control groups. The experiment was terminated soon after these observations, but it is logical to assume that if the diet had been made more deficient, the pathology would have been more severe (Horwitt, 1976).

5.5 BIOCHEMICAL AND PHYSIOLOGICAL FUNCTIONS

The general view is that vitamin E functions as a biological antioxidant by protecting the vital phospholipids in cellular and subcellular membranes from peroxidative degeneration. A structural role in maintaining membrane stability and permeability has been proposed (Lucy, 1972). Extensive peroxidation may lead to rupture of lysosomal membranes and concomitant release of destructive enzymes (Chow, 1979). Damage to mitochondrial membranes prevents the production of antibodies and of other mechanisms required for normal recovery from pathological diseases and environmental stress (Scott, 1978).

The diverse vitamin E deficiency signs observed in animals are attributable to secondary effects of the widespread damage caused to the membranes of muscle and nerve cells by lipid peroxidation. The prevention of encephalomalacia in chicks by adding low concentrations of synthetic antioxidants to vitamin E-deficient diets strongly suggests that, in preventing this particular disease, vitamin E acts solely as an antioxidant and that such action is nonspecific. Certain other diseases, such as exudative diathesis and liver necrosis, can be prevented by supplementing the diet with either vitamin E or selenium, thus demonstrating a relationship between these two micronutrients. This interrelationship, as it is currently understood, can be explained by discussing some of the biochemical reactions involved in lipid peroxidation.

The vast majority of situations involving lipid peroxidation proceed through a free radical-mediated chain reaction initiated by the abstraction of a hydrogen atom from an unsaturated lipid by a reactive free radical, followed by a complex sequence of propagative reactions. In simple terms, carbon-centered alkyl radicals (R•), produced from a molecule of polyunsaturated fatty acid (PUFA) in an initiation reaction, react with oxygen to produce highly reactive peroxyl radicals (ROO•). In the peroxidation of linolenic acid, for example, these are pentadienyl radicals (Frankel, 1991). The peroxyl radicals then attack a molecule of PUFA (represented in the scheme below by RH) to produce the corresponding lipid hydroperoxide (ROOH). The new alkyl radical yields another peroxyl radical and thus a chain reaction is propagated.

$$PUFA \xrightarrow{\text{Initiation}} R\bullet$$

$$\longrightarrow R\bullet + O_2 \xrightarrow{\text{Propagation}} ROO\bullet$$

$$ROO\bullet + RH \longrightarrow ROOH + R\bullet$$

The first line of defence against the peroxidation is the removal of superoxide radicals ($\bullet O_2^-$) and hydrogen peroxide (H_2O_2), which are produced by normal metabolic processes and are precursors of extremely reactive hydroxyl radicals ($\bullet OH$).

$$H_2O_2 + \bullet O_2^- \longrightarrow \bullet OH + OH^- + O_2$$

These precursors are largely removed by the action of superoxide dismutase, which reduces superoxide to hydrogen peroxide, and by catalase and glutathione peroxidase, which independently reduce the hydrogen peroxide to water.

Other free radicals are capable of initiating the chain reaction and so chain-breaking antioxidants are needed as a second line of defence. The only major lipid-soluble, chain-breaking antioxidant that has been found in plasma, erythrocytes and tissues is vitamin E (Burton and Traber, 1990). α-Tocopherol (represented below by α-TOH) is the predominant E vitamer found in biological membranes and is anchored in the phospholipid bilayer by means of its phytyl side chain. α-Tocopherol donates a hydrogen atom from the phenolic hydroxyl group to a peroxyl radical and converts it to a hydroperoxide product. The tocopheroxyl radical (α-TO•) that is formed is sufficiently stabilized by molecular resonance to be unable to continue the chain reaction and, instead, is removed from the cycle by reaction with another peroxyl radical to form inactive non-radical products (Burton and Ingold, 1989).

$$\alpha\text{-TOH} + \text{ROO}\bullet \longrightarrow \alpha\text{-TO}\bullet + \text{ROOH}$$

$$\alpha\text{-TO}\bullet + \text{ROO}\bullet \longrightarrow \text{nonradical products}$$

The hydroperoxides that are produced during lipid peroxidation are normally reduced by the body to harmless hydroxy acids by the action of glutathione peroxidase. If the hydroperoxides are not removed, they can undergo metal-catalysed decomposition to form alkoxyl radicals ($\text{RO}\bullet$) which indirectly propagate the free radical chain through the formation of peroxyl radicals (Frankel, 1991).

$$\text{ROOH} + \text{M}^n \longrightarrow \text{RO}\bullet + \text{OH}^- + \text{M}^{n+1}$$

$$\text{ROOH} + \text{M}^{n+1} \longrightarrow \text{ROO}\bullet + \text{H}^+ + \text{M}^n$$

The implication of selenium in these activities lies in its role as a cofactor for glutathione peroxidase, which has the dual role of reducing hydrogen peroxide and organic hydroperoxides (Hoekstra, 1975). The reducing equivalents are provided by glutathione, which is a tripeptide of cystine, glutamic acid and glycine. A constant supply of dietary selenium, in the form of sodium selenate, is essential to maintain the activity of glutathione peroxidase. [Note that a non-selenium-dependent glutathione peroxidase also reduces hydrogen peroxide, but has no activity against organic hydroperoxides (Lawrence and Burk, 1976).]

Vitamin E and selenium can act independently in preventing exudative diathesis and liver necrosis, diseases in which synthetic antioxidants have little or no effect. Evidence presented by Scott (1978) and Diplock (1985) has led to the hypothesis that vitamin E is specifically and actively incorporated within biological membranes, where it stabilizes the lipoprotein structure. Without the vitamin E, the membranes have an abnormally high permeability such as is observed in the capillary plasma membrane in the chick, resulting in exudative diathesis. Selenium acts via glutathione peroxidase in destroying fatty acid hydroperoxides before they have an opportunity to damage the capillary plasma membranes to an extent that would increase their permeability.

The sulphur amino acids, methionine and cystine, exert a sparing effect on both selenium and vitamin E as shown by their ability to delay the onset of necrotizing myopathy in the chick. This effect is presumably due to an increased *de novo* synthesis of glutathione.

5.6 PHYSICOCHEMICAL PROPERTIES

Appearance and solubility

Tocopherols and tocotrienols in the pure state are pale yellow, nearly odourless, clear viscous oils which darken on exposure to oxygen.

α-Tocopheryl acetate is of similar appearance. The hydrogen succinate ester is a white, granular, nearly odourless powder.

The nonesterified tocopherols and tocotrienols are insoluble in water and readily soluble in alcohol, other organic solvents (including acetone, chloroform and ether) and in vegetable oils. The vitamin E acetates are less readily soluble in ethanol than the nonesterified vitamers.

Spectroscopic properties

Absorption

Tocopherols and tocotrienols exhibit relatively low intensities of UV absorption. Individual vitamers are characterized by a slightly different absorption maximum within the wavelength range of 292–298 nm in ethanol. Published $A_{1\,cm}^{1\%}$ values for the tocopherols dissolved in ethanol with their absorbance maxima are: α-tocopherol, 70–73.7 at 292 nm; β-tocopherol, 86–87 at 297 nm; γ-tocopherol, 90–93 at 298 nm; and δ-tocopherol, 91.2 at 298 nm (Association of Vitamin Chemists, Inc., 1966). The absorption intensity of α-tocopheryl acetate is lower still with an $A_{1\,cm}^{1\%}$ value of only 40–44 at the λ_{max} of 285.5 nm (Machlin, 1984).

The quality of an α-tocopherol standard can be easily checked by making a solution in *n*-hexane and measuring the UV absorbance at minimum (255 nm, A_{min}) and maximum (292 nm, A_{max}) wavelengths. If the quotient A_{min}/A_{max} exceeds 0.18, the standard contains less than 90% α-tocopherol (Balz, Schulte and Thier, 1996).

Fluorescence

Nonesterified tocopherols and tocotrienols exhibit strong native fluorescence. The excitation and emission maxima of α-tocopherol are 295 and 330 nm, respectively (Duggan *et al.*, 1957). The fluorescence maxima of the other tocopherols are at slightly longer wavelengths in accordance with their absorbance spectra. The fluorescent intensities of the E vitamers differ from one another and are highly dependent on the solvent. Polar solvents such as diethyl ether and alcohols provide greater intensities compared with hexane. The fluorescence is negligible when the compounds are dissolved in chlorinated hydrocarbons (Thompson, Erdody and Maxwell, 1972). α-Tocopheryl acetate has been reported to be nonfluorescent but, with the aid of a suitable spectrofluorimeter, a weak but measurable fluorescence can be obtained. The excitation and emission maxima of 285 nm and 310 nm for the ester are closer together than those for α-tocopherol.

Stability

Tocopherols and tocotrienols are destroyed fairly rapidly by sunlight and artificial light containing wavelengths in the UV region. The vitamers are slowly oxidized by atmospheric oxygen to form mainly biologically inactive quinones; the oxidation is accelerated by light, heat, alkalinity and certain trace metals. The presence of ascorbic acid completely prevents the catalytic effect of iron(III) and copper(II) by maintaining these metals in their lower oxidation states (Cort, Mergens and Greene, 1978). The vitamers can withstand heating in acid or alkaline solution provided that oxygen and UV radiation are excluded. α-Tocopheryl acetate, which lacks the reactive hydroxyl group, is practically unaffected by the oxidizing influence of air and light.

In vitro antioxidant activity

α-Tocopherol in the nonesterified form is frequently used as an antioxidant to stabilize animal fats, which have a much lower vitamin E content than vegetable oils. In the absence of an antioxidant, unsaturated fats undergo peroxidation (autoxidation) to produce hydroperoxides. These break down further to give a variety of volatile compounds such as aldehydes and ketones, which produce the disagreeable odours and flavours of rancidity.

The relative antioxidant activities of tocopherols have been investigated using methods in which the oil or lipid substrate is heated in order to accelerate the development of rancidity. In these model systems the ability of the tocopherols to extend the induction period (i.e. the time taken for rancidity to commence) is taken to represent their antioxidant effectiveness. Using such techniques and other accelerated methods, the reported antioxidant activities of the tocopherols decreased in the order $\delta > \gamma > \beta > \alpha$ (Stern *et al.*, 1947; Griewahn and Daubert, 1948; Lea and Ward, 1959). Other reports using accelerated methods (Olcott and Emerson, 1937; Hove and Hove, 1944; Dugan and Kraybill, 1956; Olcott and van der Veen, 1968; Parkhurst, Skinner and Sturm, 1968; Cort, 1974) have shown α-tocopherol to be an inferior antioxidant compared with β-, γ- and δ-tocopherols. This order of *in vitro* antioxidant activity is the reverse of the biological activities.

Hove and Hove (1944) observed that at temperatures of 35 °C and below the antioxidant activities of the tocopherols decreased in the order $\alpha > \beta > \gamma$, the reverse of the order observed at higher temperatures. In support of this observation, Lea (1960) reported that, under suitable conditions, α-tocopherol could be shown to exert the highest and δ-tocopherol the lowest *in vitro* antioxidant activity. The conditions required were: a substrate containing fatty esters more unsaturated than linoleic; measurement of activity at a comparatively early stage of the oxidation, before the end of the induction period; and not too high a temperature of oxidation. It appears from these experiments that, under certain conditions of testing, the order of *in vitro* antioxidant activities of the tocopherols conforms

to their oxidation potentials and parallels their biological activities, i.e.
$\alpha > \beta > \gamma > \delta$. This order of antioxidant activity has been confirmed in
detailed studies carried out by Burton and Ingold (1981), although there
was no significant difference between the β and γ positional isomers. In a
study conducted under near physiological conditions (Fukuzawa *et al.*,
1982), one molecule each of α-, β-, γ- and δ-tocopherol and tocol protected,
respectively, 220, 120, 100, 30 and 20 molecules of PUFA.

In a review of the chemistry and antioxidant properties of tocochromanols, Kamal-Eldin and Appelqvist (1996) recognized that the absolute and
relative *in vitro* activities of these compounds are not only dependent on
their specific chemical reactivities toward free radicals, but also on many
other possible side reactions which may be highly propagative. These side
reactions are dramatically affected by tocochromanol concentrations, by
temperature and light, type of substrate and solvent, and by other chemical species acting as pro-oxidants and synergists in the system. Thus the
mode in which the tocochromanols react is significantly affected by the
interplay of all the chemical and physical parameters of the system.

5.7 ANALYSIS

5.7.1 Scope of analytical techniques

To evaluate the vitamin E activity of vegetable oils and the products
made from them using a nonbiological assay, it is necessary to determine
the tocopherols and tocotrienols individually, as they vary widely in
biological activity. For the analysis of those animal products which are
known to contain predominantly α-tocopherol, only this vitamer need be
determined. In vitamin E-fortified foods it is usually sufficient to determine either the added α-tocopheryl acetate or, following saponification,
the total α-tocopherol (natural plus added).

The AOAC (1990) colorimetric method for determining natural or total
α-tocopherol in foods requires the isolation of α-tocopherol by two-
dimensional thin-layer chromatography and does not account for the
other E vitamers. Gas chromatography using capillary column technology facilitates the separation of the trimethylsilyl ethers of tocopherols, cholesterol and the major plant sterols without the need to
saponify samples. The present most popular analytical technique is
HPLC, which can separate all of the E vitamers and determine them
fluorimetrically without the need for chemical derivatization.

5.7.2 Expression of dietary values

The IU for vitamin E is defined as the specific activity of 1 mg of a
standard preparation, this quantity being the average amount which,

administered orally, prevents foetal resorption in 50% of a reasonable number of female rats deprived of vitamin E (Mason, 1944). The most recent International Standard of vitamin E is totally synthetic *all-rac-α-*tocopheryl acetate (authentic *dl-α*-tocopheryl acetate), which has a defined activity of 1 IU/mg.

Various international nutrition organizations have recommended that the vitamin E value in foods be expressed as milligrams of *RRR-α*-tocopherol equivalents rather than as international units. When calculating the total vitamin E activity of mixed diets in the United States, the milligram content of β-tocopherol, γ-tocopherol and α-tocotrienol should be multiplied by factors of 0.5, 0.1 and 0.3, respectively, and added to the milligram content of α-tocopherol to give the total milligrams of α-tocopherol equivalents. If only α-tocopherol in a mixed diet is reported, the value in milligrams should be multiplied by 1.2 to account for the other vitamers that are present, thus giving an approximation of the total vitamin E activity as milligrams of α-tocopherol equivalents (National Research Council, 1989).

5.7.3 Bioassays

The true biological activity of vitamin E can be determined by its ability to prevent or reverse specific vitamin E deficiency symptoms in animals *in vivo* (Machlin, 1984). Such symptoms include foetal resorption in rats, muscular dystrophy in rabbits and encephalomalacia in chicks. The erythrocyte haemolysis test and the liver-storage assay are among the many other tests that have been used. These methods are not a direct measure of biological activity, as the measurement part of the procedure is carried out *in vitro*, but they do take into account the intestinal absorption and plasma transport of the test compounds.

Resorption–gestation assay

The resorption–gestation assay is the classic assay by which vitamin E was discovered. Female rats are bred to sexual maturity on a vitamin E-deficient diet, and then mated. Without vitamin E, the rats conceive normally, but resorb their young in the later stages of pregnancy. Vitamin E-depleted rats given the critical dose of vitamin E before the eighth day of gestation bear normal young. The resorption–gestation assay gives an all-or-none response and is recognized as the only unequivocal assay for vitamin E. The assay has been extensively employed to establish the biopotencies of the individual vitamin E vitamers and their stereoisomers (Ames, 1972a, 1979; Leth and Søndergaard, 1977; Weiser and Vecchi, 1981).

Pregnant female rats are fed a vitamin E-free basal diet during the last week of gestation and during lactation in order to minimize placental and mammary transfer of vitamin E to the offspring. The female offspring are continued on the basal diet for 30–40 days after weaning, by which time they have attained a body weight of 150 g (Ames, 1979). The vitamin E-depleted virgin female rats are tested by microscopical examination of vaginal smears for signs of a normal oestrous cycle during the week before the assay. The rats are then mated with normally fed males of proven fertility. Female rats found to be pregnant are randomly divided into test groups, each of 10–15 animals. Three to five graded dosage levels of standard and of each test sample are assigned at random, one dosage level to each test group. Each dosage level of standard and test sample is administered to the animals in equally divided daily amounts over five successive days (i.e. one-fifth of the total dose per day) commencing on the fifth day of pregnancy. The range of dosage levels of standard and samples should be such that at least two doses of each preparation will fall in the intermediate zone in which some rats in a group respond and others do not. The standard and test vitamin E preparations are added to olive oil and administered orally by stomach tube; food samples are fed instead of the basal diets. An unsupplemented negative control group is included in the assay to establish that the basal diet is not contaminated with vitamin E. The standard is 2-*ambo*-α-tocopheryl acetate (the former international standard for vitamin E) containing 1.00 IU of vitamin E activity per milligram, and the upper dosage level of standard is 0.90 IU. The test samples are given in doses that correspond to the amount of vitamin E activity in standard doses. The test doses are calculated on the basis of the chemically estimated vitamin E content of the sample.

The quantitative estimation of vitamin E activity is based on an all-or-none response to the dietary supplement, as judged by the percentage of pregnancies that culminate in live offspring at term. Determining the number of live offspring at term is difficult because some pups may die soon after birth and others may be victims of maternal cannibalism. In practice, therefore, the pregnant rats are killed on the 20th day post-insemination (i.e. one day before parturition), and the uteri are examined for the number of living foetuses and for sites of implantation of foetuses that have been resorbed. Animals which have less than four implantation sites are discarded. Animals having one or more living foetuses are recorded as positive (+), while those with no living young are recorded as negative (−).

The percentage of rats in each test group that showed a positive response is calculated and each percentage is transformed to a probit by reference to statistical tables. Groups in which no rat has a living foetus (0%) or in which every rat has one or more living foetuses (100%) contribute minimally to the statistical evaluation of an assay. Linear dose–response curves for the standard and for each test sample are obtained by plotting the probits against the logarithm of the dose levels using the regression equation, and the dose level at the 50% response point (median fertility dose) is calculated. The potency of the test sample expressed in IU of vitamin E activity is determined by calculating the ratio of the median fertility doses for the standard and sample.

Among the various bioassays that have been employed for determining vitamin E activity, the resorption–gestation assay is the most difficult, laborious and time consuming. Since only one value per test group can be obtained, statistical evaluation by probit analysis yields wide confidence limits. However, the assay is based on a biological function that is specific for vitamin E; therefore, despite its practical disadvantages, it is regarded as the ultimate reference method in assessing the vitamin E biopotency of an unknown substance (Diplock, 1985).

Erythrocyte haemolysis test

The erythrocyte haemolysis test is based on the observation that erythrocytes (red blood cells) of vitamin E-depleted rats, but not of normal rats, exhibit complete haemolysis when added *in vitro* to dilute solutions of dialuric acid (a derivative of alloxan). The effect of vitamin E deficiency on erythrocyte haemolysis was first observed by György and Rose (1948) when rats on a diet deficient in vitamin E were injected with alloxan. The manifestation of deficiency is presumed to be the result of peroxidation reactions affecting components of the cell membrane, since it can be prevented by certain synthetic antioxidants administered *in vivo* or added to the cell suspension *in vitro* (Draper, 1970). Friedman *et al.* (1958) developed a curative assay in which rats are fed a vitamin E-deficient diet until their blood samples showed complete haemolysis when treated with dialuric acid. The basal diet is then supplemented with the test sample containing vitamin E and the extent to which the degree of haemolysis is reduced is proportional to the amount of vitamin E ingested.

Male or female rats weighing about 100 g are given a vitamin E-deficient diet over a 3–4-week depletion period. At the end of this period blood samples from each rat are tested for haemolysis induced by dialuric acid. To measure the percentage haemolysis, blood samples are centrifuged, the erythrocytes are resuspended in buffered saline, and 1 ml of dialuric acid solution is added to 1 ml of the cell suspension. The tubes are incubated for 1 h at 37 °C, allowed to stand at room temperature for an additional hour, centrifuged, and the absorbance of the supernatant is measured at 415 nm. Two control tubes are used, one containing cells completely haemolysed by the addition of water, and the other containing no dialuric acid to indicate the degree of spontaneous haemolysis. The value for the experimental tube, less that for the second control, is multiplied by 100 and divided by the reading for the completely haemolysed sample to give an estimate of percentage haemolysis. Only those animals showing a degree of haemolysis of 96–99% are retained in the test. In a preliminary test with five animals per dosage level, amounts of standards and test samples are selected that give a range of 20–80% haemolysis. For the bioassay itself, two or three dosage levels are selected from within this range for the standard and for each test sample, and 15–20 rats are used for each dosage level. Each dose is dissolved in 0.2 ml of olive oil and administered orally by stomach tube. After 40–44 h of dose administration, the percentage haemolysis is again determined. The logit of the percentage haemolysis plotted against the log dose of the vitamin gives a linear plot from which the vitamin E potency of the test sample can be obtained.

After a redepletion period of about 14 days, the rats can be used again for another bioassay following the same procedure.

Compared with the resorption–gestation assay, the erythrocyte haemolysis test is less specific, but it is much simpler to conduct, has a shorter duration, and a more sensitive criterion of response. Because the results from each animal, as opposed to each group of animals, are used in the statistical analysis, the haemolysis test yields narrower confidence limits. The haemolysis test correlates well with the resorption–gestation assay and also with the chick liver-storage assay, but some caution should be used in interpreting the results, as the test is not based on an *in vivo* biological response.

Liver-storage assay

The liver-storage assay is based on the observation that the vitamin E content in the liver of rats and chicks deprived of vitamin E increases linearly in response to increasing vitamin E supplementation of the basal diet. The response is determined by absorption, transport and turnover in the liver, and the assay is particularly suitable for absorption and bio-availability studies of formulated vitamin E preparations. The commonly used assay procedures use chicks, rather than rats, because chicks are more easily rendered deficient in vitamin E. However, the activities of the various vitamers of vitamin E differ significantly between the rat and chick, and the rat is assumed to be more representative of the activity for humans (Draper, 1970).

One-day-old male chicks are fed on a vitamin E-deficient basal diet for 13 days, and birds of similar weight are selected for the assay. Groups of eight chicks are fed on the basal diet supplemented with two dosage levels of the standard and of the test sample on a restricted basis for 14 days. In the method described by Pudelkiewicz *et al.* (1960) the vitamin E content of pooled livers from each group are determined by the

Emmerie-Engel colorimetric method after careful extraction, molecular distillation and column chromatography. The vitamin E content of the liver is directly proportional to the ingested dose. The assay was modified by Dicks and Matterson (1961) into a short-term method by reducing the depletion period from 13 days to 2 days without loss of precision. In a series of six trials, both procedures gave a coefficient of variation of 9.3% (Green, 1970).

The chick liver-storage assay correlates well with the resorption–gestation assay and erythrocyte haemolysis test, and yielded similar relative biopotencies of *d-* and *dl*-α-tocopheryl acetate (Bliss and György, 1967). The use of more advanced measurement techniques, such as HPLC, would improve the usefulness of the bioassay.

5.7.4 Extraction techniques

Saponification
The recovery of α-tocopherol is nearly quantitative after saponification in the presence of an antioxidant and exclusion of light, but significant losses of tocotrienols occur (Chow, Draper and Saari Csallany, 1969). The extent of the tocotrienol losses depends on the concentration of alkali present (Piironen *et al.*, 1984) and on the temperature and duration of the digestion. The contribution of the tocotrienols to total vitamin E activity in foods is generally small and thus, for most practical purposes, the losses are of little consequence.

In the analysis of fortified foods that contain significant amounts of naturally occurring vitamin E, saponification provides a convenient means of hydrolysing the α-tocopheryl acetate to its alcohol form, allowing both the supplemental and natural α-tocopherol to be measured together as a single peak by HPLC. It should be noted that if totally synthetic *all-rac*-α-tocopheryl acetate is the supplement used, its hydrolysis product, *all-rac*-α-tocopherol, is less biologically active than naturally occurring *RRR*-α-tocopherol, making it impossible to calculate a potency value for the total vitamin E. This problem does not arise if the supplement used is *RRR*-α-tocopheryl acetate.

For the efficient extraction of tocopherols from the saponification digest using hexane, the ethanol concentration in the extraction system must be below 40% (Ueda and Igarashi, 1987).

Direct solvent extraction
For the determination of vitamin E in seed oils by HPLC, the oils can be simply dissolved in hexane and analysed directly (Speek, Schrijver and Schreurs, 1985). Solid food samples demand a more rigorous method of solvent extraction. Balz, Schulte and Thier (1993) modified the Röse-Gottlieb method to extract vitamin E from infant formulas. Dipotassium

oxalate solution (35% w/v) was substituted for ammonia to avoid alkalizing the medium, and *tert*-butylmethylether was substituted for diethyl ether because of its stability against the formation of peroxides.

5.7.5 High-performance liquid chromatography

HPLC methods used for the determination of vitamin E *per se* are summarized in Table 5.7.

Detection

HPLC methods for determining the individual tocopherols (T) and tocotrienols (T3) invariably employ fluorescence detection. Studies conducted by Thompson and Hatina (1979) have shown that, compared with absorbance detection, fluorescence detection is more selective and the sensitivity under normal-phase conditions is at least 10 times greater. The fluorescent intensities of the vitamers are markedly affected by the solvent. Fluorescence is enhanced by the presence of small amounts of alcohols or ethers in the mobile phase and abolished by even traces of chlorinated hydrocarbons (Thompson, 1982). The synthetic antioxidant butylated hydroxyanisole (BHA) is fluorescent and, if present, its peak appears between β-T and γ-T. Certain vegetable oils and plant extracts contain plastochromanol-8, which is structurally related to γ-T3 and appears as a small peak close to γ-T. The relative fluorescence responses (peak areas) of the tocopherols using a mobile phase composition of 5% diethyl ether in hexane and excitation and emission maxima of 290 nm and 330 nm, respectively, were: α-T, 100; β-T, 129; γ-T, 110; and δ-T, 122. Fortunately, the fluorescence responses of the tocotrienols are very similar to their corresponding tocopherols, and therefore tocotrienol standards are not needed for calibration purposes.

The fluorescence detector response to α-T can be increased by up to 20-fold in some instruments by using short-wavelength excitation at 205 nm (Hatam and Kayden, 1979). A disadvantage of short-wavelength excitation is a marked loss of selectivity and an aggravation of quenching effects (Thompson, 1982), and hence the longer wavelength (290 nm) is usually employed in food analysis applications.

Fluorescence detection of α-tocopheryl acetate (α-TAc) depends on the use of detectors equipped with a lamp of sufficient radiation energy to stimulate fluorescence, and narrow-band excitation and emission monochromators to overcome the problem of spectral overlap. Other essential attributes are the electronic capabilities of the photomultipliers and amplifiers (Woollard, Blott and Indyk, 1987; Balz, Schulte and Thier, 1993). Using a Shimadzu Model RF-540 (150 W xenon lamp) spectrofluorimeter with excitation (ex) and emission (em) wavelengths set at 290 nm and 330 nm (both slits at 10 nm) and normal-phase HPLC conditions, Håkansson, Jägerstad and Öste (1987) found that the fluorescence response of α-TAc was 9% of that of α-T. The detection limit for α-TAc was 0.02 µg per 20 µl injection volume, which corresponded to a supplemental level of only 0.5 µg/g of food. This detection limit is more than adequate for measuring the levels of α-TAc in fortified foods, which range from 5 to 500 µg/g (Thompson, 1982).

Woollard, Blott and Indyk (1987) employed detection wavelengths of 280 nm (ex) and 335 nm (em) for the simultaneous determination of α-TAc and indigenous α-T in infant formulas under normal-phase conditions and reported a four-fold gain in sensitivity for α-TAc compared with absorbance detection. They observed that changing the excitation wavelength to 220 nm enhanced the fluorescence signal of α-TAc by a factor of three.

The separation of vitamins A and E is necessary, otherwise vitamin A will quench the fluorescence of vitamin E by absorption of the incident light (Taylor, Lamden and Tappel, 1976). The bandpasses of a dual-monochromator spectrofluorimeter are too narrow to permit the simultaneous determination of vitamins A and E, which possess widely different excitation and emission spectra. However, with the aid of a programmable detector (an instrument that can be programmed to change the excitation and emission wavelengths according to peak retention times), this analysis can be achieved if the two vitamins are adequately separated (Rhys Williams, 1985).

Electrochemical detection can be used with reversed-phase HPLC, as the supporting electrolyte is readily soluble in semi-aqueous mobile phases. The electrochemical detection of α-T in saponified animal feeds was reported to be 20 times more sensitive than fluorescence detection (Ueda and Igarashi, 1985). Electrochemical detection cannot measure α-TAc, owing to the absence of the oxidizable hydroxyl group, but this is of no concern in the determination of total α-T after saponification. It is possible, by adding the electrolyte post-column, to use electrochemical detection with normal-phase HPLC, as demonstrated by Hiroshima *et al.* (1981).

Normal-phase systems
Normal-phase HPLC is capable of separating isocratically all of the eight nonesterified vitamin E vitamers, and has been utilized to investigate the distribution of naturally occurring vitamers in a wide variety of fats, oils and foodstuffs. The vitamers are eluted in the order of α-T, α-T3, β-T, γ-T, β-T3, γ-T3, δ-T and δ-T3. Supplemental α-TAc is eluted before α-T, and the acetate and palmitate esters of vitamin A elute separately before α-TAc.

The distribution of naturally occurring tocopherols and tocotrienols in oils and fats may be determined by simply dissolving the sample in

Table 5.7 HPLC methods used for the determination of vitamin E in food

Food	Sample preparation	Column	Mobile phase	Vitamers separated	Detection	Reference
		Normal-phase chromatography				
Seed oils	Dissolve oil in hexane	Partisil PAC 5 μm	Hexane/THF (94:6)	α-, β-, γ-, δ-T; α-, β-, γ-, δ-T3	Fluorescence: ex. 210 nm em. 325 nm	Rammell and Hoogenboom (1985)
Seed oils, margarine, butter	Dissolve sample in hexane	LiChrosorb Si-60 5 μm 250 × 4 mm i.d. (column temperature 45°C for separation of γ-T and β-T3)	Diisopropyl ether gradient of 8% to 17% in hexane	α-, β-, γ-, δ-T; α-, β-, γ-, δ-T3	Fluorescence: ex. 290 nm em. 325 nm	Syväoja et al. (1986)
Cereals, flour foods (unfortified)	Extract samples with boiling 2-PrOH, re-extract with acetone, partition into hexane	LiChrosorb Si-60 5 μm 250 × 3.2 mm i.d.	0.2% 2-PrOH or 5% diethyl ether in dry hexane/water-saturated hexane (1 + 1)	α-, β-, γ-, δ-T; α-, β-, γ-T3	Fluorescence: ex. 290 nm em. 330 nm	Thompson and Hatina (1979)
Infant formulas	Saponify (hot), extract unsaponifiables with diethyl ether	As above	As above	Total α-T, β-, γ-, δ-T; α-, β-, γ-T3	As above	As above
Infant formulas	Saponify (hot), extract unsaponifiables with hexane	LiChrosphere Si-60 5 μm 120 × 4.6 mm i.d.	1% 2-PrOH and 0.5% EtOH in hexane	Total α-T, β-, γ-, δ-T	Fluorescence: ex. 292 nm em. 320 nm	Tuan et al. (1989)
Various foodstuffs, dairy products, infant formulas	Saponify (hot), extract unsaponifiables with petroleum ether/diisopropyl ether (3 + 1)	Radial-PAK cartridge containing Resolve silica 5 μm	1% 2-PrOH in hexane	Total α-T, β-, γ-, δ-T	Fluorescence: ex. 295 nm em. 330 nm	Ircik (1988)
Infant formulas	Disperse sample in nonaqueous solvent mixture (DMSO/DMF/CHCl3 2 + 2 + 1) containing 0.1% (w/v) ascorbic acid. Partition total lipid fraction into hexane, centrifuge	Radial-PAK cartridge containing Resolve silica 5 μm	0.08% 2-PrOH in hexane	α-Tocopheryl acetate	UV 280 nm	Woollard and Blott (1986)
Meat and meat products	Saponify (ambient), extract unsaponifiables with hexane	LiChrosorb Si-60 5 μm 250 × 4 mm i.d.	Hexane/diisopropyl ether (93:7)	α-, β-, γ-T; α-, β-T3	Fluorescence: ex. 292 nm em. 324 nm	Piironen et al. (1985)

Sample	Extraction	Column	Mobile phase	Compounds	Detection	Reference
Fish and fish products	As above	As above	As above	α-, β-, γ-T	As above	Syväoja et al. (1985a)
Infant formulas	As above	As above	As above	Total α-T, β-, γ-T, α-, β-T3	As above	Syväoja et al. (1985b)
Cereal products	As above	As above	As above	α-, β-, γ-, δ-T, α-, β-, γ-T3	As above	Piironen et al. (1986a)
Rice bran	Saponify (hot), extract unsaponifiables with hexane	Supelcosil LC-Si 5 μm 250 × 4.6 mm i.d.	Isooctane/ethyl acetate/acetic acid/DMP (98.15:0.9:0.85:0.1)	α-, β-, γ-, δ-T	Fluorescence: ex. 290 nm em. 330 nm	Shin and Godber (1993)
Breakfast cereals, infant formulas	Cereals: add water and MeOH to ground sample, shake and sonicate. Extract with tBME/petroleum ether (10 + 14) Formulas: reconstitute sample with water, add dipotassium oxalate solution (35% w/v) and EtOH, extract with tBME/petroleum ether (25 + 35)	LiChrosphere 100 diol 5 μm 250 × 4 mm i.d.	Two-step gradient composed of hexane and an increasing concentration of tBME; 0–4 min hexane, 4–5 min up to hexane/tBME (97:3), 5–41 min isocratic, 41–42 min up to hexane/tBME (95:5), 42–50 min isocratic	α-Tocopheryl acetate α-, β-, γ-, δ-T α-, β-, γ-, δ-T3, plastochromanol-8	Fluorescence: ex. 295 nm em. 330 nm	Balz, Schulte and Thier (1993)
			Reversed-phase chromatography			
All food types	Saponify (ambient), extract unsaponifiables with hexane, filter through anhydrous sodium sulphate	Zorbax ODS 5 μm 250 × 4.6 mm i.d.	MeCN/CH2Cl2 containing 0.001% triethylamine/MeOH (730 + 300 + 50)	α-, (β + γ)-, δ-T	Fluorescence: ex. 290 nm em. 330 nm	Hogarty, Ang and Eitenmiller (1989)
Vegetable oils	Dilute 1 g of oil to 50 ml with THF, and further dilute 15 ml of this solution to 100 ml with MeOH	Spherisorb ODS 5 μm 250 × 4.6 mm i.d.	0.05 M aqueous sodium perchlorate/MeOH (10:90)	α-, (β + γ)-, δ-T, α-, (β + γ)-, δ-T3	Amperometric: +0.600 V	Dionisi, Prodolliet and Tagliaferri (1995)

Abbreviations: see footnote to Table 2.4.
DMSO, dimethylsulphoxide; DMF, dimethylformamide; tBME, tert-butylmethylether; DMP, 2,2-dimethoxypropane.

hexane (typically 0.5 g in 50 ml), and analysing an aliquot of the hexane solution, without concentration, by normal-phase HPLC. Fluorescence detection is usually obligatory when the total lipid fraction is analysed, as absorbance detection reveals peaks of lipid origin that interfere with the peaks of the E vitamers. If the sample is saponified, absorbance detection can be utilized, as hexane extracts of the unsaponifiable matter are usually free from interfering lipoidal material. For the analysis of saponified vitamin E-fortified foods, fluorescence detection provides a much higher sensitivity, as the liberated α-T fluoresces more strongly than α-TAc.

An improvement in the column stability and reproducibility for analysis of vitamin E on a silica column was reported using a mobile phase composition of isooctane/ethyl acetate/acetic acid/2,2-dimethoxypropane (98.15/0.90/0.85/0.10) (Shin and Godber, 1993). The acetic acid component reduced retention times of the late-eluting E vitamers presumably by competing with water and polar material for binding to silanol groups in the stationary phase. 2,2-Dimethoxypropane reacts with water to form acetone and methanol, and its inclusion stabilized retention times and reduced the need for column regeneration.

Balz, Schulte and Thier (1993) utilized a diol stationary phase and a two-step gradient composed of hexane and an increasing concentration of *tert*-butylmethylether to separate α-TAc, plastochromanol-8 and the eight nonesterified E vitamers, and applied this system to the analysis of infant formulas and breakfast cereals.

Reversed-phase systems
In reversed-phase HPLC, the vitamin E vitamers are eluted in the order of δ-T3, (γ + β)-T3, α-T3, δ-T, (γ + β)-T and α-T. The positional β and γ isomers of tocopherols and tocotrienols cannot be separated using reversed-phase columns of standard dimensions. α-TAc elutes immediately in front of α-T with baseline separation. If direct solvent extraction is the extraction technique employed, the solvent used must be compatible with the mobile phase. If it is not, the solvent extract must be evaporated to dryness, and the residue redissolved in a suitable solvent.

Reversed-phase HPLC with fluorescence detection is the preferred system for the routine determination of total α-T in fortified foods and animal feeds after saponification. The use of NARP-HPLC with a predominantly hexane mobile phase allows aliquots of hexane extracts of the unsaponifiable matter to be injected directly onto the column, thus avoiding the evaporation step necessary when a semi-aqueous mobile phase is used (Rammell, Cunliffe and Kieboom, 1983).

Dionisi, Prodolliet and Tagliaferri (1995) developed a reversed-phase HPLC method with amperometric detection for the analysis of tocopherols and tocotrienols in vegetable oils. No tocotrienols were detected in olive, hazelnut, sunflower and soybean oils, whether virgin or refined. However, relatively high levels of tocotrienols were found in palm and grapeseed oils. This method could detect small quantities (1–2%) of palm and grapeseed oils in any tocotrienol-free vegetable oil and might, therefore, help to assess authenticity of certain vegetable oils, particularly the adulteration of olive oil.

5.8 BIOAVAILABILITY

5.8.1 Physiological aspects

The absorption, transport, uptake and distribution of vitamin E have been extensively reviewed (Parker, 1989; Bjørneboe, Bjørneboe and Drevon, 1990; Drevon, 1991; Cohn *et al.*, 1992; Traber, Cohn and Muller, 1993). Biokinetic studies of vitamin E have been radically improved by the application of stable isotope labelling and GC–MS to the measurement of absorption, transport, uptake and retention of tocopherols in humans and laboratory animals (Burton and Traber, 1990). Deuterated tocopherols containing three or six atoms of deuterium per molecule may be ingested safely because deuterium is a stable isotope and has no deleterious effects. Furthermore, the deuterated tocopherols do not undergo any measurable mediated exchange at the positions of substitution. Stable isotope labelling makes it possible to evaluate, within the same subject, the simultaneous competitive uptake of two (or even three) different forms of α-tocopherol labelled with different amounts of deuterium.

Intestinal absorption

Absorption of vitamin E depends on the simultaneous digestion and absorption of dietary fat (Chapter 1). Individuals with impaired fat absorption invariably exhibit low vitamin E utilization. Before absorption, any ingested supplemental α-tocopheryl acetate is hydrolysed in the lumen of the small intestine by pancreatic esterase, which requires bile salts as cofactors. The free α-tocopherol thus formed, together with vitamin E of natural origin, is solubilized within mixed micelles for passage across the brush border membrane of the enterocyte by passive diffusion. The absorption efficiency of vitamin E is greater when the vitamin is solubilized in micelles containing medium-chain triglycerides as compared with long-chain triglycerides (Gallo-Torres, 1980). Studies in thoracic duct-cannulated rats have shown that the absorption of γ-tocopherol is not affected by the presence of a 50-fold excess of α-tocopherol (Traber *et al.*, 1986), indicating that γ-tocopherol is absorbed independently of α-tocopherol. Within the enterocyte, vitamin E is incorporated into chylomicrons and secreted into the bloodstream via the lymphatic route.

Transport, uptake and distribution

The transport of vitamin E in lipoproteins has been studied using deuterium-substituted *RRR* α-tocopheryl acetate as a metabolic tracer in human subjects (Traber *et al.*, 1988). On reaching the bloodstream, the chylomicrons are attacked by lipoprotein lipase, which hydrolyses the triglycerides present in these lipid-rich particles, producing chylomicron remnants. The vitamin E remaining in the remnants is taken up by the liver, which then secretes into the plasma the newly absorbed vitamin E in very low density lipoproteins (VLDL). These triglyceride-rich particles are also subject to lipolysis by lipoprotein lipase, producing intermediate density lipoproteins (IDL). Some IDL particles are taken up by the liver, while the remainder are converted to low density lipoproteins (LDL). The newly ingested vitamin E then equilibrates between LDL and high density lipoproteins (HDL), which are the principal carriers for transport in human plasma. Erythrocytes may also be important in transport, since substantial amounts of tocopherol are localized in the cell membrane, whence it is interchanged rapidly with plasma tocopherol. Unlike vitamins A and D, there is no evidence of a specific transport protein.

Vitamin E is taken up by all tissues of the body where it is concentrated in cellular fractions rich in membrane phospholipids, such as the mitochondria and microsomes. *In vitro* studies (Traber, Olivecrona and Kayden, 1985) have shown that lipoprotein lipase also functions as a transfer protein for vitamin E, facilitating the uptake of some of the vitamin by tissues. Tissue uptake varies directly with the logarithm of the vitamin intake. Binding proteins for α-tocopherol have been found in liver and erythrocytes.

When rats were fed deuterated *RRR*- and *SRR*-α-tocopheryl acetates (Ingold *et al.*, 1987), all of the tissues examined, except for the liver, were preferentially enriched in *RRR*-α-tocopherol. It is well documented that *SRR*- and *RRR*-α-tocopherol are absorbed equally well in rats, thus the animal's ability to discriminate between these epimers occurs at some stage after absorption. Studies involving the ingestion of deuterium-labelled *RRR*- and SRR-α-tocopherols in humans (Traber *et al.*, 1990a) and liver perfusion experiments in monkeys (Traber *et al.*, 1990b) have shown that this discrimination occurs in the liver, which secretes VLDL that is preferentially enriched in *RRR*-α-tocopherol. This preference has been attributed to a hepatic α-tocopherol-binding protein, such a protein having been isolated from rat liver cytoplasm and characterized by Catignani and Bieri (1977). The function of this protein may be to transfer *RRR*-α-tocopherol from lysosomes containing chylomicron remnants to the Golgi apparatus in which the VLDL are assembled (Burton and Traber, 1990). The erythrocytes were also found to be preferentially enriched in *RRR*-α-tocopherol. The marked preference of the erythrocyte for *RRR*-α-tocopherol over *SRR*-α-tocopherol is due to a better retention of the former compound by the cell membrane, rather than a selective uptake (Cheng *et al.*, 1987).

Behrens and Madère (1991) studied the discrimination for the natural *RRR*-α-tocopherol and the synthetic *all-rac*-α-tocopherol in several rat tissues. The results indicated that tissues have a preference for the natural diastereoisomer compared with the other seven that constitute the artificial all-racemic mixture. However, this preference was small and could only be observed at a low level of tocopherol in the diet (0.035 g/kg diet). The small discrimination disappeared when the diastereoisomers were fed at a level of 0.200 g/kg diet. This may indicate not only that *RRR*-α-tocopherol is incorporated in tissues, but all other diastereoisomers with the 2R configuration are taken up as well, probably with less affinity. Another possible explanation offered by Behrens and Madère for the absence of discrimination at high concentration is that, when *RRR*-α-tocopherol is at a relatively high concentration, it could saturate any receptor and/or specific proteins and make the binding of other stereoisomers difficult, so discrimination cannot be detected.

Despite the predominance of γ-tocopherol over α-tocopherol in the average US diet (Bieri and Poukka Evarts, 1974), several studies have shown that γ-tocopherol in normal adult human plasma averages only 15% of the concentration of α tocopherol (Behrens and Madère, 1986). In addition, Meydani *et al.* (1989) reported differences in the distribution of α- and γ-tocopherol in human postprandial lipoprotein. The mechanism for discriminating between α- and γ-tocopherol appears to reside not in the intestine, but in the liver. In humans, the liver secretes α-tocopherol in the VLDL in preference to γ-tocopherol and, furthermore, γ-tocopherol is preferentially excreted in the bile (Traber and Kayden, 1989). Catabolism of the VLDL results in the enrichment of plasma LDL and HDL with α-tocopherol, which in turn leads to an enrichment of α-tocopherol in the tissues.

The liver, blood, skeletal muscle and adipose tissue represent the major storage deposits of vitamin E. The accumulation of vitamin E in adipose tissue is due simply to its lipid solubility. If vitamin E is withdrawn from the diet, the vitamin is mobilized rapidly from the liver and plasma. Mobilization from muscle is much slower, and slower still from adipose tissue. The major route of excretion is faecal elimination.

5.8.2 Dietary sources and their bioavailability

Dietary sources

The important plant sources of vitamin E are the cereal grains and those nuts, beans and seeds that are also rich in high-potency oils. The vegetable oils extracted from these plant sources are the richest dietary sources of vitamin E. Cereal grain products, fish, meat, eggs, dairy products and green leafy vegetables also provide significant amounts. Major sources of vitamin E in the United States include margarine, mayonnaise and salad dressings, fortified breakfast cereals, vegetable shortenings and cooking oils, peanut butter, eggs, potato crisps, whole milk, tomato products and apples (Sheppard, Pennington and Weihrauch, 1993).

Vegetable oils are highly unsaturated and contain a correspondingly high concentration of vitamin E to maintain the oxidative stability of their constituent PUFA. The distribution of tocopherols and tocotrienols in different plant oils varies greatly, as shown in Table 5.8. In some vegetable oils, notably soybean oil, γ-tocopherol is the major vitamer present and in palm oil γ-tocotrienol predominates. Thus measurement of total tocopherols does not accurately represent the vitamin E activity of vegetable oils or food products containing them.

Advances in chromatographic techniques, particularly HPLC, have permitted the evaluation of the individual tocopherols and tocotrienols in a wide range of foods including processed and mixed foods (Bauernfeind, 1977; McLaughlin and Weihrauch, 1979; Lehmann et al., 1986; Hogarty, Ang and Eitenmiller, 1989). The data presented in Tables 5.8–5.10 have been obtained using HPLC with fluorimetric detection. The total vitamin E activity for each food item, expressed in milligrams of α-tocopherol equivalents (α-Teq) per 100 g, is calculated by multiplying the concentration of each vitamer by its relative biopotency.

The vitamin E content of a selection of Finnish foods is presented in Table 5.9. Among the cereal grains, wheat, maize, barley, rye, rice and

Table 5.8 Distribution of tocopherols and tocotrienols in selected vegetable oils (Syväoja *et al.*, 1986)

Oil	mg/100 g[a]									mg/g
	α-T	β-T	γ-T	δ-T	α-T3	β-T3	γ-T3	δ-T3	α-Teq[b]	α-Teq/ PUFA
Corn (maize)	25.69	0.95	75.23	3.25	1.50	–	2.03	–	32.5	0.6
Olive	11.91	–	1.34	–	–	–	–	–	12.0	1.9
Palm	6.05	–	tr	–	5.70	0.82	11.34	3.33	7.4	0.7
Peanut	8.86	0.38	3.50	0.85	–	–	–	–	9.3	0.3
Rapeseed	18.88	–	48.59	1.20	–	–	–	–	22.8	0.7
Safflower	44.92	1.20	2.56	0.65	–	–	–	–	45.6	0.6
Soybean	9.53	1.31	69.86	23.87	–	–	–	–	15.9	0.3
Sunflower	62.20	2.26	2.67	–	–	–	–	–	63.3	1.0

[a] Mean values (6–10 determinations) of each oil purchased from 3 to 5 different manufacturers.
[b] α-Tocopherol equivalents = mg α-T + (mg β-T × 0.4) + (mg γ-T × 0.08) + (mg α-T3 × 0.21) + (mg β-T3 × 0.04).
–, Not detected; tr, trace.

Table 5.9 Distribution of tocopherols and tocotrienols in selected Finnish foods

Item	α-T	β-T	γ-T	δ-T	α-T3	β-T3	γ-T3	δ-T3	α-Teq[b]	Ref.
Wheat meal	1.0	0.54	–	< 0.1	0.4	2.1	–	nd	1.38	(1)
Wheat flour										(1)
1.2–1.4% ash[c]	1.6	0.8	–	–	0.3	1.7	–	nd	2.05	
c.0.7% ash[d]	0.4	0.2	–	–	0.2	1.5	–	nd	0.58	
c.0.5% ash[e]	0.2	0.1	–	–	0.1	1.4	–	nd	0.32	
Wheat bran	1.6	0.8	–	–	1.5	5.6	–	nd	2.46	(1)
Wheat germ	22.1	8.6	–	< 0.1	0.3	1.0	–	nd	25.64	(1)
Peanut	10.89	0.27	8.39	0.17	–	nd	–	nd	11.67	(2)
Potato	0.05	–	–	–	–	nd	–	nd	0.05	(2)
Carrot	0.36	tr	tr	–	0.04	nd	–	nd	0.37	(2)
Broccoli	0.68	tr	0.14	–	–	nd	–	nd	0.69	(2)
Cauliflower	0.09	–	0.26	tr	–	nd	–	nd	0.11	(2)
White cabbage	0.04	–	–	–	–	nd	–	nd	0.04	(2)
Lettuce	0.63	tr	0.34	–	–	nd	–	nd	0.66	(2)
Spinach	1.22	–	–	–	–	nd	–	nd	1.22	(2)
Tomato	0.66	tr	0.20	tr	–	nd	–	nd	0.68	(2)
Sweet pepper	2.16	0.11	0.02	tr	–	nd	–	nd	2.21	(2)
Apple (flesh only)	0.24	–	–	–	–	nd	–	nd	0.24	(2)
Orange	0.36	tr	tr	–	–	nd	–	nd	0.36	(2)
Banana	0.21	tr	tr	–	–	nd	–	nd	0.21	(2)
Peach (flesh only)	0.96	tr	0.05	–	–	nd	–	nd	0.96	(2)
Raspberry	0.88	0.15	1.47	1.19	–	nd	–	nd	1.07	(2)
Blackcurrant	2.23	tr	0.83	tr	–	nd	–	nd	2.30	(2)
Milk, raw										(3)
summer	0.11	–	–	nd	tr	nd	nd	nd	0.11	
winter	0.06	–	–	nd	tr	nd	nd	nd	0.06	
Butter										(3)
summer	2.00	–	–	nd	0.07	nd	nd	nd	2.01	
winter	1.01	–	–	nd	0.11	nd	nd	nd	1.03	
Egg, whole[f]	1.96	0.04	0.08	nd	0.25	nd	nd	nd	2.03	(3)
Beef, raw										(4)
spring	0.34	–	0.01	nd	0.05	–	nd	nd	0.35	
autumn	0.60	tr	0.01	nd	0.04	tr	nd	nd	0.61	
Liver, cow										(4)
spring	0.25	–	0.01	nd	0.01	tr	nd	nd	0.25	
autumn	1.37	–	0.01	nd	tr	–	nd	nd	1.37	
Pork, raw, shoulder	0.47	–	0.01	nd	0.05	tr	nd	nd	0.48	(4)
Chicken, raw	0.70	tr	0.06	nd	0.03	tr	nd	nd	0.71	(4)
Cod, raw	1.05	–	–	nd	nd	nd	nd	nd	1.05	(5)
Salmon, raw	2.02	–	0.02	nd	nd	nd	nd	nd	2.02	(5)

(1) Piironen *et al.* (1986a); (2) Piironen *et al.* (1986b); (3) Syväoja *et al.* (1985c); (4) Piironen *et al.* (1985); (5) Syväoja *et al.* (1985a).

[a] Pooled samples.

[b] See footnote to Table 5.8.

[c] Milled mainly from the aleurone tissue

[d] Extraction rate c.78%.

[e] Extraction rate c.74%.

[f] Eggs from hens fed with vitamin supplements containing α-tocopheryl acetate.

–, Not detected; nd, not determined; tr, trace.

oats are important plant sources of vitamin E. The vitamin E content of cereal grains is influenced by plant genetics and is adversely affected by too much rain and humidity during harvest (Bauernfeind, 1977). The germ fraction of the cereal grains contains a far higher proportion of tocopherols, and therefore a greater vitamin E activity, than the endo sperm and other nongerm fractions in which most of the tocotrienol content of the grain is found (Grams, Blessin and Inglett, 1970; Barnes and Taylor, 1981). Thus flour, which is derived from endosperm, has a low vitamin E activity compared with milling fractions containing germ and aleurone tissue. Wheat germ is the richest source of vitamin E among the various milling products.

Most of the common non-tropical vegetables and fruits contain < 1 mg α-tocopherol equivalents/100 g fresh weight, α-tocopherol being the predominant vitamer present. Green leafy vegetables are included among the richer vegetable sources of vitamin E. The mature dark green outer leaves of brassicae, which are usually discarded, contain more vitamin E than the lighter green leaves which are consumed. The almost colourless heart of white cabbage and the florets of cauliflower contain practically no vitamin E, the determined tocopherol values for these vegetables being attributable to the green parts included (Booth and Bradford, 1963). Paradoxically, yellow senescent leaves that have lost their chlorophyll contain very much more α-tocopherol than do fresh leaves (Booth and Hobson-Frohock, 1961). Presumably α-tocopherol, which resides in the chloroplasts, protects chlorophyll from destruction by the action of oxygen produced by photosynthesis and is used up during high photosynthetic activity. In apples and pears the concentration of vitamin E is greater in the skin than in the flesh. Green cooking apples contain more tocopherol than red or yellow types (Booth and Bradford, 1963).

The concentration of vitamin E in animal tissues depends on the amount in the animal's diet. In raw muscle, fat and organs from mammals and birds most of the vitamin E is in the form of α-tocopherol. The α-tocopherol content of mammalian muscle is generally < 1 mg/100 g. There is a marked seasonal variation in the α-tocopherol content of beef and mutton, the values being about twice as high in the autumn as in the spring. Cow liver shows a much greater seasonal variation. The feeding of grass or fresh silage during the summer and dry forage and concentrates during the winter explains the higher autumn values observed in ruminants. During the same season the tocopherol concentration in different meat cuts of the same animal species increases with increasing fat content. The α-tocopherol content of cow's milk is higher in summer than in winter owing to the change in the animal's diet. In eggs, all of the vitamin E is in the yolk; the concentration varies greatly depending on the level of supplemental α-tocopheryl acetate (if any) contained in the chicken feed.

Table 5.10 Distribution of tocopherols and tocotrienols in margarines (Syväoja *et al.*, 1986)

Type of margarine	mg/100 g[a]								
	α-T	β-T	γ-T	δ-T	α-T3	β-T3	γ T3	δ-T3	α-Te[b]
Hard	7.56	–	16.80	3.13	0.44	nd	0.65	0.40	9.04
(range)	(4.00–8.96)		(6.21–28.00)	(1.37–11.07)	(0–1.75)		(0–2.34)	(0–0.78)	(4.51–11.71)
Semisoft	13.26	–	31.74	8.22	–	nd	0.10	0.14	15.88
(range)	(10.21–16.30)		(20.72–42.76)	(5.98–10.45)			(0–0.20)	(0–0.28)	(11.93–19.83)
Soft	23.96	0.79	26.84	9.07	0.04	nd	0.31	0.31	26.53
(range)	(17.57–44.62)	(0–1.44)	(2.69–43.74)	(0.67–16.56)	(0–0.18)		(0–1.01)	(0–0.62)	(17.78–48.91)

[a] Mean values of each margarine purchased from 5 different manufacturers.
[b] See footnote to Table 5.8.
–, Not detected; nd, not determined.

Contrasting values of 0.46 and 1.10 (Bauernfeind, 1980), 0.70 (McLaughlin and Weihrauch, 1979) and 1.96 (Syväoja *et al.*, 1985c) mg α-tocopherol/ 100 g whole egg have been reported. In general, fish is a better source of vitamin E than meat. Tuna and salmon canned in water contained, respectively, 0.53 (Lehmann *et al.*, 1986) and 0.7 (Hogarty, Ang and Eitenmiller, 1989) mg α tocopherol/100 g and sardines canned in tomato sauce contained 3.9 mg/100 g (Hogarty, Ang and Eitenmiller, 1989).

Margarines can be divided into three groups, hard, semisoft and soft, on the basis of their vitamin E content (Table 5.10).

Vitamin E, in the form of α-tocopheryl acetate, is sometimes added to whole milk powder (Indyk, 1983) and breakfast cereals (Widicus and Kirk, 1979) to supplement dietary requirements. The vitamin E requirement increases with an increased intake of PUFA, and hence several types of high quality dietetic margarines are enriched with vitamin E (Kläui, Hausheer and Huschke, 1970).

Bioavailability

The few reported studies in normal human subjects have shown that the absorption of vitamin E is incomplete. Estimates of the absorption of physiological oral doses of tritiated *all-rac*-α-tocopherol based on the measurement of unabsorbed faecal radioactivity gave efficiencies ranging from 51 to 86% (mean 72%) (Kelleher and Losowsky, 1970) and 55 to 79% (mean 69%) (MacMahon and Neale, 1970). The efficiency is greatest for tracer doses of vitamin E and least for pharmacological doses; thus, experimentally, absorption varies inversely with the intake (Losowsky *et al.*, 1972). This effect is probably of no nutritional importance within the relatively limited range of normal dietary intakes. There is no compensatory increase in absorption in the presence of vitamin E deficiency (Losowsky *et al.*, 1972).

There have been many conflicting reports concerning the relative absorption efficiency of α-tocopherol and its acetate ester. The situation has been clarified by Burton *et al.* (1988) who showed that there was no significant difference in efficiency when the RRR forms were adminis tered together with a meal at a daily dosage of 100 mg of total vitamin E.

Effects of polyunsaturated fats

It has been established in animals and in humans that an increase in the intake of unsaturated fat accelerates the depletion and increases the requirement for vitamin E (Witting, 1970). This is because dietary PUFA are concentrated in cellular and subcellular membranes where they have the capacity to sequester corresponding amounts of vitamin E to maintain their oxidative stability. In an initial effort to relate dietary vitamin E and PUFA, Harris and Embree (1963) suggested that a fixed ratio, termed the E:PUFA ratio, might be established to define the vit- amin E adequacy of a diet. The ratio of 0.6 proposed by Harris and Embree (1963) is now considered to be an overestimate because the values for normal intakes of vitamin E and PUFA were based on food- stuffs available for consumption rather than foods actually consumed. Bieri and Poukka Evarts (1973) calculated the average α-tocopherol intake for Americans consuming a typical mixed diet to be 9 mg/day and the average PUFA (linoleic acid) intake to be 21.2 g, giving an α-tocopherol to PUFA ratio of 0.43 mg/g of PUFA. The estimated contribution of γ- tocopherol (the most prevalent form of vitamin E in the diet) raised the ratio to 0.53 mg α-tocopherol equivalents per gram of PUFA. Other reported α-tocopherol/PUFA ratios include 0.4 (Witting and Lee, 1975) and 0.52 (Thompson *et al.*, 1973) mg/g of PUFA.

Although the total vitamin E content of animal and vegetable oils parallels their PUFA content (Shmulovich, 1994), this does not necessarily imply that the richest sources of linoleic acid are also the best sources of vitamin E activity in terms of α-tocopherol equivalents. Consider, for example, soybean oil and rapeseed oil, which are the dominant vegetable oils in the United States and Canada, respectively. The α-tocopherol equivalent:PUFA ratio for rapeseed oil is 0.7 mg/g compared with 0.3 for soybean oil (Table 5.8), even though the PUFA content of soybean oil exceeds that of rapeseed oil. This difference is due to the relatively low α-tocopherol content of soybean oil. Evidence for a nonparallel intake of α-tocopherol and PUFA was reported by Lehmann *et al.* (1986) who evaluated the vitamin E activity of typical US diets containing either 10 or 30 g of linoleic acid per day. In diets supplying 3600 or 4000 kcal energy, the vitamin E activities of the low- and high-linoleic acid diets were nearly the same. This occurred because soybean oil was the most common source of linoleic acid in the high-linoleic acid diet. In contrast, in diets containing 2400 kcal and either 10 or 30 g of linoleic acid, the vitamin E activity increased from 8.8 mg to 13.5 mg α-tocopherol equivalents per day, respectively.

There are many other practical difficulties and anomalies associated with the use of the E:PUFA ratio to define dietary vitamin E adequacy. One unsurmountable problem is that an individual's vitamin E requirement depends on the lipid composition of foods consumed over a period of months, or even years, rather than on current intake (Witting, 1975). Bieri and Farrell (1976) reviewed experimental evidence and concluded that in diets with mixed types of fat from a variety of foodstuffs, such as occur in human diets, a single fixed E:PUFA ratio cannot be used to characterize vitamin E adequacy.

Effects of dietary fibre
Pectin was fed *ad libitum* to rats at levels comprising 0, 3, 6 and 8% of the total diet (Schaus *et al.*, 1985). By the end of the 8-week study, 6% and 8% pectin reduced the bioavailability of vitamin E based on decreases in plasma and liver tocopherol as well as increased erythrocyte haemolysis. Rats fed 3% pectin did not differ in any vitamin E parameters from those fed 0% pectin. The results were consistent with the working hypothesis that pectin binds bile acids and makes them unavailable for micelle formation, thus decreasing the absorption of vitamin E. Faecal fat concentrations were not significantly different in this study, which indicated that pectin does not bind to lipids. A previous report (Omaye, Chow and Betschart, 1983) showed that pectin does not bind to α-tocopherol *in vitro*. It is unlikely that a high-fibre human diet would ever contain more than 3% pectin, therefore a typical human diet would not be expected to adversely affect vitamin E status.

A diet containing a high level of wheat bran also influences vitamin E bioavailability in rats (the fibre content of wheat bran is *c*. 50%). Feeding a diet containing 20% wheat bran for 36 days resulted in a 28% decrease in plasma tocopherol concentration compared with a diet containing 5% wheat bran (Omaye and Chow, 1984a). A similar decrease was observed between 35 and 42 days in a subsequent study (Omaye and Chow, 1984b), but after 56 days on the 20% wheat bran diet the plasma tocopherol concentrations had returned to normal. The authors concluded that the high-fibre diet induced structural and functional changes of the intestinal tract, which led to a temporary reduced bioavailability of vitamin E. The reversal of this effect after 56 days was presumably due to the ability of the intestinal tract to adapt to such changes.

Kahlon *et al.* (1986) observed a transient decrease in plasma α-tocopherol concentrations in rats fed diets containing 10% fibre (provided by cellulose or wheat bran) relative to 5% celluose fibre. This decrease had disappeared by three weeks. The addition of wheat bran to the basal diet at a level of 10% increased the total vitamin E content of the diet by

4–5%, yet similar liver α-tocopherol values were obtained in rats fed fine bran and 10% cellulose. This indicated that vitamin E in the wheat bran is not readily available. Liver α-tocopherol concentrations in rats fed 10% fibre from coarse (2 mm) wheat bran were significantly lower than those fed 10% fibre from fine (0.5 mm) wheat bran or 10% cellulose fibre for six weeks. This further suggests that coarse bran also interferes with the availability of some of the vitamin E in the diet. The apparent decrease in α-tocopherol bioavailability caused by feeding coarse bran would not have been caused by adsorption of vitamin E by bran, because the fine bran particles offered nearly twice the surface area as did the coarse bran particles in the diet. The effect of coarse bran could have resulted in morphological and physiological alterations of the intestinal mucosa.

In another study (Mongeau *et al.*, 1986), diets containing graded levels of wheat bran ranging between 4 and 20% did not affect plasma α-tocopherol concentrations after 42 days of feeding. No significant differences were found in liver α-tocopherol concentrations which showed that, if a transient decrease in plasma tocopherol concentrations had occurred before the 42nd day on a high-fibre diet, this had no effects on body stores. In white adipose tissue, a dose-related increase in γ-tocopherol with increasing wheat bran intake was observed. This suggests that wheat bran increases the efficiency of absorption of γ-tocopherol. The lack of an effect of wheat bran on γ-tocopherol concentrations in either plasma or liver may reflect efficient metabolic clearance of this vitamer except in those tissues with a low tocopherol turnover rate.

5.8.3 Effects of food processing, storage and cooking

During processing the food is exposed to the destructive influences of oxygen, light and other factors. Therefore, refined and processed foods are variable and usually less predictable sources of vitamin E than whole fresh foods (Bauernfeind, 1977). Frozen vegetables retain much of their vitamin E content, but losses in the canning of beans, peas and sweetcorn can be as high as 70–90% (Ames, 1972b). Frozen foods which have been fried in vegetable oil suffer a great loss of vitamin E during freezer storage. This loss is presumably due to destruction by hydroperoxides, which are more stable at low temperatures than at high temperatures and hence accumulate (Bunnell *et al.*, 1965). Vitamin E is very stable to heat and is not destroyed during the normal cooking of meat and vegetables. Little loss of the vitamin occurs during deep-fat frying in fresh vegetable oil, but shallow-pan frying is destructive.

Large losses of α-tocopherol occur during the milling of wheat flour, and in the processing of most grains to produce breakfast cereals (Ames, 1972b). About 25% of the α-tocopherol in the dough is lost during the baking of wheat bread (Piironen, Varo and Koivistoinen, 1987). Wheat flour and rye flour, when stored at room temperature, lost about 20% of their α-tocopherol and α-tocotrienol in two months, 50–60% in six months, and about 80% in 12 months; the corresponding losses of β-tocopherol and β-tocotrienol were about 10, 30 and 60%, respectively (Piironen, Varo and Koivistoinen, 1988).

5.9 DIETARY INTAKE

The vitamin E requirement in many industrialized countries has increased over the years as a result of an increased consumption of polyunsaturated vegetable oils and decreased consumption of saturated animal fats. Recommended Dietary Allowances (RDA) of vitamin E are based on the requirements of those individuals in the population with the highest PUFA intakes (Draper, 1993). For example, the RDA for adult males in the United States is 10 mg α-tocopherol equivalents per day (National Research Council, 1989). There are many individuals whose vitamin E intakes are substantially below the RDA value, but they have a correspondingly lower intake of PUFA, and therefore a lower vitamin E requirement.

With normal intakes of vitamin E, the plasma concentration of α-tocopherol correlates highly with the plasma concentration of total lipids or cholesterol (Bieri and Farrell, 1976). It would therefore appear that plasma tocopherol values, as a measure of an individual's vitamin E status, has little meaning if not accompanied by values for plasma lipids or cholesterol. From a survey of blood samples from infants and hospitalized adults, Horwitt *et al.* (1972) proposed that a ratio of 0.8 mg of total tocopherols per gram of total lipids be considered indicative of adequate vitamin E status in humans.

Effects of high intakes

Vitamin E, being fat-soluble, accumulates in the body, especially in the liver and pancreas. Unlike vitamins A and D, however, vitamin E is regarded as essentially nontoxic. Animal studies have shown that vitamin E is not mutagenic, carcinogenic or teratogenic. In human studies with double-blind protocols and in large population studies, oral vitamin E supplementation resulted in few side effects even at doses as high as 3200 mg/day (Bendich and Machlin, 1988).

5.10 SYNOPSIS

The naturally occurring vitamin E that abounds in vegetable oils is present as nonesterified tocopherols and tocotrienols, collectively known as tocochromanols. Of the eight tocochromanols that occur in nature, α-tocopherol is the most potent, although γ-tocopherol predominates in the Western diet. Foods are fortified with synthetic α-tocopheryl acetate, the ester being more stable than the free tocopherol. The bioavailability of vitamin E is incomplete owing to an inefficiency of intestinal absorption.

Because the tocochromanols have different biopotencies, physicochemical methods of analysis require chromatographic separation and measurement of each vitamer present. This is readily achieved by means of normal-phase HPLC using fluorimetric detection. Reversed-phase HPLC using columns of standard dimensions cannot resolve the positional β and γ isomers of tocopherols and tocotrienols. Nevertheless, the reversed-phase mode is usually preferred for determining total alpha-tocopherol in fortified foods after alkaline hydrolysis of the ester.

REFERENCES

Ames, S.R. (1972a) Tocopherols. 4. Estimation in foods and food supplements. In *The Vitamins. Chemistry, Physiology, Pathology, Methods,* 2nd edn, Vol. V (eds W.H. Sebrell Jr and R.S. Harris), Academic Press, New York, pp. 225–33.

Ames, S.R. (1972b) Tocopherols. 5. Occurrence in foods. In *The Vitamins. Chemistry, Physiology, Pathology, Methods,* 2nd edn, Vol. V (eds W.H. Sebrell Jr and R.S. Harris), Academic Press, New York, pp. 233–48.

Ames, S.R. (1979) Biopotencies in rats of several forms of alpha-tocopherol. *J. Nutr.,* **109,** 2198–2204.

AOAC (1990) Alpha-tocopherol and alpha-tocopheryl acetate in foods and feeds. Colorimetric method. Final action 1972. In *AOAC Official Methods of Analysis,* 15th edn (ed. K. Helrich), Association of Official Analytical Chemists, Inc., Arlington, VA, 971.30.

Association of Vitamin Chemists, Inc. (1966) *Methods of Vitamin Assay,* 3rd edn, Interscience Publishers, New York.

Balz, M.K., Schulte, E. and Thier, H.-P. (1993) Simultaneous determination of α-tocopheryl acetate, tocopherols and tocotrienols by HPLC with fluorescence detection in foods. *Fat Sci. Technol.,* **95,** 215–20.

Balz, M., Schulte, E. and Thier, H.-P. (1996) A new parameter for checking the suitability of α-tocopherol standards. *Z. Lebensm. u.-Forsch,* **202,** 80–1.

Barnes, P.J. and Taylor, P.W. (1981) γ-Tocopherol in barley germ. *Phytochemistry,* **20,** 1753–4.

Bauernfeind, J.C. (1977) The tocopherol content of food and influencing factors. *CRC Crit. Rev. Food Sci. Nutr.,* **8,** 337–82.

Bauernfeind, J.C. (1980) Tocopherols in foods. In *Vitamin E. A Comprehensive Treatise* (ed. L.J. Machlin), Marcel Dekker, Inc., New York, pp. 99–167.

Behrens, W.A. and Madère, R. (1986) Alpha- and gamma-tocopherol concentrations in human serum. *J. Am. Coll. Nutr.,* **5,** 91–6.

Behrens, W.A. and Madère, R. (1991) Tissue discrimination between dietary *RRR*-α- and all-*rac*-α-tocopherols in rats. *J. Nutr.,* **121,** 454–9.

Bendich, A. and Machlin, L.J. (1988) Safety of oral intake of vitamin E. *Am. J. Clin. Nutr.* **48,** 612–9.

Bieri, J.G. and Farrell, P.M. (1976) Vitamin E. *Vitams Horm.,* **34,** 31–75.

Bieri, J.G. and Poukka Evarts, R. (1973) Tocopherols and fatty acids in American diets. *J. Am. Diet. Assoc.,* **62,** 147–51.

Bieri, J.G. and Poukka Evarts, R. (1974) Gamma tocopherol: metabolism, biological activity and significance in human vitamin E nutrition. *Am. J. Clin. Nutr.*, **27**, 980–986.

Bjørneboe, A., Bjørneboe, G.-E.Aa. and Drevon, CA. (1990) Absorption, transport and distribution of vitamin E. *J. Nutr.*, **120**, 233–42.

Bliss, C.I. and György, P. (1967) Bioassays of vitamin E. In *The Vitamins. Chemistry, Physiology, Pathology, Methods*, 2nd edn, Vol. VI (eds P. György and W.N. Pearson), Academic Press, New York, pp. 304–16.

Booth, V.H. and Bradford, M.P. (1963) Tocopherol contents of vegetables and fruits. *Br. J. Nutr.*, **17**, 575–81.

Booth, V.H. and Hobson-Frohock, A. (1961) The α-tocopherol content of leaves as affected by growth rate. *J. Sci. Food Agric.*, **12**, 251–6.

Brubacher, G. and Wiss, O. (1972) Tocopherols. 8. Vitamin E active compounds, synergists, and antagonists. In *The Vitamins. Chemistry, Physiology, Pathology, Methods*, 2nd edn, Vol. V (eds W.H. Sebrell Jr and R.S. Harris), Academic Press, New York, pp. 255–8.

Bunnell, R.H., Keating, J., Quaresimo, A. and Parman, G.K. (1965) Alpha-tocopherol content of foods. *Am. J. Clin. Nutr.*, **17**, 1–10.

Bunyan, J., McHale, D., Green, J. and Marcinkiewicz, S. (1961) Biological potencies of ε- and ζ_1-tocopherol and 5-methyltocol. *Brit. J. Nutr.*, **15**, 253–7.

Burton, G.W. and Ingold, K.U. (1981) Autoxidation of biological molecules. 1. The antioxidant activity of vitamin E and related chain-breaking phenolic antioxidants *in vitro*. *J. Am. Chem. Soc.*, **103**, 6472–7.

Burton, G.W. and Ingold, K.U. (1989) Vitamin E as an *in vitro* and *in vivo* antioxidant. *Ann. NY Acad. Sci.*, **570**, 7–22.

Burton, G.W. and Traber, M.G. (1990) Vitamin E: antioxidant activity, biokinetics, and bioavailability. *Annu. Rev. Nutr.*, **10**, 357–82.

Burton, G.W., Ingold, K.U., Foster, D.O. *et al.* (1988) Comparison of free α-tocopherol and α-tocopheryl acetate as sources of vitamin E in rats and humans. *Lipids*, **23**, 834–40.

Catignani, G.L. and Bieri, J.G. (1977) Rat liver α-tocopherol binding protein. *Biochim. Biophys. Acta*, **497**, 349–57.

Century, B. and Horwitt, M.K. (1965) Biological availability of various forms of vitamin E with respect to different indices of deficiency. *Fed. Proc.*, **24**, 906–11.

Cheng, S.C., Burton, G.W., Ingold, K.U. and Foster, D.O. (1987) Chiral discrimination in the exchange of α-tocopherol stereoisomers between plasma and red blood cells. *Lipids*, **22**, 469–73.

Chow, C.K. (1979) Nutritional influence on cellular antioxidant defense systems. *Am. J. Clin. Nutr.*, **32**, 1066–81.

Chow, C.K., Draper, H.H. and Saari Csallany, A. (1969) Method for the assay of free and esterified tocopherols. *Analyt. Biochem.*, **32**, 81–90.

Cohn, W., Gross, P., Grun, H. *et al.* (1992) Tocopherol transport and absorption. *Proc. Nutr. Soc.*, **51**, 179–88.

Cort, W.M. (1974) Antioxidant activity of tocopherols, ascorbyl palmitate, and ascorbic acid and their mode of action. *J. Am. Oil Chem. Soc.*, **51**, 321–5.

Cort, W.M., Mergens, W. and Greene, A. (1978) Stability of alpha- and gamma-tocopherol: Fe^{3+} and Cu^{2+} interactions. *J. Food Sci.*, **43**, 797–8.

Dicks, M.W. and Matterson, L.D. (1961) Chick liver storage bioassay of alpha-tocopherol: Methods. *J. Nutr.*, **75**, 165–74.

Dionisi, F., Prodolliet, J. and Tagliaferri, E. (1995) Assessment of olive oil adulteration by reversed-phase high-performance liquid chromatography/amperometric detection of tocopherols and tocotrienols. *J. Am. Oil Chem. Soc.*, **72**, 1505–11.

Diplock, A.T. (1973) The interaction between vitamin E and selenium. *Acta Agric Scand.*, Suppl. 19, 113–21.

Diplock, A.T. (1985) Vitamin E. In *Fat-Soluble Vitamins. Their Biochemistry and Applications* (ed. A.T. Diplock), Heinemann, London, pp. 154–224.

Dunpin, I.I.II (1970) The tocopherols In *Fat soluble Vitamins* (ed. R.A. Morton), Pergamon Press, Oxford, pp. 333–93.

Draper, H.H. (1993) Interrelationships of vitamin E with other nutrients. In *Vitamin E in Health and Disease* (eds L. Packer and J. Fuchs), Marcel Dekker, Inc., New York, pp. 53–61.

Draper, H.H., Bergan, J.G., Chiu, M. *et al.* (1964) A further study of the specificity of the vitamin E requirement for reproduction. *J. Nutr.*, **84**, 395–400.

Drevon, C.A. (1991) Absorption, transport and metabolism of vitamin E. *Free Radical Res. Comms*, **14**, 229–46.

Dugan, L.R. Jr and Kraybill, H.R. (1956) Tocopherols as carry-through antioxidants. *J. Am. Oil Chem. Soc.*, **33**, 527–8.

Duggan, D.E., Bowman, R.L., Brodie, B.B. and Udenfriend, S. (1957) A spectrophotofluorometric study of compounds of biological interest. *Archs Biochem. Biophys.*, **68**, 1–14.

Frankel, E.N. (1991) Review. Recent advances in lipid oxidation. *J. Sci. Food Agric.*, **54**, 495–511.

Friedman, L., Weiss, W., Wherry, F. and Klien, O. (1958) Bioassay of vitamin E by the dialuric acid haemolysis method. *J. Nutr.*, **65**, 143–60.

Fukuzawa, K., Tokumara, A., Ouchi, S. and Tsukatani, H. (1982) Antioxidant activities of tococopherols on Fe^{2+}-ascorbate-induced lipid peroxidation in lecithin liposomes. *Lipids*, **17**, 511–13.

Gallo-Torres, H.E. (1980) Absorption. In *Vitamin E. A. Comprehensive Treatise* (ed. L.J. Machlin), Marcel Dekker, Inc., New York, pp. 170–92.

Grams, G.W., Blessin, C.W. and Inglett, G.E. (1970) Distribution of tocopherols within the corn kernel. *J. Am. Oil Chem. Soc.*, **47**, 337–9.

Green, J. (1970) Distribution of fat-soluble vitamins and their standardization and assay by biological methods. In *Fat-Soluble Vitamins* (ed. R.A. Morton), Pergamon Press, Oxford, pp. 71–97.

Griewahn, J. and Daubert, B.F. (1948) Delta-tocopherol as an antioxidant in lard. *J. Am. Oil Chem. Soc.*, **25**, 26–7.

György, P. and Rose, C.S. (1948) Effect of dietary factors on early mortality and hemoglobinuria in rats following administration of alloxan. *Science*, **108**, 716–8.

Håkansson, B., Jägerstad, M. and Öste, R. (1987) Determination of vitamin E in wheat products by HPLC. *J. Micronutr. Anal.*, **3**, 307–18.

Harris, P.L. and Embree, N.D. (1963) Quantitative consideration of the effect of polyunsaturated fatty acid content of the diet upon the requirements for vitamin E. *Am. J. Clin. Nutr.*, **13**, 385–92.

Harris, P.L. and Ludwig, M.I. (1949) Relative vitamin E potency of natural and of synthetic α-tocopherol. *J. Biol. Chem.*, **179**, 1111–5.

Hatam, L.J. and Kayden, H.J. (1979) A high-performance liquid chromatographic method for the determination of tocopherol in plasma and cellular elements of the blood. *J. Lipid Res.*, **20**, 639–45.

Hiroshima, O., Ikenoya, S., Ohmae, M. and Kawabe, K. (1981) Electrochemical detector for high-performance liquid chromatography. V. Application to adsorption chromatography. *Chem. Pharm. Bull.*, **29**, 451–5.

Hoekstra, W.G. (1975) Biochemical function of selenium and its relation to vitamin E. *Fed. Proc.*, **34**, 2083–9.

Hogarty, C.J., Ang, C. and Eitenmiller, R.R. (1989) Tocopherol content of selected foods by HPLC/fluorescence quantitation. *J. Food Comp. Anal.*, **2**, 200–209.

Horwitt, M.K. (1965) Role of vitamin E, selenium, and polyunsaturated fatty acids in clinical and experimental muscle disease. *Fed. Proc.*, **24**, 68–72.

Horwitt, M.K. (1976) Vitamin E: a reexamination. *Am. J. Clin. Nutr.*, **29**, 569–78.

Horwitt, M.K., Harvey, C.C., Dahm, C.H. Jr and Searcy, M.T. (1972) Relationship between tocopherol and serum lipid levels for determination of nutritional adequacy. *Ann. NY Acad. Sci.*, **203**, 223–36.

Hove, E.L. and Hove, Z. (1944) The effect of temperature on the relative antioxidant activity of α-, β-, and γ-tocopherols and of gossypol. *J. Biol. Chem.*, **156**, 623–32.

Indyk, H. (1983) The routine, simultaneous determination of vitamins A and E in fortified whole milk powders. *NZ J. Dairy Sci. Technol.*, **18**, 197–208.

Indyk, H.E. (1988) Simplified saponification procedure for the routine determination of total vitamin E in dairy products, foods and tissues by high-performance liquid chromatography. *Analyst, Lond.*, **113**, 1217–21.

Ingold, K.U., Burton, G.W., Foster, D.O. *et al.* (1987) Biokinetics of and discrimination between dietary *RRR*- and *SRR*-α-tocopherols in the male rat. *Lipids*, **22**, 163–72.

IUPAC-IUB (1974) Nomenclature of tocopherols and related compounds. *Eur. J. Biochem.*, **46**, 217–9.

Joffe, M. and Harris, P.L. (1943) The biological potency of the natural tocopherols and certain derivatives. *J. Am. Chem. Soc.*, **65**, 925–7.

Kahlon, T.S., Chow, F.I., Hoefer, J.L. and Betschart, A.A. (1986) Bioavailability of vitamins A and E as influenced by wheat bran and bran particle size. *Cereal Chem.*, **63**, 490–3.

Kamal-Eldin, A. and Appelqvist, L.-A. (1996) The chemistry and antioxidant properties of tocopherols and tocotrienols. *Lipids*, **31**, 671–701.

Kasparek, S. (1980) Chemistry of tocopherols and tocotrienols. In *Vitamin E. A Com-prehensive Treatise* (ed. L.J. Machlin), Marcel Dekker, Inc., New York, pp. 7–65.

Kelleher, J. and Losowsky, M.S. (1970) The absorption of α-tocopherol in man. *Br. J. Nutr.*, **24**, 1033–47.

Kläui, H.M., Hausheer, W. and Huschke, G. (1970) Technological aspects of the use of fat-soluble vitamins and carotenoids and of the development of stabilized marketable forms. In *Fat-soluble Vitamins* (ed. R.A. Morton), Pergamon Press, New York, pp. 113–59.

Lawrence, R.A. and Burk, R.F. (1976) Glutathione peroxidase activity in selenium-deficient rat liver. *Biochem. Biophys. Res. Commun.*, **71**, 952–8.

Lea, C.H. (1960) On the antioxidant activities of the tocopherols. II. Influence of substrate, temperature and level of oxidation. *J. Sci. Food Agric.*, **11**, 212–18.

Lea, C.H. and Ward, R.J. (1959) Relative antioxidant activities of the seven tocopherols. *J. Sci. Food Agric.*, **10**, 537–48.

Lehmann, J., Martin, H.L., Lashley, E.L. *et al.* (1986) Vitamin E in foods from high and low linoleic acid diets. *J. Am. Diet. Assoc.*, **86**, 1208–16.

Leth, T. and Søndergaard, H. (1977) Biological activity of vitamin E compounds and natural materials by the resorption–gestation test, and chemical determination of the vitamin E activity in foods and feeds. *J. Nutr.*, **107**, 2236–43.

Leth, T. and Søndergaard, H. (1983) Biological activity of all-rac-α-tocopherol and *RRR*-α-tocopherol determined by three different rat bioassays. *Int. J. Vitam. Nutr. Res.*, **53**, 297–311.

Losowsky, M.S., Kelleher, J., Walker, B.E. *et al.* (1972) Intake and absorption of tocopherol. *Ann. NY Acad. Sci.*, **203**, 212–22.

Lucy, J.A. (1972) Functional and structural aspects of biological membranes: a suggested structural role for vitamin E in the control of membrane permeability and stability. *Ann. NY Acad. Sci.*, **23**, 4–11.

Machlin, I..J. (1984) Vitamin E. In *Handbook of Vitamins, Nutritional, Biochemical and Clinical Aspects* (ed. L.J. Machlin), Marcel Dekker, Inc., New York, pp. 99–145.

MacMahon, M.T. and Neale, G. (1970) The absorption of α-tocopherol in control subjects and in patients with intestinal malabsorption. *Clin. Sci.*, **38**, 197–210.

Marusich, W.L., Ackerman, G. and Bauernfeind, J.C. (1967) Biological efficacy of d- and dl-α-tocopheryl acetate in chickens. *Poultry Sci.*, **46**, 541–8.

Mason, K.E. (1944) Physiological action of vitamin E and its homologues. *Vitams Horm.* **2**, 107–53.

McLaughlin, P.J. and Weihrauch, J.C. (1979) Vitamin E content of foods. *J. Am. Diet. Ass.*, **75**, 647–65.

Meydani, M., Cohn, J.S., Macauley, J.B. *et al.* (1989) Postprandial changes in the plasma concentration of α- and γ-tocopherol in human subjects fed a fat-rich meal supplemented with fat-soluble vitamins. *J. Nutr.*, **119**, 1252–8.

Mongeau, R., Behrens, W.A., Madère, R. and Brassard, R. (1986) Effects of dietary fiber on vitamin E status in rats: dose-response to wheat bran. *Nutr. Res.*, **6**, 215–24.

National Research Council (1989) Fat-soluble vitamins. In *Recommended Dietary Allowances*, 10th edn, National Academy Press, Washington, DC, pp. 78–114.

Nelson, J.S. (1980) Pathology of vitamin E deficiency. In *Vitamin E. A Comprehensive Treatise* (ed. L.J. Machlin), Marcel Dekker, Inc., New York, pp. 397–428.

Olcott, H.S. and Emerson, O.H. (1937) Antioxidants and the autoxidation of fats. IX. The antioxidant properties of the tocopherols. *J. Am. Chem. Soc.*, **59**, 1008–9.

Olcott, H.S. and van der Veen, J. (1968) Comparison of antioxidant activities of tocol and its methyl derivatives. *Lipids*, **3**, 331–4.

Omaye, S.T. and Chow, F.I. (1984a) Comparison between meal-eating and nibbling rats fed diets containing hard red spring wheat bran: bioavailability of vitamins A and E and effects on growth. *Cereal Chem.*, **61**, 95–9.

Omaye, S.T. and Chow, F.I. (1984b) Effect of hard red spring wheat bran on the bioavailability of lipid-soluble vitamins and growth of rats fed for 56 days. *J. Food Sci.*, **49**, 504–6.

Omaye, S.T., Chow, F.I. and Betschart, A.A. (1983) In vitro interactions between dietary fiber and ^{14}C-vitamin D or ^{14}C-vitamin E. *J. Food Sci.*, **48**, 260–1.

Parker, R.S. (1989) Dietary and biochemical aspects of vitamin E. *Adv. Food Nutr. Res.*, **33**, 157–232.

Parkhurst, R.M., Skinner, W.A. and Sturm, P.A. (1968) The effect of various concentrations of tocopherols and tocopherol mixtures on the oxidative stability of a sample of lard. *J. Am. Oil Chem. Soc.*, **45**, 641–2.

Pennock, J.F., Hemming, F.W. and Kerr, J.D. (1964) A reassessment of tocopherol chemistry. *Biochem. Biophys. Res. Commun.*, **17**, 542–8.

Piironen, V., Varo, P. and Koivistoinen, P. (1987) Stability of tocopherols and tocotrienols in food preparation procedures. *J. Food Comp. Anal.*, **1**, 53–8.

Piironen, V., Varo, P. and Koivistoinen, P. (1988) Stability of tocopherols and tocotrienols during storage of foods. *J. Food Comp. Anal.*, **1**, 124–9.

Piironen, V., Varo, P., Syväoja, E.-L. *et al.* (1984) High-performance liquid chromatographic determination of tocopherols and tocotrienols and its application to diets and plasma of Finnish men. *Int. J. Vitam Nutr. Res.*, **4**, 35–40.

Piironen, V., Syväoja, E.-L., Varo, P. *et al.* (1985) Tocopherols and tocotrienols in Finnish foods: meat and meat products. *J. Agric. Food Chem.*, **33**, 1215–18.

Piironen, V., Syväoja, E.-L. Varo, P. *et al.* (1986a) Tocopherols and tocotrienols in cereal products from Finland. *Cereal Chem.*, **63**, 78–81.

Piironen, V., Syväoja, E.-L., Varo, P. *et al.* (1986b) Tocopherols and tocotrienols in Finnish foods: vegetables, fruits, and berries. *J. Agric. Food Chem.*, **34**, 742–6.

Pudelkiewicz, W.J., Matterson, L.D., Potter, L.M. *et al.* (1960) Chick tissue-storage bioassay of alpha-tocopherol: Chemical analytical techniques and relative biopotency of natural and synthetic alpha-tocopherol. *J. Nutr.*, **71**, 115–21.

Rammell, C.G., Cunliffe, B. and Kieboom, A.J. (1983) Determination of alpha-tocopherol in biological specimens by high-performance liquid chromatography. *J. Liquid Chromat.*, **6**, 1123–30.

Rammell, C.G. and Hoogenboom, J.J.L. (1985) Separation of tocols by HPLC on an amino-cyano polar phase column. *J. Liquid Chromat.*, **8**, 707–17.

Rhys Williams, A.T. (1985) Simultaneous determination of serum vitamin A and E by liquid chromatography with fluorescence detection. *J. Chromat., Biomed. Appl.*, **341**, 198–201.

Schaus, E.E., de Lumen, B.O., Chow, F.I. *et al.* (1985) Bioavailability of vitamin E in rats fed graded levels of pectin. *J. Nutr.*, **115**, 263–70.

Scott, C.G., Cohen, N., Riggio, P.P. and Weber, G. (1982) Gas chromatographic assay of the diastereomeric composition of *all-rac-α*-tocopheryl acetate. *Lipids*, **17**, 97–101.

Scott, M.L. (1970) Studies on vitamin E and related factors in nutrition and metabolism. In *The Fat-Soluble Vitamins* (eds H.F. DeLuca and J.W. Suttie), The University of Wisconsin Press, Madison, pp. 355–68.

Scott, M.L. (1978) Vitamin E. In *Handbook of Lipid Research*, Vol. II (ed. H.F. DeLuca), Plenum Press, New York, pp. 133–210.

Scott, M.L. (1980) Advances in our understanding of vitamin E. *Fed. Proc.*, **39**, 2736–9.

Sheppard, A.J., Pennington, J.A.T. and Weihrauch, J.L. (1993) Analysis and distribution of vitamin E in vegetable oils and foods. In *Vitamin E in Health and Disease* (ed. L. Packer and J. Fuchs), Marcel Dekker, Inc., New York, pp. 9–31.

Shin, T.-S. and Godber, J.S. (1993) Improved high-performance liquid chromatography of vitamin E vitamers on normal-phase columns. *J. Am. Oil Chem. Soc.*, **70**, 1289–91.

Shmulovich, V.G. (1994) Interrelation of contents of unsaturated fatty acids and vitamin E in food product lipids. *Appl. Biochem. Microbiol.*, **30**, 547–51.

Smith, C.L., Kelleher, J., Losowsky, M.S. and Morrish, N. (1971) The content of vitamin E in British diets. *Br. J. Nutr.*, **26**, 89–96.

Speek, A.J., Schrijver, J. and Schreurs, W.H.P. (1985) Vitamin E composition of some seed oils as determined by high-performance liquid chromatography with fluorometric detection. *J. Food Sci.*, **50**, 121–4.

Stern, M.H., Robeson, C.D., Weisler, L. and Baxter, J.G. (1947) δ-Tocopherol. I. Isolation from soybean oil and properties. *J. Am. Chem. Soc.*, **69**, 869–74.

Syväoja, E.-L., Salminen, K., Piironen, V. *et al.* (1985a) Tocopherols and tocotrienols in Finnish foods: fish and fish products. *J. Am Oil Chem. Soc.*, **62**, 1245–8.

Syväoja, E.-L., Piironen, V., Varo, P. *et al.* (1985b) Tocopherols and tocotrienols in Finnish foods: human milk and infant formulas. *Int. J. Vitam. Nutr. Res.*, **55**, 159–66.

Syväoja, E.-L., Piironen, V., Varo, P. *et al.* (1985c) Tocopherols and tocotrienols in Finnish foods; dairy products and eggs. *Milchwissenschaft*, **40**, 467–9.

Syväoja, E.-L., Piironen, V., Varo, P. *et al.* (1986) Tocopherols and tocotrienols in Finnish foods: oils and fats. *J. Am. Oil Chem. Soc.*, **63**, 328–9.

Taylor, S.L., Lamden, M.P. and Tappel, A.L. (1976) Sensitive fluorometric method for tissue tocopherol analysis. *Lipids*, **11**, 530–8.

Thompson, J.N. (1982) Trace analysis of vitamins by liquid chromatography. In *Trace Analysis*, Vol. II (ed. J.F. Lawrence), Academic Press, New York, pp 1–67.

Thompson, J.N. and Hatina, G. (1979) Determination of tocopherols and tocotrienols in foods and tissues by high performance liquid chromatography. *J. Liquid Chromat.*, **2**, 327–44.

Thompson, J.N. and Scott, M.L. (1969) Role of selenium in the nutrition of the chick. *J. Nutr.*, **97**, 335–42.

Thompson, J.N. and Scott, M.L. (1970) Impaired lipid and vitamin E absorption related to atrophy of the pancreas in selenium-deficient chicks. *J. Nutr.*, **100**, 797–809.

Thompson, J.N., Erdody, P. and Maxwell, W.B. (1972) Chromatographic separation and spectrophotofluorometric determination of tocopherols using hydroxyalkoxypropyl Sephadex. *Analyt. Biochem.*, **50**, 267–80.

Thompson, J.N., Beare-Rogers, J.L., Erdödy, P. and Smith, D.C. (1973) Appraisal of human vitamin E requirement based on examination of individual meals and a composite Canadian diet. *Am. J. Clin. Nutr.*, **26**, 1349–54.

Traber, M.G. and Kayden, H.J. (1989) Preferential incorporation of α-tocopherol vs γ-tocopherol in human lipoproteins. *Am. J. Clin. Nutr.*, **49**, 517–26.

Traber, M.G., Cohn, W. and Muller, D.P.R. (1993) Absorption, transport and delivery to tissues. In *Vitamin E in Health and Disease* (eds L. Packer and J. Fuchs), Marcel Dekker, Inc., New York, pp. 35–51.

Traber, M.G., Olivecrona, T. and Kayden, H.J. (1985) Bovine milk lipoprotein lipase transfers tocopherol to human fibroblasts during triglyceride hydrolysis in vitro. *J. Clin. Invest.*, **75**, 1729–34.

Traber, M.G., Kayden, H.J., Green, J.B. and Green, M.H. (1986) Absorption of water-miscible forms of vitamin E in a patient with cholestasis and in thoracic duct-cannulated rats. *Am. J. Clin. Nutr.*, **44**, 914–23.

Traber, M.G., Ingold, K.U., Burton, G.W. and Kayden, H.J. (1988) Absorption and transport of deuterium-substituted $2R, 4'R, 8'R$-α-tocopherol in human lipoproteins. *Lipids*, **23**, 791–7.

Traber, M.G., Burton, G.W., Ingold, K.U. and Kayden, H.J. (1990a) *RRR*- and *SRR*-α-tocopherols are secreted without discrimination in human chylomicrons, but *RRR*-α-tocopherol is preferentially secreted in very low density lipoproteins. *J. Lipid Research*, **31**, 675–85.

Traber, M.G., Rudel, L.L., Burton, G.W. *et al.* (1990b) Nascent VLDL from liver perfusions of cynomolgus monkeys are preferentially enriched in *RRR*-compared with *SRR*-α-tocopherol: studies using deuterated tocopherols. *J. Lipid Res.*, **31**, 687–94.

Tuan, S., Lee, T.F., Chou, C.C. and Wei, Q.K. (1989) Determination of vitamin E homologues in infant formulas by HPLC using fluorometric detection. *J. Micronutr. Anal.*, **6**, 35–45.

Ueda, T. and Igarashi, O. (1985) Evaluation of the electrochemical detector for the determination of tocopherols in feeds by high-performance liquid chromatography. *J. Micronutr. Anal.*, **1**, 31–8.

Ueda, T. and Igarashi, O. (1987) New solvent system for extraction of tocopherols from biological specimens for HPLC determination and the evaluation of 2,2,5,7,8-pentamethyl-6-chromanol as an internal standard. *J. Micronutr. Anal.*, **3**, 185–98.

Vester, J.W. and Williams, L.R. (1963) Muscle degeneration in association with apparent vitamin E deficiency in a human. *Clin. Res.*, **11**, 180.

Weiser, H. and Vecchi, M. (1981) Stereoisomers of α-tocopheryl acetate. Characterization of the samples by physico-chemical methods and determination of biological activities in the rat resorption–gestation test. *Int. J. Vitam. Nutr. Res.*, **51**, 100–113.

Weiser, H. and Vecchi, M. (1982) Stereoisomers of α-tocopheryl acetate. II. Biopotencies of all eight stereoisomers individually or in mixtures, as determined by rat resorption–gestation tests. *Int. J. Vitam. Nutr. Res.*, **52**, 351–70.

Widicus, W.A. and Kirk, J.R. (1979) High performance liquid chromatographic determination of vitamins A and E in cereal products. *J. Ass. Off. Analyt. Chem.*, **62**, 637–41.

Witting, L.A. (1970) The interrelationship of polyunsaturated fatty acids and antioxidants *in vivo*. *Prog. Chem. Fats Lipids*, **9**, 517–53.

Witting, L.A. (1975) Vitamin E as a food additive. *J. Am. Oil Chem. Soc.*, **52**, 64–8.

Witting, L.A. and Lee, L. (1975) Dietary levels of vitamin E and polyunsaturated fatty acids and plasma vitamin E. *Am. J. Clin. Nutr.*, **28**, 571–6.

Woollard, D.C. and Blott, A.D. (1986) The routine determination of vitamin E acetate in milk-powder formulations using high-performance liquid chromatography. *J. Micronutr. Anal.*, **2**, 97–115.

Woollard, D.C., Blott, A.D. and Indyk, H. (1987) Fluorometric detection of tocopheryl acetate and its use in the analysis of infant formulas. *J. Micronutr. Anal.*, **3**, 1–14.

6

Vitamin K

6.1 INTRODUCTION

The first indication of the existence of vitamin K occurred in 1929 whilst Henrik Dam was investigating the possible essentiality of cholesterol in the diet of the chick. When the chicks were fed diets which had been extracted with nonpolar solvents to remove the sterols, they developed internal haemorrhages and blood taken from these chicks clotted slowly. In 1935, Dam proposed that the curative factor present in vegetable and animal sources was a new fat-soluble vitamin which he called vitamin K. The first isolation of vitamin K_1 from alfalfa was independently reported by the teams of Doisy, Dam and Karrer in 1939. Vitamin K_2 was isolated from putrified fish meal by Doisy's group in 1941.

6.2 CHEMICAL STRUCTURE AND NOMENCLATURE

The term 'vitamin K' is a generic one used for all compounds possessing cofactor activity for γ-glutamylcarboxylase. Two forms of vitamin K exist in nature: vitamin K_1 is synthesized by green plants in the chloroplasts and vitamin K_2 is synthesized by bacteria. The chemical structures of these compounds are shown in Figure 6.1.

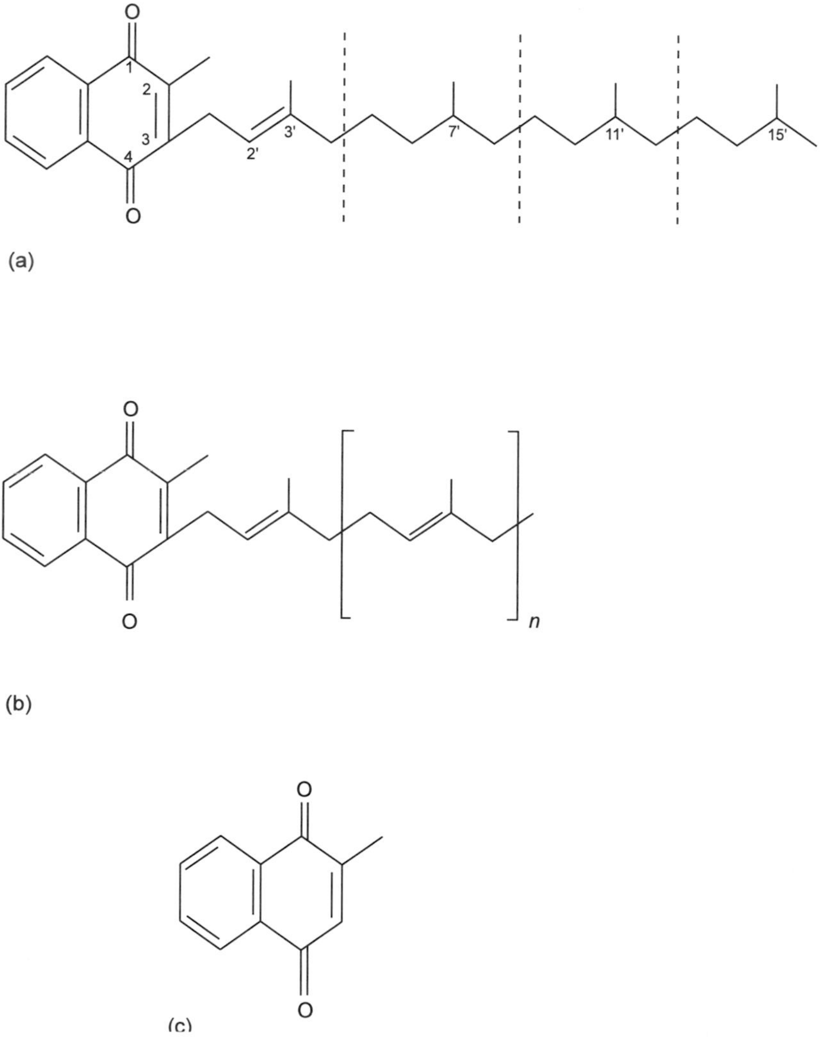

(a)

(b)

(c)

Figure 6.1 Structures of the vitamin K-active compounds. (a) Phylloquinone (vitamin K_1); (b) menaquinones (vitamin K_2); (c) menadione. The vertical dotted lines in (a) delineate the four isoprene units.

Vitamin K_1 is called phylloquinone by biochemists and nutritionists, but in the British and United States Pharmacopoeias its name is phytomenadione. Phylloquinone $(C_{31}H_{46}O_{21}; MW = 450.68)$ comprises a methyl-substituted naphthoquinone nucleus attached at C-3 to a phytyl side chain composed of three saturated and one unsaturated isoprene units, each unit having five carbon atoms. Phylloquinone therefore has a 20-carbon side chain and may be designated $K_{1(20)}$ to distinguish it from structural analogues such as $K_{1(25)}$, a synthetic compound with an extra isoprene unit. The natural form of phylloquinone is the *trans* isomer, whose stereochemistry is $2'$-*trans*, $7'R, 11'R$. Synthetic phylloquinone usually contains both the *trans* and the *cis* isomers, but only the *trans* isomer is essentially responsible for the vitamin's cofactor activity.

Vitamin K_2 refers to a family of structural analogues called menaquinones, the predominant members of which have side chains composed of 6–10 unsaturated isoprene units. They are designated MK_n (where n is the number of isoprene units), e.g. menaquinone-6 (MK-6).

The underivatized 2-methyl-1,4-naphthoquinone structure, known as menadione (vitamin K_3), is biologically active by virtue of its conversion to menaquinone-4 by the body tissues. Menadione itself is not a natural product, but menadione sodium bisulphite and other water-soluble derivatives are synthesized commercially for use as animal feed supplements. Menadione is toxic to infants at excessive dose levels (Omaye, 1984) and hence is not used in human medicine or as a food supplement.

6.3 BIOPOTENCY

There is presently little information on the relative biopotency of different K vitamers for humans, except that the hepatic turnover of long-chain menaquinones is very much slower than that of phylloquinone (Usui *et al.*, 1990).

6.4 DEFICIENCY SYNDROMES

In adult humans, a prolonged blood clotting time (e.g. prolonged prothrombin time) is the predominant, if not sole, clinical sign of vitamin K deficiency. Abnormal blood coagulation is more likely to arise from secondary causes such as malabsorption syndromes or biliary obstruction than from a dietary inadequacy of vitamin K. Subclinical deficiency has been induced in healthy adults by dietary deprivation of the vitamin (Ferland, Sadowski and O'Brien, 1993) and this may have important implications in vitamin K nutriture.

Breast-fed newborn infants (neonates), especially if premature, are susceptible to a vitamin K deficiency syndrome known as haemorrhagic

disease of the newborn in the first week of life. A case of the haemorrhagic syndrome has been defined by the British Paediatric Surveillance
Unit as: 'Any infant under six months of age with spontaneous bruising/
bleeding or intracranial haemorrhage associated with prolonged clotting
times, not due to an inherited coagulopathy or disseminated intravascu-
lar coagulation' (McNinch and Tripp, 1991). Newborn infants are thought
to be susceptible to this syndrome because (1) the placenta imposes a
barrier to the transport of vitamin K to the foetus, (2) there is inadequate
colonization of intestinal microflora during the first few days of life, and
(3) human milk, containing a concentration of around 1 μg of phylloquinone per litre (Lambert, Vanneste and De Leenheer, 1992), provides
less than the minimum requirement of vitamin K (5 μg/day). In cases of
the haemorrhagic syndrome beyond the first days of life, liver dysfunc-
tion rather than breast-feeding may be the major risk factor (Shearer,
1992). Neonates in many countries now routinely receive intramuscular
or (less effectively) oral doses of vitamin K as a prophylactic measure
against the syndrome.

6.5 BIOCHEMICAL AND PHYSIOLOGICAL FUNCTIONS

The vitamin K-dependent glutamate γ-carboxylation reaction

Vitamin K functions as a cofactor for the enzyme γ-glutamyl carboxylase,
which catalyses a unique post-translational conversion of glutamate
(Glu) residues into γ-carboxyglutamate (Gla) residues at well-defined
sites in a limited number of proteins (Figure 6.2). Because Glu is only a
weak Ca^{2+} chelator and Gla a much stronger one, the vitamin K-dependent step subsequently increases the Ca^{2+}-binding capacity of a protein
(Vermeer, 1990). Several Gla-containing proteins of hepatic origin have

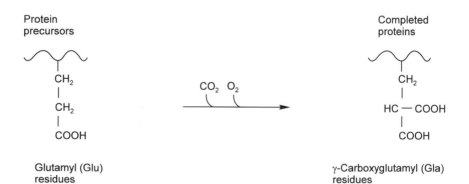

Figure 6.2 The vitamin K-dependent glutamate γ-carboxylation reaction. (From Shearer, 1993.)

a regulatory function in haemostasis; factor II (prothrombin) and factors VII, IX and X have a coagulant function, while proteins C and S have an anticoagulant function (Shearer, 1992). In bone metabolism, Gla residues facilitate the binding of calcium ions to osteocalcin, a protein synthesized by the bone-forming cells (osteoblasts) (Vermeer, Jie and Knapen, 1995).

The vitamin K epoxide cycle

In the liver, γ-glutamyl carboxylation is linked to the vitamin K epoxide cycle – a metabolic cycle which serves to conserve the pool of vitamin K available to the carboxylase (Figure 6.3). The vitamin K must be in the reduced (hydroquinone) form for carboxylation to take place. The carboxylation reaction is catalysed by γ-glutamyl carboxylase (E_1) coupled to a vitamin K epoxidase (E_2) which simultaneously converts vitamin K hydroquinone to vitamin K 2,3-epoxide. The epoxide is reduced to the quinone by vitamin K epoxide reductase (E_3). The cycle is completed by the reduction of the recycled quinone by a vitamin K reductase (K_4). The reductase enzymes E_3 and E_4 are dithiol-dependent and are inhibited by coumarin vitamin K antagonists such as warfarin. It is probable that E_3 and E_4 are one and the same enzyme. Exogenous vitamin K from the diet may enter the cycle via an NAD(P)H-dependent vitamin K reductase (E_5), which is not inhibited by warfarin (Shearer, 1992).

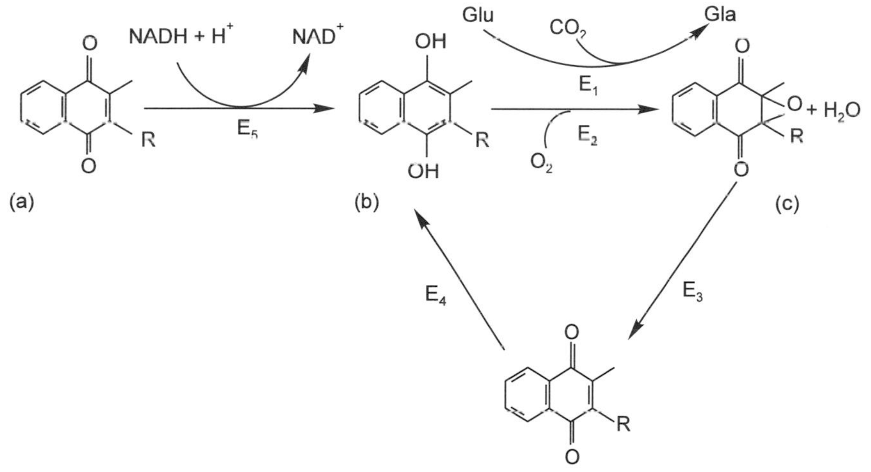

Figure 6.3 The vitamin K epoxide cycle (from Vermeer, 1990). (a) Vitamin K quinone; (b) vitamin K hydroquinone; (c) vitamin K 2,3-epoxide. The enzymes E_1 to E_5 are identified in the text. (Reproduced with permission from *Biochem. J.*, **266**, 625–36, Vermeer, Copyright 1990, Biochemical Society.)

6.6 PHYSICOCHEMICAL PROPERTIES

Appearance and solubility

Phylloquinone is a golden yellow viscous oil; the various menaquinones can be obtained in crystalline form; and menadione is a yellow crystalline powder.

Phylloquinone is insoluble in water; sparingly soluble in alcohol; and readily soluble in acetone, ether, chloroform, hexane, fat and oils. Menadione is soluble in organic solvents and only slightly soluble in water, but the bisulphite derivatives are water-soluble.

Spectroscopic properties

Absorption
Phylloquinone and the menaquinones all possess the same chromophore and exhibit identical UV absorption spectra, which contain five maxima. The ε value at the λ maximum of 248 nm is 18 900 (Dunphy and Brodie, 1971).

Fluorescence
Vitamin K compounds do not exhibit native fluorescence, but the hydro-naphthoquinone reduction products are highly fluorescent.

Stability

Vitamin K compounds are decomposed by UV radiation, alkali, strong acids, and reducing agents, but are reasonably stable to oxidizing conditions and to heat.

6.7 ANALYSIS

6.7.1 Scope of analytical techniques

Reliable methods of determining vitamin K in foods only became available in the 1980s with the development of specific HPLC techniques. Among the various vitamin K compounds, only phylloquinone is accounted for in routine food analysis. Milk-based and soy protein-based infant formulas for the full-term infant are supplemented with synthetic preparations of phylloquinone, which invariably contain about 10% of the biologically inactive *cis* isomer (Schneiderman *et al.*, 1988). Ideally, the analytical method should exclude *cis*-phylloquinone in the measurement.

HPLC techniques using gradient elution are capable of separating the individual menaquinones from one another and from phylloquinone, and such techniques have been applied to the determination of these vitamers in human (Usui *et al.*, 1989) and bovine (Hirauchi *et al.*, 1989a) liver.

6.7.2 Bioassays

Bioassays have played an important role in vitamin K research because, until the advent of direct chromatographic methods, no satisfactory physicochemical method has existed. A curative chick bioassay is primarily used to assay infant formulas and special dietary foods, but it may also be used with any food product that contains a minimum of 10 µg of phylloquinone equivalent/100 g (Robaidek, 1983). The assay is based upon the degree of lowering of elevated blood clotting times in vitamin K-depleted chicks. Blood clotting measurements (actually prothrombin times) are rapidly determined following the addition of a clotting agent (thromboplastin) and calcium chloride solution to oxalated or citrated blood. The chick is the animal of choice because its vitamin K requirement is five-fold that of the rat, it is readily depleted of vitamin K, and coprophagy is easier to control. The chick's higher requirement for vitamin K compared with the rat is at least partly attributable to the short length of its colon and rapid transit time.

A standard bioassay method has not been established, but the following describes the general procedure that is usually adopted. Day-old chicks are fed a vitamin K-deficient maintenance (basal) diet for 10–14 days, and vitamin K depletion is checked in a few chicks by determining their prothrombin times. If the blood fails to clot in 20 min, depletion is satisfactory. The remaining chicks are divided into test groups of 10 birds or more, and a reference phylloquinone standard or the material to be assayed is fed every day over a 2-week period. Two or more groups are given different known concentrations of phylloquinone to establish a standard response curve. At least two levels of the test food material are given to other groups by mixing it with the vitamin K-free maintenance diet. Another group of chicks, the positive control, is given a diet with sufficient vitamin K to meet the requirements fully. To obtain a linear response curve, the reciprocal of the prothrombin times of chicks receiving known amounts of vitamin K is plotted against the log dose. The potency of the sample is established by comparing clotting times of chicks fed the sample with those of chicks receiving known amounts of phylloquinone in their diet (Parrish, 1980).

6.7.3 Extraction techniques

Enzymatic hydrolysis
Enzymatic hydrolysis using lipase has been employed as an alternative to saponification for removing triglycerides without destroying the

alkali-labile vitamin K. When determining endogenous vitamin K in milk, it should be borne in mind that the milk fat globules are encapsulated by a membrane that may retard the action of lipase on the triglycerides. Indyk *et al* (1995) used an ultrasonic bath in which to incubate milk samples with lipase as a means of disrupting the membrane.

Direct solvent extraction
For the analysis of infant formulas, Ayi and Burgher (1985) modified the Röse-Gottlieb procedure by replacing the ammonia/ethanol treatment by acidified ethanol.

Shearer (1993) extracted vitamin K from vegetables, fruits, cereals, meats and fish by grinding these foods in a mortar with fine quartz granules before extracting with acetone. After the addition of water and hexane to the acetone extract, the phylloquinone partitioned entirely in the upper hexane phase, leaving polar impurities in the acetone/water phase.

Powdered formulas, in which the phylloquinone has been added in coated beadlet form, need to be reconstituted in water prior to solvent extraction.

Supercritical fluid extraction
Schneiderman *et al.* (1988) extracted phylloquinone from powdered infant formulas using supercritical CO_2 at 8000 psi and 60 °C for 15 min. The extracted material was recovered by depressurization of the CO_2 across an adsorbent trap, and then washed from the trap with a small volume of dichloromethane/acetone (50/50). The resultant solution was evaporated to dryness and the residue was dissolved in mobile phase for direct HPLC analysis. Trial experiments gave recoveries of 92% of phylloquinone from a Chromosorb W matrix.

6.7.4 High-performance liquid chromatography

Some of the more recent HPLC methods for determining phylloquinone in a variety of foods are summarized in Table 6.1.

Detection
Phylloquinone, with a molar absorptivity (ε) of around 19 000 in the 240–270 nm region, can be detected photometrically at levels down to 500 pg on-column. Photometric detection is satisfactory for the analysis of green leafy vegetables and vitamin K-fortified infant formulas, after purification of the sample to remove interfering lipids, but it lacks the required sensitivity for the analysis of other foods. The reversible oxidation–reduction between the quinone and hydroquinone (quinol) forms of phylloquinone allows this vitamin to be detected electrochemically with a

10-fold increase in sensitivity over photometric detection and an improved selectivity. The first combined HPLC–electrochemical measurements of vitamin K used the reductive mode, but this technique suffered from interference from the reduction of oxygen. A redox method was later developed that eliminated interference from the reduction of oxygen, and provided an increase in the selectivity and sensitivity of detection. The coulometric detector employed in this method is equipped with a dual-electrode cell in which vitamin K is first reduced upstream at the generator electrode and the hydroquinone is re-oxidized downstream at the detector electrode.

The hydroquinone reduction products of vitamin K, unlike the quinone forms, are highly fluorescent; and electrochemical (Ayi and Burgher, 1985; Langenberg *et al.*, 1986) or chemical (Booth, Davidson and Sadowski, 1994) reduction of vitamin K prior to fluorimetric detection has been utilized. Indyk (1988) reported that a commercial fluorescence detector facilitates the photochemical reduction and simultaneous fluorescence detection of phylloquinone during normal passage of the column effluent through the flow cell.

Careri *et al.* (1996) reported on the use of particle beam–mass spectrometry for the determination and unequivocal identification of phylloquinone in some vegetable samples. The proposed LC–MS method permitted phylloquinone assay at levels down to 0.1 µg/g with high specificity.

Normal-phase systems

Adsorption chromatography using silica columns facilitates the separation of the inactive *cis* isomer of phylloquinone from the active *trans* isomer. Hwang (1985) applied adsorption HPLC to the determination of phylloquinone in infant formulas using photometric detection at 254 nm. Both the *cis* and *trans* isomers could be measured in standards and in liquid formulas, but matrix interferences prevented measurement of the *cis* isomer in powdered formulas. The detection limit of the assay at a signal/noise ratio of 3 was 0.3 ng for both isomers, equivalent to 2 µg/litre of phylloquinone in the as-fed product.

Reversed-phase systems

Reversed-phase chromatography can separate phylloquinone from closely related structures, but cannot separate *cis*- and *trans*-phylloquinone.

In a method developed by Shearer (1993) for determining phylloquinone, food sample extracts are purified by silica solid-phase extraction and semi-preparative adsorption HPLC and then analysed by reversed-phase HPLC. The retention window for collecting the fraction containing the phylloquinone and internal standard excludes *cis*-phylloquinone.

Table 6.1 HPLC methods used for the determination of vitamin K in foods

Food	Sample preparation	Semi-preparative HPLC	Quantitative HPLC	Vitamers separated	Reference
Infant formulas	Mix sample with concentrated NH_4OH/MeOH, extract with CH_2Cl_2/isooctane (2 + 1). Silica column chromatography		*Normal-phase* Apex Silica 5 μm 250 × 4.6 mm i.d. Isooctane/CH_2Cl_2/2-PrOH (70 + 30 + 0.02) UV 254 nm	Cis- and *trans*-phylloquinone	Hwang (1985)
Infant formulas	Digest sample with lipase, extract with pentane		*Reversed-phase* μBondapak C_{18} 10 μm 250 × 4 mm i.d. MeOH/MeCN/THF/H_2O (39:39:16:6) UV 254 nm	Phylloquinone	Ueno and Villalobos (1983)
Soybean oil	Digest sample with lipase, extract with pentane. Alumina column chromatography		Supelcosil LC-18 5 μm 150 × 4.6 mm i.d. MeOH/MeCN/H_2O (88:10:2) UV 270 nm	Phylloquinone	Bonta and Stancher (1985)
Milk (human and bovine)	Deproteinize milk sample with EtOH at 4°C with added synthetic phylloquinone homologue (internal standard). Extract with hexane aided by sonication and vortexing. Centrifuge, evaporate hexane layer to dryness	μPorasil silica 10 μm 300 × 8 mm i.d. Heptane/ethyl acetate (99:1) UV 254 nm	Resolve C_{18} 5 μm 150 × 3.9 mm i.d. 100% MeCN UV 270 nm	Phylloquinone	Fournier et al. (1987)
Infant formulas, endogenous vitamin K in milk	Dissolve powders in warm water, add 0.8 M phosphate buffer (pH 8.0) and lipase, shake for 5 min. Incubate tubes at 37°C for 2h in an ultrasonic bath, shaking tubes every 20 min. Cool to ambient temperature. Add EtOH/MeOH (95 + 5) + solid potassium carbonate + cholesteryl phenylacetate (internal standard) and extract twice with hexane by shaking and centrifuging. Evaporate combined hexane extracts to near dryness at 40°C under vacuum and store at 4°C for 24 h if necessary. Evaporate to dryness and redissolve in hexane	Radial-PAK cartridge containing Resolve silica 5 μm Hexane/2-PrOH (99.1 + 0.1) UV 269 nm	Radial-PAK cartridge containing Resolve C_{18} 5 μm MeOH/2-PrOH/ethyl acetate/water (450 + 350 + 145 + 135) Dual-wavelength UV 269 and 277 nm	Phylloquinone, cholesteryl phenylacetate (internal standard)	Indyk et al. (1995)

Food	Extraction/preparation	HPLC purification	HPLC analysis and detection	Vitamers	Reference
Infant formulas	Supercritical fluid extraction		μBondapak C_{18} 10 μm; 150 × 3.9 mm i.d.; MeCN/CH_2Cl_2/aqueous 0.025 M sodium perchlorate (90:5:5); Amperometric detection (reductive mode), silver electrode, −1.1 V vs saturated calomel reference electrode	Phylloquinone	Schneiderman et al. (1988)
Milk (human)	Digest milk sample with lipase, extract with pentane	Nucleosil C 18 5 μm; 300 × 8 mm i.d.; MeCN/MeOH (1:1); Column temperature 35°C; UV 248 nm	Partisil ODS-2 5 μm; 250 × 4.6 mm i.d.; MeOH/EtOH/60% perchloric acid (600:400:1.2) containing 0.05 M sodium perchlorate in total solution; Dual electrochemical detection (redox mode): −450 mV (generator) +350 mV (detector)	Phylloquinone, menaquinone-4	Isshiki et al. (1988)
Milk (human)	Extract vitamin K with 3 vol of 2-PrOH/hexane (3:2), centrifuge. Evaporate three times from 10 vol of $CHCl_3$/MeOH (2:1). Silica column chromatography. Add menaquinone-7 as internal standard to purified fraction	Radial-PAK cartridge containing C_{18} 5 μm; Convex gradient of EtOH/H_2O (90:10) to EtOH/hexane (90:10); UV 254 nm	Radial-PAK cartridge containing C_{18} 10 μm; EtOH/hexane/H_2O (90:6.5:3.5) containing 0.025M TBAP; Amperometric detection (redox mode): −600 mV (reductive electrode) +200 mV (oxidative electrode)	Phylloquinone, menaquinone-7 (internal standard)	Canfield et al. (1990)
All food types	*Vegetables, fruits, cereals, meat and fish* extract sample with acetone, filter, partition the phylloquinone into hexane. *Fats, oils and dairy products*: digest sample with lipase, extract with hexane. Purify hexane extracts obtained from either extraction technique by silica solid-phase extraction. The internal standard is phylloquinone 2,3-epoxide (unlabelled for UV detection and tritium-labelled for coulometric detection)	A. Partisil silica 5 μm; 250 × 4.6 mm i.d.; 50% water-saturated CH_2Cl_2/hexane (15:85); UV 254 nm. B. Spherisorb nitrile 5 μm 250 × 4.6 mm i.d.; 50% water-saturated CH_2Cl_2 / hexane (3:97); UV 254 nm	A. Hypersil ODS; 250 × 4.6 mm i.d.; CH_2Cl_2 / MeOH (15:85); UV 270 nm. B. Spherisorb C_8 5 μm; 250 × 4.6 mm i.d.; MeOH/0.05 M acetate buffer pH 3.0 (97:3) containing 0.1 mM EDTA; Dual-electrode coulometric detection (redox mode), porous graphite electrodes, −1.5 V (generator electrode) +0.05 V (detector electrode)	Phylloquinone	Shearer (1986, 1993)

Table 6.1 Continued

Food	Sample preparation	Semi-preparative HPLC	Quantitative HPLC	Vitamers separated	Reference
Various foods (vegetable juice, whole milk, raw spinach leaves, plain bagel, raw ground beef)	*Vegetable juice*: add dihydrophylloquinone (internal standard) to sample. Extract with 2-PrOH/hexane (3 + 2) and water aided by sonication and vortexing. Silica solid-phase extraction *Milk*: as described for vegetable juice followed by liquid-phase reductive extraction for removal of lipids *Spinach*: homogenize and grind sample with sodium sulphate. Extraction and clean-up as for vegetable juice *Bread*: add $K_{1(25)}$ (internal standard) to ground sample. Extraction and clean-up as for vegetable juice *Beef*: as described for bread followed by C_{18} solid-phase extraction		Hypersil ODS 3 μm 150 × 4.6 mm i.d. $MeOH/CH_2Cl_2$ (90:10) containing 0.01 M zinc chloride, 0.005 M acetic acid and 0.005 M sodium acetate Post-column chemical reactor column packed with zinc metal Fluorescence: ex. 244 nm em. 418 nm	Phylloquinone, dihydrophylloquinone (internal standard)	Booth, Davidson and Sadowski (1994)
Egg, whole milk, yoghurt, cheese (Emmentaler), oatmeal, carrot, potato, broccoli, cauliflower, edible oils	Blend samples with appropriate solvent* and added dihydrophylloquinone (internal standard) *Cheese, oatmeal, broccoli, cauliflower:* $CH_2Cl_2/MeOH$ (2:1) *Egg, milk, yoghurt, carrot, potato:* 2-PrOH/ hexane (3:1) *Edible oils:* 100% hexane. Centrifuge. Evaporate sample extracts to dryness (except edible oils) and dissolve residue in hexane. Add equal volume of $MeOH/H_2O$ (9:1) then mix and centrifuge. Remove upper hexane layer and evaporate to dryness. Dissolve residue in 150 μl of the HPLC mobile phase		Hypersil ODS 5 μm 250 × 4.6 mm i.d. Mobile phase: 900 ml MeOH + 100 ml CH_2Cl_2 + 5 ml of a methanolic solution containing 1.37 g zinc chloride, 0.41 g sodium acetate and 0.30 g acetic acid Column temperature 40 °C Post-column chemical reactor column packed with zinc powder Fluorescence: ex. 243 nm em. 430 nm	Phylloquinone, dihydrophylloquinone (internal standard)	Jakob and Elmadfa (1996)

Sample	Sample preparation	Analysis	Analyte	Reference
Canola (rapeseed) oil	Digest oil sample + internal standard (menaquinone-4) with lipase in pH 7.7 buffer at 37 °C, extract with hexane. Silica solid-phase extraction	PartiSphere C$_{18}$ 5 μm 150 × 4.6 mm i.d. 350 ml MeOH + 150 ml MeCN + 5 ml MeOH/MeCN (85:15) solution containing 2 M zinc chloride, 1 M sodium acetate and 1 M acetic acid (pH of total solution, 3.3) Post-column chemical reduction with zinc dust column Fluorescence detection: ex. 254 nm (filter) em. 400 nm (filter)	Phylloquinone, menaquinone-4 (internal standard)	Gao and Ackman (1995)
Vegetables (raw and processed)	Homogenize with 2-PrOH, partition into hexane	Hypersil-MOS 5 μm MeOH/H$_2$O (92.5 + 7.5) containing 0.03 M sodium perchlorate Coulometric reduction: −500 mV Fluorescence detection: ex. 320 nm em. 430 nm	Phylloquinone	Langenberg et al. (1986)
Soybean oil	Shake oil sample Add K$_{1(25)}$ (internal standard) to oil sample. Shake with 0.9% NaCl solution + EtOH and partition vitamin K into hexane. Centrifuge, evaporate to dryness. Silica solid-phase extraction	Beckman XL C$_8$ 3 μm 70 × 4.7 mm i.d. cartridge MeCN/EtOH (95:5) containing 0.005 M sodium perchlorate Electrochemical reduction followed by fluorescence detection	Phylloquinone, K$_{1(25)}$ (internal standard)	Moussa et al. (1994)
Vegetables (carrot, tomato, Brussels sprouts, spinach)	Ultrasonically shake sample with MeOH, centrifuge. Mix with solid sodium carbonate and heat at 80 °C for 1 h. Partition the alkaline solution with hexane by vortex-mixing, centrifuge Extract three times with hexane. Evaporate the pooled extracts to a low volume at 35 °C under vacuum. Evaporate to dryness under N$_2$ and redissolve final residue in MeOH, filter (0.2 μm)	LiChrosorb RP-8 10 μm 250 × 4.6 mm i.d. 100% MeOH UV 247 nm or particle beam-mass spectrometry using negative-ion CI detection	Phylloquinone	Careri et al. (1996)

Abbreviations: see footnote to Table 2.4.
TBAP, tetrabutyl ammonium perchlorate.

NARP–HPLC can be used in conjunction with UV absorbance detection, but not with electrochemical detection because the mobile phase is not polar enough to dissolve the electrolyte needed to conduct a current. With electrochemical detection, therefore, a semi-aqueous mobile phase is used in conjunction with a less retentive octyl-bonded (C_8) stationary phase. The addition of 0.1 mM EDTA to the semi-aqueous mobile phase prevents the reduction of metal ions at the generator electrode. With UV detection, the phylloquinone is quantified by the method of peak height ratios using phylloquinone 2,3-epoxide as the internal standard. The epoxide is electrochemically inactive and therefore, when electrochemical detection is used, quantification is accomplished by the technique of radioisotopic dilution using tritiated phylloquinone 2,3-epoxide as the internal standard.

In a method proposed by Booth, Davidson and Sadowski (1994) for the determination of phylloquinone in various food types, extracted samples are subjected to silica solid-phase extraction, followed (in the case of meat or milk samples) by further purification by reversed-phase solid-phase extraction or liquid-phase reduction extraction, respectively. The residues obtained after evaporation of the eluates or reductive extracts are dissolved in 100% dichloromethane, followed immediately by the addition of 10 mM zinc chloride solution, 5 mM acetic acid and 5 mM sodium acetate. This final test solution is analysed by reversed-phase HPLC using a mobile phase composed of methanol/dichloromethane (90/10) containing 10 mM zinc chloride, 5 mM acetic acid and 5 mM sodium acetate. The fluorescent hydroquinone reduction products of the injected phylloquinone and internal standard are produced on-line using a post-column chemical reactor packed with zinc metal. A synthetic analogue of phylloquinone, $K_{1(25)}$, was later used as the internal standard for all food commodities.

6.8 BIOAVAILABILITY

6.8.1 Physiological aspects

Absorption and transport of dietary phylloquinone in the small intestine

Ingested phylloquinone, together with the other fat-soluble vitamins, is solubilized within mixed micelles in the duodenum and absorbed in the proximal small intestine. The absorbed phylloquinone enters the systemic circulation chemically unchanged via the lymphatic route in association with chylomicrons (Blomstrand and Forsgren, 1968). Following lipolysis of the chylomicrons in the bloodstream, the chylomicron remnants containing the phylloquinone are taken up by the liver. A fraction of the newly ingested phylloquinone is secreted back into the circulation after incorporation into very low-density lipoproteins (VLDL). Phylloquinone is also found in association with low-density lipoproteins (LDL) and high-density lipoproteins (HDL) but the dynamics of the transfer of the vitamin between the lipoprotein classes is poorly understood. Unlike vitamins A and D, no specific plasma carrier protein for vitamin K has been identified (Shearer, 1993).

Excretion

Phylloquinone is rapidly metabolized in the liver to a variety of water-soluble metabolites, which are secreted into the bile and eliminated in the faeces or, to a lesser extent, excreted in the urine (Shearer, McBurney and Barkhan, 1974).

6.8.2 Dietary sources and their bioavailability

Dietary sources

Booth, Sadowski and Pennington (1995) tabulated the phylloquinone content of 261 foods using the HPLC method of analysis described by Booth, Davidson and Sadowski (1994). Representative data from this table are presented in Table 6.2. The foods selected for analysis were those from the Food and Drug Administration's (FDA) Total Diet Study, which lists core foods in the American food supply on the basis of data from the US Department of Agriculture's 1987–1988 Nationwide Food Consumption Survey (Pennington, 1992).

The highest concentrations of vitamin K (in the form of phylloquinone) are found in green leafy vegetables, with values for cabbage, broccoli, Brussels sprouts, spinach and collards ranging between 98 and 440 µg of phylloquinone/100 g of vegetable. Such vegetables are the top contributors to vitamin K intake in the American diet (Booth, Pennington and Sadowski, 1996a). Phylloquinone contents of green leafy vegetables (but not cabbage) increase during plant maturation. Cabbage contains 3–6 times more phylloquinone in the outer leaves compared with the inner leaves. The phylloquinone content of a given vegetable differs according to the geographical growth location, suggesting that climate, soil and growing conditions may be influencing factors (Ferland and Sadowski, 1992a). Other types of vegetable (roots, bulbs and tubers), cereal grains and their milled products, fruits and fruit juices are poor sources of vitamin K. Animal products (meat, fish, milk products and eggs) contain low concentrations of phylloquinone, but appreciable amounts of menaquinones are present in liver.

Some vegetable oils, including canola (rapeseed), soybean and olive oils, are rich sources of phylloquinone, whereas peanut oil and corn

Table 6.2 Phylloquinone content of selected foods (Booth, Sadowski and Pennington, 1995)

Item	μg Phylloquinone/ 100 g mean (SD)	Average serving size (g)	μg Phylloquinone/ serving
Whole milk, fluid	0.3 (0.02)	244	0.7
Cheddar cheese	2.1 (0.2)	28	0.6
Eggs, boiled	0.3 (0.05)	50	0.2
Beef steak, loin, pan-cooked	1.8 (0.2)	85	1.5
Pork chop, pan-cooked	3.1 (0.07)	85	2.6
Pork sausage, pan-cooked	3.4 (1.4)	56	1.9
Liver, beef, fried	2.7 (0.1)	85	2.3
Chicken breast, roasted	< 0.01 (< 0.01)	85	< 0.01
Haddock, pan-cooked	5.2 (0.1)	85	4.4
Tuna, canned in oil, drained	24 (1.2)	56	14
Kidney beans, dry, boiled	8.4 (0.9)	88	7.4
Pinto beans, dry, boiled	3.7 (0.20)	86	3.2
Peanut butter, smooth	0.3 (0.02)	32	0.1
White bread	1.9 (0.1)	50	1.0
Whole wheat bread	3.4 (0.05)	57	1.9
Orange, raw	< 0.01 (< 0.01)	154	< 0.01
Apple, red, raw	1.8 (0.09)	154	2.8
Banana, raw	0.2 (0.02)	126	0.3
White potato			
baked with skin	1.1 (0.1)	140	1.5
boiled without skin	0.3 (0.02)	136	0.4
French fries, fast food	4.4 (0.2)	68	3.0
Spinach, fresh/frozen, boiled	360 (70)	90	324
Broccoli, fresh/frozen, boiled	113 (2.5)	78	88
Carrot, fresh, boiled	15 (0.5)	78	12
Tomato sauce, plain, bottled	2.9 (0.4)	61	1.8
Green beans, fresh/ frozen, boiled	16 (6.6)	62	9.7
Brussels sprouts, fresh/ frozen, boiled	289 (55)	78	225
Celery, raw	32 (4.0)	55	17
Iceberg lettuce, raw	31 (8.6)	89	28
Mayonnaise, regular, bottled	41 (1.2)	14	5.8
French salad dressing	51 (5.7)	29	15
Margarine, stick, regular	33 (4.3)	14	4.6

Table 6.3 Phylloquinone content in various vegetable oils (from Ferland and Sadowski, 1992b)

Type	Number of brands analysed	Phylloquinone (μg/100 g of oil)[a]
Peanut	3	0.65 ± 0.27
Corn (maize)	2	2.91 ± 1.28
Sunflower	2	9.03 ± 0.17
Safflower	2	9.13 ± 2.64
Olive	6	55.5 ± 6.3
Canola (rapeseed)	4	141 ± 17.0
Soybean	5	193 ± 28

[a] Average value (mean ± SEM). Each brand was analysed in triplicate.

(maize) oil are not (Table 6.3). Soybean oil is the most commonly consumed vegetable oil in the American diet. The addition of phylloquinone-rich vegetable oils in the processing and cooking of foods that are otherwise poor sources of vitamin K makes them potentially important dietary sources of the vitamin. This is particularly evident, for example, when chicken, fish, eggs and potatoes are fried in certain vegetable oils.

Those margarines, mayonnaises and regular-calorie salad dressings that are derived from phylloquinone-rich vegetable oils are second to green leafy vegetables in their phylloquinone content. The addition of these fats and oils to mixed dishes and desserts has an important impact on the amount of vitamin K in the American diet (Booth, Pennington and Sadowski, 1996a).

Various menaquinones have been found in fermented foods (Sakano *et al.*, 1988), salmon, shellfish, beef, pork, chicken, egg yolk, cheese and butter (Hirauchi *et al.*, 1989b) but the amounts may not be nutritionally significant in some of these foods. Livers of ruminant species (e.g. cow) contain significant concentrations (10–20 μg/100 g) of some menaquinones (Hirauchi *et al.*, 1989a), while cheese contains significant quantities of MK-8 (5–10 μg/100 g) and MK-9 (10–20 μg/100 g) (Shearer, Bach and Kohlmeier, 1996).

Bioavailability

Bioavailability of dietary phylloquinone and menaquinones
Metabolic studies in healthy adult humans (Shearer, McBurney and Barkhan, 1974) showed that about 20% of an oral dose of [³H]phylloquinone was excreted in the faeces unchanged, suggesting that 80% was absorbed. A further 34–40% of the radioactivity was recovered in the faeces as polar metabolites. These did not seem to be formed in the intestine, but were the result of biliary excretion of metabolites.

Little is known about the bioavailability of vitamin K from foods. The absorption efficiency of phylloquinone is likely to vary widely, being probably least efficient from green leafy vegetables, where the vitamin is intimately associated with the thylakoid membranes of chloroplasts, and being most efficient from processed foods, such as oils, margarine and dairy produce (Shearer, 1993). Both long- and short-chain menaquinones are readily absorbed by rats after oral ingestion (Groenen-van Dooren *et al.* (1995) and therefore menaquinones present in the lipid fraction of processed foods are likely to be incorporated into mixed micelles and absorbed along with phylloquinone.

Gijsbers, Jie and Vermeer (1996) investigated the effect of food composition on vitamin K absorption in human subjects by HPLC measurement of serum vitamin K concentrations. In one experiment, five healthy volunteers were subjected to three different vitamin K regimens on the first day of three successive weeks. In each case an oral dose of 1 mg (2.2 μmol) of phylloquinone was taken at 08.00 hours either in the form of a detergent-solubilized pharmaceutical concentrate (Konakion, week 1), or as 227 g boiled spinach plus 25 g butter (week 2), or as 227 g boiled spinach without added fat (week 3). Blood samples were taken every hour during the first 10 h of the experimental day and at 08.00 h next morning ($t = 24$ h). Food was restricted until 10 h after vitamin K ingestion. The data from this experiment showed that phylloquinone was readily absorbed from Konakion and circulated in blood with a peak time of 4 h. Phylloquinone from spinach (with or without added fat) was more slowly absorbed, presumably because of digestive factors, such as the rate at which the vitamin is extracted from the plant cell membranes.

The effects of the nutritional regimen on phylloquinone absorption are shown in Table 6.4. The availability of phylloquinone from spinach

Table 6.4 Effect of vitamin K from different sources on serum phylloquinone levels in human subjects (from Gijsbers, Jie and Vermeer, 1996)

Subject	AUC[a], nmol/lh (%)[b]		
	Konakion[c]	Spinach per se	Spinach + butter
A	310 (100)	5.8 (1.9)	18.4 (13.1)
B	216 (100)	9.8 (4.6)	26.4 (12.3)
C	201 (100)	6.9 (3.4)	17.6 (8.8)
D	122 (100)	13.3 (10.9)	25.6 (21.0)
E	117 (100)	4.2 (3.7)	18.2 (15.6)
Mean	193.1 (100)	8.0 (4.1)	25.8 (13.3)

[a] Area under each individual absorption curve.
[b] The AUC values for the various nutritional regimens expressed as a percentage of the corresponding Konakion experiment.
[c] Detergent-solubilized phylloquinone.

(without added fat) was only 4% that from Konakion, but the simultaneous consumption of butter improved this figure by a factor of three. The effect of fat is probably due to its stimulation of bile secretion, which facilitates the absorption of phylloquinone. Phylloquinone is absorbed from Konakion with an efficiency of about 80% (Shearer, McBurney and Barkhan, 1974) and, based on this figure, the data of Gijsbers, Jie and Vermeer (1996) indicate that less than 10% of the phylloquinone present in green vegetables is absorbed in the digestive tract.

Gut microfloral synthesis as a possible source of menaquinones
Various menaquinones ranging from MK-6 to MK-11 are synthesized by specific strains of bacteria inhabiting the colon. Among the major forms, MK-10 and MK-11 are produced by *Bacteroides* species, MK-8 by *Enterobacteria* species, MK-7 by *Veillonella* species and MK-6 by *Eubacterium lentum* (Shearer, 1993). The hydrophobic menaquinones (n > 7) form more than 95% of the total pool of intestinal menaquinones (Groenen-van Dooren *et al.*, 1993) with MK-10 predominating (Shearer, 1995).

The extent to which the menaquinones produced by the enteric bacteria can be absorbed and utilized by humans as a source of vitamin K has been the subject of debate for many years. The widely held assumption that colonic menaquinones are available to the host in significant amounts is largely based on clinical observations that patients with marginal or no vitamin K intake develop bleeding episodes associated with a prolonged prothrombin time only when they are are given broad-spectrum antibiotics (Allison *et al.*, 1987). It is generally inferred from such observations that the prolongation of prothrombin time is caused ultimately by destruction of the gut microflora and consequent vitamin K deficiency. However, many antibiotics are vitamin K antagonists and promote hypoprothrombinaemia directly (Lipsky, 1988). Therefore, the effect of antibiotics on prothrombin time is not solid evidence of menaquinone availability.

A greater wealth of evidence suggests that the menaquinones produced by colonic bacteria cannot be absorbed in amounts sufficient to prevent vitamin K deficiency under conditions of dietary deprivation. For example, the rat, which is notoriously resistant to dietary deprivation of vitamin K, displays signs of deficiency when coprophagy is prevented (Barnes and Fiala, 1959). This implies that the rat can only utilize enterically synthesized menaquinones through coprophagy and subsequent absorption from the jejunum. Another study showed that rats fed a vitamin K-deficient diet developed signs of vitamin K deficiency, despite the fact that the amounts of menaquinones in the colon had actually increased compared with a control group of rats fed a normal diet (Uchida and Komeno, 1988). The importance of sufficient dietary vitamin K consumption in rats was confirmed by Groenen-van Dooren *et al.*

(1993). In this study, rats receiving a low-fibre, vitamin-K deficient diet developed hypoprothrombinaemia when coprophagy was prevented.

Ichihashi *et al.* (1992) examined the colonic absorption of MK-9 and MK-4 using the rat *in situ* loop method. MK-9 is a typical bacterially produced menaquinone, while MK-4 is a minor bacterial product forming less than 0.5% of the total pool of intestinal menaquinones (Groenen-van Dooren *et al.*, 1993). [14C]MK-4 was shown to be absorbed via the portal bloodstream, even when bile was not present, although its absorption from the colon was slower than that from the jejunum. There was no absorption of MK-4 from the colon by the lymphatic route, even when bile was present. In contrast, when [14C]MK-9 was administered into the colonal loop without bile, most of the radioactivity was recovered from the colonal loop, and no transfer of MK-9 into the lymph or blood was observed. The data from these experiments suggest that menaquinones of short chain length, which constitute only a small and perhaps negligible part of the bacterially produced menaquinones, are absorbed from the colon via the portal pathway. Absorption rates decrease markedly with an increase in chain length and the bulk of the bacterially produced menaquinones are not absorbed from the colon by any route. The extremely poor colonic absorption of MK-4 and MK-9 in rats was also reported by Groenen-van Dooren *et al.* (1995).

In a human study (Udall, 1965), 10 patients receiving daily doses of warfarin (an antagonist of vitamin K in the blood-clotting process) were given large (500 mg) doses of phylloquinone directly into the caecum via a nasointestinal tube. No significant change in prothrombin time was observed in any patient, indicating that the vitamin was not absorbed. In contrast, one additional patient given an equal dose of phylloquinone by mouth showed the expected response, i.e. a fall in the prothrombin time.

In summary, the more recent data do not support the assumption that menaquinones synthesized by the intestinal microflora are directly absorbed in the colon. The normal bile salt-mediated mechanism of absorption is precluded because bile salts are reabsorbed in the distal ileum. Given that long-chain colonic menaquinones are integral components of the lipophilic bacterial cell membrane and even in the free form are totally insoluble in water, it is difficult to envisage how large amounts can be solubilized for absorption in a region where bile salts are lacking. However, it is well known that – at least in humans – bacteria also inhabit to some extent the distal ileum, and the possibility exists that some bacterially synthesisized menaquinones may be absorbed from this region by the bile salt-mediated lymphatic pathway. Shearer, Kries and Saupe (1992) reported that virtually all the menaquinones present in human ileal juice could be removed by simple centrifugation or by passage through a 0.2 μ filter. This suggests that the menaquinones in ileal juice are still membrane-bound and not in the micellar form which would favour the lymphatic absorption process.

6.8.3 Effects of food processing, storage and cooking

Stability studies (Ferland and Sadowski, 1992b) on the phylloquinone content of vegetable oils demonstrated that the vitamin is significantly affected by heat. When subjected to temperatures of 185–190 °C for 40 min, 7% of the original phylloquinone was lost during 20 min of heating and 11% during 40 min of heating. Phylloquinone is extremely sensitive to both fluorescent light and sunlight. After only 2 days of exposure, fluorescent light decreased the phylloquinone content of rapeseed and safflower oils by 46 and 59%, respectively, and daylight by 87 and 94%, respectively. Amber glass bottles were very effective in protecting the phylloquinone in oils from the destructive effects of light.

Richardson, Wilkes and Ritchey (1961) compared the relative vitamin K activity of raw vegetables which had been preserved by freezing, heat-processing (canning) and γ-irradiation. The frozen foods were stored at −20 °C, while the heat-processed and irradiated foods were stored at ambient temperature (24–27 °C). Vitamin K activity of the foods was determined by prothrombin times of chick plasma. The results indicated that there was no appreciable loss of vitamin K activity in the foods preserved by any of the methods examined or when stored for 15 months.

Langenberg *et al.* (1986) used an HPLC method to demonstrate that neither cooking nor γ-irradiation of vegetables lowered the phylloquinone content, nor did the phylloquinone content of commercially available vegetable products in cans or glass containers, or dried or deep-frozen, differ significantly from the content of fresh vegetables.

Hydrogenation is a common process used by the food industry to increase the oxidative stability of polyunsaturated oils and to convert liquid oils into semi-solid fats, thereby increasing their commercial applications. During the commercial hydrogenation of vegetable oils a percentage of the phylloquinone present is converted to 2′,3′-dihydrophylloquinone, owing to saturation of the 2′,3′ double bond of the phytyl side chain (Davidson *et al.*, 1996). Dihydrophylloquinone has been identified in human plasma following its ingestion in the form of margarine (Booth *et al.*, 1996), indicating that it is absorbed. Of the 261 foods analysed in the FDA's Total Diet Study, 36 foods contain dihydrophylloquinone in concentrations greater than 1.0 µg/100 g of food (Booth, Pennington and Sadowski, 1996b). All of these 36 foods contain hydrogenated fats or oils, with the highest concentrations of dihydrophylloquinone being found in assorted cookies and stick margarine. However, when expressed in micrograms per average serving size, the richest sources of the hydrogenated vitamin are commercial apple pies (27 µg) followed by fast-food french-fries (25 µg). Assuming that dihydrophyllo-quinone and phylloquinone have equivalent bioavailability, the hydro-genated compound accounts for 30% of the total vitamin K intake in two-year-old American children. The percentage contribution decreases with age and is higher among men than among women. The abundance of dihydrophylloquinone in the American diet warrants further investig-ation into its biological activity and bioavailability, neither of which are known for humans at present.

6.9 DIETARY INTAKE

The current RDA for vitamin K is set at 80 and 65 µg for adult men and women, respectively (National Research Council, 1989). This dietary recommendation was established on the basis of the amount of vitamin K required to maintain normal blood clotting times (Olson, 1987).

The analytical data for vitamin K content in foods reported by Booth, Sadowski and Pennington (1995) from the FDA's Total Diet Study were applied to the Total Diet Study's consumption model to determine how much vitamin K is in the American diet in relation to the current RDA (Booth, Pennington and Sadowski, 1996a). Of the 14 age–gender groups selected, the 25- to 30-year-old men and women consumed less than the current RDA. In contrast, formula-fed infants had estimated vitamin K intakes six times greater than the RDA. All other groups consumed amounts within the RDAs, but lower than 90 µg/day.

Ferland, Sadowski and O'Brien (1993) reported that dietary restriction of vitamin K to 10 µg/day (i.e. 12.5% of the RDA) in normal human subjects results in a decreased synthesis of Gla (as indicated by decreased urinary Gla excretion) without affecting blood coagulation. This dietary induced subclinical deficiency raises the question that a more appropri-ate criterion of vitamin K adequacy might be the daily intake that ensures that all vitamin K-dependent proteins in the body are in their fully γ-carboxylated form. Further research is necessary to identify the possible long-term consequences to health of subclinical vitamin K deficiency and this may lead to a new set of RDAs for the vitamin.

Effects of high intakes

No harmful effects attributable to the long-term ingestion of elevated amounts of the natural forms of vitamin K (phylloquinone and mena-quinones) have been reported (Shearer, 1993).

The synthetic compound menadione has been shown to cause haemolysis and liver damage in the newborn. This toxicity is attributed to the chemical reactivity of the unsubstituted 3-position in the naphthoquinone ring and its ability to combine with tissue sulphydryl groups (Shearer, 1993).

Assessment of nutritional status

Assays designed to show changes in the carboxylation status of Gla-containing proteins have replaced traditional coagulation assays. The usual approach for determining these changes is to measure levels of undercarboxylated Gla-containing proteins in blood.

6.10 SYNOPSIS

The main natural sources of vitamin K are green leafy vegetables, which contain exclusively phylloquinone. Some vegetable oils, including rapeseed and soybean oils, are important sources of phylloquinone and their use as cooking oils or in margarine, mayonnaise and salad dressing has a major impact on the amount of vitamin K in the Western diet.

The phylloquinone in plant tissues is tightly bound to the membranes of the chloroplasts, whereas in dairy produce it is dissolved in the fat fraction. The bioavailability of membrane-bound phylloquinone in vegetables appears to be very low (< 10%). In contrast, the phylloquinone content of dairy produce is dissolved in the fat fraction and consequently is absorbed with high efficiency. Factors affecting the bioavailability of dietary vitamin K include the type of food, the dietary fat content and the length of the polyprenyl side chain in the vitamin K molecule.

Another potential source of vitamin K is the menaquinones synthesized in the colon by specific strains of bacteria. The relative contribution of enteric menaquinones to vitamin K nutriture is, however, uncertain.

The vitamin K content of foods can only be reliably determined by HPLC, but the results obtained provide no information about the vitamin's bioavailability.

REFERENCES

Allison, P.M., Mummah-Schendel, L.L., Kindberg, C.G. *et al.* (1987) Effects of a vitamin K-deficient diet and antibiotics in normal human volunteers. *J. Lab. Clin. Med.*, **110**, 180–8.

Ayi, B.K. and Burgher, A.M. (1985) Determination of vitamin K$_1$ in infant formula products by high performance liquid chromatography with electrofluorometric detection. In *Production, Regulation and Analysis of Infant Formula – a Topical Conference*, May 14–16th, Association of Official Analytical Chemists, VA, pp. 83–9.

Barnes, R.H. and Fiala, G. (1959) Effects of the prevention of coprophagy in the rat. VI. Vitamin K. *J. Nutr.*, **68**, 603–14.

Blomstrand, R. and Forsgren, L. (1968) Vitamin $K_1^{-3}H$ in man – its intestinal absorption and transport in the thoracic duct lymph. *Int. Z. Vitaminforsch.*, **38**, 45–64.

Booth, S.L., Davidson, K.W. and Sadowski, J.A. (1994) Evaluation of an HPLC method for the determination of phylloquinone (vitamin K_1) in various food matrices. *J. Agric. Food Chem.*, **42**, 295–300.

Booth, S.L., Pennington, J.A.T. and Sadowski, J.A. (1996a) Food sources and dietary intakes of vitamin K-1 (phylloquinone) in the American diet: data from the FDA Total Diet Study. *J. Am. Dietetic Assoc.*, **96**, 149–54.

Booth, S.L., Pennington, J.A.T. and Sadowski, J.A. (1996b) Dihydro-vitamin K_1: primary food sources and estimated dietary intakes in the American diet. *Lipids*, **31**, 715–20.

Booth, S.L., Sadowski, J.A. and Pennington, J.A.T. (1995) Phylloquinone (vitamin K_1) content of foods in the U.S. Food and Drug Administration's total diet study. *J. Agric. Food Chem.*, **43**, 1574–9.

Booth, S.L., Davidson, K.W., Lichtenstein, A.H. and Sadowski, J.A. (1996) Plasma concentrations of dihydro-vitamin K_1 following dietary intake of a hydrogenated vitamin K_1-rich vegetable oil. *Lipids*, **31**, 709–13.

Bueno, M.P. and Villalobos, M.C. (1983) Reverse phase high pressure liquid chromatographic determination of vitamin K_1 in infant formulas. *J. Ass. Off. Analyt. Chem.*, **66**, 1063–6.

Canfield, L.M., Hopkinson, J.M., Lima, A.F. *et al.* (1990) Quantitation of vitamin K in human milk. *Lipids*, **25**, 406–11.

Careri, M., Mangia, A., Manini, P. and Taboni, N. (1996) Determination of phylloquinone (vitamin K_1) by high performance liquid chromatography with UV detection and with particle beam-mass spectrometry. *Fresenius J. Analyt. Chem.*, **355**, 48–56.

Davidson, K.W., Booth, S.L., Dolnikowski, G.G. and Sadowski, J.A. (1996) Conversion of vitamin K_1 to $2', 3'$-dihydrovitamin K_1 during the hydrogenation of vegetable oils. *J. Agric. Food Chem.*, **44**, 980–3.

Dunphy, P.J. and Brodie, A.F. (1971) The structure and function of quinones in respiratory metabolism. *Meth. Enzym.*, **18C**, 407–61.

Ferland, G. and Sadowski, J.A. (1992a) Vitamin K_1 (phylloquinone) content of green vegetables: effects of plant maturation and geographical growth location. *J. Agric. Food Chem.*, **40**, 1874–7.

Ferland, G. and Sadowski, J.A. (1992b) Vitamin K_1 (phylloquinone) content of edible oils: effects of heating and light exposure. *J. Agric. Food Chem.*, **40**, 1869–73.

Ferland, G., Sadowski, J.A. and O'Brien, M.E. (1993) Dietary induced subclinical vitamin K deficiency in normal human subjects. *J. Clin. Invest.*, **91**, 1761–8.

Fournier, B., Sann, L., Guillaumont, M. and Leclercq, M. (1987) Variations of phylloquinone concentration in human milk at various stages of lactation and in cow's milk at various seasons. *Am. J. Clin. Nutr.*, **45**, 551–8.

Gao, Z.H. and Ackman, R.G. (1995) Determination of vitamin K_1 in canola oils by high performance liquid chromatography with menaquinone-4 as an internal standard. *Food Res. Int.*, **28**, 61–9.

Gijsbers, B.L.M.G., Jie, K.-S.G. and Vermeer, C. (1996) Effect of food composition on vitamin K absorption in human volunteers. *Br. J. Nutr.*, **76**, 223–9.

Groenen-van Dooren, M.M.C.L., Soute, B.A.M., Jie, K.-S.G. *et al.* (1993) The relative effects of phylloquinone and menaquinone-4 on the blood coagulation factor synthesis in vitamin K-deficient rats. *Biochem. Pharmacol.*, **46**, 433–7.

Groenen-van Dooren, M.M.C.L., Ronden, J.E., Soute, B.A.M. and Vermeer, C. (1995) Bioavailability of phylloquinone and menaquinones after oral and colorectal administration in vitamin K-deficient rats. *Biochem. Pharmacol.*, **50**, 797–801.

Hirauchi, K., Sakano, T., Notsumoto, S. *et al.* (1989a) Measurement of K vitamins in animal tissues by high-performance liquid chromatography with fluorimetric detection. *J. Chromat. Biomed. Appl.*, **497**, 131–7.

Hirauchi, K., Sakano, T., Notsumoto, S. *et al.* (1989b) Measurement of K vitamins in foods by high-performance liquid chromatography with fluorometric detection. *Vitamins (Japan)*, **63**, 147–51 (in Japanese).

Hwang, S.-M. (1985) Liquid chromatographic determination of vitamin K_1 trans- and cis-isomers in infant formula. *J. Ass. Off. Analyt. Chem.*, **68**, 684–9.

Ichihashi, T., Takagishi, Y., Uchida, K. and Yamada, H. (1992) Colonic absorption of menaquinone-4 and menaquinone-9 in rats. *J. Nutr.*, **122**, 506–12.

Indyk, H. (1988) The photoinduced reduction and simultaneous fluorescence detection of vitamin K_1 with HPLC. *J. Micronutr. Anal.*, **4**, 61–70.

Indyk, H.E. and Woollard, D.C. (1995) The endogenous vitamin K_1 content of bovine milk: temporal influence of season and lactation. *Food Chem.*, **54**, 403–7.

Indyk, H.E., Littlejohn, V.C., Lawrence, R.J. and Woollard, D.C. (1995) Liquid chromatographic determination of vitamin K_1 in infant formulas and milk. *J. AOAC Int.*, **78**, 719–23.

Isshiki, H., Suzuki, Y., Yonekubo, A. *et al.* (1988) Determination of phylloquinone and menaquinone in human milk using high performance liquid chromatography. *J. Dairy Sci.*, **71**, 627–32.

Jakob, E. and Elmadfa, I. (1996) Application of a simplified HPLC assay for the determination of phylloquinone (vitamin K_1) in animal and plant food items. *Food Chem.*, **56**, 87–91.

Lambert, W.E., Vanneste, L. and De Leenheer, A.P. (1992) Enzymatic sample hydrolysis and HPLC in a study of phylloquinone concentration in human milk. *Clin. Chem.*, **38**, 1743–8.

Langenberg, J.P., Tjaden, U.R., de Vogel, E.M. and Langerak, D.Is. (1986) Determination of phylloquinone (vitamin K_1) in raw and processed vegetables using reversed phase HPLC with electrofluorometric detection. *Acta Alimentaria*, **15**, 187–98.

Lipsky, J.J. (1988) Antibiotic-associated hypoprothrombinemia. *J. Antimicrob. Chemother.*, **21**, 281–300.

McNinch, A.W. and Tripp, J.H. (1991) Haemorrhagic disease of the newborn in the British Isles: two year prospective study. *Br. Med. J.*, **303**, 1105–9.

Moussa, F., Depasse, F., Lompret, V. *et al.* (1994) Determination of phylloquinone in intravenous fat emulsions and soybean oil by high-performance liquid chromatography. *J. Chromat. A*, **664**, 189–94.

National Research Council (1989) Fat-soluble vitamins. In *Recommended Dietary Allowances*, 10th edn, National Academy Press, Washington, DC, pp. 78–114.

Olson, J.A. (1987) Recommended dietary intakes (RDI) of vitamin K in humans. *Am. J. Clin. Nutr.*, **45**, 587–92.

Omaye, S.T. (1984) Safety of megavitamin therapy. In *Nutritional and Toxicological Aspects of Food Safety* (ed. M. Friedman), Plenum Press, New York, pp. 169–203.

Parrish, D.B. (1980) Determination of vitamin K in foods: a review. *CRC Crit. Rev. Food Sci. Nutr.*, **13**, 337–52.

Pennington, J.A.T. (1992) Total diet studies: the identification of core foods in the United States food supply. *Food Addit. Contam.*, **9**, 253–64.

Richardson, L.R., Wilkes, S. and Ritchey, S.J. (1961) Comparative vitamin K activity of frozen, irradiated and heat-processed foods. *J. Nutr.*, **73**, 369–72.

Robaidek, E. (1983) Bioassay methods for nutrients in processed foods. *Food Technol.*, **37**(1), 81–3.

Sakano, T., Notsumoto, S., Nagaoka, T. *et al.* (1988) Measurement of K vitamins in food by high-performance liquid chromatography with fluorometric detection. *Vitamins (Japan)*, **62**, 393–8 (in Japanese).

Schneiderman, M.A., Sharma, A.K., Mahanama, K.R.R. and Locke, D.C. (1988) Determination of vitamin K_1 in powdered infant formulas, using supercritical fluid extraction and liquid chromatography with electrochemical detection. *J. Ass. Off. Analyt. Chem.*, **71**, 815–7.

Shearer, M.J. (1986) Vitamins. In *HPLC of Small Molecules. A Practical Approach* (ed. C.K. Lim), IRL Press, Oxford, pp. 157–219.

Shearer, M.J. (1992) Vitamin K metabolism and nutriture. *Blood Rev.*, **6**, 92–104.

Shearer, M.J. (1993) Vitamin K. In *Encyclopaedia of Food Science, Food Technology and Nutrition*, Vol. 7 (eds R. Macrea, R.K. Robinson and M.J. Sadler), Academic Press, London, pp. 4804–16.

Shearer, M.J. (1995) Vitamin K. *Lancet*, **345**, 229–34.

Shearer, M.J., Bach, A. and Kohlmeier, M. (1996) Chemistry, nutritional sources, tissue distribution and metabolism of vitamin K with special reference to bone health. *J. Nutr.*, **126**, 1181S–6S.

Shearer, M.J., McBurney, A. and Barkhan, P. (1974) Studies on the absorption and metabolism of phylloquinone (vitamin K_1) in man. *Vit. Horm.*, **32**, 513–42.

Shearer, M.J., Kries, R.V. and Saupe, J. (1992) Comparative aspects of human vitamin K metabolism and nutriture. *J. Nutr. Sci. Vitaminol.*, Special edition **S-13-3**, 413–6.

Uchida, K. and Komeno, T. (1988) Relationships between dietary and intestinal vitamin K, clotting factor levels, plasma vitamin K, and urinary Gla. In *Current Advances in Vitamin K Research* (ed. J.W. Suttie), Elsevier Applied Science, New York, pp. 477–92.

Udall, J.A. (1965) Human sources and absorption of vitamin K in relation to anticoagulation stability. *J. Am. Med. Ass.*, **194**, 127–9.

Usui, Y., Nishimura, N. Kobayashi, N. *et al.* (1989) Measurement of vitamin K in human liver by gradient elution high-performance liquid chromatography using platinum-black catalyst reduction and fluorimetric detection. *J. Chromat. Biomed Appl.*, **489**, 291–301.

Usui, Y., Tanimura, H., Nishimura, N. *et al.* (1990) Vitamin K concentrations in the plasma and liver of surgical patients. *Am. J. Clin. Nutr.*, **51**, 846–52.

Vermeer, C. (1990) γ-Carboxyglutamate-containing proteins and the vitamin K-dependent carboxylase. *Biochem. J.*, **266**, 625–36.

Vermeer, C., Jie, K.S.G. and Knapen, M.H.J. (1995) Role of vitamin K in bone metabolism. *Annu. Rev. Nutr.*, **15**, 1–22.

Zonta, F. and Stancher, B. (1985) Quantitative analysis of phylloquinone (vitamin K_1) in soy bean oils by high-performance liquid chromatography. *J. Chromat.*, **329**, 257–63.

7

Thiamin (vitamin B_1)

7.1 INTRODUCTION

At one time, the disease beriberi was believed to be caused by a micro-organism or toxin. The first indication of a nutritional aetiology was the virtual elimination of beriberi in the Japanese Navy in 1885, brought about by increasing the proportion of meat and vegetables in the staple rice diet. In 1890, Eijkman, a Dutch medical officer stationed in Java, discovered that feeding chickens on polished rice induced a polyneuritis closely resembling human beriberi, which could be prevented by the addition of rice bran to the avian diet. A few years later, Grijns extracted a water-soluble 'polyneuritis preventive factor' from rice bran and correctly concluded that beriberi is the result of a dietary lack of an essential nutrient. By 1926, two Dutch chemists, Jansen and Donath, had succeeded in isolating the factor in crystalline form from rice bran extracts. Ten years later, R.R. Williams elucidated the chemical structure of the

factor and proposed the name 'thiamine' (nowadays spelt thiamin).
Thiamin has also been known as aneurin(e), indicative of its role in
preventing neurological symptoms.

7.2 CHEMICAL STRUCTURE AND NOMENCLATURE

The thiamin molecule (Figure 7.1) comprises substituted pyrimidine and
thiazole moieties linked by a methylene bridge and systematically named
in accordance with the IUPAC-IUB Recommendations as 3-(4'-amino-
2'-methylpyrimidin-5'-ylmethyl)-5-(2-hydroxyethyl)-4-methylthiazolium.
In living tissues thiamin, in the form of the diphosphate ester, thiamin
pyrophosphate (TPP) (Figure 7.1), functions as the prosthetic group
of several enzymes in which it has a tight but noncovalent binding
affinity for the apoenzyme (enzyme protein). Small amounts of the
monophosphate (TMP) and triphosphate (TTP) esters of thiamin occur
in animal tissues, but these have no direct coenzyme activity. For food
supplementation purposes thiamin is chiefly used in the form of its
chloride hydrochloride double salt (known commercially as thiamin
hydrochloride) $C_{12}H_{17}N_4OSCl$. HCl; $MW = 337.28$. The hydrochloride
salt is hygroscopic and usually exists as the hemihydrate containing the
equivalent of about 4% water. Another commercial form of the vitamin,
thiamin mononitrate, is practically nonhygroscopic and is especially
recommended for the enrichment of flour mixes.

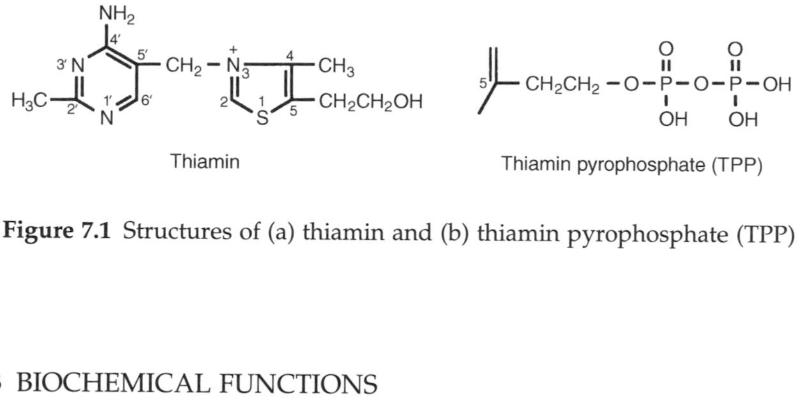

Figure 7.1 Structures of (a) thiamin and (b) thiamin pyrophosphate (TPP).

7.3 BIOCHEMICAL FUNCTIONS

Thiamin, in the form of TPP, participates in carbohydrate metabolism by
functioning as a coenzyme for the enzyme involved in the oxidative
decarboxylation of pyruvate to the important energy-rich thioester
acetyl-CoA in the tricarboxylic acid cycle. α-Ketoglutarate undergoes a
similar decarboxylation to form the analogous succinyl-CoA. TPP is also
a coenzyme for the transketolase reaction of the pentose phosphate pathway. This reaction involves the transfer of an α-keto group from xylulose-

5-phosphate to ribose-5-phosphate to form sedo-heptulose-7-phosphate and glyceraldehyde-3-phosphate. The pentose phosphate pathway provides the major source of ribose for nucleotide and nucleic acid synthesis, as well as reduced nicotinamide adenine dinucleotide phosphate (NADPH) for the synthesis of fatty acids and other compounds. Furthermore, TPP is required for the oxidative decarboxylation of the branched-chain keto acids formed after deamination of the amino acids leucine, isoleucine and valine.

7.4 DEFICIENCY SYNDROMES

The classic pathological condition resulting from a gross deficiency of thiamin in humans is beriberi, which is prevalent in Far Eastern populations where unfortified polished rice is the staple diet. The symptoms of beriberi vary according to circumstances. Wet beriberi, which is manifested as a generalized oedema and cardiac malfunction, results from severe physical exertion and high carbohydrate intake. Dry beriberi is characterized by a pronounced muscle wasting accompanied by a variety of neurological disorders; it results from relative inactivity with caloric restriction during the chronic deficiency.

A deficiency of thiamin in humans can also result in Wenicke-Korsakoff's encephalopathy, which is characterized by mental confusion and deterioration of nerve function leading, at worst, to coma. This disease manifests particularly in alcoholics in industrialized countries. Alcoholics are particularly prone to thiamin deficiency because of decreased intake of the vitamin due to decreased food consumption. In addition, decreased liver function impairs the activation and, consequently, the utilization of thiamin (Leevy and Baker, 1968).

Subclinical thiamin deficiencies are characterized by mental disturbances, fatigue and loss of weight resulting from reduced appetite and digestive problems.

The neurological and cardiovascular disorders that arise from a deficiency of thiamin are attributable to the dependence of the brain and heart upon thiamin for their energy supply. The involvement of thiamin in carbohydrate metabolism explains why the requirement varies with the proportion of carbohydrate in the diet and is increased under conditions that elevate the metabolic rate. Both dietary lipid and protein exert a thiamin-sparing action.

7.5 PHYSICOCHEMICAL PROPERTIES

Solubility and other properties

Thiamin hydrochloride is a colourless crystalline substance with a yeast-like odour and a salty nut-like taste. It is readily soluble in water

(100 g/100 ml), less soluble in glycerol (5 g/100 ml), sparingly soluble in 95% ethanol (1 g/100 ml) and absolute ethanol (0.3 g/100 ml) and insoluble in fat solvents. Since the thiamin molecule contains a quaternary nitrogen, it is a strong base and will be completely ionized over the entire range of pH normally encountered in foods.

Spectroscopic properties

The absorption spectrum of thiamin hydrochloride is pH-dependent. At pH 2.9 a single maximum at 246 nm occurs. At pH 5.5 two maxima occur at 234 nm and 264 nm which correspond to the substituted pyrimidine and thiazole moieties, respectively. Although thiamin exhibits a rather low molar absorptivity ($\epsilon = 11\,305$ at λ_{max} 246 nm, acid pH), absorbance detection has adequate sensitivity for fortified foods (Kamman, Labuza and Warthesen, 1980; Ayi *et al.*, 1985) and also for foods that are relatively rich in the vitamin, such as legumes and pork muscle (Vidal-Valverde and Reche, 1990). For other food commodities, absorbance detection is inadequate, and it is necessary to employ fluorescence detection after oxidation of the thiamin to thiochrome by pre- or post-column reaction with alkaline hexacyanoferrate(III) (potassium ferricyanide). The fluorescence excitation and emission spectra of thiochrome possess wavelength maxima at 375 nm and 432–435 nm, respectively.

Stability

As the pH of the medium is increased, thiamin becomes increasingly susceptible to deactivation by heat and chemical reaction. Mild oxidation produces thiamin disulphide without loss of thiamin activity; more vigorous oxidation with alkaline potassium hexacyanoferrate(III) produces the biologically inactive thiochrome (Figure 7.2). Thiamin phosphate esters are oxidized to thiochrome compounds in a similar manner without splitting the phosphate bonds. Thiamin is deactivated by sulphite ions even at room temperature, owing to methylene bridge cleavage. Solutions of thiamin are degraded by UV irradiation (Dwivedi and Arnold, 1973).

The effects of chemical interactions on thiamin stability in complex food systems has been reviewed (Clydesdale *et al.*, 1991).

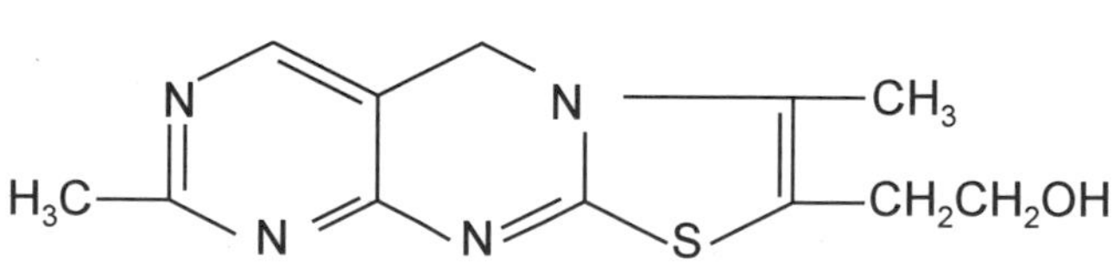

Figure 7.2 Structure of thiochrome.

7.6 ANALYSIS

7.6.1 Scope of analytical techniques

The major form of thiamin in meat products is protein-bound thiamin pyrophosphate (TPP). In cereals and cereal products, including white flour made from wheat, free thiamin predominates, but wheat germ, bran and also white bread contain TPP. For the analysis of these products it is customary to determine the combined thiamin activity of the various forms present. Thiamin can be determined microbiologically, or by fluorimetry after conversion to thiochrome, or by gas chromatography. The principal technique is HPLC using either absorbance detection of thiamin itself or fluorescence detection of thiochrome. The use of a common extraction procedure allows both thiamin and riboflavin to be assayed by HPLC either simultaneously or successively. Thiamin activity can be determined by either the curative growth test or the brachycardia test in rats.

The International Unit (IU) of vitamin B_1 activity is defined as the activity of 3 µg of crystalline thiamin hydrochloride, i.e. 1 g corresponds to 333 000 IU. The IU for thiamin is no longer used in practice, and analytical results are expressed in terms of weight (mg) of pure thiamin hydrochloride.

7.6.2 Extraction techniques

The thiamin content of foods is protected to some extent by the matrix and further stabilized through protein binding. The phosphorylated and protein-bound thiamin present in most foods is quite stable when food samples in a medium of 0.1 N hydrochloric acid (HCl) are heated to 95–100 °C or autoclaved at 121–123 °C for 30 min, but higher concentrations of acid are likely to decompose thiamin. The nonphosphorylated thiamin that occurs in cereal grain products is partly destroyed when autoclaved under the above conditions, but it can resist autoclaving at 108–109 °C for 20 min (McRoberts, 1954).

The extraction procedure generally used for the determination of total thiamin by fluorimetry, GC, HPLC and microbiological assay involves hot acid digestion to release the thiamin and thiamin phosphate esters from their association with proteins, followed by enzymatic hydrolysis of the phosphate esters to complete the liberation of thiamin. Food samples of animal origin can be autoclaved at 121 °C for 30 min with 0.1 N HCl, as the phosphorylated forms of thiamin present in such samples are not degraded under these conditions. For the majority of cereals and cereal products, which contain mostly nonphosphorylated thiamin, it is necessary to lower the autoclaving temperature to 108 °C in order to avoid vitamin loss.

A commercial diastatic enzyme preparation of fungal origin (e.g. Taka-diastase, Claradiastase or Mylase) is suitable for the hydrolysis step, as such preparations contain phosphatase activity in addition to α-amylase and other enzymes. The enzyme treatment can be omitted for the analysis of those grain products that do not contain phosphorylated thiamin. For proteinaceous samples such as meat, the proteolytic enzyme papain is sometimes added to the diastase in order to dissolve the proteins that have been denatured during the previous acid digestion. The commercial availability of diastatic enzymes has been a problem, and Defibaugh (1987) has evaluated selected enzymes for suitability as substitutes in thiamin assays.

Instead of using an enzyme hydrolysis procedure for thiamin extraction prior to HPLC, Ohta *et al.* (1993) refluxed rice flour samples with a mixture of hydrochloric acid and methanol (0.1 M HCl–40% aqueous methanol) for 30 min at 60 °C. The results obtained using this method were comparable to those obtained using the AOAC fluorimetric method and a high correlation coefficient ($r = 0.958$) was obtained between the two methods.

For the analysis of milk, the extraction procedure simply entails precipitation of the protein by acidification at room temperature, and filtration. This nonhydrolytic extraction procedure has the advantage of leaving the biologically inactive TMP intact, so this compound can be excluded from measurement.

7.6.3 Microbiological assay

Two *Lactobacilli* species have been widely used as assay organisms for the determination of thiamin, namely *L. fermentum* (ATCC No. 9338) and *L. viridescens* (ATCC No. 12706). Of the two, the latter is generally preferred as it is less susceptible to inhibitory or stimulatory substances (Scholes, 1960). *L. viridescens* requires the intact thiamin molecule for growth; neither the thiazole nor pyrimidine moiety is active singly or together. Equimolar concentrations of TPP are approximately 60% as active as thiamin (Deibel, Evans and Niven, 1957). The *L. viridescens* assay was compared with manual and semi-automated fluorimetric methods in terms of their ability to recover added thiamin hydrochloride from commercially processed foods whose natural thiamin content was destroyed with thiaminase (Defibaugh, Smith and Weeks, 1977). The microbiological method gave the best results for eight products tested, with a recovery of $99.9 \pm 1.03\%$.

In order to ensure the complete utilization of total thiamin by *L. viridescens*, the extraction procedure involves both acid and enzymatic hydrolysis as a means of liberating thiamin from all bound forms. The enzyme treatment is omitted for the analysis of grain products and milk, and for the determination of the added thiamin hydrochloride in fortified foods.

7.6.4 AOAC fluorimetric method

This method is based on the conversion of thiamin to its fluorescent oxidation product, thiochrome, by reaction with alkaline potassium hexacyanoferrate(III). The method has been adapted to the determination of thiamin in foods, grain products, bread and milk-based infant formulas. In the procedure described for foods containing TPP (AOAC, 1990a) the food sample and a standard solution of thiamin hydrochloride are taken through the following steps: acid digestion, enzymatic hydrolysis, purification by open-column chromatography, oxidation of thiamin to thiochrome, extraction of the thiochrome into isobutanol, and measurement of the fluorescence. TMP is insoluble in isobutanol, so it will not be measured in this assay. Some analysts prefer to run an internal standard by adding a known amount of thiamin to a second sample. Alyabis and Simpson (1993) modified the AOAC method by using a reversed-phase $C_{18}(50\,\mu m)$ material for the open-column chromatography in place of the Bio-Rex 70 cation exchange resin.

For the analysis of grain products such as wheat flour, macaroni, and noodle products, which do not contain significant amounts of phosphorylated or protein-bound thiamin, the enzymatic hydrolysis and chromatographic purification steps have been omitted (AOAC, 1990b). The autoclaving conditions prescribed for grain products are 20 min at 108–109 °C. The enzymatic hydrolysis step, but not the chromatography, is essential for bread and wheat germ, which both contain phosphorylated thiamin (AOAC, 1990c).

Segmented-flow analysis has been used to automate the oxidation reaction and subsequent steps of the AOAC fluorimetric method at a sampling rate of 30 per hour (Kirk, 1974, 1977; Ribbron, Stevenson and Kirk, 1977; Pelletier and Madère, 1975). Partial separation of thiamin from extraneous substances was achieved by on-line dialysis (Roy and Conetta, 1976), and interferences from the remaining impurities were corrected by measuring the blank of the sample solution. Pelletier and Madère (1977) showed that acid digestion carried out before enzymatic hydrolysis can result in low yields of thiamin from fish and from certain nonprocessed meat such as beefsteak. Losses of thiamin incurred during the separation of insoluble matter from certain fruits was prevented by addition of ethylene glycol monoethyl ether (Pelletier and Madère, 1977). Kirk (1974) observed that the chromatographic purification step, normally used for sample clean-up, was not required for most samples. When high blank values interfered with the assay results, interference was eliminated by extracting the sample with water-saturated isobutanol before automated analysis. Only sample extracts containing chocolate were found to require the use of column chromatography (Kirk, 1977). Pelletier and Madère (1975) included the chromatographic purification step in the extraction procedure, although they observed that with certain foods it appeared to be unnecessary. Soliman (1981) used neither column chromatography nor isobutanol washing to clean up sample extracts. Instead, extracts were analysed before and after addition of benzene-sulphonyl chloride, which inhibited thiochrome formation and provided a more representative blank based on the fluorescence of all the reactants except thiochrome.

7.6.5 High-performance liquid chromatography

HPLC methods used for determining thiamin *per se* are summarized in Table 7.1. Thiamin can also be determined together with riboflavin, as discussed in Chapter 2.

Pre-column derivatization allows the relatively nonpolar thiochrome to be determined using reversed-phase chromatography with its attendant ease of operation and long-term stability. The addition of orthophosphoric acid 45 s after treatment with alkaline hexacyanoferrate(III) minimizes the formation of thiamin disulphide, a pH-dependent side reaction of the oxidation reaction (Fellman *et al.*, 1982). An alternative approach is to selectively extract the thiochrome into isobutanol, and then to inject an aliquot of the organic solution onto a column of underivatized silica eluted with chloroform/methanol (90:10) (Ang and Moseley, 1980).

If the derivatization is carried out post-column, it is actually thiamin that is being chromatographed, and this compound in the ionized state is not retained under simple reversed-phase conditions. However, reversed-phase columns can be utilized for thiamin by means of ion interaction chromatography using hexane (or heptane) sulphonic acid as the ion interaction reagent, either after post-column derivatization of thiamin and fluorescence detection, or without derivatization using UV detection. Ion suppression chromatography is a possible alternative (Ohta *et al.*, 1984, 1993), but this application requires adjustment of the mobile phase pH to 2.2 or 2.5 with perchloric acid, which is approaching the limit of stability (pH 2) of silica-based column packing materials.

7.7 BIOAVAILABILITY

7.7.1 Physiological aspects

Intestinal absorption

Dietary sources of TPP are completely hydrolysed to free thiamin in the intestinal lumen by different phosphatases, including alkaline phosphatase. In the rat, the main site of absorption of thiamin is the proximal 22 cm of the small intestine, i.e. the duodenum and proximal jejunum (Sklan and Trostler, 1977). The mechanism of thiamin absorption is not fully understood but some aspects of the process will be discussed and a possible scheme of cellular events proposed.

The uptake of thiamin by rat small intestine has been studied *in vivo* using intestinal loops and *in vitro* using everted jejunal sacs (Hoyumpa *et al.*, 1975). At very low (0.06–2.0 µM) concentrations of thiamin, uptake was shown to be a saturable process and a competitive inhibitory effect of pyrithiamin implied the presence of a specific carrier-mediated transport system. Sodium replacement, anoxia, low temperature and various metabolic inhibitors led to a fall in thiamin uptake, indicating a sodium-dependent, active (energy-requiring) process. Uptake of higher concentrations (> 2 µM) of thiamin took place by simple diffusion. Only the saturable part of the thiamin uptake process will be discussed in the present text.

Komai, Kawai and Shindo (1974) studied thiamin uptake by a tissue accumulation method using everted rat intestine cut into segments of 3–5 mm length. After 30 min incubation in a medium containing 0.2 µg/ml of ^{35}S-thiamin, 42% of the total thiamin taken up was in the free form. The intracellular to extracellular concentration ratio was calculated to be 2.27 with respect to free thiamin, indicating a transport of free thiamin against a concentration gradient. All of the other requirements for the concept of active transport, i.e. the saturation kinetics, the temperature dependency, the energy requirement and the competitive inhibition by structural analogues, were also satisfied for the thiamin transport.

During the absorption process, thiamin is phosphorylated to TPP by the specific enzyme thiamin pyrophosphokinase (E.C. 2.7.6.2) (TPKase), which is localized almost exclusively in the cytoplasm (Casirola *et al.*, 1988). Human jejunal tissue also accumulates thiamin in the phosphorylated form during absorption (Rindi and Ferrari, 1977).

There is indirect evidence to support the hypothesis that thiamin must be phosphorylated by the jejunal epithelial cells for its absorption to take place. Structural analogues of thiamin inhibited both thiamin phosphorylation and thiamin uphill transport in a proportional manner (Rindi and Ventura, 1967). However, the net transport was always more inhibited than thiamin phosphorylation, suggesting that the phosphorylation may not be the sole mechanism involved. Patrini *et al.* (1981) demonstrated a highly significant ($P < 0.001$) direct correlation ($r = 0.93$) of labelled thiamin transport with labelled thiamin phosphoester content in rat jejunal tissue. Ferrari, Patrini and Rindi (1982) showed that not only thiamin, but also TMP, is transferred to the serosal side of everted jejunal sacs during absorption. The transport and accumulation of these compounds in the serosal fluid only proceeded efficiently when the

Table 7.1 HPLC methods used for the determination of thiamin

Food	Sample preparation	Column	Mobile phase	Compounds separated	Detection	Reference
		Normal-phase chromatography				
Liver, semi-synthetic animal diet	Digest homogenized sample with 0.1 M (0.1 N) HCl at 95–100 °C for 1 h. Cool, adjust pH to 4.5 and incubate with Takadiastase and papain at 45–50 °C for 3 h. Cool, dilute to volume with 0.1 M HCl, centrifuge and filter. *Derivatization:* oxidize thiamin to thiochrome with alkaline $K_3Fe(CN)_6$, partition into isobutanol and centrifuge	LiChrosorb Si-60 5 μm 250 × 4 mm i.d.	$CHCl_3$/MeOH (80:20)	Thiochrome	Fluorescence: ex. 375 nm em. 430 nm	Bailey and Finglas (1990)
		Ion interaction chromatography				
Infant formula, milk, low-fat yoghurt, eggs, salad dressing	Stir sample with water. Deproteinize by pH adjustment to 1.7–2.0 and then to 4.0–4.2. Dilute to volume with water and filter	μBondapak C_{18} 10 μm 300 × 3.9 mm i.d. Column temperature 50 °C	Water/MeOH (80:20) containing 0.15% sodium hexane sulphonate, 0.75% acetic acid and 0.1% EDTA	Thiamin	UV 248 nm	Nicolas and Pfender (1990)
Legumes, pork muscle, full cream milk powder	Autoclave ground sample with dilute HCl at 121 °C for 15 min. Cool, adjust pH to 4.0–4.5 and incubate with Takadiastase at 48 °C for 3 h. Cool, dilute to volume with water and filter	μBondapak C_{18} 10 μm 300 × 3.9 mm i.d. Column temperature 30 °C (legumes) or 50 °C (pork and milk products)	Water/MeOH (69:31) containing 5 mM sodium hexane sulphonate/5 mM sodium heptane sulphonate (75 + 25) and 0.5% acetic acid	Thiamin	UV 254 nm	Vidal-Valverde and Reche (1990)

| | | | UV 245 nm | Ayi et al. (1985) |
| | | | | |

Thiamin

Cyanopropyl-bonded phase chromatography

Clean-up and concentration: readjust pH of extract to 5.0–5.5 and keep at 0 °C. Pass aliquot of extract through an Amberlite CG-50 ion exchange column, wash column bed with water, and elute thiamin with 0.15 M HCl. Evaporate eluate to dryness, dissolve residue in water and adjust pH to 5.5–6.0. Pass aliquot of solution through a C₁₈ Sep-Pak solid-phase extraction cartridge, wash cartridge with aqueous 5 mM sodium hexane sulphonate and elute thiamin with methanolic 5 mM sodium hexane sulphonate

Infant formula products | Stir reconstituted sample with water. Deproternize by pH adjustment to 1.7, and then to 4.6. Dilute to volume with water and filter. *Clean-up and concentration*: pass filtrate through strong cation exchange solid-phase extraction cartridge and wash cartridge sequentially with water, MeOH, water. Elute thiamin with 2 M KCl/MeOH (60 : 40) at 65–75 °C. Cool and dilute to volume with cooled elution solvent | Zorbax-CN 6 μm 250 × 4.6 mm i.d. | 0.04 M triethyl-ammonium phosphate buffer (pH 7.7)/MeOH (90 + 10) |

Abbreviations: see footnote to Table 2.4.

intracellular phosphorylation of thiamin to TPP was operating. In control experiments, when there was no thiamin phosphorylation and when TPP dephosphorylation was prevailing, endogenous thiamin accumulated steadily in the tissue and only a small amount passed into the serosal fluid together with TMP.

Other data suggest that phosphorylation of thiamin is not essential for its absorption. Komai, Kawai and Shindo (1974) reported that the longitudinal distribution of TPKase activity along the digestive tract did not coincide with that of the activity of thiamin uptake. Also, the activity of TPKase was found primarily in the soluble fraction of rat intestinal mucosa, with very little activity in the mitochondrial and microsomal fractions which contain most of the microvillous membrane. Furthermore, chloroethylthiamin inhibited the uptake of thiamin by rat jejunal segments but had no effect on TPKase activity. Ferrari, Ventura and Rindi (1971) reported that intracellular phosphorylation in rat everted jejunal sacs was not prevented by the absence of sodium in the incubation medium, whereas the net transport of thiamin was critically dependent on the presence of sodium.

The mechanism of thiamin uptake has also been studied using isolated brush-border membrane vesicles. This preparation eliminates any interference from cellular organelles and metabolism, including thiamin phosphorylation by TPKase. Using vesicles prepared from rat small intestine, Casirola *et al.* (1988) found that at low concentrations ($< 1.25 \, \mu M$) thiamin was taken up with a saturable mechanism that was independent of sodium and from phosphorylation. In addition, the lack of a transitory accumulation of thiamin indicated that its crossing of the brush-border membrane is not active transport. The saturation kinetics imply the presence of carrier-mediated transport, therefore it can be inferred that the initial uptake mechanism is facilitated diffusion. In the light of this knowledge, it is evident that the pronouncement of an active transport mechanism in intact intestinal preparations was based on the dependency on cellular energy metabolism.

Rindi (1984) proposed a possible scheme of the cellular events involved in the transport of thiamin across the enterocyte (Figure 7.3). In the original scheme, the thiamin crosses the brush-border membrane by simple diffusion, but, as just mentioned, this process is now believed to be one of facilitated diffusion. The intracellular thiamin is partly phosphorylated to TPP, a process which acts to concentrate thiamin in the enterocyte so that it may subsequently leave the cell down a concentration gradient. Because the remaining free thiamin exists as a monovalent cation at physiological pH, its accumulation is further aided by the electronegativity of the cell interior (Rose, 1991). Intracellular TPP may subsequently be dephosphorylated by microsomal phosphatases to thiamin or TMP, which finally pass into the plasma.

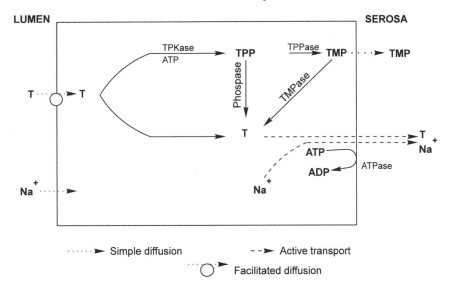

Figure 7.3 Possible scheme of cellular events involved in the transport of thiamin by the small intestine. T, thiamin; TMP, thiamin monophosphate; TPP, thiamin pyrophosphate; TPKase, thiamin pyrophosphokinase; TPPase, thiamin pyrophosphatase; TMPase, thiamin monophosphatase. (According to Rindi, 1984, and amended by Rose, 1991.)

The molecular aspects of the exit of thiamin from the enterocyte are unknown. An active transport process is indicated on the grounds that net thiamin transport proceeds against a serosa-positive electrical potential difference in everted jejunal sacs (Rose, 1988). The complete removal of sodium or the inhibition of the Na^+–K^+-ATPase by ouabain inhibits the net transport of thiamin by blocking its exit from the enterocyte, without affecting its entry and intracellular phosphorylation. This led Rindi (1984) to postulate that thiamin exit is mediated by the Na^+–K^+-ATPase located on the basolateral membrane. Further studies are needed to verify this cotransport of thiamin and sodium out of the cell.

The absorption of thiamin is enhanced in thiamin-deficient rats (Hoyumpa *et al.*, 1976; Patrini *et al.*, 1981), indicating adaptive regulation of uptake.

Other metabolic considerations

The absorbed thiamin, on entering the portal bloodstream, is bound to plasma albumin and conveyed to the liver. Uptake of thiamin by the liver takes place by a sodium-dependent active transport process (Lumeng *et al.*, 1979; Yoshioka, 1984).

Within the tissues, thiamin is converted to phosphate esters by three intracellular enzymes (Tanphaichitr, 1994). In all tissues thiamin is phosphorylated to its coenzyme form, TPP, by the catalytic action of thiamin diphosphokinase.

$$\text{Thiamin} + \text{ATP} \longrightarrow \text{TPP} + \text{AMP(adenosine monophosphate)}$$

In the brain and other nervous tissue, some of the TPP is converted to TTP by thiamin diphosphate-ATP phosphoryltransferase:

$$\text{TPP} + \text{ATP} \longrightarrow \text{TTP} + \text{ADP(adenosine diphosphate)}$$

Nervous tissue also contains thiamin diphosphatase which converts small amounts of TPP to TMP and inorganic phosphate:

$$\text{TPP} \longrightarrow \text{TMP} + \text{PPi}$$

Vitamin B_1 has a relatively high turnover rate in the body and is not stored in large amounts in any tissue. About half of the body stores are found in skeletal muscles, the remainder being distributed mainly in the liver, heart, kidney and brain. Of the total vitamin B_1 content of the body, about 80% is TPP, 10% is TTP, and the remainder is TMP and free thiamin.

Thiamin in excess of binding and storage capacity is rapidly excreted in the urine together with small amounts of its metabolites. Only negligible amounts of thiamin are excreted in the bile. Urinary excretion falls to near zero during thiamin deficiency, indicating a renal conservation mechanism (McCormick, 1988).

Because of its low storage capacity, a continuous supply of dietary thiamin is necessary. During thiamin deprivation there is a rather rapid loss of the vitamin from all tissues except the brain.

7.7.2 Dietary sources and their bioavailability

Dietary sources

All plant and animal tissues contain thiamin and it is therefore present in all natural unprocessed foods. The thiamin content of a selection of foods is presented in Table 7.2. Rich sources of thiamin include yeast and yeast extract, wheat bran, oatmeal, whole-grain cereals, pulses, nuts, lean pork, heart, kidney and liver. Beef, lamb, chicken, eggs, vegetables and fruits contain intermediate amounts, while milk contains a relatively low amount. In cereal grains the thiamin is very unevenly distributed, being relatively low in the starchy endosperm and high in

Table 7.2 Thiamin content of various foods (from Holland *et al.*, 1991)

Food	mg Thiamin/100 g
Cow's milk, whole, pasteurized	0.04
Egg, chicken, whole, raw	0.09
Wheat flour, wholemeal	0.47[a]
Rice, brown, raw	0.59
Rice, white, easy cook, raw	0.41
Beef, lean only, raw, average	0.07
Lamb, lean only, raw, average	0.14
Pork, lean only, raw, average	0.89
Chicken, meat only, raw	0.10
Liver, ox, raw	0.23
Cod, raw, fillets	0.08
Potato, maincrop, old, average, raw	0.21
Soya beans, dried, raw	0.61
Red kidney beans, dried, raw	0.65
Peas, raw	0.74
Brussels sprouts, raw	0.15
Apples, eating, average, raw	0.03
Orange juice, unsweetened	0.08
Peanuts, plain	1.14
Yeast extract (Marmite)	3.10

[a] This level is for fortified flour.

the germ. The milling of cereals removes most of the thiamin, so white flour, breakfast cereals and, in certain countries, polished rice are enriched by addition of the vitamin. Other frequently enriched food-stuffs are breads, macaroni, spaghetti and milk modifiers (chocolate, malt, etc.).

The major form of thiamin in meat products is TPP, which exists as a protein complex. Small amounts of TMP, TTP and nonphosphorylated thiamin have been detected in the liver, heart, kidney and brain of the rat (Ishii *et al.*, 1979). Of the total thiamin present in milk, 50–70% is free thiamin, 5–17% is protein-bound thiamin, and 18–45% is TMP, which itself has no thiamin activity (Hartman and Dryden, 1974). The natural thiamin content of most cereals and cereal products, including white flour made from wheat, is present almost entirely in the form of non-phosphorylated thiamin; exceptions are wheat germ and bran which contain TPP (Andrews and Nordgren, 1941; Hoffer, Alcock and Geddes, 1943). There is evidence that the absence of phosphorylated thiamin in plant material is caused by phosphatase activity when the material is extracted for assay (Obermeyer, Fulmer and Young, 1944). White bread contains TPP due to the phosphorylation of thiamin added for enrichment during the yeast fermentation of the dough as well as the contribution from the yeast itself (McRoberts, 1957).

Bioavailability

Warnick *et al.* (1956) reported that human subjects excreted less thiamin ʷʰᵉⁿ ⁱᵒⁱᵃᵗᵒᵉˢ ᶠᵘʳⁿⁱˢʰᵉᵈ ｒ ᵒⁿᵉ third of the dietary thiamin than when the same amount was furnished by brown rice, ᵗʰⁱᵃᵐⁱⁿ, ᵗᵃᵐᵇ ᵃⁿᵈ ᵖᵒʳᵏ together, or pure thiamin. To determine whether the lower excretion of thiamin during the potato test periods was due to low availability of the thiamin in potatoes or to increased utilization of the vitamin in the tissues, pig feeding studies were carried out followed by thiamin assays of the animal tissues. The results indicated that the pigs fed potatoes had a higher concentration of thiamin in the muscle tissues than did the pigs fed the rice diet. Thiamin concentration in the heart, liver and kidney did not vary with the type of diet eaten. The decrease in urinary excretion and the increase in tissue storage of thiamin when potatoes were the important source of this vitamin suggested that unidentified factors associated with the source of dietary thiamin may influence the utilization and deposition of thiamin in the tissues.

Thiamin appears to be well utilized when supplied by diets containing wheat or animal protein (Edwards *et al.*, 1971). Girija, Sharada and Pushpamma (1982) determined the bioavailability of thiamin from curries prepared with two green leafy vegetables commonly consumed in rural areas in India. Using the balance technique and dose–response curves obtained from human urinary excretion data, the bioavailability values were 59 and 62% from amaranth and drumstick curries, respectively.

Thiamin is considered to be totally available in a wide variety of food products (Gregory and Kirk, 1978), and therefore a chemical assay, used in conjunction with an effective extraction procedure, can substitute for a biological assay as a measure of vitamin B₁ activity, at least for certain foods. This was illustrated in a study conducted by Gregory and Kirk (1978), in which a semi-automated fluorimetric method was compared with a rat bioassay for the determination of thiamin in green beans. Calculation of available thiamin using the various rat assay dose–response curves indicated a mean value of $7.30 \pm 2.46 \, \mu g/g$, as thiamin hydrochloride, which compared favourably with the mean chemically determined value of $7.46 \pm 0.14 \, \mu g/g$.

Anti-thiamin factors
Anti-thiamin factors refer to naturally occurring compounds that act on thiamin in such a way that the products formed no longer possess the biological activity of the vitamin *in vivo*. They may be divided into two categories: thermolabile enzymes (thiaminases) and thermostable substances such as tannins in plants. Thiaminases will be discussed first.

Thiaminase I (E.C. 2.5.1.2) has been found in shellfish, the viscera of freshwater fish, ferns, and in some bacteria (e.g. *Bacillus thiaminolyticus*).

It cleaves the thiamin molecule by an exchange reaction with an organic base or a sulphydryl compound via a nucleophilic displacement on the methylene group of the pyrimidine moiety. Thiaminase II (E.C. 3.5.99.2), which has been found only in microorganisms, is a hydrolytic enzyme that hydrolyses the methylene-thiazole-N bond (Evans, 1975). Thiaminases are usually inaccessible to thiamin in living cells, but become accessible when tissues are disrupted at pH 4–8 (Murata, 1982). The enzymes can act during food storage and preparation and during passage through the gastrointestinal tract (Tanphaichitr, 1994).

Thiaminase activity of raw fermented fish in the diet of rural Thais was estimated to contain 4.5 units per gram. The amount consumed daily was found to be about 50 g of the fish, which is equivalent to 225 units of the enzyme (Vimokesant *et al.*, 1982). Habitual intakes of raw freshwater fish (with or without fermentation) and raw shellfish are risk factors for the development of thiamin deficiency. Since thiaminases are inactivated by heat, the cooking of such foods destroys the enzymes completely.

High anti-thiamin activity attributable to thermostable factors has been found in tea, coffee, rice bran, betel nuts, ferns and a variety of fruits and vegetables including blueberries, blackcurrants, red chicory, beetroot, red cabbage and Brussels sprouts (Hilker and Somogyi, 1982). Taungbodhitham (1995) analysed 14 types of vegetable commonly consumed by people in southern Thailand and found anti-thiamin factors in 10 types. Samples with a high anti-thiamin activity tended to have a low thiamin content.

The thermostable anti-thiamin activity in plants is associated with polyphenols such as hydroxylated derivatives of cinnamic acid, flavonoids and tannins. The chemical structures of some of these compounds are shown in Figure 7.4. In plant tissues, caffeic acid occurs mainly in conjugated forms such as chlorogenic acid and is released from such forms by hydrolysis. Roasted coffee has a high content of chlorogenic acid and contains smaller quantities of caffeic acid and other diphenols. In tests using pure solutions of plant polyphenols, two cinnamic acid derivatives (caffeic acid and chlorogenic acid) showed a greater degradation of thiamin than did two flavonoids (quercetin and dihydroquercetin), although they were less effective as antioxidants (Yang and Pratt, 1984). The major product of thiamin degradation was found to be thiamin disulphide.

Tannic acid or tannin is a generic term to describe a complex and non-uniform group of polyhydroxyphenols, which occur in fruits and vegetables commonly consumed by people in various parts of the world (Rungruangsak *et al.*, 1977). Tannins may be divided into two groups: derivatives of flavonols – so-called condensed tannins; and hydrolysable tannins, which are esters of a sugar, usually glucose, with one or more trihydroxybenzenecarboxylic acids. Tannins comprise 7–14% of tea leaf

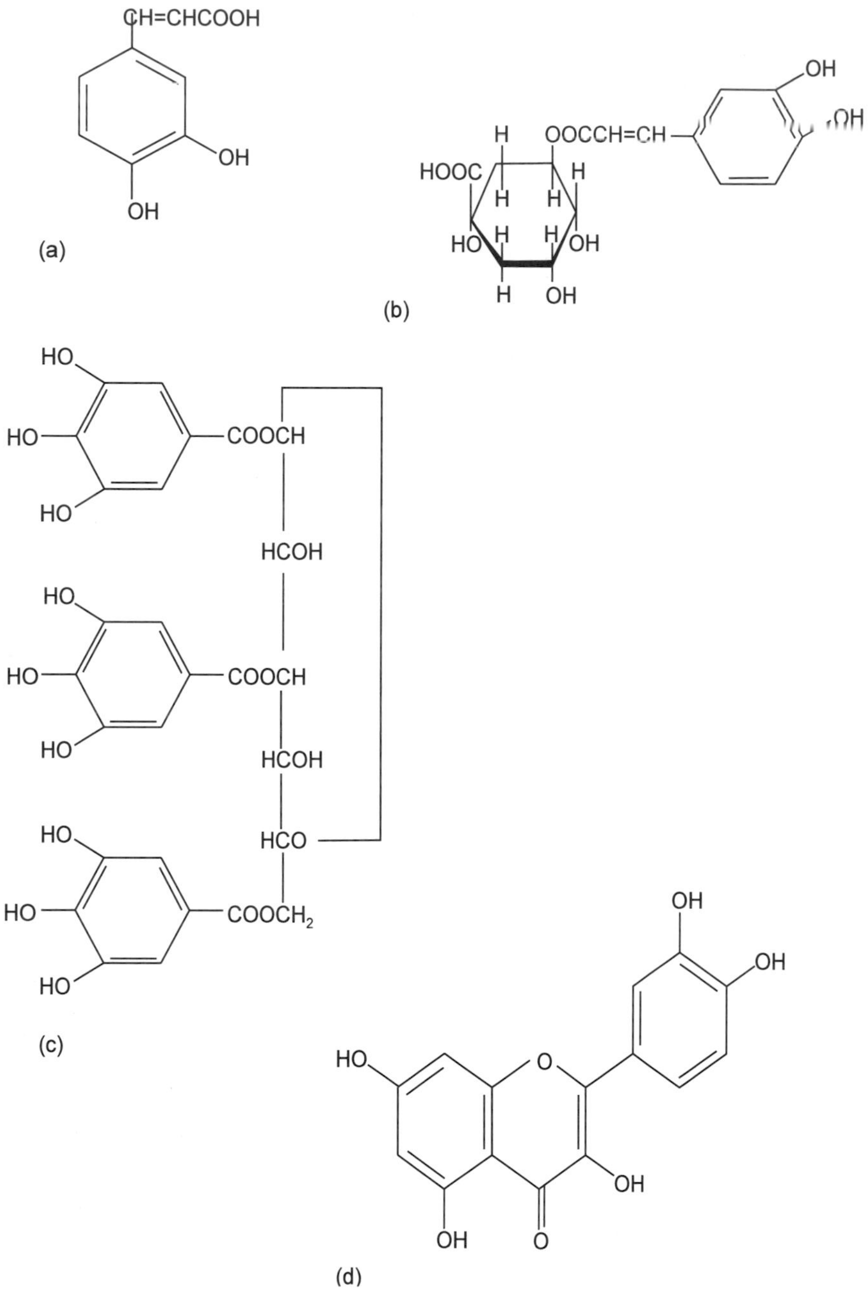

Figure 7.4 Polyphenolic compounds possessing anti-thiamin activity. (a) Caffeic acid (3,4-dihydroxycinnamic acid); (b) chlorogenic acid; (c) corilagin (a tannin); (d) quercetin (a flavone).

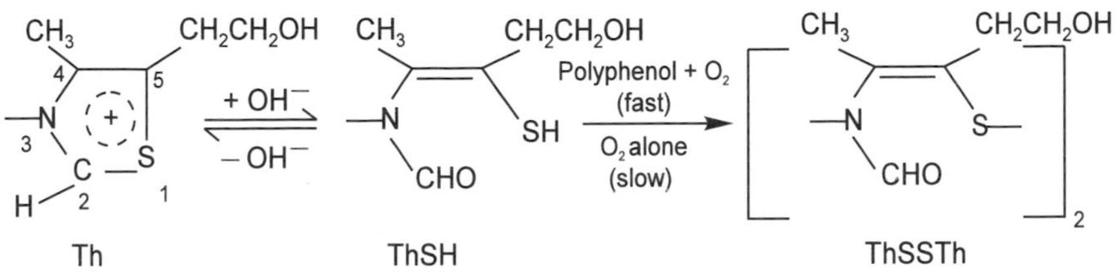

Figure 7.5 A scheme showing the effect of OH⁻ on the opening of the thiazole ring of thiamin (Th) to the sulphydryl (ThSH) form leading to the formation of thiamin disulphide (ThSSTh).

dry weight, which gives tea the astringent taste. Betel nuts are also good sources of tannins.

In all polyphenolic anti-thiamin factors the position of the hydroxyl groups is important for activity. Compounds with *ortho* hydroxyl groups have the highest activity, followed by *para*-hydroxylated compounds with medium activity and *meta*-hydroxylated compounds with no activity (Hilker and Somogyi, 1982). Panijpan and Ratanaubolchai (1980) proposed a broad scheme to explain the mechanism for thiamin–polyphenol interaction. In Figure 7.5 only the thiazole moiety and changes thereon are depicted. At a sufficiently high pH (> 6.5) the polyphenol ionizes and the OH⁻ groups attack the C-2 to open the ring and yield a sulphydryl derivative (ThSH) which exists in equilibrium with thiamin (Th). In the presence of oxygen, *ortho*- and *para*-polyphenolic compounds undergo oxidation and polymerization to yield active quinones and less active polymerized products. The quinones interact with the sulphydryl compound to give thiamin disulphide (ThSSTH) as the major reaction product. In the absence of the polyphenol compound the oxidation of the sulphydryl compound to thiamin disulphide is slow. Thiamin disulphide shows the full activity of the vitamin itself because it is easily reduced to thiamin *in vivo* by reducing agents such as cysteine, ascorbic acid and glutathione (Evans, 1975). However, in the absence of reducing agents, thiamin disulphide is not the ultimate major product of thiamin degradation by polyphenolic compounds. Further hydrolysis of the disulphide and subsequent oxidation yield inactive degradation products.

Rungruangsak *et al.* (1977) showed that ascorbic acid prevents thiamin modification by tannic acid *in vitro*. Under the conditions used, ascorbic acid appeared not only to protect the remaining thiamin from further modification by tannic acid, but also to be capable of partially reversing the reaction. Cysteine and reduced glutathione were found to be less effective than ascorbic acid.

In rural parts of Thailand the chewing of fermented tea leaves and betel nut as stimulants is common practice. Vimokesant *et al.* (1982) reported on studies of the effects of tea drinking and chewing of fermented tea leaves on the thiamin status of Thai subjects. Thiamin status was meas-ured as the effect of TPP stimulation of blood transketolase activity – a high effect indicates thiamin deficiency. After drinking of tea or chewing of fermented tea leaves, the TPP effect was significantly increased and thereafter decreased to normal by thiamin supplementation. Abstention from chewing betel nuts reduced a high TPP effect to a normal level, which was again increased significantly when the subjects resumed their chewing habits. A high TPP effect could be prevented by a delay in tea-leaf chewing after meals or consuming foods high in ascorbic acid along with the meals. The consumption of coffee, either caffeinated or decaffein-ated, also produced an anti-thiamin effect in human subjects as shown by a decreased urinary excretion of thiamin (Hilker and Somogyi, 1982).

Definite effects of anti-thiamin factors on thiamin status have not been demonstrated in rat feeding trials (Hilker and Somogyi, 1982). However, experiments using rat everted intestinal sacs showed that thiamin pass-age through the intestinal wall was diminished in the presence of caffeic acid (Schaller and Höller, 1976). These investigators suggested that in the case of a limited supply of thiamin, deficiency could arise from a dis-turbance in the saturable transport process.

Effects of alcohol
The acute effect of ethanol in rats is to significantly reduce the absorption of thiamin by inhibiting in a reversible manner the low concentration, saturable component of the dual transport process. More specifically, ethanol appears to allow the initial uptake of thiamin across the apical membrane of the enterocyte but impairs the subsequent movement of the vitamin across the basolateral membrane into the serosa. The fall in the rate of cellular thiamin exit is associated with inhibition of the basolateral Na^+–K^+-ATPase activity (Hoyumpa, 1986).

Studies in rats may not be completely relevant to humans since jejunal perfusion studies in chronic alcoholic patients demonstrated that neither alcoholism nor acute exposure to alcohol (5% concentration of ethanol) significantly affected thiamin absorption (Breen *et al.*, 1985).

Effects of dietary fibre
Omaye, Chow and Betschart (1982) demonstrated certain *in vitro* interac-tions of thiamin with selected dietary fibre constituents which could influence vitamin bioavailability in long-term high fibre dietary regimes. The interactions were consistent with binding of thiamin to the fibre matrix or trapping by interstitial water. Ranhotra *et al.* (1985) reported that the bioavailability of thiamin present in whole wheat bread was slightly higher than that in thiamin-restored white bread as determined by erythrocyte transketolase (ETK) activity and liver thiamin content in the rat; thus the higher fibre content of whole wheat bread did not impair

the bioavailability of the thiamin present. Yu and Kies (1993) determined the bioavailability of thiamin to humans from wet and dry milled maize brans which were coarsely or finely ground. Using a double cross-over design, the nine subjects were fed laboratory controlled diets containing unsupplemented bread or bread supplemented with each type of bran to supply 20 g fibre/day. Urinary excretion values were not significantly affected by the experimental treatment.

7.7.3 Effects of food processing, storage and cooking

Storage losses of thiamin in foods depend upon conditions such as pH, temperature and moisture content. Frozen foods and dehydrated foods incur little loss of thiamin during storage. An alkaline environment during processing or cooking will promote loss of thiamin; for instance, the use of baking powder in cake making can destroy over 50% of the original thiamin content of the flour. In the baking of bread with yeasts the medium is slightly acidic and the loss is reduced to 15–30%, mostly in the crust. The addition of sulphiting agents to fruit and minced meat to inhibit enzymatic and nonenzymatic browning can cause total destruction of thiamin, but these foods are not important sources of the vitamin. In the processing of milk the following losses have been reported: pasteurization, 9–20%; sterilization, 30–50%; spray-drying, 10%; roller drying, 15%; and condensing (canning), 40% (Gubler, 1991).

7.8 DIETARY INTAKE

Because TPP is essential as a coenzyme for the release of energy from carbohydrates, alcohol and fats, there is an increased requirement for thiamin in situations where metabolism is heightened, e.g. during high muscular activity, pregnancy and lactation, and also during protracted fever and hyperthyroidism. In practice, the requirement for thiamin is expressed in terms of total caloric intake.

In 1985 adult men in the United States consumed an average of 1.75 mg of thiamin/day, equivalent to 0.68 mg/1000 kcal. The corresponding intakes for adult women and children aged 1 to 5 years were 1.05 mg (0.69 mg/1000 kcal) and 1.12 mg (0.79 mg/1000 kcal), respectively.

The current US recommendations (National Research Council, 1989) are based on 0.5 mg of thiamin per 1000 kcal for healthy children, adolescents and adults, and 0.4 mg/1000 kcal for infants. This leads to RDAs ranging from 1.1 mg/day for females aged between 11 and 50 years and 1.5 mg/day for males aged between 15 and 50 years. A minimum intake of 1.0 mg/day is advised for adults, irrespective of energy intake. This applies particularly to the elderly, who may be unable to utilize the vitamin efficiently. The allowance for infants up to 6 months is 0.3 mg/day, increasing to 1.0 mg/day by 10 years of age. An additional allowance of 0.4 mg/day is recommended throughout pregnancy to accomodate maternal and foetal growth and increased maternal energy intake. To account for both thiamin secretion in milk and increased energy intake during lactation, an additional 0.5 mg/day is recommended throughout lactation.

In the United Kingdom, the Reference Nutrient Intakes are based upon 0.4 mg of thiamin per 1000 kcal for healthy people aged over 1 year. Typical RNIs are 1.0 mg/day for males aged between 19 and 50 years, and 0.8 mg/day for females aged from 15 years upwards, with a daily increment of 0.1 mg during the last trimester of pregnancy and 0.2 mg throughout lactation (Department of Health, 1991).

Effects of high intakes

Thiamin is of extremely low toxicity when administered orally to humans, because excess amounts are rapidly excreted in the urine. Large parenteral doses administered over a long period have been reported to produce clinical manifestations and, in some cases, even death (Cumming, Briggs and Briggs, 1981).

Assessment of nutritional status

The most commonly used procedure for assessing thiamin nutritional status in humans has been the measurement of erythrocyte transketolase (EC 2.2.1.1) activity (ETKA) and its stimulation *in vitro* by the addition of TPP (TPP effect) (Sauberlich, 1984). The TPP effect represents the difference between basal and stimulation values expressed as a percentage of the basal value. As thiamin deficiency progresses, the TPP effect increases. A TPP effect > 20% indicates thiamin deficiency, 15–20% indicates marginal deficiency, and < 15% indicates sufficiency. The TPP effect reflects a decrease in dietary intake of thiamin before any other signs of thiamin inadequacy are detectable. However, in chronic deficiency, the TPP added *in vitro* cannot restore ETKA fully; under such conditions, the TPP effect may be in the normal range of 0–15% (Tanphaichitr, 1994).

7.9 SYNOPSIS

The major form of thiamin in meat products is protein-bound TPP. In cereals and cereal products, including white flour made from wheat, free thiamin predominates, but wheat germ, bran and also white bread contain TPP. For the analysis of these products it is customary to determine the combined thiamin activity of the various forms present.

Extraction of the free and protein-bound thiamin in milk and milk-based foods is quite straightforward and requires only pH adjustment to induce protein precipitation, and filtration. When determining the total thiamin content of a food commodity, the test material is extracted by autoclaving with dilute mineral acid followed by enzymatic hydrolysis in order to convert protein-bound and phosphorylated forms of the vitamin to free thiamin. The principal techniques for determining thiamin are direct fluorimetric measurement after conversion to thiochrome, and HPLC with fluorimetric detection of thiochrome. The use of a common extraction procedure allows thiamin and riboflavin to be assayed by HPLC either simultaneously or successively.

Few studies have been conducted on the bioavailability of thiamin present naturally in foods, but it is generally considered to be high. The presence of thermostable polyphenols possessing anti-thiamin activity has been demonstrated in a wide variety of fruits and vegetables as well as in tea and coffee.

REFERENCES

Alyabis, A.M. and Simpson, K.L. (1993) Comparison of reverse-phase C-18 open column with the Bio-Rex 70 column in the determination of thiamin. *J. Food Comp. Anal.*, **6**, 166–71.

Andrews, J.S. and Nordgren, R. (1941) The application of the thiochrome method to the thiamin analysis of cereals and cereal products. *Cereal Chem.*, **18**, 686 95.

Ang, C.Y.W. and Moseley, F.A. (1980) Determination of thiamin and riboflavin in meat and meat products by high-pressure liquid chromatography. *J. Agric. Food Chem.*, **28**, 483–6.

AOAC (1990a) Thiamin (vitamin B_1) in foods. Fluorometric method. Final action. In *AOAC Official Methods of Analysis*, 15th edn (ed. K. Helrich), Association of Official Analytical Chemists, Inc., Arlington, VA, 942.23.

AOAC (1990b) Thiamine (vitamin B_1) in grain products. Fluorometric (rapid) method. Final action. In *AOAC Official Methods of Analysis*, 15th edn (ed. K. Helrich), Association of Official Analytical Chemists, Inc., Arlington, VA, 953.17.

AOAC (1990c) Thiamine (vitamin B_1) in bread. Fluorometric method. Final action 1960. In *AOAC Official Methods of Analysis*, 15th edn (ed. K. Helrich), Association of Official Analytical Chemists, Inc., Arlington, VA, 957.17.

Ayi, B.K., Yuhas, D.A., Moffett, K.S. *et al.* (1985) Liquid chromatographic determination of thiamine in infant formula products by using ultraviolet detection. *J. Ass. Off. Analyt. Chem.*, **68**, 1087–92.

Bailey, A.L. and Finglas, P.M. (1990) A normal phase high performance liquid chromatographic method for the determination of thiamin in blood and tissue samples. *J. Micronutr. Anal.*, **7**, 147–57.

Breen, K.J., Buttigieg, R., Iossifidis, S. *et al.* (1985) Jejunal uptake of thiamin hydrochloride in man: influence of alcoholism and alcohol. *Am. J. Clin. Nutr.*, **42**, 121–6.

Casirola, D., Ferrari, G., Gastaldi, G. *et al.* (1988) Transport of thiamine by brush-border membrane vesicles from rat small intestine. *J. Physiol.*, **398**, 329–39.

Clydesdale, F.M., Ho, C.-T., Lee, C.Y. *et al.* (1991) The effects of postharvest treatment and chemical interactions on the bioavailability of ascorbic acid, thiamin, vitamin A, carotenoids, and minerals. *Crit. Rev. Food Sci. Nutr.*, **30**, 599–638.

Cumming, F., Briggs, M. and Briggs, M. (1981) Clinical toxicology of vitamin supplements. In *Vitamins in Human Biology and Medicine* (ed. M.H. Briggs), CRC Press, Inc., Boca Raton, Florida, pp. 187–243.

Defibaugh, P.W. (1987) Evaluation of selected enzymes for thiamine determination. *J. Ass. Off. Analyt. Chem.*, 70, 514–17.

Defibaugh, P.W., Smith, J.S. and Weeks, C.E. (1977) Assay of thiamin in foods using manual and semiautomated fluorometric and microbiological methods. *J. Ass. Off. Analyt. Chem.*, 60, 522–7.

Deibel, R.H., Evans, J.B. and Niven, C.F. (1957) Microbiological assay for thiamin using *Lactobacillus viridescens*. *J. Bact.*, 74, 818–21.

Department of Health (1991) *Dietary Reference Values for Food Energy and Nutrients for the United Kingdom.* Report on Health and Social Subjects, No. 41, HM Stationery Office, London.

Dwivedi, B.K. and Arnold, R.G. (1973) Chemistry of thiamine degradation in food products and model system: a review. *J. Agric. Food Chem.*, 21, 54–60.

Edwards, C.H., Booker, L.K., Rumph, C.H. *et al.* (1971) Utilization of wheat by adult man: excretion of vitamins and minerals. *Am. J. Clin. Nutr.*, 24, 547–55.

Evans, W.C. (1975) Thiaminases and their effects on animals. *Vitamins Horm.*, 33, 467–504.

Fellman, J.K., Artz, W.E., Tassinari, P.D. *et al.* (1982) Simultaneous determination of thiamin and riboflavin in selected foods by high-performance liquid chromatography. *J. Food Sci.*, 47, 2048–50, 2067.

Ferrari, G., Patrini, C. and Rindi, G. (1982) Intestinal thiamin transport in rats. Thiamin and thiamin phosphoester content in the tissue and serosal fluid of everted jejunal sacs. *Pflügers Arch.*, 393, 37–41.

Ferrari, G., Ventura, U. and Rindi, G. (1971) The Na$^+$-dependence of thiamin intestinal transport *in vitro*. *Life. Sci.*, 10, 67–75.

Girija, V., Sharada, D. and Pushpamma, P. (1982) Bioavailability of thiamine, riboflavin and niacin from commonly consumed green leafy vegetables in the rural areas of Andhra Pradesh in India. *Int. J. Vitam., Nutr. Res.*, 52, 9–13.

Gregory, J.F. III and Kirk, J.R. (1978) Comparison of chemical and biological methods for determination of thiamin in foods. *J. Agric. Food Chem.*, 26, 338–41.

Gubler, C.J. (1991) Thiamin. In *Handbook of Vitamins*, 2nd edn (ed. L.J. Machlin), Marcel Dekker, Inc., New York, pp. 233–81.

Hartman, A.M. and Dryden, L.P. (1974) Vitamins in milk and milk products. In *Fundamentals of Dairy Chemistry*, 2nd edn (eds B.H. Webb, A.H. Johnson and J.A. Alford), AVI Publishing Co., Inc., Westport, CT, pp. 325–441.

Hilker, D.M. and Somogyi, J.C. (1982) Antithiamins of plant origin: their chemical nature and mode of action. *Ann. NY Acad. Sci.*, 378, 137–45.

Hoffer, A., Alcock, A.W. and Geddes, W.F. (1943) A rapid method for the determination of thiamine in wheat and flour. *Cereal Chem.*, 20, 717–29.

Holland, B., Welch, A.A., Unwin, I.D. *et al.* (1991) *McCance and Widdowson's The Composition of Foods*, 5th edn, Royal Society of Chemistry and Ministry of Agriculture, Fisheries and Food.

Hoyumpa, A.M. Jr (1986) Mechanisms of vitamin deficiencies in alcoholism. *Alcoholism Clin. Exp. Res.*, 10, 573–81.

Hoyumpa, A.M. Jr, Middleton, H.M., Wilson, F.A. and Schenker, S. (1975) Thiamine transport across the rat intestine. 1. Normal characteristics. *Gastroenterology*, 68, 1218–27.

Hoyumpa, A.M. Jr, Nichols, S., Schenker, S. and Wilson, F.A. (1976) Thiamine transport in thiamine-deficient rats: role of the unstirred water layer. *Biochim. Biophys. Acta*, 436, 438–47.

Ishii, K., Sarai, K., Sanemori, H. and Kawasaki, T. (1979) Concentrations of thiamine and its phosphate esters in rat tissues determined by high-performance liquid chromatography. *J. Nutr. Sci. Vitaminol.*, **25**, 517–23.

Kamman, J.F., Labuza, T.P. and Warthesen, J.J. (1980) Thiamin and riboflavin analysis by high performance liquid chromatography. *J. Food Sci.*, **45**, 1497–9, 1504.

Kirk, J.R. (1974) Automated methods for the analysis of thiamine in milk, with application to other selected foods. *J. Ass. Off. Analyt. Chem.*, **57**, 1081–4.

Kirk, J.R. (1977) Automated analysis of thiamine, ascorbic acid, and vitamin A. *J. Ass. Off. Analyt. Chem.*, **60**, 1234–7.

Komai, T., Kawai, K. and Shindo, H. (1974) Active transport of thiamine from rat small intestine. *J. Nutr. Sci. Vitaminol.*, **20**, 163–77.

Leevy, C.H. and Baker, H. (1968) Vitamins and alcoholism. Introduction. *Am. J. Clin. Nutr.*, **21**, 1325–8.

Lumeng, L., Edmondson, J.W., Schenker, S. and Li, T.-K. (1979) Transport and metabolism of thiamin in isolated rat hepatocytes. *J. Biol. Chem.*, **254**, 7265–8.

McCormick, D.B. (1988) Thiamin. In *Modern Nutrition in Health and Disease*, 7th edn (eds M.E. Shils and V.R. Young), Lea & Febiger, Philadelphia, pp. 355–61.

McRoberts, L.H. (1954) Report on the determination of thiamin in enriched flour. Comparison of fluorometric methods. *J. Ass. Off. Agric. Chem.*, **37**, 757–70.

McRoberts, L.H. (1957) Report on thiamine in enriched cereal and bakery products. *J. Ass. Off. Agric. Chem.*, **40**, 843–53.

Murata, K. (1982) Actions of two types of thiaminases on thiamin and its analogues. *Ann. NY Acad. Sci.*, **378**, 146–56.

National Research Council (1989) Water-soluble vitamins. In *Recommended Dietary Allowances*, 10th edn, National Academy Press, Washington, DC, pp. 115–73.

Nicolas, E.C. and Pfender, K.A. (1990) Fast and simple liquid chromatographic determination of nonphosphorylated thiamine in infant formula, milk, and other foods. *J. Ass. Off. Analyt. Chem.*, **73**, 792–8.

Obermeyer, H.G., Fulmer, W.C. and Young, J.M. (1944) Cocarboxylase hydrolysis by a wheat phosphatase. *J. Biol. Chem.*, **154**, 557–9.

Ohta, H., Baba, T., Suzuki, Y. and Okada, E. (1984) High-performance liquid chromatographic analysis of thiamine in rice flour with fluorimetric post-column derivatization. *J. Chromat.*, **284**, 281–4.

Ohta, H., Maeda, M., Nogata, Y. *et al.* (1993) A simple determination of thiamine in rice (*Oryza sativa* L.) by high-performance liquid chromatography with post-column derivatization. *J. Liquid Chromat.*, **16**, 2617–29.

Omaye, S.T., Chow, F.I. and Betschart, A.A. (1982) In vitro interaction of 1-^{14}C-ascorbic acid and 2-^{14}C-thiamin with dietary fiber. *Cereal Chem.*, **59**, 440–3.

Panijpan, B. and Ratanaubolchai, K. (1980) Kinetics of thiamine-polyphenol interactions and mechanism of thiamine disulphide formation. *Int. J. Vitam. Nutr. Res.*, **50**, 247–53.

Patrini, C., Cusaro, G., Ferrari, G. and Rindi, G. (1981) Thiamine transport by rat small intestine 'in vitro': influence of endogenous thiamine content of jejunal tissue. *Acta Vitaminol. Enzymol.*, **3** n.s., 17–26.

Pelletier, O. and Madère, R. (1975) Comparison of automated and manual procedures for determining thiamine and riboflavin in foods. *J. Food Sci.*, **40**, 374–9.

Pelletier, O. and Madère, R. (1977) Automated determination of thiamin and riboflavin in various foods. *J. Ass. Off. Analyt. Chem.*, **60**, 140–6.

Ranhotra, G., Gelroth, J., Novak, F. and Bohannon, F. (1985) Bioavailability for rats of thiamin in whole wheat and thiamin-restored white bread. *J. Nutr.*, **115**, 601–6.

Ribbron, W.M., Stevenson, K.E. and Kirk, J.R. (1977) Comparison of semiautomated and manual methods for the determination of thiamine in baby cereals
and infant and dietary formulas. *J. Ass. Off. Analyt. Chem.*, **60**, 737–8.

Rindi, G. (1984) Thiamin absorption by small intestine. *Acta Vitaminol. Enzymol.*,
6, 47–55.

Rindi, G. and Ferrari, G. (1977) Thiamine transport by human intestine in vitro.
Experientia, **33**, 211–13.

Rindi, G. and Ventura, U. (1967) Phosphorylation and uphill intestinal transport
of thiamine, in vitro. *Experientia*, **23**, 175–6.

Rose, R.C. (1988) Transport of ascorbic acid and other water-soluble vitamins.
Biochim. Biophys. Acta, **947**, 335–66.

Rose, R.C. (1991) Intestinal transport of water-soluble vitamins. In *Handbook of
Physiology, Section 6: The Gastrointestinal System, Vol. 4. Intestinal Absorption and
Secretion* (eds M. Field and R.A. Frizzell), American Physiological Society,
Bethesda, MD, pp. 421–35.

Roy, R.B. and Conetta, A. (1976) Automated analysis of water-soluble vitamins in
food. *Food Technol.*, **30**(10), 94, 95, 98, 100, 103, 104.

Rungruangsak, K., Tosukhowong, P., Panijpan, B. and Vimokesant, S.L. (1977)
Chemical interactions between thiamin and tannic acid. I. Kinetics, oxygen
dependence and inhibition by ascorbic acid. *Am. J. Clin. Nutr.*, **30**, 1680–5.

Sauberlich, H.E. (1984) Newer laboratory methods for assessing nutriture of
selected B-complex vitamins. *Ann. Rev. Nutr.*, **4**, 377–407.

Schaller, K. and Höller, H. (1976) Thiamine absorption in the rat. IV. Effects of
caffeic acid (3,4-dihydroxycinnamic acid) upon absorption and active transport
of thiamine. *Int. J. Vitam. Nutr. Res.*, **46**, 143–8.

Scholes, P.M. (1960) The microbiological assay of thiamine. Two methods modified for use with small amounts of test material. *Analyst, Lond.*, **85**, 883–9.

Sklan, D. and Trostler, N. (1977) Site and extent of thiamin absorption in the rat. *J.
Nutr.*, **107**, 353–6.

Soliman, A.-G.M. (1981) Comparison of manual and benzenesulphonyl chloride-
semiautomated thiochrome methods for determination of thiamine in foods. *J.
Ass. Off. Analyt. Chem.*, **64**, 616–22.

Tanphaichitr, V. (1994) Thiamin. In *Modern Nutrition in Health and Disease*, 8th
edn, Vol. 1 (eds M.E. Shils, J.A. Olson and M. Shike), Lea & Febiger, Philadelphia, pp. 359–65.

Taungbodhitham, A.K. (1995) Thiamin content and activity of antithiamin factor
in vegetables of southern Thailand. *Food Chem.*, **52**, 285–8.

Vidal-Valverde, C. and Reche, A. (1990) An improved high performance liquid
chromatographic method for thiamin analysis in foods. *Z. Lebensmittelunters
u.-Forsch.*, **191**, 313–18.

Vimokesant, S., Kunjara, S., Rungruangsak, K. *et al.* (1982) Beriberi caused by
antithiamin factors in food and its prevention. *Ann. NY Acad. Sci.*, **378**, 123–36.

Warnick, K.P., Zaehringer, M.V., Bring, S.V. and Woods, E. (1956) Physiological
availability of thiamine from potatoes and from brown rice. *J. Nutr.*, **59**, 121–33.

Yang, P.-F. and Pratt, D.E. (1984) Antithiamin activity of polyphenolic antioxidants. *J. Food Sci.*, **49**, 489–92.

Yoshioka, K. (1984) Some properties of the thiamine uptake system in isolated rat
hepatocytes. *Biochim. Biophys. Acta*, **778**, 201–9.

Yu, B.H. and Kies, C. (1993) Niacin, thiamin, and pantothenic acid bioavailability
to humans from maize bran as affected by milling and particle size. *Plant Foods
Human Nutr.*, **43**, 87–95.

8

Riboflavin and other flavins (vitamin B₂)

8.1 INTRODUCTION

The existence of a heat-stable growth factor in yeast extracts was first discovered by Emmett and Luros in 1920. In 1932, Warburg and Christian isolated from yeast a yellow enzyme which contained a non-protein component subsequently shown to be a phosphate ester of an alloxazine derivative. The structure of riboflavin was determined by Kuhn's group in 1933 and the vitamin was first synthesized by Karrer's group in 1935.

In bound coenzymic forms, riboflavin participates in oxidation–reduction reactions in numerous metabolic pathways and in energy production in the mitochondrial electron transport chain.

8.2 CHEMICAL STRUCTURE AND NOMENCLATURE

The principal vitamin B_2-active flavins found in nature are riboflavin, riboflavin-5'-phosphate (= flavin mononucleotide, FMN) and riboflavin-5'-adenosyldiphosphate (= flavin adenine dinucleotide, FAD). The structures of these compounds are depicted in Figure 8.1. The parent riboflavin molecule comprises a substituted isoalloxazine moiety with a ribitol side chain and is systematically named as 7,8-dimethyl-10-(1'-D-ribityl)isoalloxazine ($C_{17}H_{20}N_4O_6$; MW = 376.36). The 'mononucleotide' and 'dinucleotide' designations for FMN and FAD, respectively, are actually incorrect but are nevertheless still accepted. FMN is not a nucleotide, as the sugar group is not ribose, and the isoalloxazine ring is neither a purine nor a pyrimidine. FAD is composed of a nucleotide (adenosine monophosphate, AMP) and the so-called flavin pseudonucleotide. In biological tissues, FAD and, to a lesser extent, FMN occur almost entirely as prosthetic groups for a large variety of flavin enzymes (flavoproteins). In most flavoproteins the flavins are bound tightly but noncovalently to the apoenzyme. In mammalian tissues less than 10% of the FAD is covalently attached to specific amino acid residues of four important apoenzymes. These are found within succinate and sarcosine dehydrogenases, monoamine oxidase and L-gluconolactone oxidase in which FAD is peptide-linked to an N-histidyl or S-cysteinyl residue via the 8-methyl group (McCormick, 1994).

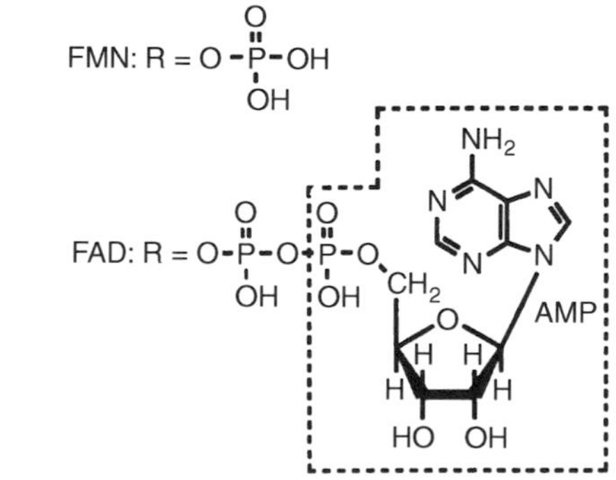

Figure 8.1 Structures of riboflavin, FMN and FAD.

The term riboflavin is confusing as it may be used in two different contexts. It is either synonymous with vitamin B_2 (a generic descriptor for all biologically active flavins) or it refers specifically to the parent riboflavin molecule (MW = 376.36) described above. In this text, the names riboflavin, FMN and FAD will be used in most cases as specific terms and flavins or vitamin B_2 will be used as generic terms.

8.3 DEFICIENCY SYNDROMES

Unlike the dramatic effects of gross thiamin deficiency, no pathologically severe symptoms attributed to vitamin B_2 deficiency have been observed in humans. Symptoms of vitamin B_2 deficiency have been induced experimentally in volunteers whose diets were lacking only in the vitamin or who were fed vitamin B_2 antagonists. Deficiency symptoms in humans usually include lesions of the lips (cheilosis) and angles of the mouth (angular stomatitis), a fissured and magenta-coloured tongue (glossitis), seborrhoeic follicular keratosis of the nose and forehead, and dermatitis of the anogenital region. Ophthalmic symptoms are a superficial vascularization of the cornea accompanied by intense photophobia.

8.4 PHYSICOCHEMICAL PROPERTIES

Solubility

The riboflavin compound is only sparingly soluble in water (100–250 µg/ml at room temperature and 2337 µg/ml at 100 °C). It is more soluble in dilute hydrochloric acid and in alkali (but unstable in alkali); slightly soluble in absolute ethanol; and insoluble in acetone, diethyl ether, chloroform or benzene. The solubility of riboflavin in water is increased in the presence of aromatic compounds such as nicotinamide. The sodium salt of FMN is very soluble in water, and FAD is freely soluble in water (Yagi, 1962).

Spectroscopic properties

Absorption
Riboflavin in aqueous solution exhibits a UV-visible spectrum containing four major bands centred around 223, 266, 373 and 445 nm. The positions of the maxima and their absorbance coefficients vary somewhat according to the nature of the solvent in which the compound is dissolved. The spectra of riboflavin and FMN are practically identical to one another under similar conditions, but the spectrum of FAD is slightly different. All three flavins lose their absorbance in the visible region when they are reduced to their colourless 1,5-dihydro forms known as leuco bases.

Table 8.1 The molar absorptivity (ε) values of flavins dissolved in 0.1 M phosphate buffer, pH 7.0[a]

Flavin	Wavelength (nm)		
	260	375	450
Riboflavin	27 700	10 600	12 200
FMN	27 100	10 400	12 200
FAD	37 000	9 300	11 300

[a] From Yagi, K. (1962) *Methods of Biochemical Analysis*, Vol. 10 (ed. D. Glick). Copyright © 1962, by John Wiley & Sons, Inc. Reprinted by permission of John Wiley & Sons, Inc.

Molar absorptivity (ε) values of flavins dissolved in 0.1 M phosphate buffer, pH 7.0, are given in Table 8.1.

Fluorescence
Riboflavin, FMN and FAD in aqueous solution exhibit an intense yellow–green fluorescence with an emission maximum at around 530 nm when excited at 440–500 nm. The fluorescence of flavins is a characteristic of uncharged neutral forms of isoalloxazines; anions and cations do not fluoresce. Riboflavin shows a maximum and equal fluorescence intensity in the pH range between 3.5 and 7.5. The same is true for FMN, but FAD displays maximal intensity at pH 2.7–3.1. In equimolar neutral aqueous solutions, riboflavin and FMN exhibit practically the same fluorescence intensity, whereas that of FAD is only about 14% of that of riboflavin. The binding of flavins to proteins generally results in a dramatic quenching of fluorescence. Reduced forms of flavins do not fluoresce (Koziol, 1971).

Stability
Crystalline riboflavin is stable in dry form under normal lighting. In solution, vitamin activity is destroyed by exposure to both UV radiation and visible light, the rate of destruction increasing with an increase in temperature and pH. Sattar, deMan and Alexander (1977) reported that the wavelength range of 350–520 nm was generally damaging to riboflavin in aqueous solution, with the 415–455 nm range being responsible for the greatest destruction of the vitamin. The principal photodegradation product under neutral and acidic conditions is lumichrome (7,8-dimethylalloxazine), while under alkaline conditions the major product is lumiflavin (7,8-trimethylisoalloxazine) (Figure 8.2). The latter degradation product exhibits a strong blue fluorescence, which has been used as a means of detection in many assay procedures.

If protected from light, aqueous solutions of riboflavin in the pH range between 2 and 5 are heat-stable up to 120 °C; above pH 7 the isoalloxazine

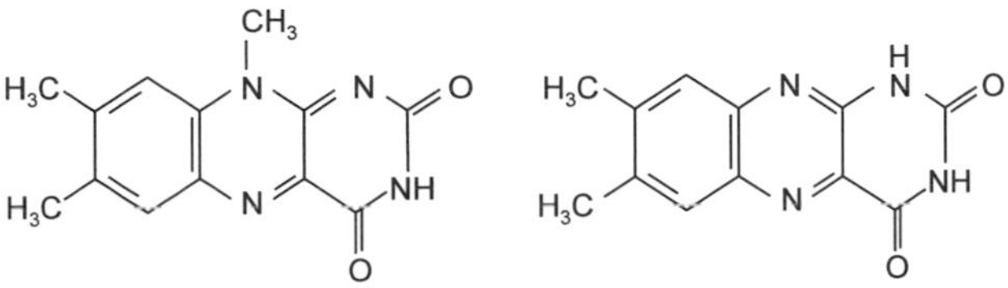

Figure 8.2 Photodegradation products from riboflavin: (a) Lumiflavin; (b) Lumichrome.

ring is rapidly destroyed at elevated temperatures. Optimal stability is obtained at pH 3.5–4.0. Riboflavin is stable towards oxygen and many oxidizing agents in the absence of light.

8.5 ANALYSIS

8.5.1 Scope of analytical techniques

Vitamin B_2 occurs naturally in biological tissues as protein-bound FAD and FMN, while riboflavin itself is the major form in bovine milk. It is usually required to determine the total riboflavin content (less ambiguously termed 'vitamin B_2 activity') of the sample, which is defined in the nutritional context as the sum of FAD, FMN and riboflavin. Current techniques for determining total riboflavin are microbiological assay, fluorimetry, and HPLC using fluorimetric detection. HPLC can also be used to determine the individual flavins. The biological activity of preparations containing vitamin B_2 can be examined in the curative rat growth test using crystalline riboflavin as the reference standard.

Riboflavin, FMN and FAD are equally active as vitamins in human nutrition (Yagi, 1962). No IU of vitamin B_2 activity has been defined, and analytical results are expressed in terms of weight (mg) of chemically pure riboflavin.

8.5.2 Extraction techniques

The flavin cofactors naturally present in biological tissues are stabilized through protein binding, and riboflavin added to foods is protected to some extent by the food matrix. FMN is partially hydrolysed to riboflavin when food samples are heated to 95–100 °C or autoclaved at 121–123 °C in a medium of 0.1 N HCl for 30 min. A substantial fraction of the FMN is converted to the biologically active isomeric 2'-, 3'- and 4'-phosphates (Nielsen, Rauschenbach and Bacher, 1986). FAD is totally hydrolysed to FMN when food samples are heated in a medium of 0.1 N HCl under the above conditions or incubated in a medium of 10% (w/v) trichloroacetic acid (TCA) at 38 °C overnight. No destruction of FAD occurs at 0 °C for 30 min in 10% TCA (Koziol, 1971).

When carrying out physicochemical or microbiological assays for ribo flavin, it is necessary to release the flavins from their intimate association with proteins and to completely convert the FAD to FMN. Both of these requirements are readily accomplished (for noncovalently bound flavins) by autoclaving food samples at 121 °C for 30 min with dilute mineral acid (usually 0.1 N HCl) at a pH of < 3. During acid digestion some of the FMN is hydrolysed to riboflavin, and a small fraction of the FMN (ribo-flavin-5'-phosphate) is converted to the isomeric 2'-, 3'- and 4'-phos-phates. The complete conversion of FMN to riboflavin can only be achieved by subsequent enzymatic hydrolysis, for which a standardized diastatic enzyme preparation such as Takadiastase or Claradiastase is used. Watada and Tran (1985) reported that Mylase was as effective as Takadiastase, the latter being unavailable at that time. These are rela-tively inexpensive and crude preparations that contain varying degrees of phosphatase activity. In practice, the complete enzymatic conversion of FMN to riboflavin may not always be achieved, the degree of hydro-lysis depending on the source and batch-to-batch phosphatase activity of the enzyme and on the incubation conditions.

For the analysis of milk, eggs and dairy products, it is common practice to determine the riboflavin specifically, on the assumption that free or loosely bound riboflavin is the predominant naturally occuring flavin present. In this case, the extraction procedure simply entails precipitation of the protein by acidification and filtration, omitting the acid and enzyme digestion steps. Rashid and Potts (1980) removed the protein from milk and milk products by filtration after treatment with acidified lead acetate solution.

Acid and enzymatic hydrolysis carried out successively are incapable of liberating the covalently bound FAD of certain enzymes, and hence this source of FAD will not be measured. This is perhaps fortuitous when the nutritional value of the food sample is under assessment, as there is evidence that covalently bound FAD is largely unavailable to the host.

8.5.3 Microbiological assay

The organism traditionally used for determining vitamin B_2 is *Lactobacil-lus casei* subsp. *rhamnosus* (ATCC No. 7469). Lactic acid bacteria cannot utilize FAD, and the growth response of *L. casei*, measured turbi-dimetrically, differs significantly between riboflavin and FMN (Langer and Charoensiri, 1966). As most of the vitamin B_2 activity in foods is present in the form of FMN after acid extraction, it would be more accurate to use FMN as the standard in the microbiological assay instead of riboflavin.

L. casei is stimulated by starch (Andrews, Boyd and Terry, 1942), and either stimulated or inhibited by long-chain free fatty acids (e.g. palmitic, stearic, oleic and linoleic acids) and other lipids, including lecithin (Bauernfeind, Sotier and Boruff, 1942; Strong and Carpenter, 1942; Kodicek and Worden, 1945). Kornberg, Langdon and Cheldelin (1948) proposed the use of *Enterococcus faecalis* (ATCC No. 10100) which, with a sensitivity to 0.1 ng riboflavin/ml (Barton-Wright, 1952), is 50 times more sensitive compared with *L. casei*. Lipids also stimulate the growth of *E. faecalis*, but its higher sensitivity to riboflavin allows samples to be prepared for assay at a higher dilution, resulting in negligible amounts of interfering lipids and other extraneous matter in the assay tubes. Barton-Wright (1967) used a basal medium that is more synthetic in composition than the original medium used by Kornberg, Langdon and Cheldelin (1948). Comparative assay values using *E. faecalis* and *L. casei* (Table 8.1) show excellent agreement for many of the foods tested. The higher value of 0.33 µg/g obtained for white flour using *L. casei* is apparently due to the stimulatory effect of lipids, as a prior ether extraction lowered this value to 0.20 µg/g, making it comparable to the value of 0.21 µg/g using *E. faecalis*. The same ether treatment produced no lowering of the *E. faecalis* assay value. *E. faecalis*, being practically unaffected by lipids and other substances, is preferred to *L. casei* as an assay organism (Barton-Wright, 1967). Bell (1971) pointed out that there are no dehydrated media for this assay, and media have to be prepared from the ingredients.

The extraction procedure for the *L. casei* assay necessitates autoclaving the food sample with 0.1 N HCl at 121 °C for 30 min. During acid hydrolysis, FAD, which cannot be utilized by lactic acid bacteria, is completely degraded to FMN and riboflavin, and some of the FMN is also degraded to riboflavin. The complete conversion of FMN to riboflavin is not necessary, since FMN, riboflavin and the various isomeric riboflavin monophosphates are all nutrients for the growth of *L. casei*. The acid digestion eliminates the troublesome starch, but it liberates interfering free fatty acids.

The general extraction procedure used in *L. casei* assays for analysing foods of very low fat content (such foods include many cereals) involves acid hydrolysis, followed by precipitation of the denatured proteins at pH 4.5. The precipitated proteins, together with the small amount of lipoidal material and any non-hydrolysed starch, is removed by simple filtration through paper. Omission of the filtration step produced a pronounced drift in riboflavin values for whole wheat flour, which was typical of cereal products in general (Strong and Carpenter, 1942). Ether extraction could be used to remove interfering lipids, but, for samples of negligible fat content, the filtration step results in a valid assay (i.e. free from drift), and is much simpler and quicker to accomplish. High-fat foodstuffs should be given a preliminary extraction with petroleum ether to remove the bulk of the lipids before the acid hydrolysis step. This initial extraction does not completely remove the stimulatory lipids (Wegner, Kemmerer and Fraps, 1942) and a further extraction with diethyl ether is necessary after filtering off the precipitated proteins. Milk should be separated by centrifugation, and the serum shaken in a separating funnel with diethyl ether. Petroleum ether is not a suitable fat solvent with milk, as the mixture tends to form an emulsion (Barton-Wright and Booth, 1943).

High-fat foods, such as wheat germ, maize, oats, soya beans, meat, cheese and mixed diets, require a preliminary extraction with petroleum ether (b.p. 40–60 °C) before acid hydrolysis to remove neutral fats. This involves extracting the dried, finely ground sample with petroleum ether for 16–18 h in a Soxhlet apparatus. The defatted material is autoclaved, adjusted to pH 4.5, diluted and filtered. A 50 ml aliquot of the filtrate is shaken with two or three 30 ml portions of diethyl ether in a separating funnel. The combined ether extracts are washed two or three times with water, and the washings are added to the bulked aqueous layers. Finally, the pH of the extract is adjusted to 6.8, and the extract is filtered, if necessary, and diluted to 100 ml (or other suitable volume) for direct assay (Barton-Wright, 1952).

The published extraction procedure using *E. faecalis* (Kornberg, Langdon and Cheldelin, 1948) entailed the addition of 20 ml of water and 3 ml of 1 N H_2SO_4 to 1 g of the dry material to be assayed, and autoclaving for 30 min. The pH was adjusted to 4.5–5.0, and the extract diluted to contain $c.0.001$–0.002 μg riboflavin/ml.

8.5.4 AOAC fluorimetric method

The native fluorescence exhibited by riboflavin enables this vitamin to be assayed fluorimetrically without the need for chemical derivatization. The approach taken for the determination of vitamin B_2 using direct fluorimetry is dictated by the relative fluorescence intensities of the three major flavins. FMN and riboflavin exhibit equal fluorescent intensity on a molar basis, whereas the fluorescence of FAD is much less intense. It is therefore necessary to completely convert the FAD to FMN and riboflavin by autoclaving the food sample at 121 °C for 30 min with 0.1 N HCl. A fluorimetric method has been adopted as final action by the AOAC for the determination of vitamin B_2 in foods, including ready-to-feed milk-based infant formulas. The general procedure (AOAC, 1990a) involves the following steps: acid digestion, precipitation of proteinaceous material, oxidation, and measurement of the fluorescence.

In a rapid procedure for determining added riboflavin in dried milk products (Woodrow, Torrie and Henderson, 1969), samples were deproteinized by acidification and fluorescence readings of the filtrate were obtained without further purification.

The fluorimetric method has been semi-automated using segmented-flow analysis to perform in-line dialysis and permanganate oxidative clean-up after manual acid digestion of food samples (Egberg and Potter, 1975). Excess permanganate was reduced with sodium bisulphite, and metaphosphoric acid was added as a manganese-sequestering agent to prevent the precipitation and build-up of manganese dioxide in the reagent lines. A collaborative study (Egberg, 1979) led to the adoption of the method by the AOAC as Final Action in 1982 (AOAC, 1990b).

Russell and Vanderslice (1992a) applied flow-injection analysis to the standard AOAC fluorimetric method (AOAC, 1990a) using sodium dithionite ($Na_2S_2O_4$) dissolved in 0.4% sodium acetate as the blank determination, as specified in the AOAC semi-automated method.

8.5.5 High-performance liquid chromatography

HPLC methods for determining vitamin B_2 are complicated by the fact that the FMN and riboflavin liberated during acid hydrolysis are chromatographically separated, and both of these flavins have to be measured. The biologically active isomeric riboflavin monophosphates are also separated from FMN and riboflavin, and the non-measurement of these isomers constitutes an unavoidable source of error.

One technique for determining vitamin B_2 is to convert the bound flavins completely to riboflavin using both acid and enzymatic hydrolysis. Some workers heat the hydrolysed extract with 50% w/w trichloroacetic acid to deactivate and precipitate the previously added enzyme. The vitamin activity is then calculated by measuring the riboflavin peak in the chromatogram. In practice, the complete conversion of FMN to riboflavin cannot always be guaranteed.

The alternative approach is to omit the enzymatic step, and measure separately the FMN and riboflavin peaks resulting from acid hydrolysis and protein precipitation. For acid extracts containing significant amounts of FMN, i.e. where the FMN peak height is > 25% of the sum of the FMN and riboflavin peak heights (such as were obtained from raw beef, corned beef, fresh and cooked liver, and canned mushrooms), FMN and riboflavin were calculated separately, using their corresponding response factors (ratio of concentration of standard to peak height), and the results were summed to obtain a value for the vitamin B_2 activity expressed as riboflavin. For smaller amounts of FMN the total vitamin B_2 could be obtained by summation of FMN and riboflavin peak heights without significant error (Reyes *et al.*, 1988). The necessity of applying a correction factor to compensate for the lack of purity of commercial FMN preparations has been stressed by Russell and Vanderslice (1992b). It should be noted that the quantity of FMN found does not represent the original FMN content of the food sample, as the FMN peak originates largely from hydrolysed FAD.

Table 8.2 HPLC methods used for the determination of riboflavin and other flavins

Food	Sample preparation	Column	Mobile phase	Compounds separated	Detection	Reference
			Reversed-phase chromatography			
Dairy products (milk, whole milk powder, cheese)	Homogenize sample with 6% formic acid containing 2 M urea and then centrifuge. Mix a 2 ml aliquot with sorboflavin (internal standard). *Clean-up and concentration:* pass solution through C_{18} solid-phase extraction column, wash column bed with 10% formic acid and elute the flavins with 10% formic acid/MeOH (4:1)	LC-18 3 μm (Supelco) 75 × 4.6 mm i.d.	14% MeCN in 0.1 M KH_2PO_4 (final pH 2.9)	FAD, FMN, sorboflavin (internal standard), riboflavin	Fluorescence: ex. 450 nm em. 530 nm	Bilic and Sieber (1990)
Dairy products, meat products, fish, fruit, vegetables, flour, baked products, beer, coffee	Autoclave homogenized sample with 0.05 M (0.1 N) H_2SO_4 at 121°C for 20 min. Cool, adjust pH to 4.5 and incubate with Claradiastase at 45°C overnight. Cool, dilute to volume and filter. *Clean-up and concentration:* pass aliquot of filtrate through C_{18} Sep-Pak solid-phase extraction cartridge, wash cartridge with water and elute the riboflavin with 40–70% MeOH	Spherisorb ODS-2 5 μm 250 × 4.6 mm i.d.	Water/MeOH (65:35)	Riboflavin (representing total vitamin B_2 activity)	Fluorescence: ex. 445 nm em. 525 nm	Ollilainen et al. (1990)
Bovine milk	Boil 1.0 g milk sample to inactivate pyrophosphatase. Digest with buffered Pronase (pH 6.8) for 1 h at 45°C to simultaneously release the flavins bound to milk proteins and deproteinize. Cool, adjust volume with phosphate buffer then adjust pH to 5.5. Centrifuge and filter	Capcell Pak C_{18} 5 μm 250 × 4.6 mm i.d. Column temperature 40°C	90% MeOH in water (Solvent A) and 0.01 M phosphate buffer (pH 5.5) (Solvent B). Linear gradient from 35% of A and 65% of B to 95% of A and 5% of B over 8 min and 5 min	FAD, FMN, riboflavin	Fluorescence: ex. 462 nm em. 520 nm	Ianno, Shirahuji and Hoshi (1991)

Sample	Sample preparation	Column	Mobile phase	Analytes	Detection	Reference
Raw and cooked meats, dairy products, eggs, cereal products	Homogenize sample with MeOH/CH$_2$Cl$_2$(9 + 10) and 7-ethyl-8-methyl-riboflavin (internal standard). Add 0.1 M citrate-phosphate buffer, pH 5.5 containing 0.1% sodium azide, and rehomogenize and centrifuge	2 PLRP-S 5 μm (Polymer Laboratories) columns in series 250 × 4.6 mm i.d. + 150 × 4.6 mm i.d. Column temperature 40 °C	MeCN/0.1% sodium azide in 0.01 M citrate-phosphate buffer, pH 5.5 in multistep gradient elution programme	FAD, FMN, riboflavin, 7-Et-8-Me-RF (internal standard)	Fluorescence: ex. 450 nm em. 522 nm	Russell and Vanderslice (1992b) (Figure 8.3)
Cereals, milk (whole, semi-skimmed, skimmed, evaporated)	On-line system consisting of microwave extraction followed by dialysis and trace enrichment with a C$_{18}$ mini-column	Spherisorb ODS-2 5 μm 250 × 4.6 mm i.d.	Initial: 94% of 0.1 M sodium acetate buffer (pH 4.8) and 6% of water–MeCN–MeOH (50 + 40 + 10) Linear gradient elution: Proportion of water–MeCN–MeOH mixture increased from 6 to 100% over 30 min	FMN, riboflavin	Fluorescence: ex. 450 nm em. 520 nm (cut-off filter 400 nm)	Greenway and Kometa (1994)

Ion interaction chromatography

Sample	Sample preparation	Column	Mobile phase	Analytes	Detection	Reference
Flour, bread, raw beef, corned beef, chicken liver, mushrooms, milk and milk products, cereals	Defat high fat samples with hexane. Autoclave homogenized sample with 0.1 M (0.1 N) HCl at 121 °C for 30 min. Adjust pH to 6.0 and then to 4.5. Dilute to volume and filter	LiChrosorb RP-8 10 μm 250 × 4.0 mm i.d.	Water/MeOH (60:40) containing 5 mM sodium hexane sulphonate	FMN, riboflavin (representing total vitamin B$_2$ activity)	Fluorescence: ex. 440 nm (filter) em. 565 nm (filter)	Reyes et al. (1988)

Abbreviations: see footnote to Table 2.4.

Most published HPLC methods for determining vitamin B_2 utilize ODS stationary phases either with aqueous/organic mobile phases in the simple reversed-phase mode or with eluents containing hexane or heptane sulphonic acid as ion interaction agents (Table 8.2). Fluorescence monitoring is the preferred means of detection in most cases. The detection sensitivity can be increased by irradiating the riboflavin at high pH to form lumiflavin (Ang and Moseley, 1980), but this technique has not been widely adopted. For samples with a low vitamin B_2 content ($< 10\,\mu g$ riboflavin/100 g) Lumley and Wiggins (1981) employed a trace enrichment technique in which successive 100 µl aliquots of the test extract were injected into the liquid chromatograph whilst using pure water as the mobile phase. The riboflavin which concentrated at the head of the ODS guard column could then be eluted as a tight band by changing to the methanol–water mobile phase.

The use of a 254 nm fixed-wavelength absorbance detector provided a minimum on-column detection limit of 0.4 ng riboflavin, which was adequate for the analysis of legumes and full cream milk powder. It was necessary to employ adsorption chromatography on a Florisil column followed by C_{18} solid-phase extraction as clean-up procedure to remove interfering substances (Vidal-Valverde and Reche, 1990). Absorbance monitoring at 270 nm was also sufficiently sensitive for the analysis of milk and milk products after clean-up and concentration using C_{18} solid-phase extraction (Ashoor *et al.*, 1985) and has the advantage in simultaneously detecting riboflavin decomposition products (Toyosaki, Yamamoto and Mineshita, 1986). Stancher and Zonta (1986) utilized visible absorbance detection at 446 nm to avoid detection interference in the analysis of Italian cheeses. The on-column detection limit of 2.5 ng (cf. 0.1 ng by fluorescence) was more than adequate for determining the relatively high concentration of riboflavin (at least $100\,\mu g/100\,g$) present in the cheese commodities analysed.

Russell and Vanderslice (1992b) employed a nondegradative two-step extraction procedure for the simultaneous quantification of riboflavin, FMN and FAD in a variety of foods. An internal standard, 7-ethyl-8-methyl-riboflavin, was added at the start of the extraction procedure, and separation of the flavins and internal standard was accomplished with a polymer-based column packing and a multistep gradient elution programme (Figure 8.3). Appropriate correction factors were applied to account for the impurities in the commercial FMN and FAD standards.

Bilic and Sieber (1990) extracted riboflavin, FMN and FAD from dairy products by homogenization with 6% formic acid containing 2 M urea in the presence of sorboflavin added as an internal standard. Sorboflavin contains a glucityl side chain instead of a ribityl chain on the isoalloxazine ring.

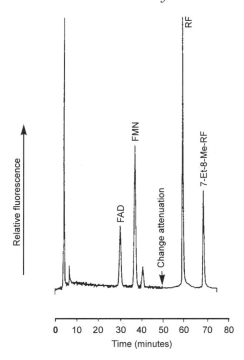

Figure 8.3 Reversed-phase HPLC of the flavins extracted from hamburger. Conditions as given in Table 8.2. (From Russell and Vanderslice, 1992b. Copyright © 1991 by Department of Agriculture, Government of Canada.)

Greenway and Kometa (1994) developed an on-line sample preparation method for the determination of riboflavin and FMN in milk and cereal samples by reversed-phase HPLC. The on-line system consisted of microwave extraction followed by dialysis and trace enrichment with a C_{18} mini-column. Sample preparation was minimal, with milk samples being directly introduced into the system and cereal samples only needing to be ground prior to analysis. During the microwave extraction all the FAD was converted into FMN and 15% of the FMN was converted into riboflavin. The full analysis time on the ground samples was about 20 min. Results were found to be in agreement with the AOAC fluorimetric method and a previously reported HPLC method.

8.6 BIOAVAILABILITY

8.6.1 Physiological aspects

Of the vitamin B_2-active flavins that occur in the diet, the coenzymes FMN and FAD predominate in plant and animal tissues. Protein-bound riboflavin is naturally present in milk and eggs and free riboflavin is obtainable from fortified processed foods.

Intestinal absorption

The FMN and FAD present in the ingested food are released from non-covalent binding to flavoproteins as a consequence of acidification in the stomach and gastric and intestinal proteolysis. Riboflavin is similarly released from its association with binding proteins (Merrill *et al.*, 1981).

The flavin coenzymes are hydrolysed in the upper small intestine to free riboflavin, which is then absorbed. Hydrolysis of both FMN and FAD is effected by alkaline phosphatase (E.C. 3.1.3.1) which has a broad specificity and is located on the brush-border membrane of the enterocyte (Daniel, Binninger and Rehner, 1983). Two additional brush-border enzymes, FMN phosphatase and FAD pyrophosphatase, participate in the degradation of the flavin coenzymes (Akiyama, Selhub and Rosenberg, 1982). The considerably smaller amounts of covalently bound flavins are released as 8α-(peptidyl) riboflavins, which are absorbed along with the free riboflavin (Chia, Addison and McCormick, 1978).

Experimental studies using rat everted jejunal sacs have led to the conclusion that the intestinal uptake of physiological amounts of ribo-flavin is an active, carrier-mediated transport process that is dependent on the presence of sodium ions in the incubation medium (Daniel, Wille and Rehner, 1983; Said, Hollander and Duong, 1985; Middleton, 1990). When larger amounts of riboflavin are administered experimentally, the carrier-mediated transport system becomes saturated and absorption by simple diffusion becomes quantitatively more important. This dual process of absorption has been confirmed under *in vivo* conditions (Feder, Daniel and Rehner, 1991).

The absorbed riboflavin is phosphorylated to FMN within the enterocyte by flavokinase (E.C. 2.7.1.26) from the intestinal mucosa. Kasai *et al.* (1988) examined the correlation between the absorption of various radioactive riboflavin analogues and their chemical structures and concluded that this phosphorylation is one of the specific processes for the absorption of riboflavin. Further studies (Kasai *et al.*, 1990) supported this conclusion because analogues that were absorbed at low concentrations through a process specific for riboflavin were phosphorylated effectively by flavokinase, whereas those that were absorbed solely through simple diffusion at all concentrations were not phosphorylated, or only phosphorylated weakly.

Uptake measurements using intact intestinal tissue preparations (e.g. everted sacs and perfused intestinal segments) are subject to interference by the intracellular metabolism of riboflavin. Such interference can be eliminated by the use of isolated brush-border membrane vesicles because this preparation lacks intracellular components. Using vesicles prepared from rat (Daniel and Rehner, 1992) and human (Said and Arianas, 1991) small intestine, it appeared that the riboflavin uptake system was an electrogenic Na^+–riboflavin cotransport process. However, in similar preparations from the rabbit, the uptake process appeared to be independent of sodium and electroneutral in nature (Said, Mohammadkhani and McCloud, 1993). Carrier-mediated transport of riboflavin across the basolateral membrane of rabbit intestine was also Na^+-independent and electroneutral (Said, Hollander and Mohammadkhani, 1993).

The absorption of riboflavin is efficient along the length of the small intestine (i.e. jejunum and ileum) in the guinea pig (Hegazy and Schwenk, 1983) and the rabbit (Said, Mohammadkhani and McCloud, 1993). The capacity for uptake is much smaller than that for the uptake of sugars, but the affinity of the riboflavin transport carrier for its substrate is several thousand times higher than that of the sugar carrier for its substrate (Hegazy and Schwenk, 1983). Absorption is enhanced by the presence of bile salts in the intestinal lumen (Mayersohn, Feldman and Gibaldi, 1969). A role of bile salts is also indicated by impaired absorption in cases of biliary obstruction in children (Jusko *et al.*, 1971). Bile salts may increase the absorption of riboflavin in the following possible ways. The detergent action of bile salts enhances the solubility of riboflavin in the intestinal lumen and increases the permeability of the brush-border membrane to the vitamin. Bile salts inhibit gastric emptying and proximal intestinal transit, resulting in an increased residence time of riboflavin at the absorption sites.

Said and Ma (1994) examined the uptake of radioactive riboflavin by an *in vitro* cell culture system comprising confluent monolayers of the human-derived Caco-2 intestinal epithelial cells. At the confluent stage, Caco-2 cells differentiate spontaneously to become enterocyte-like absorptive cells, having well-defined apical and basolateral membranes and intercellular junctional complexes. The involvement of an active carrier-mediated process in the initial phase of riboflavin uptake (a 3 min incubation time) was confirmed by the demonstration of energy dependence, temperature dependence, saturation kinetics and inhibition by unlabelled riboflavin and related analogues. Riboflavin uptake was Na^+- and pH-independent and the initial phase occurred without metabolic alteration of the transported riboflavin. Inhibitors of anion transport did not produce inhibition of riboflavin uptake by Caco-2 cells. Thus, contrary to its apparent behaviour in renal transport, riboflavin does not appear to act as an anion with regard to its intestinal transport.

Growing Caco-2 monolayers in a riboflavin-deficient and oversupplemented medium caused significant enhancement and suppression of riboflavin uptake, respectively (Said and Ma, 1994). This adaptive up- or downregulation of riboflavin uptake was further studied by Said, Ma and Grant (1994) who established a role for the 'second messenger' cAMP (adenosine $3',5'$-cyclic monophosphate) in the regulation mechanism. They found that compounds that increased intracellular cAMP concentration through different mechanisms caused a significant and concentration-dependent inhibition (downregulation) in riboflavin uptake. This inhibition was in contrast to other findings of stimulation of intestinal uptake of D-glucose by these compounds (Sharp and Debnam, 1994), indicating that the effect of cAMP-stimulating compounds on riboflavin intestinal uptake is not generalized in nature. Furthermore, the inhibitory effect could not be attributed to changes in the intracellular trapping mechanism, since only 19% of the total radioactivity taken up by Caco-2 cells following 7 min of incubation with [^{3}H]riboflavin was metabolized. Inhibition of riboflavin uptake by one of the cAMP-stimulating compounds was mediated through a significant decrease in the number and/or activity of the riboflavin carriers and not through a decreased affinity of the carrier for its substrate.

To summarize, the small intestine is well adapted to completely extracting the small amounts of riboflavin that are largely bound within the ingested flavin coenzymes. The coenzymes are dephosphorylated in the lumen and the liberated riboflavin is extracted very efficiently by a high-affinity, carrier-mediated transport system, which is distributed along the entire length of the small intestine. The uptake mechanism is controversial but, according to the most recent evidence, appears to be sodium-independent and electroneutral in nature. After uptake, some of the riboflavin is metabolically trapped within the enterocyte as FMN. At the serosal surface most of the FMN is probably dephosphorylated to riboflavin, which enters the portal system *en route* to the liver.

Other metabolic considerations

Circulatory transport of riboflavin involves loose association with plasma albumin and tight associations with some globulins. The extent to which flavins are bound to plasma proteins is not believed to be crucial in regulating tissue availability of the vitamin, except that the proteins may decrease losses of the vitamin during glomerular filtration (White and Merrill, 1986). Uptake of riboflavin by hepatocytes (liver cells) appears to be a Na^+-independent facilitated diffusion process (Aw, Jones and McCormick, 1983). Cellular entry of riboflavin in all tissues is followed by metabolic trapping by flavokinase-catalysed phosphorylation to FMN. Release of riboflavin from cells requires hydrolysis of FMN by nonspecific phosphatases (McCormick, 1994).

Conversion of FMN to FAD and binding to specific flavoproteins occurs in most tissues, but particularly in the small intestine, liver, heart, kidney and brain. The liver is the major storage site of the vitamin and contains about one-third of the total body flavins, 70–90% of which is in the form of FAD. Free riboflavin constitutes less than 5% of the stored flavins. Other storage sites are the spleen, kidney and cardiac muscle. These depots maintain significant amounts of the vitamin even in severe deficiency states.

At least 40% of plasma riboflavin is considered not to be bound in humans (Rose, 1988); thus reabsorption by the kidney is necessary to prevent urinary loss of the vitamin. There is some evidence for a carrier-mediated transport system for riboflavin in the kidney (Spector, 1982; Lowy and Spring, 1990) but the uptake mechanism remains to be investigated. For normal adults eating varied diets, riboflavin accounts for 60–70% of flavin compounds in the urine; the remainder are riboflavin metabolites (McCormick, 1994). Urinary excretion studies carried out in humans have suggested that any riboflavin secreted into the bile is almost fully reabsorbed, i.e. the vitamin is subject to enterohepatic cycling (Jusko and Levy, 1967).

8.6.2 Dietary sources and their bioavailability

Dietary sources

Living cells require FMN and FAD as the prosthetic groups of a variety of enzymes, and hence the flavins are found, at least in small amounts, in all natural unprocessed foods. The riboflavin content of a selection of foods is presented in Table 8.3. Yeast extract is exceptionally rich in vitamin B_2, and liver and kidney are also rich sources. Wheat bran, eggs, meat, milk and cheese are important sources in diets containing these foods. Cereal grains contain relatively low concentrations of flavins, but are important sources in those parts of the world where cereals constitute the staple diet. The milling of cereals results in considerable loss (up to 60%) of vitamin B_2, so white flour is enriched by addition of the vitamin. The enrichment of bread and breakfast cereals contributes significantly to the dietary supply of vitamin B_2. Polished rice is not usually enriched, since the yellow colour of the vitamin would make the rice visually unacceptable. However, most of the flavin content of the whole brown rice is retained if the rice is steamed prior to milling. This process drives the water-soluble vitamins in the germ and aleurone layers into the endosperm (Cooperman and Lopez, 1991).

In most foods the vitamin B_2 is present in the form of protein-bound flavins, predominantly as FAD. Several investigators have reported that from 54 to 95% of the total vitamin B_2 in bovine milk is present as free riboflavin; the rest is present as FMN and FAD (Hartman and Dryden, 1974). In milk a small proportion (14% in whole milk) of the flavins is

Table 8.3 Riboflavin content of various foods (from Holland *et al.*, 1991)

Food	mg Riboflavin/100 g
Cow's milk, whole, pasteurized	0.17
Cheese, cheddar, average	0.40
Egg, chicken, whole, raw	0.47
Wheat flour, wholemeal	0.09
Rice, brown, raw	0.07
Rice, white, easy cook, raw	0.02
Beef, lean only, raw, average	0.24
Lamb, lean only, raw, average	0.28
Pork, lean only, raw, average	0.25
Chicken, meat only, raw	0.16
Liver, ox, raw	3.10
Cod, raw, fillets	0.07
Potato, maincrop, old, average, raw	0.02
Soya beans, dried, raw	0.27
Peas, raw	0.02
Brussels sprouts, raw	0.11
Apples, eating, average, raw	0.02
Orange juice, unsweetened	0.02
Peanuts, plain	0.10
Yeast extract (Marmite)	11.00

bound noncovalently to specific proteins (Kanno *et al.*, 1991). Egg white and egg yolk contain specialized riboflavin-binding proteins that are required for storage of free riboflavin in the egg for use by the developing embryo (Merrill *et al.*, 1981).

Bioavailability

The bioavailability of riboflavin has not been extensively studied. In a human study involving eight women subjects, Everson *et al.* (1948) compared the rise in urinary output of riboflavin when an equal amount (1 mg) of the vitamin was supplied as a pure chemical and as a test food. The relative bioavailability values for ice cream, green peas and almonds were calculated to be 90, 41 and 39%, respectively. Further studies (Everson, Pearson and Matteson, 1952) revealed that the lower bioavailabilities of green peas and almonds were at least partly due to their incomplete digestibilities, ice cream being completely digested. Girija, Sharada and Pushpamma (1982) determined the bioavailability of riboflavin from curries prepared with three green leafy vegetables commonly consumed in rural areas in India. Using the balance technique and dose–response curves obtained from human urinary excretion data, the bioavailability values were 47, 46 and 51% from amaranth, gogu and drumstick leaf curries, respectively.

The covalently bound FAD, which makes up $< 10\%$ of the FAD in mammalian tissues, is largely unavailable as a source of vitamin B_2 in the rat (Chia, Addison and McCormick, 1978).

Effects of alcohol
Bioavailability is impaired by the excessive consumption of alcohol, which appears to inhibit the intestinal FMN phosphatase and FAD pyrophosphatase necessary to release riboflavin for absorption (Pinto, Huang and Rivilin, 1984).

Effects of dietary fibre
Roe *et al.* (1978) examined the influence of various dietary fibre sources on riboflavin absorption as measured by urinary excretion of the vitamin in healthy male subjects. Coarse and fine wheat bran, cellulose and cabbage were incorporated as single fibre sources into the diet on an isofibrous basis (12 g fibre/day). Compared with a control group receiving a low-fibre diet, each fibre source increased urinary excretion when 15 mg FMN was ingested with the breakfast meal. It was concluded that the dietary fibre sources enhanced riboflavin absorption, probably by slowing the passage of chyme in the intestine and thereby increasing the duration of vitamin exposure at the absorption sites.

8.6.3 Effects of food processing and cooking

Significant losses of vitamin B_2 are caused in milk by photodegradation. As much as 85% of the vitamin content of milk in glass bottles may be destroyed after only 2 h exposure to bright sunlight (Horwitt, 1972). The breakdown product, lumichrome, is a much stronger oxidizing agent than is riboflavin and catalyses massive destruction of the ascorbic acid content. The formation of highly reactive singlet oxygen during the photodegradation of flavins promotes the rapid autoxidation of unsaturated fatty acids in the milk fat, resulting in an unpleasant 'off' flavour (Coultate, 1989). The pasteurization, evaporation and condensing of milk has little effect on the vitamin B_2 content.

Vitamin B_2 is stable during the heat processing and normal cooking of foods if light is excluded.

8.7 DIETARY INTAKE

Because of the limited storage capacity for all forms of vitamin B_2, the margin between dietary intake resulting in deficiency (0.55 mg/day) and that resulting in tissue saturation (1.1 mg/day) is very small (Horwitt *et al.*, 1950). In 1985 adult men in the United States consumed an average of 2.08 mg of riboflavin/day; the intake was 1.34 mg for adult women, and 1.57 mg for their children aged 1 to 5 years (National Research Council, 1989).

In the United States, the riboflavin allowance for healthy people of all ages is based on 0.6 mg/1000 kcal, even though there is no evidence that the requirement is correlated directly to energy expenditure. This leads to RDAs ranging from 0.4 mg/day for early infants to 1.7 mg/day for young adult males. However, for elderly people and others whose daily caloric intake may be less than 2000 kcal, a minimum of 1.2 mg/day is recommended to ensure tissue saturation. During pregnancy an additional 0.3 mg/day is recommended. An additional daily intake of 0.5 mg is recommended for the first 6 months of lactation and 0.4 mg for the second 6 months (National Research Council, 1989).

In the United Kingdom, recommendations for riboflavin intake are not based on energy intake. The Reference Nutrient Intake is 1.3 mg/day for men and 1.1 mg/day for women, with daily increments of 0.3 mg and 0.5 mg for pregnancy and lactation, respectively (Department of Health, 1991).

Effects of high intakes

The low solubility of riboflavin and the limited capacity of intestinal absorption mechanisms probably account for the lack of toxicity following oral doses of up to 20 mg of riboflavin daily (Cumming, Briggs and Briggs, 1981).

Assessment of nutritional status

The most commonly used current method for assessing riboflavin status is based upon the change in activity of the FAD-dependent enzyme erythrocyte glutathione reductase (EGR) as measured by the oxidation of NADPH to NADP. The *in vitro* addition of FAD to freshly lysed erythrocytes, substrate and NADPH provides an activity coefficient, which is inversely proportional to the urinary excretion of riboflavin. Riboflavin deficiency is indicated by an activity coefficient > 1.2, while tissue saturation of riboflavin with adequate FAD results in no additional stimulation and an activity coefficient of 1.0 (Cooperman and Lopez, 1991).

8.8 SYNOPSIS

Vitamin B_2 comprises riboflavin, FAD and FMN. In most foods the vitamin is present in the form of protein-bound flavins, predominantly as FAD. Vitamin B_2 is present in a wide variety of natural unprocessed foods but, apart from its abundance in liver, occurs in very small amounts (typically $< 0.5\,\mathrm{mg}/100\,\mathrm{g}$). The bioavailability of vitamin B_2 is related to the digestibility of the food, being generally high for meat and dairy products and lower for plant products.

Physicochemical and microbiological methods of analysis involve high-temperature acid digestion of the food sample to break down the protein and convert the FAD to FMN and riboflavin. The complete conversion of FMN to riboflavin can only be achieved by subsequent enzymatic hydrolysis. HPLC is a suitable assay technique becaused it allows fluorimetric measurement of the riboflavin with a high degree of sensitivity.

REFERENCES

Akiyama, T., Selhub, J. and Rosenberg, I.H. (1982) FMN phosphatase and FAD pyrophosphatase in rat intestinal brush borders: role in intestinal absorption of dietary riboflavin. *J. Nutr.*, **112**, 263–8.

Andrews, J.S., Boyd, H.M. and Terry, D.E. (1942) Riboflavin analysis of cereals. Application of the microbiological method. *Ind. Engng Chem. Analyt. Edn*, **14**, 271–4.

Ang, C.Y.W. and Moseley, F.A. (1980) Determination of thiamin and riboflavin in meat and meat products by high-pressure liquid chromatography. *J. Agric. Food Chem.*, **28**, 483–6.

AOAC (1990a) Riboflavin (vitamin B_2) in foods and vitamin preparations. Fluorometric method. Final action 1971. In *AOAC Official Methods of Analysis*, 15th edn (ed. K. Helrich), Association of Official Analytical Chemists, Inc., Arlington, VA, 970.65.

AOAC (1990b) Riboflavin in foods and vitamin preparations. Automated method. Final action 1982. In *AOAC Official Methods of Analysis*, 15th edn (ed. K. Helrich), Association of Official Analytical Chemists, Inc., Arlington, VA, 981.15.

Ashoor, S.H., Knox, M.J., Olsen, J.R. and Deger, D.A. (1985) Improved liquid chromatographic determination of riboflavin in milk and dairy products. *J. Ass. Off. Analyt. Chem.*, **68**, 693–6.

Aw, T.Y., Jones, D.P. and McCormick, D.B. (1983) Uptake of riboflavin by isolated rat liver cells. *J. Nutr.*, **113**, 1249–54.

Barton-Wright, E.C. (1952) *The Microbiological Assay of the Vitamin B-Complex and Amino Acids*, Sir Isaac Pitman & Sons, Ltd, London.

Barton-Wright, E.C. (1967) The microbiological assay of certain vitamins in compound feeding stuffs. *J. Ass. Publ. Analysts*, **5**, 8–23.

Barton-Wright, E.C. and Booth, R.G. (1943) The assay of riboflavin in cereals and other products. *Biochem. J.*, **37**, 25–30.

Bauernfeind, J.C., Sotier, A.L. and Boruff, C.S. (1942) Growth stimulants in the microbiological assay for riboflavin and pantothenic acid. *Ind. Engng Chem. Analyt. Edn*, **14**, 666–71.

Bell, J.G. (1971) The microbiological assay of B-complex vitamins. In *Proceedings of the University of Nottingham Residential Seminar on Vitamins* (ed. M. Stein), Churchill Livingstone, Edinburgh and London, pp. 165–87.

Bilic, N. and Sieber, R. (1990) Determination of flavins in dairy products by high-performance liquid chromatography using sorboflavin as internal standard. *J. Chromat.*, **511**, 359–66.

Chia, C.P., Addison, R. and McCormick, D.B. (1978) Absorption, metabolism, and excretion of 8α-(amino acid) riboflavins in the rat. *J. Nutr.*, **108**, 373–81.

Cooperman, J.M. and Lopez, R. (1991) Riboflavin. In *Handbook of Vitamins*, 2nd edn (ed. L.J. Machlin), Marcel Dekker, Inc., New York, pp. 283–310.

Coultate, T.P. (1989) *Food. The Chemistry of its Components*, 2nd edn, Royal Society of Chemistry, London.

Cumming, F., Briggs, M. and Briggs, M. (1981) Clinical toxicology of vitamin supplements. In *Vitamins in Human Biology and Medicine* (ed. M.H. Briggs), CRC Press, Inc., Boca Raton, Florida, pp. 187–243.

Daniel, H. and Rehner, G.I. (1992) Sodium-dependent transport of riboflavin in brush border membrane vesicles of rat small intestine is an electrogenic process. *J. Nutr.*, **122**, 1454–61.

Daniel, H., Binninger, E. and Rehner, G. (1983) Hydrolysis of FMN and FAD by alkaline phosphatase of the intestinal brush-border membrane. *Int. J. Vitam. Nutr. Res.*, **53**, 109–14.

Daniel, H., Wille, U. and Rehner, G. (1983) In vitro kinetics of the intestinal transport of riboflavin in rats. *J. Nutr.*, **113**, 636–43.

Department of Health (1991) *Dietary Reference Values for Food Energy and Nutrients for the United Kingdom*. Report on Health and Social Subjects, No. 41, HM Stationery Office, London.

Egberg, D.C. (1979) Semiautomated method for riboflavin in food products: collaborative study. *J. Ass. Off. Analyt. Chem.*, **62**, 1041–4.

Egberg, D.C. and Potter, R.H. (1975) An improved automated determination of riboflavin in food products. *J. Agric. Food Chem.*, **23**, 815–20.

Everson, G., Pearson, E. and Matteson, R. (1952) Biological availability of certain foods as sources of riboflavin. *J. Nutr.*, **46**, 45–53.

Everson, G., Wheeler, E., Walker, H. and Caulfield, W.J. (1948) Availability of riboflavin of ice cream, peas, and almonds judged by urinary excretion of the vitamin by women subjects. *J. Nutr.*, **35**, 209–23.

Feder, S., Daniel, H. and Rehner, G. (1991) In vivo kinetics of intestinal absorption of riboflavin in rats. *J. Nutr.*, **121**, 72–9.

Girija, V., Sharada, D. and Pushpamma, P. (1982) Bioavailability of thiamine, riboflavin and niacin from commonly consumed green leafy vegetables in the rural areas of Andhra Pradesh in India. *Int. J. Vitam. Nutr. Res.*, **52**, 9–13.

Greenway, G.M. and Kometa, N. (1994) On-line sample preparation for the determination of riboflavin and flavin mononucleotides in foodstuffs. *Analyst, Lond.*, **119**, 929–35.

Hartman, A.M. and Dryden, L.P. (1974) Vitamins in milk and milk products. In *Fundamentals of Dairy Chemistry*, 2nd edn (eds B.H. Webb, A.H. Johnson and J.A. Alford), AVI Publishing Co., Inc., Westport, CT, pp. 325–441.

Hegazy, E. and Schwenk, M. (1983) Riboflavin uptake by isolated enterocytes of guinea pigs. *J. Nutr.*, **113**, 1702–7.

Holland, B., Welch, A.A., Unwin, I.D. *et al.* (1991) *McCance and Widdowson's The Composition of Foods*, 5th edn, Royal Society of Chemistry and Ministry of Agriculture, Fisheries and Food.

Horwitt, M.K. (1972) Riboflavin, V. Occurrence in food. In *The Vitamins. Chemistry, Physiology, Pathology, Methods*, 2nd edn, Vol. V (eds W.H. Sebrell, Jr and R.S. Harris), Academic Press, New York, pp. 46–9.

Horwitt, M.K., Harvey, C.C., Hills, O.W. and Liebert, E. (1950) Correlation of urinary excretion of riboflavin with dietary intake and symptoms of ariboflavinosis. *J. Nutr.*, **41**, 247–64.

Jusko, W.J. and Levy, G. (1967) Absorption, metabolism, and excretion of riboflavin-5'-phosphate in man. *J. Pharm. Sci.*, **56**, 58–62.

Jusko, W.J., Levy, G., Yaffe, S.J. and Allen, J.E. (1971) Riboflavin absorption in children with biliary obstruction. *Am. J. Dis. Child.*, **121**, 48–52.

Kanno, C., Shirahuji, K. and Hoshi, T. (1991) Simple method for separate determination of three flavins in bovine milk by high performance liquid chromatography. *J. Food Sci.*, **56**, 678–81, 700.

Kanno, C., Kanehara, N., Shirafuji, K. *et al.* (1991) Binding form of vitamin B₂ in bovine milk: its concentration, distribution and binding linkage. *J. Nutr. Sci. Vitaminol.*, **37**, 15–27.

Kasai, S., Nakano, H., Kinoshita, T. *et al.* (1988) Intestinal absorption of riboflavin, studied by an *in situ* circulation system using radioactive analogues. *J. Nutr. Sci. Vitaminol.*, **34**, 265–80.

Kasai, S., Nakano, H., Maeda, K. and Matsui, K. (1990) Purification, properties, and function of flavokinase from rat intestinal mucosa. *J. Biochem.*, **107**, 298–303.

Kodicek, E. and Worden, A.N. (1945) The effect of unsaturated fatty acids on *Lactobacillus helveticus* and other Gram-positive micro-organisms. *Biochem. J.*, **39**, 78–85.

Kornberg, H.A., Langdon, R.S. and Cheldelin, V.H. (1948) Microbiological assay for riboflavin. *Analyt. Chem.*, **20**, 81–3.

Koziol, J. (1971) Fluorometric analyses of riboflavin and its coenzymes. *Methods Enzymol.*, **18B**, 253–85.

Langer, B.W. Jr and Charoensiri, S. (1966) Growth response of *Lactobacillus casei* (ATCC 7469) to riboflavin, FMN, and FAD. *Proc. Soc. Exp. Biol. Med.*, **122**, 151–2.

Lowy, R.J. and Spring, K.R. (1990) Identification of riboflavin transport by MDCK cells using quantitative fluorescence video microscopy. *J. Membr. Biol.*, **117**, 91–9.

Lumley, I.D. and Wiggins, R.A. (1981) Determination of riboflavin and flavin mononucleotide in foodstuffs using high-performance liquid chromatography and a column-enrichment technique. *Analyst, Lond.*, **106**, 1103–8.

Mayersohn, M., Feldman, S. and Gibaldi, M. (1969) Bile salt enhancement of riboflavin and flavin mononucleotide absorption in man. *J. Nutr.*, **98**, 288–96.

McCormick, D.B. (1994) Riboflavin. In *Modern Nutrition in Health and Disease*, 8th edn, Vol. 1 (eds M.E. Shils, J.A. Olson and M. Shike), Lea & Febiger, Philadelphia, pp. 366–75.

Merrill, A.H. Jr, Lambeth, J.D., Edmondson, D.E. and McCormick, D.B. (1981) Formation and mode of action of flavoproteins. *Annu. Rev. Nutr.*, **1**, 281–317.

Middleton, I.I.M. III (1990) Uptake of riboflavin by rat intestinal mucosa in vitro. *J. Nutr.*, **120**, 588–93.

National Research Council (1989) Water-soluble vitamins. In *Recommended Dietary Allowances*, 10th edn, National Academy Press, Washington, DC, pp. 115–73.

Nielsen, P., Rauschenbach, P. and Bacher, A. (1986) Preparation, properties, and separation by high-performance liquid chromatography of riboflavin phosphates. *Methods Enzymol.*, **122G**, 209–20.

Ollilainen, V., Mattila, P., Varo, P. *et al.* (1990) The HPLC determination of total riboflavin in foods. *J. Micronutr. Anal.*, **8**, 199–207.

Pinto, J., Huang, Y. and Rivlin, R. (1984) Selective effects of ethanol and acetaldehyde upon intestinal enzymes metabolizing riboflavin: mechanism of reduced flavin bioavailability due to ethanol. *Am. J. Clin. Nutr.*, **39**, 685 (abstr).

Rashid, I. and Potts, D. (1980) Riboflavin determination in milk. *J. Food Sci.*, **45**, 744–5.

Reyes, E.S.P., Norris, K.M., Taylor, C. and Potts, D. (1988) Comparison of paired-ion liquid chromatographic method with AOAC fluorometric and microbiological methods for riboflavin determination in selected foods. *J. Ass. Off. Analyt. Chem.*, **71**, 16–19.

Roe, D.A., Wrick, K., McLain, D. and van Soest, P. (1978) Effects of dietary fiber sources on riboflavin absorption. *Fed. Proc.*, **37**, 756 (abstr).

Rose, R.C. (1988) Transport of ascorbic acid and other water-soluble vitamins. *Biochim. Biophys. Acta*, **947**, 335–66.

Russell, L.F. and Vanderslice, J.T. (1992a) Comments on the standard fluorometric determination of riboflavin in foods and biological tissues. *Food Chem.*, **43**, 79–82.

Russell, L.F. and Vanderslice, J.T. (1992b) Non-degradative extraction and simultaneous quantitation of riboflavin, flavin mononucleotide, and flavin adenine dinucleotide in foods by HPLC. *Food Chem.*, **43**, 151–62.

Said, H.M. and Arianas, P. (1991) Transport of riboflavin in human intestinal brush border membrane vesicles. *Gastroenterology*, **100**, 82–8.

Said, H.M. and Ma, T.Y. (1994) Mechanism of riboflavine uptake by Caco-2 human intestinal epithelial cells. *Am. J. Physiol.*, **266**, G15–21.

Said, H.M., Hollander, D. and Duong, Y. (1985) A dual, concentration-dependent transport system for riboflavin in rat intestine *in vitro*. *Nutr. Res.*, **5**, 1269–79.

Said, H.M., Hollander, D. and Mohammadkhani, R. (1993) Uptake of riboflavin by intestinal basolateral membrane vesicles: a specialized carrier-mediated process. *Biochim. Biophys. Acta*, **1148**, 263–8.

Said, H.M., Ma, T.Y. and Grant. K. (1994) Regulation of riboflavin intestinal uptake by protein kinase A: studies with Caco-2 cells. *Am. J. Physiol.*, **267**, G955–9.

Said, H.M., Mohammadkhani, R. and McCloud, E. (1993) Mechanism of transport of riboflavin in rabbit intestinal brush border membrane vesicles. *Proc. Soc. Exp. Biol. Med.*, **202**, 428–34.

Sattar, A., deMan, J.M. and Alexander, J.C. (1977) Light-induced degradation of vitamins. I. Kinetic studies on riboflavin decomposition in solution. *Can. Inst. Food Sci. Technol.*, **10**, 61–4.

Sharp, P.A. and Debnam, E.S. (1994) The role of cyclic AMP in the control of sugar transport across the brush border and basolateral membranes of rat jejunal enterocytes. *Exp. Physiol.*, **79**, 203–14.

Spector, R. (1982) Riboflavin transport by rabbit kidney slices: characterization and relation to cyclic organic acid transport. *J. Pharmacol. Exp. Ther.*, **221**, 394–8.

Stancher, B. and Zonta, F. (1986) High performance liquid chromatographic analysis of riboflavin (vitamin B_2) with visible absorbance detection in Italian cheeses. *J. Food Sci.*, **51**, 857–8.

Strong, F.M. and Carpenter, L.E. (1942) Preparation of samples for microbiological determination of riboflavin. *Ind. Engng Chem. Analyt. Edn*, **14**, 909–13.

Toyosaki, T., Yamamoto, A. and Mineshita, T. (1986) Simultaneous analysis of riboflavin and its decomposition products in various milks by high-performance liquid chromatography. *J. Micronutr. Anal.*, **2**, 117–23.

Vidal-Valverde, C. and Reche, A. (1990) Reliable system for the analysis of riboflavin in foods by high performance liquid chromatography and UV detection. *J. Liquid Chromat.*, **13**, 2089–101.

Watada, A.E. and Tran, T.T. (1985) A sensitive high-performance liquid chromatography method for analyzing riboflavin in fresh fruits and vegetables. *J. Liquid Chromat.*, **8**, 1651–62.

Wegner, M.I., Kemmerer, A.R. and Fraps, G.S. (1942) Influence of the method of preparation of sample on microbiological assay for riboflavin. *J. Biol. Chem.*, **146**, 547–51.

White, H.B. and Merrill, A.H. Jr (1986) Riboflavin-binding proteins. *Ann. Rev. Nutr.*, **8**, 279–99.

Woodrow, I.L., Torrie, K.M. and Henderson, G.A. (1969) A rapid method for the determination of riboflavin in dried milk products. *J. Inst. Can. Technol. Aliment.*, **2**, 120–2.

Yagi, K. (1962) Chemical determination of flavins. In *Methods of Biochemical Analysis*, Vol. 10 (ed. D. Glick), John Wiley & Sons, New York, pp. 319–56.

9

Niacin and tryptophan

9.1 INTRODUCTION

The human disease of pellagra was first described in Spain by Casal in 1735 after the introduction of maize into Europe from the Americas. In the 1920s, Goldberger (USA) reported that pellagra and black tongue in dogs responded to treatment with animal protein and also to boiled protein-free extracts of yeast. In 1937, Elvehjem found that the active component in liver extracts used to successfully treat canine black tongue was nicotinamide, and reports that nicotinic acid cured pellagra soon followed.

The nicotinamide nucleotide coenzymes, NAD and NADP, function as proton and electron carriers in a wide variety of oxidation–reduction reactions. Examples are reactions concerned with the release of energy from carbohydrates, fatty acids and amino acids, and with the synthesis of amino acids, fatty acids and pentoses for nucleotide and nucleic acid production.

Niacin can be synthesized *in vivo* from the essential amino acid L-tryptophan. An average intake of protein will probably provide more than enough tryptophan to meet the body's requirement for niacin without the need for any preformed niacin in the diet.

9.2 CHEMICAL STRUCTURE AND NOMENCLATURE

Niacin

Niacin is the generic descriptor for two vitamers, nicotinic acid and nicotinamide, which are systematically named as pyridine 3-carboxylic acid ($C_6H_5O_2N$; MW = 123.11) and pyridine 3-carboxamide ($C_6H_6ON_2$; MW = 122.12), respectively (Figure 9.1). This nomenclature is not to be confused with American usage of the term niacin to denote specifically

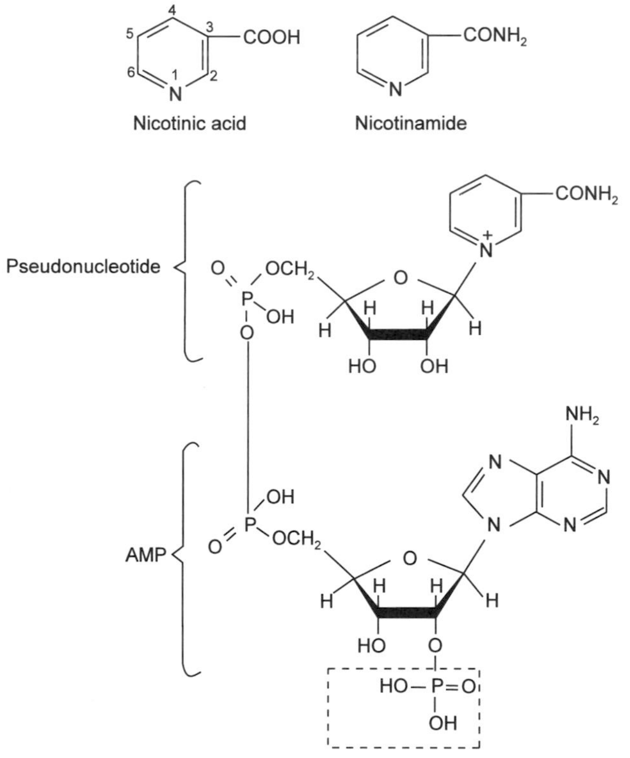

NAD$^+$ and NADP$^+$ (phosphate group within dotted area is absent in NAD$^+$)

Figure 9.1 Structures of niacin compounds and the nicotinamide nucleotides.

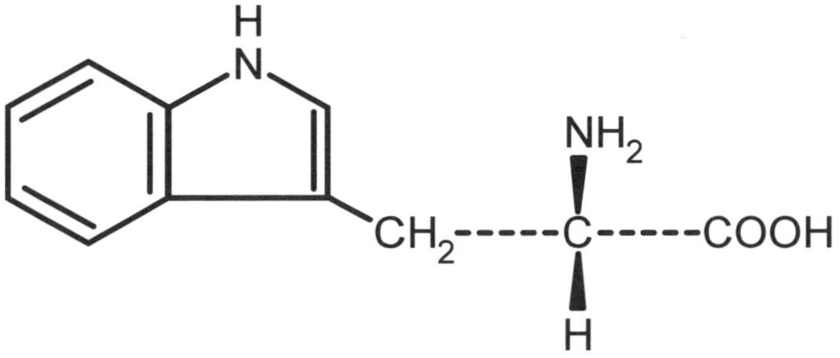

Figure 9.2 Structure of L-tryptophan.

nicotinic acid, and niacinamide to denote the amide. In living tissues nicotinamide is the reactive moiety of the coenzymes NAD and NADP. The coenzymes are composed of a nucleotide (AMP) and a pseudo-nucleotide containing the nicotinamide moiety (Figure 9.1). The oxidized forms of the coenzymes carry a positive charge (NAD^+ and $NADP^+$); reduced forms carrying two electrons and one proton (and associated with an additional proton) are represented as NADH and NADPH. Although the coenzymes are only loosely associated with the apoenzyme during catalysis, most cellular NAD and NADP is stored in the cytoplasm bound to protein (Weiner and van Eys, 1983).

Tryptophan

L-Tryptophan ($C_{11}H_{12}N_2O_7$; MW $= 204.22$) is an essential amino acid whose structure (Figure 9.2) contains an indole ring. D-Tryptophan is not utilized by humans and many domestic animals; it is, however, fully utilized by the rat (Nielsen *et al.*, 1985).

9.3 DEFICIENCY SYNDROMES

A deficiency in niacin results in pellagra, which is a nutritional disease endemic among poor communities who subsist chiefly on maize. The prognosis is complicated by signs of protein-energy malnutrition and by an imbalance of amino acid intake, particularly low levels of tryptophan and high levels of leucine. The disease is diagnosed in the early stages by a sunburn-like dermatitis affecting skin areas exposed to light. The dermatitis may progress to a scaled and cracked condition, and in chronic cases the skin becomes rough and thickened with a brown pigmentation. Gastrointestinal disturbances (abdominal pain, diarrhoea and loss of appetite) may also appear, accompanied by oral lesions similar to those described for riboflavin deficiency. Early neurological symptoms include tremor, irritability, anxiety and depression, with delirium and dementia

sometimes occurring in severe and chronic cases. The response to nicotinamide therapy is rapid and dramatic: untreated pellagra results
eventually in death.

Because most proteins contain at least 1.0% tryptophan, it is theoretically possible to maintain adequate niacin status on a diet devoid of
niacin but containing > 100 g of protein. Primary deficiencies are rare
(at least in industrialized countries), but secondary deficiencies may arise
from gastrointestinal disorders or alcoholism.

9.4 PHYSICOCHEMICAL PROPERTIES OF NICOTINIC ACID, NICOTINAMIDE AND TRYPTOPHAN

Solubility and other properties

Nicotinic acid crystallizes into colourless needles (m.p. 235 °C) which are
odourless and have a tart taste; the crystals of nicotinamide (m.p. 129°C)
are similar, but have a bitter taste. Nicotinic acid is sparingly soluble in
water (1.67 g/100 ml at 25 °C) and ethanol (0.73 g/100 ml) and insoluble
in acetone and diethyl ether. It is, however, freely soluble in boiling
water. The pH of a saturated aqueous solution is 3 and the pK_a is 4.8
(25 °C). The sodium salt of nicotinic acid is readily soluble in water (71 g/
100 ml) and yields a solution of about pH 7. Nicotinamide is readily
soluble in water (100 g/100 ml) and in ethanol (67 g/100 ml); there is
solubility in acetone and very slight solubility in diethyl ether and chloroform. An aqueous solution of nicotinamide yields a pH of 6 and the pK_a is
3.3 (20 °C). Both nicotinic acid and nicotinamide have the properties of
bases and form quaternary ammonium salts when in acid solution.
Nicotinic acid, being amphoteric, forms carboxylic acid salts when in a
basic solution, but nicotinamide possesses no acidic properties.

The solubility of L-tryptophan in water is 11.4 g/l at 25 °C; pK_{a1} is 2.38
and pK_{a2} is 9.39.

Spectroscopic properties

Nicotinic acid and its amide exhibit similar absorption spectra in the UV
region. The absorptivity is strongly affected by pH, being higher in an
acidic than in an alkaline solution, but the λ_{max} remains almost
unchanged at 261 nm. The $A_{1\,cm}^{1\%}$ value for nicotinamide in 0.1 N H_2SO_4
at 261 nm is 478 (Merck & Co., Inc., 1983). The presence of electrolytes
also has a marked effect on the absorbances of the solutions (Strohecker
and Henning, 1966).

Tryptophan exhibits native fluorescence by virtue of its indole group
and has excitation and emission maxima at 285 nm and 365 nm, respectively, when dissolved in 0.1 M ammonium hydroxide (pH 11) (Duggan *et
al.*, 1957).

Stabilities of nicotinic acid, nicotinamide and tryptophan

Both nicotinic acid and nicotinamide are unaffected by atmospheric oxygen, light and heat in the dry state and in neutral aqueous solution. Nicotinamide is hydrolysed quantitatively to nicotinic acid when autoclaved in a medium of 1 N mineral acid or in alkaline solution, the nicotinic acid being resistant to these conditions. The oxidized coenzymes (NAD^+ and $NADP^+$) are labile to alkali (Guilbert and Johnson, 1977), whereas the reduced coenzymes (NADH and NADPH) are labile to acid (Chaykin, 1967).

In the presence of air, free tryptophan is degraded when heated at 100 °C in the pH range 2–7 and in both strongly acidic and strongly basic solutions. When oxygen is removed by evacuation, the rate of tryptophan decomposition drops sharply. In the complete absence of oxygen, alkaline solutions of tryptophan are stable when heated (Friedman and Cuq, 1988).

9.5 ANALYSIS

9.5.1 Scope of analytical techniques

The niacin in mature cereal grains exists largely as chemically bound forms of nicotinic acid that are nutritionally unavailable, while in pulses and other noncereal plant foods it is present mainly as free nicotinic acid. In meat the niacin content is mainly in the form of nicotinamide, and this is the form used in the fortification of processed foods.

It is customary when employing *in vitro* analytical methods to measure the total niacin (i.e. free plus bound) content of the food sample; however, this provides a gross overestimate of the biologically available niacin of several staple cereal-based foods. Several nutritionists have suggested that the measurement of free niacin in foods would provide an accurate estimate of the content of available niacin but, in view of the apparent partial utilization of bound nicotinic acid present in mature cereal grains, this may underestimate the niacin value in cereal products. Nevertheless, the measurement of free niacin would seem to be the more appropriate for such products as a means of ensuring an adequate level of enrichment. The terms 'total' and 'free' niacin are defined by the extraction methods employed in the analysis. Total niacin generally refers to the niacin that is extractable by autoclaving the sample with alkali or 1 N acid; free niacin is frequently defined as the niacin extractable by autoclaving with 0.1 N acid (Gregory, 1985). Both total and free niacin can be measured microbiologically, colorimetrically or by HPLC. HPLC also permits the separate measurement of nicotinamide and nicotinic acid.

The standard microbiological assay procedure for determining total niacin does not account for tryptophan, because the mild acid digestion used to extract the niacin does not hydrolyse the proteins, and the lactic acid bacterium employed as the test organism cannot utilize peptide-bound tryptophan. Apart from using an animal bioassay, or possibly a protozoan assay, an estimate of the niacin equivalent content would necessitate the determination of L-tryptophan in a separate assay.

The amino acid composition of foods and feedstuffs is routinely determined by means of an automatic amino acid analyser (Gehrke *et al.*, 1985). In this equipment the amino acids present in protein hydrolysates are separated by ion exchange chromatography and measured colorimetrically after reaction with ninhydrin (Spackman, Stein and Moore, 1958). The preceding acid hydrolysis (i.e. 6 N HCl, 24 h, sealed evacuated tubes, 110 °C) unfortunately results in extensive degradation of the tryptophan, particularly in the presence of carbohydrate (Miller, 1967), and hence it is necessary to determine this amino acid in a separate assay. The most satisfactory method of analysis involves HPLC after release of the peptide-bound tryptophan by alkaline or enzymatic hydrolysis.

Nicotinic acid and nicotinamide possess equal vitamin activity, the free acid being converted to the amide in the body. No IU of niacin activity has been defined, and analytical results for niacin are expressed in weight units (mg) of pure nicotinamide. To calculate the niacin equivalent of a diet, one adds one-sixtieth of the weight of tryptophan present to the weight of nicotinamide.

9.5.2 Extraction techniques for niacin

Autoclaving food samples in the presence of 1 N (0.5 M) H_2SO_4 for 30 min at 121 °C liberates nicotinamide from its coenzyme forms, but does not completely liberate the bound nicotinic acid from cereal products. The general extraction procedure for cereal products is autoclaving in the presence of calcium hydroxide solution, which readily liberates the nicotinic acid from its chemically bound forms. During the autoclaving with 1 N acid or calcium hydroxide solution, the liberated nicotinamide is hydrolysed to nicotinic acid.

When alkaline hydrolysis is applied to immature cereal grains (such as sweet corn), chemical or microbiological assay values for niacin are lower than those obtained by biological assay. This is because much of the niacin in immature cereal grains is contained in NAD, and this compound is destroyed by alkali. Preliminary boiling of immature cereal grain samples in water at neutral pH releases the nicotinamide from the NAD with the result that chemical or microbiological assay values for niacin are higher, and similar to those obtained by biological assay. Therefore, if it is necessary to use calcium hydroxide to extract food samples, it may be prudent always to test whether a preliminary boiling with water at neutral pH increases the values for niacin (Wall and Carpenter, 1988).

Hepburn (1971) observed that extraction of wheat products with water and dilute acid yielded niacin contents that tended to increase with both time and temperature of treatment. This indicates that, although bound niacin is reported to resist hydrolysis under such mild conditions, gradual breakdown does occur during extraction. Hepburn therefore decided to use the least stringent conditions possible for estimating free niacin, namely suspension of the samples in distilled water with continuous stirring for 30 min. Niacin values so obtained tended to be lower than those obtained after autoclaving with 0.1 N acid, but results were consistent in replication. Furthermore, niacin added in enrichment mixtures appeared to be fully recovered by the method. Other extraction techniques used to estimate free niacin have included blending with 50% aqueous ethanol (Carter and Carpenter, 1981) and boiling (Ghosh, Sarkar and Guha, 1963) or autoclaving (Vidal-Valverde and Reche, 1991) with 0.1 N HCl. The latter authors found that treatment of the acid hydrolysate with Takadiastase was absolutely necessary in the case of legume samples, because of the high starch content.

For the determination of added nicotinic acid or nicotinamide as a colour fixative in fresh meat, meat samples have been extracted by boiling with 96% ethanol (Gorin and Schütz, 1970) and blending with water (van Gend, 1973; Takatsuki *et al.*, 1987; Hamano *et al.*, 1988), acetonitrile (Tanaka *et al.*, 1989), methanol (Oishi *et al.*, 1988) or methanol after addition of a small amount of phosphoric acid (Tsunoda *et al.*, 1988). To extract the added nicotinamide from fortified foods, Rees (1989) heated the samples with $2 N$ H_2SO_4 on a boiling water bath for 30 min. In contrast to autoclaving with $1 N$ H_2SO_4 for 30 min, which completely hydrolyses the amide function, this somewhat milder acid treatment caused only a 6% conversion of nicotinamide to nicotinic acid.

Roy and Merten (1983) found that autoclaving meat samples with $1 N$ HCl in the presence of urea resulted in a significant increase in the niacin content when compared with extraction with $1 N$ acid alone. This suggests the release of niacin from non-ester conjugates by the acid–urea combination, possibly from amide-linked forms.

9.5.3 Microbiological assays for niacin

Determination of total niacin

Microbiological techniques currently employed for determining total niacin are based on the assay method developed by Snell and

Wright (1941) using *Lactobacillus plantarum* (ATCC No. 8014). *L. plantarum* responds equally well on a molar basis to nicotinic acid, nicotinamide, nicotinuric acid (an inactive metabolite) and NAD without preliminary hydrolysis (Snell and Wright, 1941); hence this organism cannot be used to differentiate between nicotinic acid and nicotinamide. The growth of *L. plantarum* on a nicotinic acid basal medium is not affected by free fatty acids or phospholipids, apart from an inhibitory effect of linoleic acid at a relatively high concentration (Kodicek and Pepper, 1948a), so there are no problems of growth stimulation or inhibition due to lipids. Nevertheless, Barton-Wright (1945, 1952) recommended that high-fat samples be Soxhlet-extracted with light petroleum ether for 16–18 h before acid or alkaline hydrolysis to prevent the formation of oily emulsions which may hinder complete extraction.

The extraction procedure recommended by the Association of Vitamin Chemists, Inc. (1966) for the microbiological determination of total niacin for all foodstuffs, including cereals and cereal products, entails autoclaving the sample at 121 °C for 30 min in the presence of 1 N H_2SO_4. This treatment liberates nicotinamide from its coenzyme forms and simultaneously hydrolyses it to nicotinic acid. The treatment does not completely liberate the bound nicotinic acid from cereal products; alkaline hydrolysis is necessary to do this. However, extraction of cereal products with 1 N H_2SO_4 yielded similar nicotinic acid values obtained microbiologically as did extraction with 1 N NaOH (Melnick, 1942), and microbiological values obtained on wheat bran, corn bran and rice bran were uniformly higher than chemical values when 1 N H_2SO_4 was used for extraction (Sweeney and Parrish, 1954). These findings support the experimental evidence presented by Krehl and Strong (1944) that *L. plantarum* is able to utilize the bound nicotinic acid in cereals to a considerable extent. Although alkaline hydrolysis has been used to ensure the complete liberation of bound nicotinic acid in microbiological assays (Kodicek and Pepper, 1948b), the use of 1 N H_2SO_4 is preferred for practical reasons because alkaline extraction of fatty samples gives extracts that are difficult to clarify.

A radiometric–microbiological assay has been developed for the determination of total niacin in foods based on the measurement of $^{14}CO_2$ generated from the metabolism of L-[1-^{14}C]methionine by *Kloeckera apiculata* (ATCC No. 9774; formerly called *K. brevis*) (Guilarte and Pravlik, 1983). Sample preparation entailed autoclaving with 1 N HCl for 90 min at 121 °C and, after cooling, pH adjustment to 4.5. Removal of precipitated debris was not necessary. Radiometric assay sensitivity, defined as the lowest concentration of vitamin producing a response of at least twice background (no vitamin), was 2 ng per vial (5.2 ml).

Determination of bound nicotinic acid

In mature cereal grains, most of the niacin present exists in the form of bound nicotinic acid, which is biologically unavailable to humans unless pretreated with alkali. The general approach to estimating the bound nicotinic acid content of a cereal sample is to treat the sample with weak (0.1 N) acid, which extracts the free niacin but does not liberate nicotinic acid from the bound forms. The resultant extract therefore contains both free and bound nicotinic acid, but provides a value only for the free form when analysed microbiologically using a selective organism. A portion of the acid extract is subsequently treated with 1 N NaOH to liberate bound nicotinic acid, and the resultant solution is assayed for total niacin. The difference between the results obtained for total and free niacin is a measure of bound nicotinic acid. This approach depends on the assay organism being unresponsive to bound nicotinic acid; thus the usual organism, *L. plantarum*, is unsuitable. The two methods described below utilize organisms that fulfil this requirement.

In a method described by Clegg (1963), food samples weighing 2–4 g are extracted with 50 ml of 0.1 N HCl on a boiling water bath for 30 min. After cooling, the pH of the solution is adjusted to 3.5 with 0.1 N NaOH with constant stirring to avoid localized alkalinity. The extract is made up to 100 ml, filtered, and an aliquot of the filtrate is washed with an equal volume of chloroform. An aliquot of the fat-free extract is heated with 1 N NaOH in a boiling water bath for 30 min to liberate the bound niacin. The alkali-treated solution and an untreated aliquot of the fat-free extract at pH 3.5 are adjusted to pH 6.5 and diluted to contain *c*.30 ng of total or free niacin/ml, respectively. In order to avoid the liberation of bound nicotinic acid during the subsequent autoclaving of medium plus sample, aliquots of the alkali-untreated solution are added aseptically to previously autoclaved assay tubes containing the medium. The alkali-treated solution is dispensed into assay tubes before autoclaving in accordance with standard procedure. The difference between the values for total and free niacin, using *L. casei* as the assay organism, represents the bound niacin content of the sample.

A similar procedure for determining the bound nicotinic acid in natural materials was reported by Ghosh, Sarkar and Guha (1963) using *Leuconostoc mesenteroides* subsp. *mesenteroides* (ATCC No. 9135) as the assay organism. In this procedure the homogenized sample was extracted with 0.1 N HCl in a boiling water bath for 45 min, and the extract was cooled and centrifuged. The residue was treated twice with 0.1 N HCl in the cold and then centrifuged. The combined centrifugate was used for the assay of free nicotinic acid, bound nicotinic acid and nicotinamide. Free nicotinic acid was determined by adjusting an aliquot of the acid extract to pH 6.2–6.4 and centrifuging off the precipitated material. The centrifugate was adjusted to pH 3.5 then sterilized by steaming at atmospheric pressure for 15 min. It was found that a portion of the bound nicotinic acid was hydrolysed when the materials were sterilized at 10 lb/in² for 15 min at pH 6.2–6.4, but when the sterilization was effected at pH 3.5–4 by steaming, only traces of nicotinic acid were released from the bound form. Bound nicotinic acid, together with free nicotinic acid, was determined by treating a second aliquot of the extract with 0.5 N NaOH for 10 min at room temperature, and assaying the nicotinic acid content after removal of the material precipitated at pH 6.2–6.4. Nicotinamide, together with free and bound nicotinic acid, was estimated by hydrolysing the bound nicotinic acid with 0.5 N NaOH, then hydrolysing the amide with 1 N HCl at 15 lb/in² for 1 h. The total nicotinic acid was assayed after removal of the material precipitated at pH 6.2–6.4. The difference in *Leuc. mesenteroides* assay results between the nicotinic acid content of the extract before and after alkaline hydrolysis represented the bound nicotinic acid content of the sample. The difference between the total nicotinic acid content and the nicotinic acid content obtained after alkaline hydrolysis represented the nicotinamide content of the sample.

Determination of added nicotinic acid

Nicotinic acid added to fresh meat can be determined directly using *Leuc. mesenteroides*. Gorin and Schütz (1970) suspended the minced meat sample in 0.2 M acetate buffer (pH 4.8) and shook the suspension for 10 min at room temperature. Nicotinic acid was extracted while the buffer flocculated the meat proteins. This mild extraction process does not convert nicotinamide to nicotinic acid. The results obtained were in good agreement with a spectrophotometric method.

9.5.4 AOAC colorimetric methods for determining niacin

The AOAC colorimetric method for the determination of niacin in foods and feeds (AOAC, 1990a) is based on the König reaction, in which pyridine derivatives are reacted with cyanogen bromide and an aromatic amine, sulphanilic acid. The pyridine ring is opened up, and the intermediate product is coupled with the amine to form a yellow dye, whose absorbance can be measured photometrically.

There are two different procedures: one for noncereal foods and feeds, and the other for cereal products. Noncereal foods and feeds are extracted by autoclaving with 1 N H_2SO_4 for 30 min at 15 lb (104 kPa) pressure in order to liberate nicotinamide from its coenzyme forms and hydrolyse it to nicotinic acid. The reaction with cyanogen bromide and sulphanilic acid is carried out at room temperature, and the resulting colour is measured at 450 nm. Cereal products are autoclaved with calcium hydroxide solution for 2 h at 15 lb pressure to liberate the nicotinic acid from its chemically bound forms. The reaction with cyanogen bromide and sulphanilic acid is carried out in the cold under somewhat different conditions, and the colour is measured at 470 nm.

The colorimetric method has been semi-automated using segmented-flow analysis with on-line dialysis for the determination of nicotinic acid (representing total niacin) released from food samples by autoclaving with calcium hydroxide solution (Egberg, Potter and Honold, 1974). A reference flow cell was employed to eliminate blank colour interference. The semi-automated procedure was shown to compare favourably with a microbiological assay for 63 different food products ($r = 0.9937$). The method has been adopted as Final Action by the AOAC for the determination of niacin in cereal products (Gross, 1975; AOAC, 1990b) and foods and feeds (Egberg, 1979; AOAC, 1990c). The on-line generation of cyanogen chloride using the Technicon AutoAnalyzer II system has been investigated by Ge, Oman and Ebert (1986) and shown to represent a definite improvement in the safe handling of the reagents.

9.5.5 High-performance liquid chromatography

Niacin

Selected HPLC methods for determining niacin using absorbance (UV) detection are summarized in Table 9.1.

Chase *et al.* (1993) determined nicotinic acid (representing total niacin) in purified extracts of fortified foods using anion exchange HPLC and UV detection (254 nm). Comparative results obtained using the AOAC microbiological assay (*L. plantarum*) were 11% higher than the HPLC results and the correlation coefficient (r) was 0.990. In addition to fortified products, several foods containing naturally low levels of niacin were evaluated. Beef soup gave the same results by both HPLC and microbiological methods, whereas the HPLC result for tuna was 17% higher. When canned sweet corn and frozen lima beans were extracted (Table 9.1), the nicotinic acid levels in the final dilution approached the detection limit of 0.11 µg nicotinic acid/ml. Because of sample matrix interferences at these low levels, niacin could not be accurately quantified by the HPLC method

The free nicotinamide and nicotinic acid naturally present in fresh meat, as well as that possibly added, can be extracted quantitatively with water or methanol and determined by HPLC with reasonable precision. Strong cation exchange chromatography permits the separation of the two vitamers, together with ascorbic acid and sorbic acid, in aqueous extracts of meat (Hamano *et al.*, 1988). Comparison of results obtained by

Table 9.1 HPLC methods used for the determination of niacin

Food	Sample preparation	Column	Mobile phase	Compounds separated	Detection	Reference
Anion exchange chromatography						
Fortified food products (bread, spaghetti, egg noodles, macaroni, cereals, infant formulas)	Add 50 ml water to suitably prepared sample followed by 6 ml H_2SO_4/water $(1+1)$. Autoclave at 121 °C for 45 min and then cool. Adjust pH to 6.0–6.5 and then immediately adjust to pH 4.5. Dilute with water and filter. *Clean-up*: readjust pH of extract to 0.5–1.0 and transfer to Florisil (magnesium silicate) column. Wash column bed with 0.25 M H_2SO_4 and elute nicotinic acid with 25 ml 0.5 M NaOH into a flask containing 1ml glacial acetic acid. Dilute to volume with water and filter	PRP-X100 250 × 4.1 mm i.d.	2% acetic acid in water	Nicotinic acid (representing total niacin)	UV 254 nm	Chase et al. (1993)
Strong cation exchange chromatography						
Fresh meat (to test for added niacin)	Homogenize sample in water then boil, cool and filter	Partisil SCX 10 μm 250 × 4.6 mm i.d.	0.05 M phosphate buffer (pH 3.0)	Nicotinic acid, nicotinamide, ascorbic acid, sorbic acid	UV 260 nm	Hamano et al. (1988)
Ion interaction chromatography						
Beef, semolina, cottage cheese	Blend sample with water and $Ca(OH)_2$ mixture, then autoclave at 121 °C for 15 min. Cool to 0 °C, filter, adjust to pH 6.5–7.0 with oxalic acid, filter. Pass 10 ml of filtrate through C_{18} Sep-Pak solid-phase extraction cartridge. Discard first 6 ml of eluate and collect next 3.5 ml. Add 1 drop 85% H_3PO_4 and mix	C_{18} LC-18-DB 5 μm (Supelco) 150 × 4.6 mm i.d.	23% MeCN, 0.10% H_3PO_4, 0.10% sodium dodecylsulphate in water (final pH 2.8)	Nicotinic acid (representing total niacin)	UV 254 nm	Tyler and Genzale (1990) (Figure 9.3)

Sample	Extraction	Column	Mobile phase	Compound	Detection	Reference
Legumes, lyophilized pork muscle	Autoclave ground sample with dilute HCl at 121 °C for 15 min and cool. Adjust pH to 4.0–4.5 and incubate with Takadiastase at 48 °C for 3h. Cool, filter and dilute to volume with water. *Clean-up and concentration:* readjust pH of extract to 4.7 ±0.02 and pass an aliquot through a Dowex 1-X8 (Bio-Rad) anion exchange column. Wash column with water, and elute nicotinic acid with 0.15 M HCl. Evaporate eluate to dryness and dissolve residue in MeOH/0.1 M acetate buffer of the required pH to obtain a final pH of 4.7–4.9	μBondapak C_{18} 10 μm 300 × 3.9 mm i.d.	0.01 M sodium acetate buffer (pH 4.66)/MeOH (9:1) containing 0.005 M tetrabutylammonium bromide (final pH 4.72)	Nicotinic acid (representing free niacin)	UV 254 nm	Vidal-Valverde and Reche (1991)
Instant coffee	Dissolve sample in hot water, cool, dilute to volume and filter. Pass through Sep-Pak C_{18} solid-phase extraction cartridge	Spherisorb ODS-2 150 × 5 mm i.d.	MeOH/water (8:92) containing 5 mM tetrabutylammonium hydroxide (final pH adjusted to 7.0)	Nicotinic acid	UV 254 nm	Trugo, Macrae and Trugo (1985)
Fresh beef and pork, fresh fish, fish products (to test for added niacin)	Homogenize sample in water and dilute to volume. Centrifuge and filter. Deproteinize aliquot of the filtrate with saturated zinc sulphate solution followed by 1 M NaOH. Dilute to volume with water, let stand for 30 min and filter	Radial-PAK μBondapak C_{18} 10 μm 100 × 8 mm	(i) 5 mM tetrabuty-ammonium phosphate (PIC-A) in MeOH/water (1 + 9) (ii) 10 mM heptane sulphonic acid (PIC-B7) in water	(i) Nicotinic acid (ii) Nicotinamide	UV 263 nm	Takatsuki et al. (1987)
Beef, pork, tuna (to test for added niacin)	Add H_3PO_4 to homogenized sample, then extract with MeOH and filter	Shim-Pack FLC ODS 50 × 4.6 mm i.d. Column temperature 50 °C	1 mM sodium dodecylsulphate and 0.02 M H_3PO_4/MeOH (7:3) pH 2.4	Nicotinic acid, nicotinamide	UV 261 nm	Tsunoda et al. (1988)

Table 9.1 Continued

Food	Sample preparation	Column	Mobile phase	Compounds separated	Detection	Reference
Fresh meat (to test for added niacin)	Extract homogenized sample with MeOH. *Clean-up:* Sep-Pak alumina N solid-phase extraction cartridge eluting nicotinamide with MeOH and nicotinic acid with 0.1 M NaHCO₃	LiChrosorb RP-18	MeOH/acetate buffer (pH 5) (2:10) containing 0.1 M sodium acetate and 10 mM tetrabutyl-ammonium hydroxide	Nicotinic acid, nicotinamide (separate chromatograms)	UV 261 nm	Oishi *et al.* (1988)
		Reversed-phase chromatography				
Cereal products, mushrooms	*Acid extraction:* autoclave sample with 0.1 M (0.2 N) H₂SO₄ at 1 bar for 1 h and cool. Incubate with buffered Clarase at 45 °C for 3 h. Cool, dilute to volume and filter. *Alkaline extraction:* heat sample with Ca(OH)₂ suspension at 95–100 °C for 30 min, then autoclave at 1 bar for 30 min. Cool and dilute to volume. Refrigerate overnight and centrifuge	Nicotinic acid fraction transferred from reversed-phase column (Nucleosil C₁₈ 5 µm) to anion exchange column (Nucleosil SB 5 µm) using automatic column switching. Both columns 150 × 4.6 mm i.d.	A. 0.57% acetic acid adjusted to pH 3.0 with NaOH to elute the nicotinic acid from the C₁₈ column and to place it on to the anion exchange column. B. Mobile phase A/MeOH (5:95) to flush the C₁₈ column. C. 2.28% acetic acid adjusted to pH 3.0 with NaOH to elute the nicotinic acid from the anion exchange column	Nicotinic acid (representing total niacin)	UV 254 nm	van Niekerk *et al.* (1984)

Abbreviations: see footnote to Table 2.4.

the HPLC method proposed by Hamano *et al.* with those obtained by a standard colorimetric method (König reaction) indicated poor agreement, attributed to deficiencies in the latter method.

Other investigators have used ion interaction chromatography for determining the niacin compounds. Takatsuki *et al.* (1987) found it necessary to use two different mobile phases for determining each compound in deproteinized aqueous extracts of meat; methanol/water $(1 + 9)$ containing buffered tetrabutylammonium ion was used for nicotinic acid, and water containing buffered heptane sulphonic acid for nicotinamide. The detection limit for a $10\,g$ sample was $1\,mg/100\,g$ for nicotinic acid and nicotinamide. Values greater than $20\,mg/100\,g$ indicated the illegal addition of either vitamer. Tsunoda *et al.* (1988) reported the simultaneous determination of the two vitamers in methanolic extracts of meat and fish using dodecylsulphate as the ion interaction agent. Oishi *et al.* (1988) obtained nicotinic acid and nicotinamide fractions from methanolic extracts of fresh meat by selective elution from an alumina solid-phase extraction cartridge, and chromatographed each fraction using tetrabutylammonium ion in the mobile phase.

In determining the naturally occurring niacin in foodstuffs, it must be decided whether to estimate total niacin or free ('available') niacin. In addition to the extraction problem, it is necessary to purify sample extracts prior to HPLC because the absorbance detection usually employed is not highly selective. Kral (1983) used pulsed amperometric detection for the analysis of fruit juices subjected to acid hydrolysis but, despite the improved selectivity, it was still necessary to employ cation exchange chromatography as a clean-up step.

Tyler and Genzale (1990) reported a simple, yet efficient, means of purifying alkaline digests of three major food representatives (beef, semolina and cottage cheese). After autoclaving with calcium hydroxide solution, digests were cooled in an ice bath, filtered, and then adjusted to pH 6.5–7.0 with oxalic acid to precipitate the excess calcium. An aliquot $(10\,ml)$ of the filtered suspension was loaded onto a C_{18} solid-phase extraction cartridge, the first $6\,ml$ of effluent was discarded, and the remaining effluent was collected and acidified with phosphoric acid. Analysis of the purified extracts was performed by ion interaction chromatography at acidic pH using sodium dodecylsulphate as the ion interaction agent. A feature of this method is the unusually sharp nicotinic acid peak, which resulted from the injection of the analyte in a large volume $(200\,\mu l)$ of dilute phosphoric acid, and which led to a detection limit of $c.0.5\,mg/100\,g$. The high efficiency of the purification step is evident from the lack of irrelevant peaks in the chromatogram obtained for semolina (Figure 9.3). The high efficiency of the single clean-up step is attributable to the amphoteric nature of nicotinic acid. At pH 7, nicotinic acid is not retained on the C_{18} cartridge, but a large number of pigmented

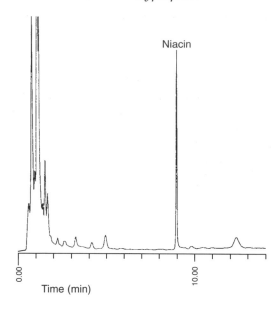

Time (min)

Figure 9.3 Ion interaction HPLC of nicotinic acid extracted from semolina after Sep-Pak C_{18} clean-up and acidification of Sep-Pak effluent. Conditions as given in Table 9.1. (From Tyler and Genzale, 1990. Reprinted from *The Journal of AOAC International*, 1990 **73**, No. 03, p. 468. Copyright 1986 by AOAC International, Inc.)

compounds are retained. Conversely, nicotinic acid is strongly retained on the C_{18} analytical column at acidic pH in the presence of an ion interaction agent, whereas many polar interferences elute in the dead volume. Tyler and Genzale (1990) compared HPLC results with results obtained from a microbiological assay. Agreement between the two methods was obtained for the analysis of beef, semolina and cottage cheese, but certain food samples (e.g. instant coffee) gave higher niacin values using HPLC.

Van Niekerk *et al.* (1984) did not include a clean-up step in their assay procedure for the analysis of cereal products and mushrooms subjected to acid or alkaline hydrolysis. They employed, instead, on-line two-dimensional HPLC, whereby the nicotinic acid fraction was transferred from a reversed-phase column to an anion exchange column. The HPLC method gave lower values for the acid extracts than for the alkaline extracts (except for the standard), whereas results obtained microbiologically on both acid and alkaline extracts showed no difference. This suggests that part of the nicotinic acid in the acid extract was present in a bound form that was available to the assay organism (*L. plantarum*). The HPLC method had a detection limit of 0.5 mg nicotinic acid/100 g of sample when a 5 g sample was extracted and diluted to 100 ml.

For the estimation of free ('available') niacin in legumes and meat (Vidal-Valverde and Reche, 1991), samples (1–10 g, equivalent to 10–100 μg of niacin) were mixed with hydrochloric acid (30 ml of 0.1 N + 1 ml of 6 N HCl) and autoclaved at 121 °C for 15 min. The acid digest was then adjusted to pH 4.0–4.5 and incubated with Takadiastase to hydrolyse the starch present in the legumes. The extracts were purified by strong anion exchange chromatography, and then analysed by ion interaction HPLC. The shape and retention times of the nicotinic acid peak were extremely sensitive to the pH of the mobile phase and, consequently, to column temperature, because of the amphoteric nature of nicotinic acid. Each different food required a trial-and-error procedure to ascertain the column temperature required for optimum separation.

Tryptophan

A typical HPLC procedure for the determination of tryptophan in foods involves the following steps:

(1) alkaline hydrolysis of the sample at 110–125 °C in an air-deprived medium for 16–20 h;
(2) dilution of hydrolysate and neutralization with 6 M HCl;
(3) clarification of the dilute hydrolysate;
(4) analysis by reversed-phase HPLC with ion suppression and fluorimetric detection (Jones, Hitchcock and Jones, 1981; Nielsen and Hurrell, 1985; Landry, Delhaye and Viroben, 1988; Hagen and Augustin, 1989; Bech-Andersen, 1991).

Because hot alkali attacks glass, alkali-inert and autoclavable plastics such as polypropylene should be used for the hydrolysis. The prolonged alkaline treatment causes complete racemization of L-tryptophan in proteins (Friedman and Cuq, 1988), but this is of no concern when reversed-phase HPLC is used.

The alkalis that have been investigated for protein hydrolysis are sodium hydroxide, lithium hydroxide and barium hydroxide. Delhaye and Landry (1993) reported that hydrolysis of wheat flour and soybean meal in the presence 1.35 M (2.7 N) $Ba(OH)_2$ at 110 °C for 16 h, at 125 °C for 8 h or at 140 °C for 4 h led to a complete release of tryptophan. There was no degradation of tryptophan when the hydrolysis time was extended to 24 h at 125 °C or 16 h at 140 °C. A 10 h hydrolysis at 125 °C allowed the autoclaving to be started at any time of the working day and stopped at any time of the night. The autoclave could then cool for the rest of the night (or weekend, considering the great stability of tryptophan in barium hydroxide).

The time-consuming steps of neutralization of alkaline hydrolysates with 6 M HCl followed by quantitative transfer, dilution to volume and centrifuging were replaced in a simplified procedure (Landry and Delhaye, 1992, 1994) by diluting a very small (10 μl) aliquot of cooled hydrolysate with 1 ml of acid solution (0.01 M HCl, 0.1 M acetic acid). The acid treatment destroys any traces of barium carbonate and the co-precipitation of sample debris results in a clear solution ready for injection into the liquid chromatograph. The dilution step can be carried out in a sample vial when an autosampler is used. Barium hydroxide octahydrate has a limited solubility in water and must be added to the sample in solid form. In the conventional procedure the amount of added alkali must be accurate as it dictates the amount of acid required for neutralization. This does not apply to the simplified procedure since the hydrolysate aliquot for HPLC is taken from a solution saturated in barium hydroxide. Consequently, an approximate amount of powdered barium hydroxide close to 4.2 g may be added to the sample with a calibrated scoop.

Using the simplified procedure (Table 9.2 and Figure 9.4), Landry and Delhaye (1994) reported a tryptophan recovery of 95.3 to 106.8% (mean of $102 \pm 3.7\%$) corrected from losses of 5-methyltryptophan added as an internal standard. This quantitative recovery depended on the removal of air from the autoclave and thorough removal of atmospheric oxygen from the hydrolysis medium and sample particles by allowing the vapour to escape from the boiling liquids at 100 °C for 5 min before the rise in pressure. If this purging treatment was omitted or incomplete, barytic hydrolysis did not lead to a quantitative recovery. Hydrolysis with 4.2 N NaOH under the same conditions and with autoclave purging did not provide a complete recovery of tryptophan. The difference between $Ba(OH)_2$ and NaOH with respect to tryptophan recovery could be due to the extent of oxygen removal from the hydrolysis medium and sample. The NaOH solution, being more viscous than the $Ba(OH)_2$ solution, promotes a more rapid swelling of the sample during solvation, thereby impeding the release of entrapped air when the autoclave was purged. These considerations for NaOH hydrolysis would also apply to LiOH and no significant improvements were found by changing from NaOH to LiOH (Nielsen and Hurrell, 1985).

Proteases have been considered as an alternative to alkalis but the use of such enzymes has resulted in incomplete hydrolysis. However, Garcia and Baxter (1992) reported the complete hydrolysis of soy- and milk-based infant formulas using Pronase in a medium of 0.1 M trizma buffer (pH 8.5, 50 °C, 6 h). The commercial preparation of Pronase used by these authors contained c.25% (w/w) of added calcium acetate as stabilizer to retard the loss of protease activity (loss of zinc). The use of phosphate buffer instead of trizma buffer

Table 9.2 HPLC method used for the determination of tryptophan

Food	Sample preparation	Column	Mobile phase	Compounds separated	Detection	Reference
		Reversed-phase chromatography				
Maize, wheat, sorghum, whey, bran, pea, soybean, fishmeal	Weigh an amount of sample containing 50 mg of protein into a 30 ml polypropylene tube. Add 4.2 g of Ba(OH)$_2$, 8H$_2$O, 5-methyltryptophan (3 µmol, internal standard), 2.5 ml of water and 2 or 3 carborundum grains. Vortex-mix and add 4 ml of water. Place tubes in racks and cap with beakers. Place rack in autoclave when water boils and bolt. When the pressure reaches 1 bar, allow the vapour to escape for 5 min to drive off the oxygen. Heat to raise the pressure to 1.4 bar (125°C) and leave for 10 h, the autoclave being left to cool during the night. Cool the hydrolysates to 0°C. Dilute 10 µl of hydrolysate with 1 ml of acid solution (0.01 M HCl, 0.1 M acetic acid)	Nova-Pak C$_{18}$ 4 µm 150 × 3.9 mm i.d. Column temperature 45°C	0.07 M sodium acetate buffer (pH 4.5)/MeOH (80 + 20)	Tryptophan, 5-methyltryptophan (internal standard)	Fluorescence: ex. 285 nm em. 345 nm	Landry and Delhaye (1992, 1994) (Figure 9.4)

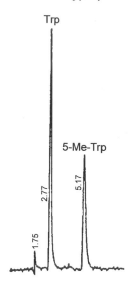

Figure 9.4 Reversed-phase HPLC of tryptophan in a wheat hydrolysate containing added 5-methyltryptophan as internal standard. Conditions as given in Table 9.2. (Reprinted with permission from *J. Agric. Food. Chem.*, **40**, 776–9, Landry and Delhaye, Copyright 1992, American Chemical Society.)

would precipitate the calcium as calcium phosphate, and this removal of stabilizer could explain other reports of long digestion times and incomplete recoveries.

It has been recommended that samples containing more than 5% fat should be defatted by ether extraction prior to hydrolysis to maximize recovery of tryptophan (Sato *et al.*, 1984; Rogers and Pesti, 1990). Starch or maltodextrin reduces the loss of tryptophan by an unknown mechanism when the hydrolysis is carried out in the near absence of oxygen (Friedman and Cuq, 1988). Starch undergoes oxidative degradation in alkaline solution in the presence of oxygen. Apparently, the polysaccharide is capable of removing the residual traces of oxygen in the hydrolysis mixture by incorporating the oxygen into products that do not react with tryptophan. The protective effect of starch or maltodextrin is not sufficient, however, to permit the hydrolysis to be conducted without the removal of most of the oxygen by evacuation (Hugli and Moore, 1972). The amount of starch required depends on the carbohydrate content of the sample. Samples that are rich in carbohydrate, such as corn meal and wheat flour, do not require the addition of starch. Bech-Andersen (1991) found that commercially hydrolysed starch contains a component which coelutes with tryptophan during analytical HPLC and used lactose to protect the amino acid.

Analytical methods involving alkaline hydrolysis only accurately predict the bioavailable tryptophan when *in vivo* protein digestibility and amino acid absorption are close to completion. In a study of processed and stored foods, methods using alkaline hydrolysis predicted smaller losses of tryptophan than those measured by a rat growth assay (Nielsen, Klein and Hurrell, 1985). For most samples, a good agreement was found between the rat assay value and the chemical value when the latter was multiplied by true protein digestibility.

9.6 BIOAVAILABILITY

9.6.1 Physiological aspects

Protein digestibility

Protein digestibility is discussed here because it influences the bioavailability of the niacin precursor, tryptophan.

The digestibility of a protein represents that proportion which is not excreted in the faeces and which is, therefore, assumed to be absorbed by the animal. Protein is the only major dietary constituent that contains nitrogen, and therefore protein digestibility can be determined by measuring the nitrogen content of the protein consumed and the nitrogen content of the faeces. If the faecal nitrogen (designated F) is subtracted from the nitrogen intake (I) and expressed as a proportion of the intake, $(I - F)/I$, the result is termed apparent digestibility. Determinations are carried out in rats, which digest a variety of food proteins in a similar manner to humans (Bodwell, Satterlee and Hackler, 1980).

Faecal nitrogen is derived not only from undigested and unabsorbed dietary protein, but also from nondietary sources. These endogenous sources constitute sloughed-off cells of the intestinal lining, microbial proteins and unabsorbed residues of gastrointestinal secretions. Hence faecal nitrogen is not a correct measurement of undigested protein. When allowance is made for the nondietary component, referred to as metabolic faecal nitrogen (M), the true digestibility can be calculated from $I - (F - M)/I$. The metabolic faecal nitrogen is determined from the faecal nitrogen output from rats fed on a protein-free diet. The true protein digestibility of food is always higher than the apparent digestibility because of the metabolic faecal nitrogen factor. This factor is relatively constant and more or less independent of food intake. Apparent digestibility deviates more from true digestibility at low levels of protein intake than at high levels of intake because the metabolic faecal nitrogen makes up a larger proportion of the faecal nitrogen when the protein intake is low (Hopkins, 1981). Therefore, true protein digestibility, being independent of protein intake, is generally regarded as being

Table 9.3 True digestibility by adults of protein in some common foods (from Hopkins, 1981)

Protein source	Number of reports	Digestibility, %	
		Mean	Range
Animal protein sources			
Meat, poultry, fish	10	94	90–99
Milk, whole	3	94	90–98
Egg, whole	9	98	93–106
Cereal products			
Wheat, whole	6	86	80–93
Wheat flour, white	2	96	96–97
Bread, white wheat	5	97	95–101
Bread, whole wheat	2	92	91–92
Wheat, ready-to-eat cereals	9	77	53–88
Maize, whole	4	87	84–92
Maize, ready-to-eat cereals	5	70	62–78
Rice, polished	4	89	82–93
Rice, ready-to-eat cereals	2	70	63–77
Oatmeal	4	86	76–92
Oats, ready-to-eat cereals	4	72	63–89
Legume products			
Soybean flour	5	86	75–92
Soybean protein isolate	3	95	93–97
Peanut flour	3	93	91–98
Peanut butter	1	95	–

more accurate than apparent digestibility for comparing different protein sources. Apparent digestibilities are satisfactory for making comparisons provided the levels of protein intake are similar. Results for protein digestibility are expressed either as a percentage or, more recently, as a ratio. Thus 60% digestibility is expressed as 0.6 (Bender, 1993).

Values obtained from human studies for true protein digestibility in some common foods are given in Table 9.3. Animal protein sources and refined (low-fibre) plant protein sources such as white wheat flours or bread and soy protein isolate are highly digestible (mean values 94–99%). Whole-grain cereals and oatmeal are less digestible (c.86%) and highly processed ready-to-eat cereals have lower digestibilities (70–77%). High-fibre vegetable protein sources including sorghum, millet and legumes (beans, lentils, peas and other pulses) have true protein digestibilities of about 80% or lower. The seemingly low protein digestibility of legumes may be explained by the increased excretion of endogenous nitrogen by their consumption (Shurpalekar, Sundaravalli and Narayana Rao, 1979).

Typical North American whole diets based on animal and/or refined vegetable protein sources gave an average true protein digestibility of

92%, with values of individual studies ranging from 88% to 96%. Diets from other cultures based on coarse and less refined vegetable protein have low protein digestibilities. For example, two studies carried out in India showed that the protein of diets containing mainly rice, dahl, milk powder and vegetables had true digestibilities of 73% and 77%. In a third Indian study, diets based on the cereal, ragi, were even more poorly digested (54%). There is experimental evidence that the low digestibility of diets from certain cultures is due to an inherent characteristic of the foods and not to genetic differences in the population or to pathological problems with the digestive tract (Hopkins, 1981).

Differences in protein digestibility of various foods may arise from inherent differences in the configuration of the protein molecule, specific amino acid bonding, and the combination of protein with lipids, various metals, nucleic acids or carbohydrates (Kies, 1981). Processing techniques influence the digestibility of proteins – some positively and some negatively. In the milling of flour, reduction of the grain size, breakage of the hard outer kernel and possibly removal of the bran layer increase the surface area, making enzymatic digestion much more efficient. For some foods, thermal processing can improve protein digestibility by destroying protease inhibitors and opening the protein structure through denaturation (Rupnow, 1992).

The method of rat faecal analysis for determining protein digestibility might not be valid in the case of legumes. The observed increase in faecal nitrogen, rather than representing a failure to absorb dietary nitrogen, might be at least partly due to the excretion of endogenous nitrogen above the level expected in normal metabolism. The increased excretion of endogenous nitrogen could be due to a greatly increased turnover of intestinal epithelial cells, resulting in the excretion of cellular protein in amounts that overwhelm the normal reabsorption process (Bender and Mohammadiha, 1981; Sandaradura and Bender, 1984). However, experiments conducted by Fairweather-Tait, Gee and Johnson (1983), in which intestinal mucosal cell turnover in rats was monitored by observing the uptake and subsequent loss of tritium from the labelled DNA precursor, thymidine, indicated that mucosal protein loss could not fully account for the increased faecal nitrogen excretion. Alternatively, the increased endogenous nitrogen could originate from bacterial cells. Sarwar (1987) proposed that the high concentrations of insoluble dietary fibre in legumes stimulate fermentation in the large intestine, thereby increasing the amount of bacterial protein in the faeces.

The true digestibility of crude protein is a reasonable approximation of true digestibilities of amino acids (including tryptophan) in diets based on animal protein sources, cereals and oilseed products. In most mixed diets consumed in North America the differences between true digestibilities of protein and tryptophan are less than 10%. However, true digestibility of crude protein does not accurately predict digestibilities of limiting amino acids in diets based on legumes. In beans, peas and lentils, values for the true digestibility of tryptophan (reflecting tryptophan bioavailability) have been reported to be 25% lower than those of crude protein (Sarwar, 1987). This lower digestibility may be due to the presence of tryptophan in the less digestible parts of the grain.

In general, amino acid bioavailability is of practical concern only at marginal or inadequate intakes of total protein (Kies, 1981).

Methods of determining the bioavailability of tryptophan

Methods such as faecal analysis, animal growth assay and microbiological assay for determining the bioavailability of amino acids have been critically discussed (Meade, 1972; Elwell and Soares, 1975).

Faecal analysis
When referring to amino acids, the term bioavailability is synonymous with true digestibility (Sarwar, 1987) and hence the calculation used in faecal analysis is similar to that for estimating true protein digestibility:

$$\% \text{ Trp bioavailability} = \frac{\text{Trp intake} - (\text{faecal Trp} - \text{metabolic faecal Trp})}{\text{Trp intake}} \times 100$$

Rat growth assay
In this assay, the growth responses of rats fed test food diets are compared with growth responses of rats fed basal diets containing graded levels of L-tryptophan. The reference protein included in the basal diet is casein in which tryptophan is 100% available. Zein, which is devoid of tryptophan, is used as a basal ingredient to provide a source of natural protein. For the standard curve, mean weight gains are plotted against total tryptophan intake determined by chemical analysis. Rat growth assays have been used to determine tryptophan availability in different food model systems (Nielsen *et al.*, 1985) and selected foods (McDonough *et al.*, 1989).

Microbiological assay
Streptococcus zymogenes is commonly employed as the test organism for the measurement of available amino acids based on the methodology of Ford (1962, 1964). This bacterium is vigorously proteolytic and has an absolute requirement for exogenous tryptophan among certain other amino acids. Pronase has been found to be more satisfactory as a predigesting enzyme than papain (Hewitt and Ford, 1985). Wells *et al.* (1989) compared the microbiological assay with the rat growth assay as a means

Table 9.4 Comparison of estimates of percentage available tryptophan from *Streptococcus zymogenes* and rat assays (from Wells *et al.*, 1989)

Protein source	S. zymogenes method			Rat assay
	Digest 1	Digest 2	Mean	
Casein	123	114	119	(100)[a]
Non-fat milk	–	130	130	105
Non-fat milk (heated)	96	104	100	90
Tuna	112	103	108	99
Chicken frankfurters (defatted)	94	99	97	98
Beef salami	98	104	102	93
Sausage	152	114	133	92
Beef stew	106	100	103	102
Macaroni and cheese	104	86	95	109
Chick peas	88	90	89	87
Pinto beans	92	79	86	61
Pea protein concentrate	105	105	105	97
Soy isolate	130	116	123	114
Peanut butter	109	91	100	95
Whole wheat cereal	123	114	119	80
Rice and wheat gluten cereal	101	101	101	100
Rolled oats	74	74	74	98

[a] Standard curve.

of measuring tryptophan bioavailability in the same commercial food products (Table 9.4). Microbial values for availability were generally higher than values obtained by rat assay, but the two methods were significantly correlated ($P = 0.05$) in a paired *t*-test.

Protein digestion

Digestion of dietary protein begins in the stomach with the action of pepsin, a complex of several proteolytic enzymes which breaks the polypeptide chain at many sites, especially the peptide bond on the amino side of aromatic amino acids (phenylalanine and tyrosine) (Kies, 1981). Pepsin thus acts on proteins to produce short peptide chains without liberating a significant quantity of amino acids. The gastric phase accounts for $< 10\%$ of total protein digestion in humans (Steele and Harper, 1990).

The major site of protein digestion is the small intestine. All but a negligible fraction of ingested protein is broken down completely and absorbed as amino acids and small peptides. Between 20 and 30 g of protein contained in digestive juices and about 30 g of protein derived from desquamated cells of the small intestine are added to ingested protein, and these too are mostly digested and absorbed. In humans, these endogenous sources comprise *c*.30–50% of the total protein digested (Alpers, 1987). The intestinal proteases are secreted as proenzymes from the exocrine pancreas in response to hormonal and vagal stimulation. A brush border enzyme, enterokinase, which is released from the intestinal mucosa by the action of bile acids, activates trypsinogen to trypsin. Trypsin in turn activates the other proenzymes to produce an array of proteases.

The final stages of protein digestion are accomplished by a large number of peptidases secreted by the brush border and cytoplasm of the enterocyte. Small peptides containing up to five amino acid residues can be absorbed by the enterocyte where they are largely broken down to free amino acids (Emery, 1993). The absorbed amino acids enter the portal vein and are transported directly to the liver. The protein contained in the faeces is not food protein, but comes from colonic bacteria, mucoproteins and cell debris.

Intestinal absorption of tryptophan

L-Amino acids are actively transported across the brush-border membrane of enterocytes by a variety of carriers according to whether the amino acids are acidic, basic or neutral in character. Neutral amino acids are so called because they have a relatively low net electrical charge by virtue of their aromatic ring (e.g. tryptophan, phenylalanine, tyrosine), long side chain (e.g. methionine) or branched side chain (e.g. valine, leucine, isoleucine).

Little is known about the mechanism of intestinal tryptophan transport. Munck and Rasmussen (1975) studied the transport characteristics in rat jejunum and concluded that tryptophan is transported across the brush-border membrane by a carrier of neutral amino acids alone. Tryptophan, tyrosine and phenylalanine competitively inhibited the brush-border transport of each other (Cohen and Huang, 1964), suggesting that these three amino acids share a common carrier. It has been established that brush-border uptake of tryptophan occurs by a sodium-independent transport system with a broad selectivity for certain neutral amino acids. A similar system operates for the basolateral exit of tryptophan from the enterocyte (Stevens, Kaunitz and Wright, 1984; Hopfer, 1987).

Intestinal absorption of niacin

Much of the available niacin in the diet will be in the form of the nicotinamide nucleotides, with meat and milk containing free nicotinamide. Ingested nicotinamide nucleotides are hydrolysed by enzymes in the small intestine and the liberated nicotinamide is rapidly absorbed and conveyed to the liver.

Gross and Henderson (1983) showed that NAD degradation in rat intestine commences with the formation of nicotinamide ribonucleotide by the action of a pyrophosphatase found in intestinal secretions or released from desquamated cells. The ribonucleotide is rapidly hydrolysed to nicotinamide riboside, which is then hydrolysed more slowly to nicotinamide. Both of these reactions are catalysed by enzymes that are membrane-bound or intracellular.

Absorption of nicotinamide and nicotinic acid across the brush-border membrane occurs by simple diffusion, followed by a rapid sequence of metabolic steps to form NAD primarily within the cytosolic compartment of the enterocyte. The NAD and intermediate metabolites are not freely permeable to the cell membrane, resulting in the metabolic trapping of nicotinamide and nicotinic acid. The metabolic trapping creates the concentration gradient across the brush-border membrane, which is necessary for the passive absorption of the two vitamers (Schuette and Rose, 1983; Stein *et al.*, 1994).

Renal reabsorption of niacin

Schuette and Rose (1986) studied brush-border transport and renal metabolism of nicotinic acid in rat brush-border membrane vesicles and renal cortical slices exposed to a physiological concentration of the vitamin. The data suggested that reabsorption of nicotinic acid at the brush border takes place by an active sodium-dependent mechanism. The Na^+–nicotinic acid co-transport appeared to be electroneutral and consequently membrane potential effects on transport were a result of effects on the carrier itself. Within the absorptive cell, most of the nicotinic acid is rapidly metabolized to intermediates in the Preiss–Handler pathway for NAD biosynthesis. The transporter moves nicotinic acid into the cell faster than the enzyme system metabolizes it – a situation which allows an expression of active transport.

Other metabolic considerations

In the liver, nicotinic acid and nicotinamide, together with tryptophan, are converted via a common intermediary (nicotinic acid ribonucleotide) into NAD, some of which is utilized by the liver itself. The surplus NAD is hydrolysed in the liver to free nicotinamide, which is then released into the general circulation accompanied by the nicotinic acid that was not metabolized. On reaching the tissues, the niacin vitamers are used for the intracellular synthesis of NAD and NADP. There is a continuous turnover of these nucleotides in the body and very little storage. Excess niacin is converted in the liver to methylated derivatives, which are excreted into the urine (McCormick, 1988).

9.6.2 Dietary sources of niacin and tryptophan and their bioavailability

Dietary sources

Because of the contribution of tryptophan, foods containing balanced protein are important contributors to total niacin equivalent intake. The niacin, tryptophan and total niacin equivalent contents of a selection of foods are presented in Table 9.5.

Lean red meat, poultry and liver contain high levels of both niacin and tryptophan and, together with legumes, are important sources of the vitamin. In meat the niacin is largely in the form of nicotinamide formed by *post mortem* hydrolysis of NAD. Peanut butter is an excellent source of niacin. Cheese and eggs are relatively poor sources of preformed niacin, but these high-protein foods contain ample amounts of tryptophan and therefore have a high niacin equivalent. Fruits and vegetables provide useful amounts, depending on the dietary intake. Other useful sources are whole grain cereals, bread, tea and coffee.

In mature cereal grains most of the niacin is present as bound nicotinic acid and is concentrated in the aleurone and germ layers (Figure 9.5).

Table 9.5 Niacin, tryptophan and total niacin equivalents of some representative foods (mg/100 g) (from Holland *et al.*, 1991)

Food	Niacin	Tryptophan	Niacin equivalent from tryptophan[a]	Total niacin equivalents
Cow's milk, whole, pasteurized	0.1	42	0.7	0.8
Cheese, cheddar, average	0.1	360	6.0	6.1
Egg, chicken, whole, raw	0.1	222	3.7	3.8
Wheat flour, wholemeal	5.7[b]	150	2.5	8.2
Rice, brown, raw	5.3	90	1.5	6.8
Rice, white, easy cook, raw	4.2	96	1.6	5.8
Beef, lean only, raw, average	5.2	258	4.3	9.5
Lamb, lean only, raw, average	6.0	264	4.4	10.4
Pork, lean only, raw, average	6.2	228	3.8	10.0
Chicken, meat only, raw	7.8	228	3.8	11.6
Liver, ox, raw	13.4	270	4.5	17.9
Cod, raw, fillets	1.7	192	3.2	4.9
Potato, maincrop, old, average, raw	0.6	30	0.5	1.1
Soyabeans, dried, raw	2.2	342	5.7	7.9
Peas, raw	2.5	66	1.1	3.6
Brussels sprouts, raw	0.2	42	0.7	0.9
Apples, eating, average, raw	0.1	6	0.1	0.2
Orange juice, unsweetened	0.2	6	0.1	0.3
Peanuts, plain	13.8	330	5.5	19.3
Yeast extract (Marmite)	58.0	540	9.0	67.0
Coffee, instant	22.00[c]	174	2.90	24.9

[a] Niacin equivalent = mg tryptophan ÷ 60.
[b] This level is for fortified flour.
[c] Can be as high as 39 mg/100 g. Decaffeinated instant coffee contains about the same amount.

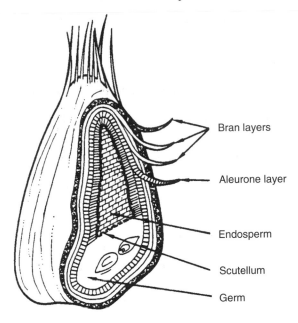

Bran layers

Aleurone layer

Endosperm

Scutellum

Germ

Figure 9.5 Longitudinal section of a wheat grain.

Milling to produce white flour removes most of the vitamin with the bran. In the UK it is compulsory by law to add niacin to white flour (mostly 70% extraction rate) at 16 mg/kg. All flour other than wholemeal (100% extraction) must be enriched (Bender, 1978).

As discussed below, some plant-derived foods contain niacin in chemically bound forms that result in their bioavailabilities being low. In mature cereal grains, for example, as much as 70% of the niacin may be biologically unavailable after conventional cooking. Most food composition tables give total niacin, and are compiled from the results of analyses in which nicotinic acid is liberated from unavailable bound forms by hydrolysis with acid or alkali. Therefore, tabulated niacin contents for many plant foods, particularly mature cereals, overestimate their value in providing biologically available niacin.

Bioavailability

In mature cereal grains (those examined included maize, wheat, rice, barley and sorghum) 85–90% of the total niacin content exists in chemically bound forms of nicotinic acid, the remainder being present as free nicotinic acid and nicotinamide (Ghosh, Sarkar and Guha, 1963). The majority of this bound nicotinic acid is biologically unavailable after conventional cooking (Wall and Carpenter, 1988). At one time there

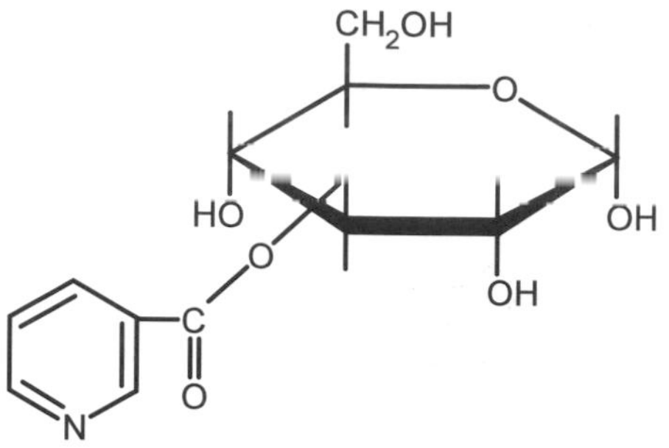

Figure 9.6 Structure of β-3-*O*-nicotinoyl-D-glucose.

were two views concerning the nature of the bound nicotinic acid in cereal grains. The nicotinic acid was contained either within a polysaccharide, referred to as niacytin (Kodicek and Wilson, 1960) or within a polypeptide, which was called niacinogen (Das and Guha, 1960). Investigations by Mason, Gibson and Kodicek (1973) revealed that the bound nicotinic acid of wheat bran is not present as a single substance, but is incorporated in a number of macromolecules that are both polysaccharide and glycopeptide in character. Mason and Kodicek (1973) subjected preparations of bound nicotinic acid of wheat bran to partial acid hydrolysis and identified a subunit as nicotinoyl glucose (Figure 9.6). The nicotinoyl ester bond linking the nicotinic acid and glucose moieties most probably exists in the polysaccharide and glycopeptide fractions isolated from wheat bran, and is so far the only established linkage in bound nicotinic acid.

Nicotinoyl glucose itself is readily utilized, so why should this compound be unavailable when present in plant tissues? Mason and Kodicek (1973) suggested that its incorporation within indigestible celluloses and hemicelluloses prevents access of the gastrointestinal esterases to the nicotinoyl ester bonds. Alternatively, esterase activity may be poor: the methyl ester of nicotinic acid was only 15% as effective as the free acid in supporting the growth of rats (Wall and Carpenter, 1988). About 10% of the total niacin was released as free nicotinic acid after extraction of sorghum meal with 0.1 N HCl (Magboul and Bender, 1982). This suggests that a small proportion of bound nicotinic acid can be hydrolysed by gastric juice and made available.

Pellagra, the disease caused by a deficiency of both niacin and tryptophan, has commonly been found in population groups having maize as their staple food. The generally accepted explanation for this association is the unavailability of niacin in maize, coupled with a very low proportion of tryptophan in zein (the major protein in maize). Mexican and Central American peasants, and also Hopi Indians in Arizona, rely upon maize as a staple food and yet do not experience pellagra. The explanation for this paradox lies in the way in which these people prepare the maize for bread-making. In the traditional preparation of Mexican tortillas, the maize is soaked at alkaline pH in lime-water before baking and this process releases the nicotinic acid from its bound forms. In the making of piki bread the Hopi Indians use wood ash, which is alkaline and also results in the liberation of nicotinic acid. The availability of nicotinic acid in tortillas baked from maize treated with lime-water has been demonstrated in pigs by Kodicek *et al.* (1959).

Wall, Young and Carpenter (1987) established that niacin exists in different forms during the development of the maize grain, the changes being consistent with the biological needs of the seed. In the early milky and dough stages (21 to 28 days after pollination), the predominant niacin-containing compounds are the pyridine nucleotides NAD and NADP. These coenzymes are required by the plant to supply energy for producing polysaccharides, proteins and fats for deposition and storage within the seed. The oxidized forms of these coenzymes (NAD^+ and $NADP^+$) are biologically available as a source of niacin and release nicotinamide on heating. The reduced coenzymes (NADH and NADPH), which may also be present in small amounts, are unavailable, because they are acid-labile and break down in the gastric juice. As the grain matures, some free nicotinamide and nicotinic acid are present and, finally, nutritionally unavailable carbohydrate-bound nicotinic acid and trigonelline (Figure 9.7) are formed for storage. Small amounts of pyridine nucleotides are retained in the germ and aleurone layer to facilitate germination.

Carpenter *et al.* (1988) confirmed in rat growth assays the high available niacin value of maize harvested at the milky stage, whether it was of a sweet or starchy type, and the much lower value of each type at maturity. The higher level of tryptophan in the maize protein at the milky stage also contributes to its overall niacin equivalent value. Therefore, the niacin of sweet corn (corn-on-the-cob) and also immature

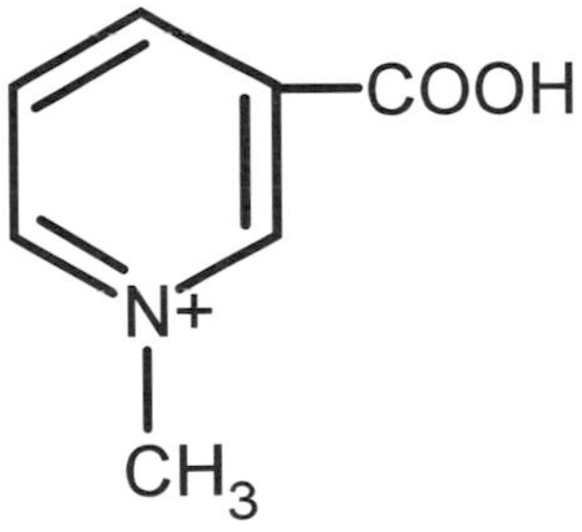

Figure 9.7 Structure of trigonelline (*N*-methylnicotinic acid).

Table 9.6 Bioavailability of niacin in cereal and noncereal foods (from Carter and Carpenter, 1982)

Food	Total niacin (chemical) mg/kg	Free niacin[a] (chemical) mg/kg	Available niacin (rat assay) mg/kg	Bioavailability[b] %
Cereal foods				
Maize, boiled	18.8	1.1	7 ± 1.8	37
Tortilla	12.6	11.7	14 ± 1.8	111
Sweet corn, raw	51.3	Tr	40 ± 6.4	78
Sweet corn, steamed	56.4	45	48 ± 7.2	85
Sorghum, boiled	45.5	1.6	15 ± 3.8	33
Rice, boiled	70.7	17.3	29 ± 6.6	26[c]
Whole wheat, raw	51.4	ND	16 ± 3.0	31
Whole wheat, boiled	57.3	ND	18 ± 5.4	31
Concentrate from wheat bran	2830	Tr	480 ± 240	17
Non-cereal foods				
Potato, baked	50.7	12	32 ± 6.9	52[c]
Liver, baked beef	306	297	314 ± 30	103
Beans, autoclaved	26.2	22	31 ± 3.1	118
Peanut flour	255	Tr	117 ± 20	46
Coffee extract	597	315	420 ± 67	70

[a] Free nicotinic acid + nicotinamide.
[b] Bioavailability = (available niacin ÷ total niacin) × 100%.
[c] Estimate of the bioavailability of the (total − free niacin) assuming that the free niacin is fully available to the rat.
ND, none detected; Tr, trace.

starchy corn has a high bioavailability both raw, when it is present mostly in NAD, and cooked (at neutral pH), when the NAD has largely broken down to yield free nicotinamide.

Carter and Carpenter (1982) compared the available niacin values of foods determined by the weight gain in young rats with the contents of free and total niacin determined by chemical analysis (Table 9.6). The bioavailability of niacin in each food is calculated by dividing the rat assay value by the total niacin content (both in units of mg/kg) and multiplying by 100. The niacin in liver and beans was present largely, if not entirely, as free niacin and was fully bioavailable. In mature cereals (maize, sorghum, rice and wheat) only a small proportion of free niacin, or none, was detected and bioavailabilities ranged from 26% to 37%. The complete bioavailability of niacin in tortilla demonstrated the effect of alkali in liberating free nicotinic acid from its bound forms in maize. Potato and peanut flour contained little or no free niacin and exhibited bioavailabilities of 52% and 46%, respectively. The coffee extract had about half its niacin in free form and the bioavailability was 70%. High bioavailabilities were obtained for both raw and steamed sweet corn owing to their niacin being mainly in the form of NAD and free nicotinamide, respectively.

Estimates of tryptophan bioavailability in selected foods using a rat growth assay were 90–100% for most products tested (McDonough *et al.*, 1989). One notable exception was pinto beans whose low value of 59% could be explained by a low protein digestibility. Kies (1981) reported that the protein contained in the bran fraction of wheat is not available to the human, presumably because of interference of absorption by dietary fibre.

The utilization of tryptophan as a precursor of NAD depends on an adequate dietary supply of other essential amino acids. If the supply falls, less tryptophan is utilized in protein synthesis and more becomes available for conversion to NAD. A high dietary intake of leucine inhibits the synthesis of NAD from tryptophan, without affecting the utilization of niacin obtained directly from the diet. Leucine is found in high concentration in sorghum. In parts of India, where sorghum is the staple food, leucine may be pellagragenic at times of food shortage when intakes of both tryptophan and niacin are very low (Bender and Bender, 1986).

9.6.3 Effects of food processing, storage and cooking

Niacin

Niacin is generally stable during the commercial processing, storage and domestic cooking of foods. There is no significant loss of niacin during milk processing (Tannenbaum, Young and Archer, 1985).

The steaming of rice prior to milling does not significantly increase the niacin content of polished rice as, unlike other B-group vitamins, the nicotinic acid in its bound form does not diffuse into the endosperm (Koetz and Neukom, 1977).

Trigonelline, an abundant constituent of green coffee beans, is demethylated to nicotinic acid during the roasting process. The amount of nicotinic acid formed is 80–150 mg/kg in medium roasted coffee and up to 500 mg/kg in a dark roast (Hoffmann–La Roche Inc., 1992).

Niacin is present in uncooked foods mainly as NAD and NADP, but these nucleotides may undergo some degree of hydrolysis during cooking to yield nicotinamide. The use of baking powder liberates much of the bound vitamin from cereal flours during baking.

Tryptophan

From experiments with different food model systems, Nielsen *et al.* (1985) concluded that the tryptophan residues in food proteins are relatively stable during processing and storage. They were not easily oxidized and were relatively resistant to reactions with oxidizing lipids, alkali, quinones and reducing sugars. Additional studies conducted by de Weck,

Nielsen and Finot (1987) showed that, although free tryptophan is readily oxidized in the presence of 0.1 M hydrogen peroxide at 50 °C, protein-bound tryptophan is comparatively stable. Peroxide-treated casein produced no significant decrease in tryptophan bioavailability when fed to rats in a growth assay. Thus, in normal methods of food manufacture and storage, appreciable loss of tryptophan by oxidation is not very likely and any such loss will have little nutritional significance. Only severe heat treatments in the presence of oxygen cause a significant degradation of protein-bound tryptophan (Cuq, Vié and Cheftel, 1983).

Tryptophan residues in proteins are susceptible to UV radiation, which leads to the formation of off-flavours in milk exposed to sunlight (Friedman and Cuq, 1988).

9.7 DIETARY INTAKE

Requirements for niacin are related to energy intake because of the involvement of NAD and NADP as coenzymes in the oxidative release of energy from food. Estimation of niacin requirement is complicated by the conversion of tryptophan to the vitamin. The efficiency of the conversion is affected by a variety of influences, including the amounts of tryptophan and niacin ingested, protein and energy intake, hormonal status, and vitamin B_6 and riboflavin nutriture. In humans, c.60 mg of L-tryptophan yield 1 mg of niacin. One niacin equivalent (NE) is therefore equal to either 1 mg of available niacin or 60 mg of L-tryptophan (FAO/WHO, 1967). A notable exception to the 60 : 1 conversion ratio is the state of pregnancy, in which the conversion is about twice as efficient. This increased conversion is presumably due to the stimulation by oestrogen of tryptophan oxygenase, which is a rate-limiting enzyme in the biosynthetic pathway. Conversion is also increased when contraceptive pills are used.

All the information used in estimating niacin requirements for humans comes from studies conducted in the United States during the 1950s on adult men and women (National Research Council, 1989). In terms of energy intake, the allowance for adults of both sexes and all ages is 6.6 NE/1000 kcal (4200 kJ). As a safeguard it is also suggested that a minimum daily intake of 13 NE be maintained regardless of energy intake. An additional 2 NE/day to the RDA is recommended for pregnant women in the United States but not in the UK, where enhanced tryptophan conversion during pregnancy is deemed sufficient. Both countries recommend an additional allowance during lactation.

When estimating dietary intakes of niacin, the following approximations of tryptophan content may be used: maize products, 0.6%; other grains, fruits and vegetables, 1.0%; meats, 1.1%; milk, 1.4%; and eggs, 1.5% of the protein in each food. In the average US diet, 65% of the protein comes from meat, milk and eggs. The calculated average daily intakes of niacin in the United States are 41 mg NE for men and 27 mg NE for women, which are well above the respective RDAs of 15–20 and 13–15 mg NE (National Research Council, 1989).

There are no data on the niacin requirements of children from infancy through adolescence. Human milk from a well nourished mother supplies about 7 NE/1000 kcal, which provides the basis for the RDA of 8 NE/1000 kcal for formula-fed infants up to 6 months of age. The niacin allowances for children more than 6 months of age are based on the same standard as for adults, i.e. 6.6 NE/1000 kcal.

Effects of high intake

Nicotinic acid administered orally at doses as low as 100 mg/day causes peripheral vasodilatation, with the appearance of skin flushing. In high doses, nicotinic acid competes with uric acid for excretion, leading to an increase in the incidence of gouty arthritis. Of greatest concern is possible liver damage, and in one report severe jaundice occurred at doses of 750 mg/day for only 3 months. Nicotinamide does not cause vasodilatation, but is otherwise two to three times as toxic as the acid (Miller and Hayes, 1982; Alhadeff, Gualtieri and Lipton, 1984).

Assessment of nutritional status

Biochemical methods for evaluating niacin status are not well established and no reliable blood test has been demonstrated. The most widely used test is measurement of the urinary excretion of N^1-methylnicotinamide by fluorimetry or HPLC (Bender and Bender, 1986). A value (mg/g creatinine) of < 0.5 indicates niacin deficiency; 0.5–1.59 indicates marginal deficiency; and > 1.6 indicates adequate status.

9.8 SYNOPSIS

The vitamin activity of niacin is provided by nicotinic acid and nicotinamide, the latter being the reactive moiety of the coenzymes NAD and NADP. Humans are able to synthesize niacin from tryptophan and, for an adult in nitrogen balance, a typical protein intake can provide enough tryptophan to meet niacin requirements.

Nicotinic acid and nicotinamide are both stable and either may be used in the fortification of foods. In meat, a rich source of niacin, the vitamin is present mainly in the form of free nicotinamide due to *post mortem* degradation of NAD. In mature cereal grains, most of the total niacin content exists as chemically bound forms of nicotinic acid, which are nutritionally unavailable. However, the niacin in such grains can be made available by treating the grain with alkali before or during cooking as a means of releasing the nicotinic acid. Immature cereal grains such as sweetcorn contain mainly NAD as a source of niacin, and this is degraded to nicotinamide after conventional cooking at neutral pH Pulses and other noncereal plant foods contain mainly free niacin.

The determination of niacin is complicated by the various forms of the vitamin that occur in foods. It has been customary to determine the total niacin (i.e. free plus bound) content of the food sample by chemical or microbiological assay using an alkaline or strong acid extraction procedure. However, this approach provides a gross overestimate of available niacin for several staple cereal-based foods. A better prediction of available niacin is to measure just the free niacin, yet this may result in a degree of undervaluation because of the apparent partial utilization of bound nicotinic acid. Nevertheless, it is prudent to determine free niacin for cereal-based foods as a means of ensuring an adequate level of enrichment.

The niacin content of foods is usually expressed as niacin equivalents. Apart from using an animal bioassay, an estimation of the niacin equivalent content would necessitate the determination of L-tryptophan in a separate assay. This is best achieved by means of HPLC.

REFERENCES

Alhadeff, L., Gualtieri, T. and Lipton, M. (1984) Toxic effects of water-soluble vitamins. *Nutr. Rev.*, **42**, 33–40.
Alpers, D.H. (1987) Digestion and absorption of carbohydrates and proteins. In *Physiology of the Gastrointestinal Tract*, 2nd edn (ed. L.R. Johnson), Raven Press, New York, pp. 1469–87.
AOAC (1990a) Niacin and niacinamide in drugs, foods and feeds. Colorimetric method. Final action 1962. In *AOAC Official Methods of Analysis*, 15th edn (ed. K. Helrich), Association of Official Analytical Chemists, Inc., Arlington, VA, 961.14.
AOAC (1990b) Niacin and niacinamide in cereal products. Automated method. Final action 1976. In *AOAC Official Methods of Analysis*, 15th edn (ed. K. Helrich), Association of Official Analytical Chemists, Inc., Arlington, VA, 975.41.
AOAC (1990c) Niacin and niacinamide in foods, drugs and feeds. Automated method. Final action 1982. In *AOAC Official Methods of Analysis*, 15th edn (ed. K. Helrich), Association of Official Analytical Chemists, Inc., Arlington, VA, 981.16.
Association of Vitamin Chemists, Inc. (1966) *Methods of Vitamin Assay*, 3rd edn, Interscience Publishers, New York.
Barton-Wright, E.C. (1945) The theory and practice of the microbiological assay of the vitamin B-complex; together with the assay of selected amino acids and potassium. *Analyst, Lond.*, **70**, 283–94.
Barton-Wright, E.C. (1952) *The Microbiological Assay of the Vitamin B-Complex and Amino Acids*, Sir Isaac Pitman & Sons, Ltd, London.
Bech-Andersen, S. (1991) Determination of tryptophan with HPLC after alkaline hydrolysis in autoclave using α-methyl-tryptophan as internal standard. *Acta Agric. Scand.*, **41**, 305–9.

Bender, A.E. (1978) *Food Processing and Nutrition*, Academic Press, London.

Bender, A.E. (1993) Protein quality. In *Encyclopaedia of Food Science, Food Technology and Nutrition*, Vol. 6 (eds R. Macrae, R.K. Robinson and M.J. Sadler), Academic Press, London, pp. 3820–24.

Bender, A.E. and Mohammadiha, H. (1981) Low digestibility of legume nitrogen. *Proc. Nutr. Soc.*, **40**, 66A.

Bender, D.A. and Bender, A.E. (1986) Niacin and tryptophan metabolism: the biochemical basis of niacin requirements and recommendations. *Nutr. Abstr. Rev. (Ser. A)*, **56**, 695–719.

Bodwell, C.E., Satterlee, L.D. and Hackler, L.R. (1980) Protein digestibility of the same protein preparations by human and rat assays and by in vitro enzymic digestion methods. *Am. J. Clin. Nutr.*, **33**, 677–86.

Carpenter, K.J., Schelstraete, M., Vilicich, V.C. and Wall, J.S. (1988) Immature corn as a source of niacin for rats. *J. Nutr.*, **118**, 165–9.

Carter, E.G.A. and Carpenter, K.J. (1981) Bound niacin in sorghum and its availability. *Nutr. Res.*, **1**, 571–9.

Carter, E.G.A. and Carpenter, K.J. (1982) The available niacin values of foods for rats and their relation to analytical values. *J. Nutr.*, **112**, 2091–103.

Chase, G.W. Jr, Landen, W.O. Jr, Soliman, A.-G.M. and Eitenmiller, R.R. (1993) Liquid chromatographic analysis of niacin in fortified food products. *J. AOAC Int.*, **76**, 390–3.

Chaykin, S. (1967) Nicotinamide coenzymes. *Annu. Rev. Biochem.*, **36**, 149–70.

Clegg, K.M. (1963) Bound nicotinic acid in dietary wheaten products. *Br. J. Nutr.*, **17**, 325–9.

Cohen, L.L. and Huang, K.C. (1964) Intestinal transport of tryptophan and its derivatives. *Am. J. Physiol.*, **206**, 647–52.

Cuq, J.C., Vié, M. and Cheftel, J.C. (1983) Tryptophan degradation during heat treatments: Part 2 – degradation of protein-bound tryptophan. *Food Chem.*, **12**, 73–88.

Das, M.L. and Guha, B.C. (1960) Isolation and chemical characterization of bound niacin (niacinogen) in cereal grains. *J. Biol. Chem.*, **235**, 2971–6

Delhaye, S. and Landry, J. (1993) Quantitative determination of tryptophan in food and feedstuffs: practical considerations on autoclaving samples for hydrolysis. *J. Agric Food Chem.*, **41**, 1633–4.

de Wek, D., Nielsen, H.K. and Finot, P.-A. (1987) Oxidation rate of free and protein-bound tryptophan by hydrogen peroxide and the bioavailability of the oxidation products. *J. Sci. Food Agric.*, **41**, 179–85.

Duggan, D.E., Bowman, R.L., Brodie, B.B. and Udenfriend, S. (1957) A spectrophotofluorometric study of compounds of biological interest. *Arch. Biochem. Biophys.*, **68**, 1–14.

Egberg, D.C. (1979) Semiautomated method for niacin and niacinamide in food products: collaborative study. *J. Ass. Off. Analyt. Chem.* **62**, 1027–30.

Egberg, D.C., Potter, R.H. and Honold, G.R. (1974) The semiautomated determination of niacin and niacinamide in food products. *J. Agric. Food Chem.*, **22**, 323–6.

Elwell, D. and Soares, J.H. Jr (1975) Amino acid bioavailability: a comparative evaluation of several assay techniques. *Poultry Sci.*, **54**, 78–85.

Emery, P.W. (1993) Digestion and absorption of protein and nitrogen balance. In *Encyclopaedia of Food Science, Food Technology and Nutrition*, Vol. 6 (eds R. Macrae, R.K. Robinson and M.J. Sadler), Academic Press, London, pp. 3824–7.

Fairweather-Tait, S.J., Gee, J.M. and Johnson, I.T. (1983) The influence of cooked kidney beans (*Phaseolus vulgaris*) on intestinal cell turnover and faecal nitrogen excretion in the rat. *Br. J. Nutr.*, **49**, 303–12.

FAO/WHO (1967) Requirements of vitamin A, thiamine, riboflavine and niacin. *World Health Organisation Technical Report Series No. 362*, World Health Organisation, Geneva.

Ford, J.E. (1962) A microbiological method for assessing the nutritional value of proteins. 2. The measurement of 'available' methionine, leucine, isoleucine, arginine, histidine, tryptophan and valine. *Br. J. Nutr.*, **16**, 409–25.

Ford, J.E. (1964) A microbiological method for assessing the nutritional value of proteins. 3. Further studies on the measurement of available amino acids. *Br. J. Nutr.*, **18**, 449–60 (and plate 1 facing p. 460).

Friedman, M. and Cuq, J.-L. (1988) Chemistry, analysis, nutritional value, and toxicology of tryptophan in food. A review. *J. Agric. Food Chem.*, **36**, 1079–93.

Garcia, S.E. and Baxter, J.H. (1992) Determination of tryptophan content in infant formulas and medical nutritionals. *J. AOAC Int.*, **75**, 1112–9.

Ge, H., Oman, G.N. and Ebert, F.J. (1986) On-line generation of cyanogen chloride in semiautomated determination of niacin and niacinamide in food products. *J. Ass. Off. Analyt. Chem.*, **69**, 560–2.

Gehrke, C.W., Wall, L.L. Sr, Absheer, J.S. *et al.* (1985) Sample preparation for chromatography of amino acids: acid hydrolysis of proteins. *J. Ass. Off. Analyt. Chem.*, **68**, 811–21.

Ghosh, H.P., Sarkar, P.K. and Guha, B.C. (1963) Distribution of the bound form of nicotinic acid in natural materials. *J. Nutr.*, **79**, 451–3.

Gorin, N. and Schütz, G.P. (1970) Comparison of a microbiological and a spectrophotometric method for the determination of nicotinic acid in fresh meat. *J. Sci. Food Agric.*, **21**, 423–5.

Gregory, J.F. III (1985) Chemical changes of vitamins during food processing. In *Chemical Changes in Food during Processing* (eds T. Richardson and J.W. Finley), Van Nostrand Reinhold Co., New York, pp. 373–408.

Gross, A.F. (1975) Automated method for the determination of niacin and niacinamide in cereal products: collaborative study. *J. Ass. Off. Analyt. Chem.*, **58**, 799–803.

Gross, C.J. and Henderson, L.M. (1983) Digestion and absorption of NAD by the small intestine of the rat. *J. Nutr.*, **113**, 412–20.

Guilarte, T.R. and Pravlik, K. (1983) Radiometric-microbiologic assay of niacin using *Kloeckera brevis*: analysis of human blood and food. *J. Nutr.*, **113**, 2587–94.

Guilbert, C.C. and Johnson, S.L. (1977) Investigation of the open ring form of nicotinamide adenine dinucleotide. *Biochemistry*, **16**, 335–44.

Hagen, S.R. and Augustin, J. (1989) Determination of tryptophan in foods by isocratic reversed-phase high-performance liquid chromatography. *J. Micronutr. Anal.*, **5**, 303–9.

Hamano, T., Mitsuhashi, Y., Aoki, N. *et al.* (1988) Simultaneous determination of niacin and niacinamide in meats by high-performance liquid chromatography. *J. Chromat.*, **457**, 403–8.

Hepburn, F.N. (1971) Nutrient composition of selected wheats and wheat products. VII. Total and free niacin. *Cereal Chem.*, **48**, 369–72.

Hewitt, D. and Ford, J.E. (1985) Nutritional availability of methionine, lysine and tryptophan in fish meals, as assessed with biological, microbiological and dye-binding assay procedures. *Br. J. Nutr.*, **53**, 575–86.

Hoffmann–La Roche Inc. (1992) Vitamins, Part XV: Niacin. In *Encyclopedia of Food Science and Technology*, Vol. 4 (ed. Y.H. Hui), John Wiley & Sons, Inc., New York, pp. 2776–83.

Holland, B., Welch, A.A., Unwin, I.D. *et al.* (1991) *McCance and Widdowson's The Composition of Foods*, 5th edn, Royal Society of Chemistry and Ministry of Agriculture, Fisheries and Food.

Hopfer, U. (1987) Membrane transport mechanisms for hexoses and amino acids in the small intestine. In *Physiology of the Gastrointestinal Tract*, 2nd edn (ed. L.R. Johnson), Raven Press, New York, pp. 1499–1526.

Hopkins, D.T. (1981) Effects of variation in protein digestibility. In *Protein Quality in Humans: Assessment and In Vitro Estimation* (eds C.E. Bodwell, J.S. Adkins and D.T. Hopkins), AVI Publishing Co. Inc., Westport, CT, pp. 169–193.

Hugli, T.E. and Moore, S. (1972) Determination of the tryptophan content of proteins by ion exchange chromatography of alkaline hydrolysates. *J. Biol. Chem.*, **247**, 2828–34.

Jones, A.D., Hitchcock, C.H.S. and Jones, G.H. (1981) Determination of tryptophan in feeds and feed ingredients by high-performance liquid chromatography. *Analyst*, **106**, 968–73.

Kies, C. (1981) Bioavailability: a factor in protein quality. *J. Agric. Food Chem.*, **29**, 435–40.

Kodicek, E. and Pepper, C.R. (1948a) A critical study of factors influencing the microbiological assay of nicotinic acid. *J. Gen. Microbiol.*, **2**, 292–305.

Kodicek, E. and Pepper, C.R. (1948b) The microbiological estimation of nicotinic acid and comparison with a chemical method. *J. Gen. Microbiol.*, **2**, 306–14.

Kodicek, E. and Wilson, P.W. (1960) The isolation of niacytin, the bound form of nicotinic acid. *Biochem. J.*, **76**, 27P.

Kodicek, E., Braude, R., Kon, S.K. and Mitchell, K.G. (1959) The availability to pigs of nicotinic acid in *tortilla* baked from maize treated with lime-water. *Br. J. Nutr.*, **13**, 363–84.

Koetz, R. and Neukom, H. (1977) Nature of bound nicotinic acid in cereals and its release by thermal and chemical treatment. In *Physical, Chemical and Biological Changes in Food caused by Thermal Processing* (eds T. Høyem and O. Kvåle), Applied Science Publishers Ltd, London, pp. 305–10.

Kral, K. (1983) Determination of nicotinic acid in fruit juices by HPLC with amperometric detection at a SMDA. *Z. Analyt. Chem.*, **314**, 479–82 (in German).

Krehl, W.A. and Strong, F.M. (1944) Studies on the distribution, properties, and isolation of a naturally occurring precursor of nicotinic acid. *J. Biol. Chem.*, **156**, 1–12.

Landry, J. and Delhaye, S. (1992) Simplified procedure for the determination of tryptophan of foods and feedstuffs from barytic hydrolysis. *J. Agric. Food Chem.*, **40**, 776–9.

Landry, J. and Delhaye, S. (1994) Determination of tryptophan in feedstuffs: comparison of sodium hydroxide and barium hydroxide as hydrolysis agents. *Food Chem.*, **49**, 95–7.

Landry, J., Delhaye, S. and Viroben, G. (1988) Tryptophan content of feedstuffs as determined from three procedures using chromatography of barytic hydrolysates. *J. Agric. Food Chem.*, **36**, 51–2.

Magboul, B.I. and Bender, D.A. (1982) The nature of niacin in sorghum. *Proc. Nutr. Soc.*, **41**, 50A.

Mason, J.B., Gibson, N. and Kodicek, E. (1973) The chemical nature of the bound nicotinic acid of wheat bran: studies of nicotinic acid-containing macromolecules. *Br. J. Nutr.*, **30**, 297–311.

Mason, J.B. and Kodicek, E. (1973) The chemical nature of the bound nicotinic acid of wheat bran: studies of partial hydrolysis products. *Cereal Chem.*, **50**, 637–46.

McCormick, D.B. (1988) Niacin. In *Modern Nutrition in Health and Disease*, 7th edn (eds M.E. Shils and V.R. Young), Lea & Febiger, Philadelphia, pp. 320–75.

McDonough, F.E., Bodwell, C.E., Wells, P.A. and Kamalu, J.A. (1989) Bioavailability of tryptophan in selected foods by rat growth assay. *Plant Foods Human Nutr.*, **39**, 85–91.

Meade, R.J. (1972) Biological availability of amino acids. *J. Animal Sci.*, **35**, 713–23.

Melnick, D. (1942) Collaborative study of the applicability of microbiological and chemical methods to the determination of niacin (nicotinic acid) in cereal products. *Cereal Chem.*, **19**, 550 67.

Merck & Co., Inc. (1983) *Merck Index*, 10th edn (ed. M. Windholz), Merck & Co., Inc., Rahway, NJ.

Miller, D.R. and Hayes, K.C. (1982) Vitamin excess and toxicity. In *Nutritional Toxicology*, Vol. I (ed. J.N. Hathcock), Academic Press, New York, pp. 81–133.

Miller, E.L. (1967) Determination of the tryptophan content of feedingstuffs with particular reference to cereals. *J. Sci. Food Agric.*, **18**, 381–6.

Munck, B.G. and Rasmussen, S.N. (1975) Characteristics of rat jejunal transport of tryptophan. *Biochim. Biophys. Acta*, **389**, 261–80.

National Research Council (1989) Water-soluble vitamins. In *Recommended Dietary Allowances*, 10th edn, National Academy Press, Washington, DC, pp. 115–73.

Nielsen, H.K. and Hurrell, R.F. (1985) Tryptophan determination of food proteins by h.p.l.c. after alkaline hydrolysis. *J. Sci. Food Agric.*, **36**, 893–907.

Nielsen, H.K., Klein, A. and Hurrell, R.F. (1985) Stability of tryptophan during food processing and storage. 2. A comparison of methods used for the measurement of tryptophan losses in processed foods. *Br. J. Nutr.*, **53**, 293–300.

Nielsen, H.K., de Weck, D., Finot, P.A. *et al.* (1985) Stability of tryptophan during food processing and storage. 1. Comparative losses of tryptophan, lysine and methionine in different model systems. *Br. J. Nutr.*, **53**, 281–92.

Oishi, M., Amakawa, E., Ogiwara, T. *et al.* (1988) Determination of nicotinic acid and nicotinamide in meats by high performance liquid chromatography and conversion of nicotinamide in meats during storage. *J. Food Hygiene Soc. Japan*, **29**, 32–7 (in Japanese).

Rees, D.I. (1989) Determination of nicotinamide and pyridoxine in fortified food products by HPLC. *J. Micronutr. Anal.*, **5**, 53–61.

Rogers, S.R. and Pesti, G.M. (1990) Determination of tryptophan from feedstuffs using reverse phase high-performance liquid chromatography. *J. Micronutr. Anal.*, **7**, 27–35.

Roy, R.B. and Merten, J.J. (1983) Evaluation of urea-acid system as medium of extraction for the B-group vitamins. Part II. Simplified semi-automated chemical analysis for niacin and niacinamide in cereal products. *J. Ass. Off. Analyt. Chem.*, **66**, 291–6.

Rupnow, J.H. (1992) Proteins: biochemistry and applications. In *Encyclopedia of Food Science and Technology*, Vol. 3 (ed. Y.H. Hui), John Wiley & Sons, Inc., New York, pp. 2182–91.

Sandaradura, S.S. and Bender, A.E. (1984) The effect of cooked legumes on mucosal cell turnover in the rat. *Proc. Nutr. Soc.*, **43**, 89A.

Sarwar, G. (1987) Digestibility of protein and bioavailability of amino acids in foods. *Wld Rev. Nutr. Diet.*, **54**, 26–70.

Sato, H., Seino, T., Kobayashi, T. *et al.* (1984) Determination of the tryptophan content of feed and feedstuffs by ion exchange liquid chromatography. *Agric. Biol. Chem.*, **48**, 2961–9.

Schuette, S.A. and Rose, R.C. (1983) Nicotinamide uptake and metabolism by chick intestine. *Am. J. Physiol.*, **245**, G531–8.

Schuette, S. and Rose, R.C. (1986) Renal transport and metabolism of nicotinic acid. *Am J. Physiol.*, **250**, C694–C703.

Shurpalekar, K.S., Sundaravalli, O.E. and Narayana Rao, M. (1979) Effect of legume carbohydrates on protein utilization and lipid levels in rats. *Nutr. Rep. Int.*, **19**, 119–24.

Snell, E.E. and Wright, L.D. (1941) A microbiological method for the determination of nicotinic acid. *J. Biol. Chem.*, **139**, 675–86.

Spackman, D.H., Stein, W.H. and Moore, S. (1958) Automatic recording apparatus for use in the chromatography of amino acids. *Analyt. Chem.*, **30**, 1190–1206.

Steele, R.D. and Harper, A.E. (1990) Proteins and amino acids. In *Present Knowledge in Nutrition*, 6th edn (ed. M.L. Brown), International Life Sciences Institute Nutrition Foundation, Washington, DC, pp. 67–79.

Stein, J., Daniel, H., Whang, E. *et al.* (1994) Rapid postabsorptive metabolism of nicotinic acid in rat small intestine may affect transport by metabolic trapping. *J. Nutr.*, **124**, 61–66.

Stevens, B.R., Kaunitz, J.D. and Wright, E.M. (1984) Intestinal transport of amino acids and sugars: advances using membrane vesicles. *Ann. Rev. Physiol.*, **46**, 417–33.

Strohecker, R. and Henning, H.M. (1966) *Vitamin Assay – Tested Methods*, Verlag Chemie, Weinheim.

Sweeney, J.P. and Parrish, W.P. (1954) Report on the extraction of nicotinic acid from naturally-occurring materials. *J. Ass. Off. Agric. Chem.*, **37**, 771–7.

Takatsuki, K., Suzuki, S., Sato, M. *et al.* (1987) Liquid chromatographic determination of free and added niacin and niacinamide in beef and pork. *J. Ass. Off. Analyt. Chem.*, **70**, 698–702.

Tanaka, A., Iijima, M., Kikuchi, Y. *et al.* (1989) Gas chromatographic determination of nicotinamide in meats and meat products as 3-cyanopyridine. *J. Chromat.*, **466**, 307–17.

Tannenbaum, S.R., Young, V.R. and Archer, M.C. (1985) Vitamins and minerals. In *Food Chemistry*, 2nd edn (ed. O.R. Fennema), Marcel Dekker, Inc., New York, pp. 477–544.

Trugo, L.C., Macrae, R. and Trugo, N.M.F. (1985) Determination of nicotinic acid in instant coffee using high-performance liquid chromatography. *J. Micronutr. Anal.*, **1**, 55–63.

Tsunoda, K., Inoue, N., Iwasaki, H. *et al.* (1988) Rapid simultaneous analysis of nicotinic acid and nicotinamide in foods, and their behaviour during storage. *J. Food Hyg. Soc. Japan*, **29**, 262–6 (in Japanese).

Tyler, T.A. and Genzale, J.A. (1990) Liquid chromatographic determination of total niacin in beef, semolina, and cottage cheese. *J. Ass. Off. Analyt. Chem.*, **73**, 467–9.

van Gend, H.W. (1973) An automated colorimetric method for the determination of free nicotinic acid in minced meat. *Z. Lebensmittelunters. u.-Forsch.*, **153**, 73–7.

van Niekerk, P.J., Smit, S.C.C., Strydom, E.S.P. and Armbruster, G. (1984) Comparison of a high-performance liquid chromatographic and microbiological method for the determination of niacin in foods. *J. Agric. Food Chem.*, **32**, 304–7.

Vidal-Valverde, C. and Reche, A. (1991) Determination of available niacin in legumes and meat by high-performance liquid chromatography. *J. Agric. Food Chem.*, **39**, 116–21.

Wall, J.S. and Carpenter, K.J. (1988) Variation in availability of niacin in grain products. *Food Technol.*, **42**(10), 198–202, 204.

Wall, J.S., Young, M.R. and Carpenter, K.J. (1987) Transformation of niacin-containing compounds in corn during grain development: relationship to niacin nutritional availability. *J. Agric. Food Chem.*, **35**, 752–8.

Weiner, M. and van Eys, J. (1983) *Nicotinic Acid. Nutrient–Cofactor–Drug*, Marcel Dekker, Inc., New York.

Wells, P., McDonough, F., Bodwell, C.E. and Hitchens, A. (1989) The use of *Streptococcus zymogenes* for estimating tryptophan and methionine bioavailability in 17 foods. *Plant Foods Human Nutr.*, **39**, 121–7.

10

Vitamin B₆

10.1 INTRODUCTION

In 1934 György observed the appearance of a scaly dermatitis (acrodynia) in rats fed on diets free from the whole vitamin B complex and supplemented with thiamin and riboflavin. This observation led to the establishment of a 'rat acrodynia-preventative factor' and its designation as vitamin B₆. The isolation of the pure crystalline vitamin was first reported by Lepkovsky in 1938, and the synthesis of pyridoxine was accomplished by Harris and Folkers in the following year. Discovery of the existence of pyridoxal and pyridoxamine and the recognition of their phosphorylated forms as coenzymes is largely credited to Esmond E. Snell during 1944–1948.

Vitamin B₆ functions as a coenzyme for many enzymes involved in amino acid metabolism, including the biosynthesis of niacin from tryptophan. It is also the essential coenzyme for glycogen phosphorylase, the enzyme responsible for the utilization of muscle and liver glycogen reserves. Biogenic amines such as dopamine, serotonin, histamine and γ-aminobutyric acid, which are implicated as neurotransmitters, are either synthesized or metabolized with the aid of vitamin B_6-dependent enzyme reactions. Vitamin B_6 is involved in the synthesis of sphingosine, and a deficiency of the vitamin leads to impaired development of brain lipids and incomplete myelination of nerve fibres in the central nervous system (Dakshinamurti, Paulose and Siow, 1985).

10.2 CHEMICAL STRUCTURE AND NOMENCLATURE

Vitamin B_6 is the generic descriptor for all 3-hydroxy-2-methylpyridine derivatives which exhibit qualitatively in rats the biological activity of pyridoxine (Snell, 1986). Six B_6 vitamers are known, namely pyridoxine or pyridoxol (PN), pyridoxal (PL) and pyridoxamine (PM), which possess, respectively, alcohol, aldehyde and amine groups in the 4-position; their 5'-phosphate esters are designated as PNP, PLP and PMP (Figure 10.1). Pyridoxine is systematically named as 3-hydroxy-4,5-bis(hydroxy-methyl)-2-methylpyridine and is available commercially as its hydrochloride salt, PN.HCl ($C_8H_{11}O_3$ N.HCl; MW = 205.65). PN.HCl is the only form of vitamin B_6 used in the fortification of foods.

In its role as a coenzyme, PLP is bound tightly to the apoenzyme by a Schiff base (aldimine) linkage formed through condensation of the

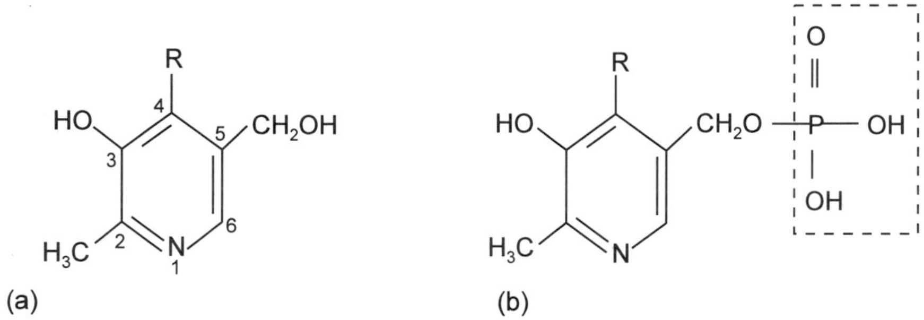

R	Vitamers
CH₂OH	PN, PNP
CHO	PL, PLP
CH₂NH₂	PM, PMP

Figure 10.1 Structures of vitamin B_6 compounds showing (a) nonphosphorylated and (b) phosphorylated forms.

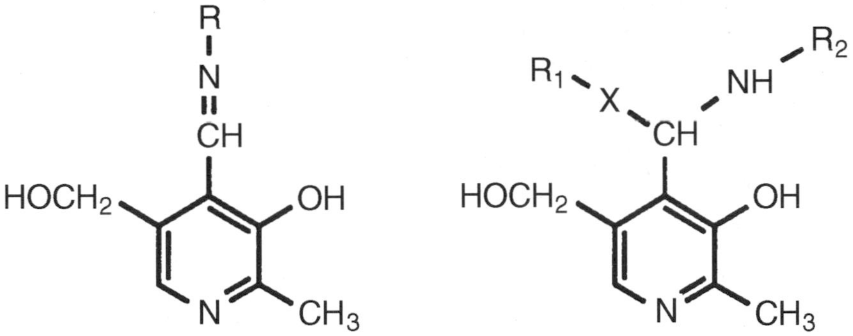

Figure 10.2 Potential interactions involving pyridoxal and amino groups of food proteins. (a) Schiff base; (b) substituted aldamine (X = amino, sulphydryl or imidazole). Analogous reactions would occur with pyridoxal phosphate and proteins.

4-carbonyl group with the ε-amino group of specific lysine residues (Morino and Nagashima, 1986). The resultant Schiff base compound may be subject to nucleophilic attack by a neighbouring amino, sulphydryl or imidazole group to form a substituted aldamine (Figure 10.2) (Gregory and Kirk, 1977).

A ubiquitous bound form of PN that occurs in plant tissues is a glucoside conjugate, 5'-O-(β-D-glucopyranosyl) pyridoxine (Figure 10.3), designated in this text as PN-glucoside (Gregory, 1988). Two minor derivatives in which an organic acid is esterified to the C-6 position of the glucosyl moiety of PN-glucoside have been identified in legume seedlings (Gregory, 1988). A more complex derivative of PN-glucoside containing cellobiose and 5-hydroxydioxindole-3-acetic acid moieties has been identified as a major form of vitamin B_6 in rice bran and a minor form in wheat bran and legumes (Tadera and Orite, 1991).

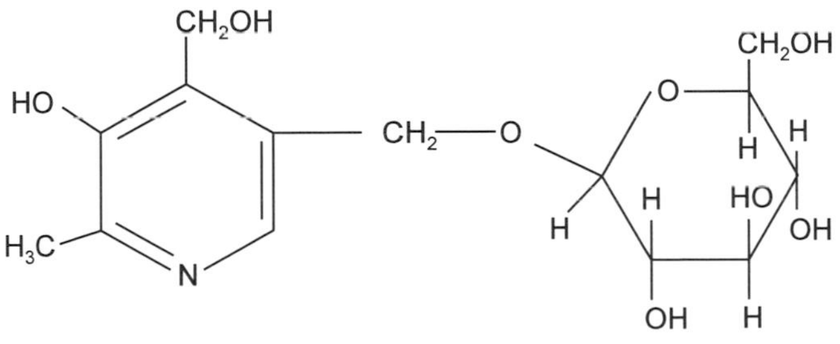

Figure 10.3 Structure of 5'-O-(β-D-glucopyranosyl) pyridoxine, the primary form of glycosylated vitamin B_6 in foods.

10.3 DEFICIENCY SYNDROMES

Vitamin B$_6$ is widely distributed in foods, and any diet so poor as to be insufficient in this vitamin would most likely lack adequate amounts of other B-group vitamins. For this reason, a primary clinical deficiency of B$_6$ in the adult human is rarely encountered. In the 1950s, an occurrence of convulsions in infants was traced to an unfortified liquid milk-based canned formula that had undergone autoclaving in manufacture (Coursin, 1954). There is some evidence that the convulsions were caused by an insufficient production of γ-aminobutyric acid, the major inhibitory neurotransmitter in the brain (Ebadi, 1978). The administration of the antagonist deoxypyridoxine to adult volunteers receiving diets low in vitamin B$_6$ resulted in lesions of the skin and mouth that resembled those of riboflavin and niacin deficiency. These symptoms responded to B$_6$ therapy, but did not respond to thiamin, riboflavin or niacin (National Research Council, 1980).

10.4 PHYSICOCHEMICAL PROPERTIES

Solubility

PN.HCl is a white, odourless, crystalline powder with a salty taste and a melting point of 204–206 °C (with decomposition). It is readily soluble in water (1 g/5 ml), sparingly soluble in ethanol (1 g/100 ml) and very slightly soluble in diethyl ether and chloroform. The pH of a 5% aqueous solution is 2.3–3.5; pK_a values are 5.0 and 9.0 (25 °C). The free base is readily soluble in water and slightly soluble in acetone, chloroform and diethyl ether. In aqueous solutions the B$_6$ vitamers exist in a variety of equilibrium forms, depending upon the pH (Snell, 1958).

Spectroscopic properties

Absorption
The absorption spectrum of PN.HCl is strongly dependent upon the pH of the solution. At pH 1.8 there is a single maximum at 290 nm which diminishes in intensity at pH 4.5, with the appearance of a new maximum at 324 nm. At pH 7.0 the 324 nm maximum increases in intensity, the 290 nm maximum disappears and a new maximum appears at 253 nm. The spectra of PL.HCl at corresponding pH values are similar to those of PN.HCl, the only obvious difference being at pH 7.0, where the maximum occurs at 316 nm. The $A_{1cm}^{1\%}$ values of acidified solutions (pH about 2) of PN.HCl and PL.HCl are, respectively, *c.*430 at λ_{max} 290.5 nm and 440 at λ_{max} 286.5 (Strohecker and Henning, 1966).

Fluorescence

PN, PL (hemiacetal) and PM all display native fluorescence whose intensity is pH-dependent. PM is the most highly fluorescent compound of the three. The phosphorylated forms are only weakly fluorescent.

Stability

Saidi and Warthesen (1983) observed that no significant degradation of PN.HCl took place when aqueous solutions protected from light were held at 40 °C and 60 °C for up to 140 days at pH levels ranging from 4 to 7. Under the same conditions, PM.2HCl showed a trend of increasing loss with increasing pH, while PL.HCl showed a marked loss at pH 5, but only a moderate loss above and below that pH value. Ang (1979) showed that PN.HCl was the most stable and PM.2HCl the least stable of the three vitamers after exposure of aqueous solutions to normal laboratory light at different pH values. Low-actinic amber glassware or yellow or gold fluorescent lighting protected the vitamers from photodegradation. The principal photodegradation product of PLP is 4-pyridoxic acid 5'-phosphate (Reiber, 1972).

Shephard and Labadarios (1986) investigated the degradation of vitamin B_6 standard solutions after experiencing difficulties in the reproducibility of B_6 vitamer standard determinations in an HPLC method. Of the crystalline vitamer standards used (PN.HCl, PM.2HCl, PL.HCl, PLP and PMP.HCl), all but PLP were stable when stored individually in the dark. Solutions of PLP in water were stable when stored frozen (-20 °C) at a concentration of 1 mg/ml (pH 3.3). However, storage at room temperature in the dark for 24 h of a laboratory working solution (1 µg/ml) either in sodium acetate buffer (pH 5.5) or in distilled water (pH 6.1) resulted in a 20% or 95% loss of PLP, respectively, due to hydrolysis to PL. When prepared in 0.01 M HCl, PLP was stable for at least 2 days at room temperature. It was shown that Schiff base formation and transamination reactions can occur between the vitamers themselves at room temperature or below. In order to prevent these reactions, vitamer solutions must be stored separately and compound standards must be prepared before use in a fairly acid medium (pH 3.0).

10.5 ANALYSIS

10.5.1 Scope of analytical techniques

Vitamin B_6 exists as pyridoxine (PN), pyridoxal (PL) and pyridoxamine (PM), together with their 5'-phosphorylated derivatives, PNP, PLP and PMP.PNP does not occur to any significant extent in natural products. PN.HCl is used in the fortification of foods. The Schiff base forms of PLP and PL that predominate in foods of animal origin are presumed to be totally available. PN and PNP are virtually absent in animal tissues, one exception being liver tissue, in which they are detectable at very low levels. Plant-derived foods contain mostly PN, a significant proportion of which may be present as PN glucoside and/or other conjugated forms which appear to be largely unavailable to humans.

The total available vitamin B_6 activity of a food or diet containing all forms of the vitamin is measured most conclusively by an animal assay. By feeding the material directly, both the free and bound forms can exert their combined effect. The free vitamers are equally active mole for mole when fed to animals as separate supplements in solution. However, when mixed in the ration, PM and PL are less active than PN (Bliss and György, 1967). Both the rat and the chick are sensitive to influences of diet composition, especially with respect to fermentable carbohydrate which provides the means for vitamin B_6 production by microflora in the large intestine. The utilization of microbially produced vitamin B_6 via coprophagy or direct intestinal absorption can bias quantitative bioassays.

Total vitamin B_6 activity is usually estimated microbiologically using a turbidimetric yeast assay; the radiometric–microbiological assay is a more recent innovation. With the aid of HPLC, it is possible to measure all of the B_6 vitamers present, and also PN-glucoside, after nondestructive extraction. A good HPLC technique is capable of isolating the inactive metabolite 4-pyridoxic acid, which may be present in foods, from the B_6 vitamers. The various bound forms of vitamin B_6 can be assayed microbiologically or by HPLC using selective extraction procedures. Gregory (1988) has critically reviewed methods for determination of vitamin B_6 in foods and other biological materials.

No IU of vitamin B_6 activity has been defined, and analytical results are expressed in weight units (mg) of pure PN.HCl. All six vitamers are considered to have approximately equivalent biological activity in humans as a result of their enzymatic conversion to the major coenzyme form, PLP (Brin, 1978).

10.5.2 Extraction techniques

The aldehydic B_6 vitamers PL and PLP, which are reversibly bound to proteins as Schiff bases and substituted aldamines, and the PN-glucosides, which occur exclusively in plant tissues, are readily liberated by acid extraction with autoclaving (Vanderslice *et al.*, 1985). The conjugated form of vitamin B_6 identified in brans and legumes (Tadera and Orite, 1991) released PN only when hydrolysed with alkali and then with β-glucosidase (Tadera, Kaneko and Yagi, 1986).

Present techniques for extracting vitamin B_6 from food matrices are based largely on the investigations reported by Atkin *et al.* (1943) and Rabinowitz and Snell (1947a). Both research groups found that autoclaving samples with 0.055 N HCl for up to 5 h gave an efficient extraction for a wide range of foods, but not for cereals, which required a higher concentration of acid. This requirement was correctly attributed to the presence of an unidentified bound form of vitamin B_6, now known to be PN-glucoside. The efficiency of extraction depended not only on the strength of the acid, but also on the volume of acid used. Another factor to consider is that PL is liberated from PLP much more rapidly than PM is liberated from PMP (Rabinowitz and Snell, 1947b). The possibility of interaction between PL or PLP with amino acids during the extraction procedure of Rabinowitz and Snell (autoclaving with 0.055 N HCl at 20 lb pressure for 5 h) was investigated by Rabinowitz, Mondy and Snell (1948). No loss of activity for *S. carlsbergensis* was observed when PL or PLP was autoclaved in the presence of a relatively high concentration of glutamic acid, which indicated that transamination does not occur under these conditions.

Because animal and plant tissues differ greatly with respect to the forms of vitamin B_6 contained in them, there is no single set of conditions that can quantitatively extract vitamin B_6 from both plant and animal products. In the AOAC microbiological method (AOAC, 1990), animal-derived foods are autoclaved with 0.055 N HCl for 5 h at 121 °C. This treatment hydrolyses phosphorylated forms of vitamin B_6, whilst also liberating PL from its Schiff base and substituted aldamine bound forms. Plant-derived foods are autoclaved with 0.44 N HCl for 2 h at 121 °C, the stronger acid environment being necessary to liberate PN from its glycosylated form. Polansky, Reynolds and Vanderslice (1985) proposed that food composites be autoclaved with 0.2 N HCl for 5 h at 121 °C. Polansky, Murphy and Toepfer (1964) reported that, for whole wheat samples, autoclaving with 0.055 N HCl, instead of 0.44 N HCl, gave a similar PL value, but lower values of PN and PM. Conversely, autoclaving meat products with 0.44 N HCl, instead of 0.055 N HCl, gave approximately the same PN and PL values, but only about half of the PM (Polansky and Toepfer, 1969a). The superiority of the lower concentration of acid for extraction of vitamin B_6 from animal products does not result from destruction of the vitamin by the stronger acid, since subsequent enzymatic hydrolysis of 2 N acid hydrolysates liberated as much B_6 as the direct enzyme treatment (Rubin and Scheiner, 1946). Rather, it is due to the incomplete liberation of the vitamin by the more concentrated acid; the optimum release occurs between pH 1.5 and 2.0, with a maximum at pH 1.7–1.8 (Rubin, Scheiner and Hirschberg, 1947). To satisfy these strict pH criteria, one must always ensure that the acid is added in amounts that exceed the buffering capacity of the sample. Analyses based on acid hydrolysis would tend to overestimate the biologically available vitamin B_6 in those plant-derived foods that contain significant quantities of β-glucoside conjugates.

The AOAC acid hydrolysis procedures have no effect upon the peptide-bound ε-pyridoxyllysine and its 5'-phosphate derivative, which are formed during the heat-sterilization of evaporated milk and other animal derived canned foods (section 10.0.3). These conjugates, which possess anti-vitamin B$_6$ activity under certain conditions, exhibit 75–80% stability when subjected to 6 N HCl at 105 °C for 48 h (Gregory, Ink and Sartain, 1986).

Recent advances in HPLC permit the simultaneous separation of all six B$_6$ vitamers, plus pyridoxic acid, when used in conjunction with an extraction procedure that preserves the integrity of the vitamers. Treatment of samples with deproteinizing agents such as metaphosphoric, perchloric, trichloroacetic or sulphosalicylic acid at ambient temperature readily hydrolyses Schiff bases, whilst preserving the phosphorylated vitamers. These acids do not hydrolyse PN-glucoside, and hence results obtained using these extractants provide a better estimate of available vitamin B$_6$ than results obtained using acid hydrolysis. The high efficiency of extraction using these acidic reagents is partly due to the conversion of the pyridine bases to quaternary ammonium salts, thereby increasing their solubility in water. Their use as extracting agents also prevents enzymatic interconversion of B$_6$ vitamers during homogenization of samples. In such procedures it is usually necessary to remove excess reagent, which might otherwise interfere with the analytical chromatography. Trichloroacetic acid can be removed by extraction with diethyl ether; perchloric acid by reaction with 6 M KOH and precipitation as insoluble potassium perchlorate; and sulphosalicylic acid by chromatography on an anion exchange column. An extraction procedure using 5% sulphosalicylic acid has been successfully applied to such complex foods as pork, dry milk and cereals (Vanderslice *et al.*, 1980). Recoveries of B$_6$ vitamers added to samples were 95–105% for all vitamers except for PNP, where the recovery was 85%.

Free B$_6$ vitamers are extractable from foods by homogenization in 0.01 M sodium acetate buffer, pH 6.8 (Addo and Augustin, 1988). Selective hydrolysis of phosphorylated B$_6$ vitamers and PN-glucoside can be achieved enzymatically by incubation with acid phosphatase and β-glucosidase, respectively.

Kabir, Leklem and Miller (1983a) examined the vitamin B$_6$ value of 22 varied foods using the level of glycosylated vitamin B$_6$ as an index of bioavailability. Homogenized food samples were stirred in duplicate flasks with 0.1 M phosphate buffer (pH 6.8) for 2 h at room temperature, after which the pH was adjusted to 5.0 with 1 N HCl. To one of the duplicate flasks was added β-glucosidase; the other flask received no enzyme. All flasks were then incubated at 37 °C for 2 h with shaking. Enzyme action was stopped by addition of 1 N HCl to all flasks, followed by steaming for 5 min. The extracts, after pH adjustment, dilution and filtration, were assayed microbiologically using *S. carlsbergensis* as the assay organism. In addition, the total vitamin B_6 of each food was determined microbiologically following acid hydrolysis. The percentage of glycosylated vitamin B_6 in each sample was calculated as follows:

$$\% \text{ Glycosylated vitamin } B_6 = (G - F)/T \times 100$$

where $G = $ mg $B_6/$g sample treated with β-glucosidase, $F = $ mg $B_6/$g sample without enzyme, and $T = $ mg total $B_6/$g sample as determined after acid hydrolysis.

Gregory (1989) pointed out that the above procedure, although generally capable of providing accurate results, can seriously underestimate the concentration of glycosylated vitamin B_6 in raw plant tissues which contain significant β-glucosidase activity.

Ekanayake and Nelson (1986) devised an *in vitro* method for estimating available vitamin B_6 in processed model foods using a two-stage enzymatic digestion (pepsin/pancreatin) of the food matrix, followed by treatment of the hydrolysate with acidified methanol, and determination of the hydrolysed vitamers by HPLC. The results obtained showed good correlation with the available vitamin B_6 content as determined by rat bioassay. The method was subsequently applied to the analysis of natural foods (Ekanayake and Nelson, 1988).

10.5.3 Microbiological assays

The ideal assay organism for the microbiological determination of total vitamin B_6 should exhibit an equal growth response to PN, PL and PM, as these vitamers have equal biological activities in humans (Brin, 1978). Lactic acid bacteria have no utility, because they do not respond to PN (Snell, 1981). The yeast *Saccharomyces carlsbergensis (uvarum)* (ATCC No. 9080) responds vigorously to all three free bases and is the most widely used assay organism for determining vitamin B_6. Acid hydrolysis of food samples is necessary for the determination of total vitamin B_6 because *S. carlsbergensis* utilizes only the unbound non-phosphorylated forms of the vitamin (Morris, Hughes and Mulder, 1959). The growth response of *S. carlsbergensis* to PL relative to that to PN is practically equal, or somewhat less, but the response to PM is markedly less than that to PN and is dose-dependent within the working concentration range of 2–10 ng molar equivalents of PN per tube (Guilarte, McIntyre and Tsan, 1980; Gregory, 1982). This unequal response leads to an underestimation of the total vitamin B_6 content if the sample extract contains predominantly PM (e.g. a processed meat product), but is of little concern in plant-derived foods or in foods that are substantially fortified with PN. Morris, Hughes and Mulder (1959) pointed out that cultures of *S. carlsbergensis* obtained from different sources, but reputed to be derived from the same parent strain, possess different nutritional requirements, particularly in their response to thiamin and vitamin B_6. It is therefore important to study the growth requirements of any strain of *C. carlsbergensis* before it is used as an assay organism.

The problem of differential response in the *S. carlsbergensis* assay can be overcome by separating PN, PL and PM chromatographically and assaying each vitamer individually. Toepfer and Polansky (1970) reported the results of a collaborative study using such a technique, which was adopted as Final Action by the AOAC in 1975 (AOAC, 1990). In this method, acid-hydrolysed food samples are adjusted to pH 4.5 with 6 N KOH to precipitate the denatured proteins, then diluted with water and filtered. An aliquot of the filtrate is applied to an open column packed with Dowex AG 50W-X8 cation exchange resin, and fractions containing PL, PN and PM are obtained by elution with a sequence of boiling buffer solutions. Each vitamer is then assayed using its own standard curve. For a number of animal products, total vitamin B_6 values, obtained by adding the results of the individual vitamers, were statistically higher than non-chromatographed values calculated using a PN standard curve (Polansky and Toepfer, 1969a). An additional benefit of the chromatographic technique is the removal from high-starch plant products of acid-treated glucose compounds which would otherwise stimulate yeast growth (Toepfer and Polansky, 1964). Data obtained by the chromatographic procedure compared well with data obtained by rat growth assay for total vitamin B_6 in a few selected food samples (Toepfer *et al.*, 1963). For routine purposes, the chromatographic step is omitted, and aliquots of the filtrate are diluted according to the expected vitamin B_6 content before addition to the assay tubes. A more efficient and reproducible fractionation of B_6 vitamers can be obtained using HPLC, as performed by Gregory, Manley and Kirk (1981).

Guilarte (1984) used a modified version of the AOAC (1990) procedure in which 200 ml of 0.5 N HCl or 0.05 N HCl, for plant and animal products, respectively, was added to 1–2 g of the dry product, and the mixtures were autoclaved for 90 min at 121 °C. The cooled extracts were adjusted to pH 4.5 using 6 N NaOH, diluted to 250 ml with water, centrifuged or filtered, and finally diluted to the appropriate concentration for turbidimetric assay.

Guilarte, McIntyre and Tsan (1980) compared *S. carlsbergensis* with another yeast, *Kloeckera apiculata* (ATCC No. 9774; formerly called *K. brevis*), for their ability to utilize PN, PL and PM in the concentration range needed for the measurement of vitamin B_6 in biological materials. The results showed that, unlike *S. carlsbergensis*, *K. apiculata* responded equally to all three vitamers at a concentration range of 2–10 ng molar equivalents of PN per tube. A practically equal response of *K. apiculata* to

PN, PL and PM was previously reported by Barton-Wright (1971) and Daoud, Luh and Miller (1977). Guilarte, McIntyre and Tsan (1980) proposed that *K. apiculata* should be used instead of *S. carlsbergensis* as the standard turbidimetric microbiological assay organism for vitamin B_6 in biological materials. However, this proposal has not found acceptance among certain other research groups. Gregory (1982) conducted a study under conditions comparable to those employed by Guilarte, McIntyre and Tsan (1980), and found that *K. apiculata* exhibited an even lower relative response to PM than that obtained with *S. carlsbergensis*. A similar disparity in the response to PM with *K. apiculata* was reported by Polansky (1981). These conflicting data (Gregory, 1983; Guilarte and Tsan, 1983) suggest that subtle environmental factors or culturing variables affect the specificity of *K. apiculata*. This as yet unresolved discrepancy between results from different research groups illustrates the importance of checking the growth response to PN, PL and PM of any assay organism before proceeding to routine determinations.

The growth response of *S. carlsbergensis* and *K. apiculata* is influenced by KCl and NaCl formed as a result of pH adjustment of acid-hydrolysed food samples with $6 N$ KOH or NaOH (Guilarte, 1984). This potential source of interference can be reduced or eliminated by treating the standard in a similar fashion as the food sample to be assayed.

Barton-Wright (1971) encountered the occasional problem of excessive growth in the blanks using *K. apiculata*, which could not be reduced using the PN-depletion technique of Gare (1968). The problem was overcome by maintaining the organism in liquid stock culture containing PN.HCl instead of on the conventional malt agar slope. The basal medium was modified in several respects, notably by substituting a 10% charcoal-treated malt extract solution for vitamin-free casein hydrolysate, which often proved difficult to free completely from vitamin B_6.

A radiometric–microbiological assay developed for the determination of total vitamin B_6 in foods, and based upon the measurement of $^{14}CO_2$ generated from the metabolism of L-$[1-^{14}C]$valine by *K. apiculata* (Guilarte, Shane and McIntyre, 1981; Guilarte, 1983, 1986), gave an assay sensitivity of 0.25 ng of PN, PL or PM per vial (3.2 ml total volume). The method described is subject to the same controversy encountered in the turbidimetric assay with regard to the specificity of *K. apiculata* towards PN, PL and PM.

10.5.4 Fluorimetric method

The first published methods for the fluorimetric determination of vitamin B_6 in foods (Fujita, Fujita and Fujino, 1955a,b; Fujita, Matsuura and Fujino, 1955) involved acid hydrolysis of the food samples, chromatographic purification, chemical conversion of the eluted vitamers to 4-pyridoxic acid, and acid treatment of this intermediate to form the lactone deriva-tive. Different pre-treatment procedures were necessary for selectively determining each of the vitamers. The procedure described by Fujita, Matsuura and Fujino (1955) for determining PN was adapted by Hen-nessy *et al.* (1960) to the analysis of white flour enriched by the addition of PN.HCl, as well as bread made from this flour. Modifications included an additional enzymatic (Mylase) digestion step after acid hydrolysis. A simplified modification of the Hennessy method was applied to PN.HCl-enriched foods in general (Strohecker and Henning, 1966).

The Strohecker and Henning (1966) method was modified by Šebečić and Vedrina-Dragojević (1992) for the determination of total vitamin B₆ in foods. Soya bean samples, which have a complex composition and are notoriously difficult to analyse, were chosen to test the applicability of the suggested procedure. Sample extraction involved the following steps: autoclaving in the presence of sulphuric acid, buffering to pH 4.5, diges-tion with Claradiastase at 45 °C for 30 min, dilution and filtration. PM was converted to PN by boiling the filtrate with sulphuric acid/nitrous acid solution and the resultant solution was neutralized and filtered. As PL is an intermediate product in the oxidation of PN to 4-pyridoxic acid by permanganate, a separate procedure of PL oxidation was not carried out. The filtrate was applied to an open column containing Permutit-T ion exchange resin and the column bed was washed with distilled water to remove unwanted material. Both PN and PL were eluted in one step with warm sulphuric acid and the eluate was diluted with acid. The PN was oxidized to 4-pyridoxic acid by the addition of ice-cold potassium permanganate solution and surplus permanganate was removed by the dropwise addition of 3% hydrogen peroxide. Lactonization was ac-complished by the addition of hydrochloric acid and boiling for 12 min. After cooling, EDTA was added and the solution was diluted with ammonia solution and filtered. The fluorescence intensity of the 4-pyridoxic acid lactone produced was measured at 350 nm (excitation) and 430 nm (emission). Total vitamin B₆ was calculated on the basis of the difference in fluorescence of the sample and fluorescence of the sample with added known amount of B₆ vitamers (method of standard additions). To prepare a sample blank, a duplicate sample was taken through the procedure up to the oxidation of PN with permanganate, and the 4-pyridoxic acid thus formed was destroyed by incubating for 12 min in boiling water (without HCl). EDTA was then added to the cooled solution, and the solution was diluted with ammonia solution and filtered.

10.5.5 High-performance liquid chromatography

Selected HPLC methods for determining vitamin B₆ are summarized in Table 10.1. The HPLC method for vitamin B₆ recommended by COST 91

(Brubacher, Müller-Mulot and Southgate, 1985) advocates autoclaving the homogenized test material in a medium of $0.2\,N$ H_2SO_4 for 30 min at 121 °C. Standard solutions of PN, PL and PM are subjected to the same conditions (external standardization). HPLC is performed under reversed-phase ion suppression conditions using an ODS stationary phase, a mobile phase of $0.08\,N$ H_2SO_4 and a fluorescence detector. PM, PL and PN are separated (in that order of elution) within 10 min and quantified, and the amounts of each vitamer are totalled to give the vitamin B_6 content expressed in mg PN/100 g of sample. The small molecular weight differences (1 mg of PN $\equiv$ 1.008 mg of PM $\equiv$ 1.012 mg of PL) are ignored. The detection limits are 10 μg of PM, 30 μg of PL and 30 μg of PN in a 100 g sample.

For the analysis of human milk, Morrison and Driskell (1985) hydrolysed the phosphorylated B_6 vitamers with phosphohydrolase, and then determined the PL, PN and PM by means of ion interaction chromatography with a gradient elution programme. The vitamin B_6 antagonist 4'-deoxypyridoxine (4-dPN), which is commercially available as the hydrochloride salt, was used as an internal standard. With the aid of fluorescence detection, the minimum detectable limits for PM and PN were 1 ng/injection, which was equivalent to 6 μg/litre of milk. Excellent correlations between HPLC and microbiologically derived data were obtained for PL and total vitamin B_6, but not for PM and PN, where several values were below minimum detectable limits.

Ekanayake and Nelson (1988) estimated the biologically available vitamin B_6 content of foods by subjecting partially defatted samples to a two-stage enzymatic digestion (pepsin/pancreatin), followed by treatment with trichloroacetic acid and then methanol to terminate the hydrolysis reaction and precipitate the proteins. The hydrolysates were centrifuged at $28\,000 \times g$ and the supernatant fraction was transferred to a hypodermic syringe. The pellet was resuspended in HPLC mobile phase and centrifuged. The supernatant was collected in the same syringe, and the pooled fractions were passed through a C_{18} solid-phase extraction cartridge. The purified extracts were analysed by reversed-phase HPLC with ion suppression, using post-column reaction with buffered sodium bisulphite to form fluorescent hydroxysulphonate derivatives of PLP and PL. To calculate the available vitamin B_6, non-phosphorylated vitamer equivalents were calculated from the phosphorylated forms by using the molarity conversion factors of 0.6764 for PLP to PL and 0.6777 for PMP to PM. Transamination of PLP to PMP and PL to PM occurred on digestion and was taken into account when calculating the recoveries. Total vitamin B_6 contents of the food samples were determined after acid hydrolysis (AOAC microbiological procedure) and HPLC analysis. The validity of the *in vitro* enzymatic digestion procedure as a true measure of available vitamin B_6 was evaluated by comparing the ratio of available to

Table 10.1 HPLC methods used for the determination of vitamin B_6 compounds

Food	Sample preparation	Column	Mobile phase	Compounds separated	Detection	Reference
			Strong cation exchange chromatography			
Milk (bovine, caprine, human)	Deproteinize with 20% TCA, centrifuge, extract supernatant with diethyl ether and discard the ether. Filter the aqueous portion (0.45 μm membrane). Add 3,5-diaminobenzoic acid as internal standard to an aliquot of filtrate immediately before injection	Vydac 401TP-B 10 μm silica-based cation exchanger 300 × 4.6 mm i.d.	Ternary gradient elution Solvent A: 0.02 M HCl Solvent B: NaH₂PO₄ 10 g/l adjusted to pH 3.3 with conc. H₃PO₄ Solvent C: NaH₂PO₄ 48 g/l adjusted to pH 5.9 with anhydrous Na₂HPO₄ Linear gradient from 100% A at 10 min to 100% B at 20 min, then a linear gradient from 100% B to 100% C at 30 min. At 38 min programme returns to 0 time and 100% A	PLP, 4-PA, PMP, internal standard, PL, PN, PM	Fluorescence: ex. 330 nm em. 400 nm after post-column addition of buffered sodium bisulphite	Aahuren and Coburn (1990)
			Anion exchange chromatography			
Raw and fried chicken	Homogenize sample with 5% SSA and 3-hydroxypyridoxine (HOP) as internal standard. Add CH₂Cl₂, homogenize, centrifuge and filter (0.45 μm followed by 0.22 μm) *Clean-up*: Dowex AG2-X8, Cl⁻ form anion exchange column (340 × 10 mm) eluted with 0.1 M (0.1 N) HCl. Fluorescence detection	Two 75 × 7.5 mm Bio-Rad anion exchange columns containing diethylaminoethyl groups bound to G5000 power white supports	0.12 M NaCl + 0.02 M glycine buffer, pH 9.8. A second buffer is used to clean the columns after each analysis	PM, PMP, PN, PL, HOP (internal standard), PLP	Programmable fluorescence after post-column addition of buffered sodium bisulphite For PM and PMP: ex. 330 nm em. 400 nm For PN, PL and HOP: ex. 310 nm em. 400 nm For PLP: ex. 330 nm em. 400 nm	Olds, Vanderslice and Brochetti (1993)

Reversed-phase chromatography

Sample	Sample preparation	Column	Mobile phase	Vitamers	Detection	Reference
Food composites, vegetables, meat and meat products, complete meals	Homogenize sample in a medium of 0.1 M (0.2 N) H_2SO_4 and autoclave at 121 °C for 30 min. Cool, dilute to volume with water and filter	Spherisorb RB ODS 5 μm 250 × 4 mm i.d.	0.04 M H_2SO_4	PM, PL, PN	Fluorescence: ex. 290 nm em. 395 nm	Brubacher, Müller-Mulot and Southgate (1985)
Fortified breakfast cereals	Sonicate sample with 0.5 M potassium acetate (pH 4.5) and centrifuge. Deproteinize the supernatant by heating at 50 °C with 33.3% w/v TCA and centrifuging	μBondapak C_{18} 10 μm 300 × 3.9 mm i.d.	0.033 M KH_2PO_4 buffer (pH 2.2)	PM, PL, PN (Only PN present at any significant level in breakfast cereals)	Fluorescence: ex. 295 nm em. 405 nm (filters)	Gregory (1980a)
Human milk	Dilute sample with 0.033 M KH_2PO_4 buffer. Deproteinize by treating with 60% TCA and centrifuging. Dilute supernatant with equal volume of buffer. Remove remaining lipid and TCA by washing with diethyl ether. Remove ether with nitrogen	Biosphere ODS 5 μm (Bioanalytical Systems)	0.033 M KH_2PO_4 buffer (pH 2.9) containing 3% MeOH	PMP, PM, PLP, PL, PN	Fluorescence: ex. 330 nm em. 400 nm after post-column reaction with buffered sodium bisulphite	Hamaker et al. (1985)
Chicken tissues (raw and cooked)	Homogenize with HPO_3 solution, heat to 100 °C, then cool to 0 °C and centrifuge	BioSil ODS-5S 5 μm 250 × 4 mm i.d.	0.066 M KH_2PO_4 buffer (pH 3.0)	PM, PMP, PL, PLP	Fluorescence: ex. 290 nm em. 395 nm (raw samples) ex. 280 nm em. 395 nm (cooked samples)	Ang, Cenciarelli and Eitenmiller (1988)
Ground beef, lima beans, whole wheat flour, non-fat dry milk	Partially defat sample by cold extraction with $CHCl_3$/petroleum ether (1:2), air dry, store overnight at −40 °C. Slurry in water, acidify to pH 2.0, digest with pepsin (37 °C for 3h), adjust pH to 8.0, digest with pancreatin (37 °C for 12 h). Add TCA and then MeOH, mix and centrifuge. Resuspend pellet in HPLC mobile phase and recentrifuge. Pass combined supernatants through C_{18} Sep-Pak solid-phase extraction cartridge	μBondapak C_{18} 10 μm 300 × 3.9 mm i.d.	0.075 M KH_2PO_4 containing monochloroacetic acid at 1.5 g/l and acidified to pH 2.75 with H_3PO_4	PMP, PM, PLP, PL, PN representing biologically available vitamin B_6	Fluorescence: ex. 310 nm em. 390 nm after post-column addition of buffered sodium bisulphite	Ekanayake and Nelson (1988)

Table 10.1 *Continued*

Food	Sample preparation	Column	Mobile phase	Compounds separated	Detection	Reference
Beef, liver, milk	Homogenize beef and liver with 0.1 M KH_2PO_4 buffer (pH 7). Deproteinize the homogenate or the milk sample with 3 M perchloric acid and centrifuge. Resuspend the pellet in 1 M perchloric acid and recentrifuge. Adjust pH of combined supernatants to 6–7. Heat aliquot with 0.5 M sodium glyoxylate to convert PM and PMP to PL and PLP, respectively. *Derivatization:* heat with semicarbazide to form semicarbazones	Ultrasphere IP C_{18} 5 μm 250 × 4.6 mm i.d.	0.033 M KH_2PO_4 buffer (pH 2.2) containing 2.5% MeCN	Semicarbazones of PL and PLP. Quantification of PL, PLP, PM and PMP is based on assays with and without glyoxylate treatment	Fluorescence: ex. filter 365 nm transmission max; em. filter > 400 nm transmission	Gregory, Manley and Kirk (1981)

Ion interaction chromatography

Food	Sample preparation	Column	Mobile phase	Compounds separated	Detection	Reference
Human milk	Add 4'-deoxypyridoxine (4-dPN) as internal standard to milk sample. Incubate with buffered phosphohydrolase at 37 °C for 1 h. Deproteinize by heating with TCA. Defat with CH_2Cl_2. Centrifuge and adjust pH to 5.2	μBondapak C_{18} 10 μm 300 × 3.9 mm i.d.	Gradient elution programme. *Solvent A.* MeOH/water (85:15) *Solvent B.* 1% acetic acid containing 5 mM sodium heptane sulphonate (PIC-B7)	PL, PN, PM, 4-dPN (internal standard)	Fluorescence: ex. 300 nm em. 375 nm	Morrison and Driskell (1985)
Pork, broccoli, milk	Homogenize pork and broccoli samples with 5% sulphosalicylic acid (SSA) and 4'-deoxypyridoxine (4-dPN) as internal standard, add CH_2Cl_2, homogenize, centrifuge. Blend milk with 20% SSA and 4-dPN, add CH_2Cl_2, blend, centrifuge *Preparative HPLC:* AG2-X8, Cl⁻ form (Bio-Rad) anion exchange column (250 × 9 mm i.d.) eluted with 0.1 M HCl. Fluorescence detection	C_{18} (Perkin-Elmer) 3 μm 30 × 4.6 mm i.d.	Linear gradient with 0.033 M KH_2PO_4 buffer (pH 2.2) containing 8 mM octanesulphonic acid and increasing percentage of 2-PrOH	PLP, PMP, PL, PN, 4-dPN (internal standard), PM	Fluorescence: ex. 330 nm em. 400 nm after post-column reaction with buffered sodium bisulphite	Gregory and Feldstein (1985)

Sample	Sample preparation	Column	Mobile phase	Vitamers determined	Detection	Reference
Plant-derived foods (raw broccoli, peanut butter, raw green beans, raw carrots, bananas, orange juice)	Homogenize sample with 5% (w/v) SSA with 4-dPN added as internal standard. Add CH_2Cl_2, homogenize and centrifuge. *Purification and removal of SSA*: apply portion of extract to a disposable column packed with AG2-X8 anion exchanger (Bio-Rad) and elute B_6 vitamers, 4-dPN and PN-glucoside with 0.1 M HCl. Either (i) analyse eluate directly by HPLC or (ii) incubate with buffered (pH 5.0) β-glucosidase for 5 h at 37°C followed by deproteinization with TCA	As above. Later changed to Ultrasphere IP 5 μm 250 × 4.6 mm i.d. (as reported by Gregory and Sartain, 1991)	As above. Later changed from a ternary to a binary gradient (as reported by Gregory and Sartain, 1991)	*Before enzymatic hydrolysis* PLP, PMP, PL, PN-glucoside, PN, 4-dPN (internal standard), PM *After enzymatic hydrolysis* PLP, PMP, PL, PN, 4-dPN, PM	As above	Gregory and Ink (1987)
Potatoes	Mix sample with 5% sulphosalicylic acid (SSA) plus 4'-deoxypyridoxine (4-dPN) as internal standard and centrifuge. *Clean-up*: SSA removed by passage through a 40 × 8 mm i.d. 100–120 mesh AG1-X8 anion exchange column (Bio-Rad)	Radial-PAK C_{18} 4 μm 100 × 8 mm	Binary step gradient. *Solvent A*. Water containing 0.033 M H_3PO_4, 4 mM octane- and heptanesulphonic acid (pH 2.2)/2-PrOH (97.5:2.5). *Solvent B*. Water containing 0.33 M H_3PO_4 (pH 2.2)/2-PrOH (82.5:17.5). Equilibrate column with B for 1 h; switch to A 10 min before injection, then to B 3 min after injection	PLP, PMP, PN-glucoside, PL, PN, 4-dPN (internal standard), PM	Fluorescence: ex. 338 nm em. 425 nm after post-column reaction with buffered sodium bisulphite	Addo and Augustin (1988) (Figure 10.4)
Pork liver (raw), milk	To minced liver sample or to milk add 4'-deoxypyridoxine (4-dPN) as internal standard and homogenize with ice-cold 0.1–0.5 M perchloric acid. Centrifuge, adjust pH to 7.5, filter and readjust pH to 4	LiChrosphere RP-18 5 μm 125 × 4 mm i.d.	Gradient elution programme. *Solvent B*. MeOH. *Solvent B*. 0.03 M KH_2PO_4 buffer (pH 2.7) containing 4 mM octaresulphonic acid. Linear gradient: 90% (B) to 60% (B) to 90% (B)	PLP, 4-PA, PMP, PL, PN, 4-dPN (internal standard), PM	Fluorescence: ex. 330 nm em. 400 nm after post-column reaction with buffered sodium bisulphite	Bitsch and Möller (1989)

Table 10.1 Continued

Food	Sample preparation	Column	Mobile phase	Compounds separated	Detection	Reference
Wheat	Add 4-dPN as internal standard to ground sample and homogenize. Deproteinize by addition of 5% HPO_3 and centrifugation	Ultremex C_{18} 3 µm 150 × 4.6 mm i.d.	Gradient elution programme *Solvent A.* 0.033 M phosphoric acid + 8 mM 1-octanesulphonic acid adjusted to pH 2.2 with 6 M KOH *Solvent B.* 0.033 M phosphoric acid + H_2O /MeCN (90:10) adjusted to pH 2.2. Linear gradient from 100% A to 100% B during 10 min after injection, hold at 100% B for 15 min, then return to 100% A in 4.5 min	PLP, 4-PA, PMP, PN-glucoside, PL, PN, 4-dPN (internal standard), PM	Fluorescence: ex. 311 nm em. 360 nm after post-column reaction with buffered sodium bisulphite	Simpson et al. (1995)
Yeast, wheat germ, breakfast cereal (unfortified), muesli	*Dephosphorylation and conversion of PM to PL:* To ground samples add 0.05 M acetate buffer (pH 4.5), 1 M glyoxylic acid (adjusted to pH 4.5), ferrous sulphate and acid phosphatase. Incubate with continuous shaking at 37 °C, dilute to volume with water and filter *Reduction of PL to PN:* Add aliquot of filtrate to a solution containing 0.2 M NaOH and 0.1 M sodium borohydride, shake and filter	LiChrospher 60 RP Select B (C_8) 5 µm 250 × 5 mm i.d.	0.05 M KH_2PO_4/MeCN (96:4) containing 0.5 mM sodium heptane sulphonate. Total solution adjusted to pH 2.50 with H_3PO_4	PN (representing total vitamin B_6)	Fluorescence: ex. 290 nm em. 395 nm	Reitzer-Bergaentzlé, Marchioni and Hasselmann (1993)

| All food types | Homogenize sample with 5% (w/v) TCA and 4-dPN (internal standard), dilute with 5% (w/v) TCA and shake vigorously for 30 min. Centrifuge and filter. *Dephosphorylation*: dilute aliquot of filtrate with 4 M acetate buffer (pH 6.0) and incubate with Takadiastase at 45 °C for 3 h. Cool and add 15.7% (w/v) TCA to precipitate the enzyme protein. Centrifuge and filter | Hypersil ODS 3 μm 125 × 4.6 mm i.d. | 3% MeOH and 1.25 mM 1-octanesulphonic acid (PIC-B8) in 0.1 M KH_2PO_4 adjusted to pH 2.15 with H_3PO_4 | PL, PN, PM, 4-dPN (internal standard) | Fluorescence: ex. 333 nm em. 375 nm Post-column eluent mixed with 1 M K_2HPO_4 to elevate the pH before detection | van Schoonhoven et al. (1994) |

Abbreviations: see footnote to Table 2.4

SSA, sulphosalicylic acid; HPO_3, metaphosphoric acid.

total vitamin B_6 obtained by the methods described, with the same ratio obtained by rat bioassay and yeast growth assay taken from published data. When the 95% confidence intervals for each of the ratios for a given food were compared, they were found to be comparable for ground beef, nonfat dry milk and whole wheat flour, but not for lima beans. The latter discrepancy may have been due to varietal differences in lima beans.

For the determination of PL, PLP, PM and PMP, the major B_6 vitamers in foods of animal origin (Gregory, Manley and Kirk, 1981), beef and liver homogenates or milk samples were deproteinized by treatment with 3 N perchloric acid, followed by centrifugation. Duplicate portions of each extract were transferred to test tubes and diluted with phosphate buffer. Water (0.1 ml) or aqueous 0.5 N sodium glyoxylate was added, and the tubes were incubated in a boiling water bath for 15 min. The purpose of the glyoxylate treatment was to convert PM and PMP to PL and PLP, respectively. After cooling, semicarbazide (0.5 ml of a 0.2 M aqueous solution) was added to each tube, followed by a 5 min incubation in a boiling water bath. Treatment with semicarbazide converts the aldehydic B_6 vitamers to their semicarbazone derivatives. Semicarbazide treatment is also capable of breaking the Schiff base linkage between PLP and protein lysine residues, allowing the estimation of both free and protein-bound PLP (Dakshinamurti and Chauhan, 1981). The derivatization was employed because PLP-semicarbazone exhibits greater retention and fluorescence than underivatized PLP. The semicarbazone derivatization procedure does not allow the determination of PN and PNP. However, data obtained by microbiological assay show that tissue PN and PNP levels are minimal and insignificant. The derivatized solutions were filtered (0.45 μm) prior to HPLC analysis. The separation of the semicarbazones of PL and PLP was accomplished using reversed-phase HPLC with ion suppression. Quantification of the individual B_6 vitamers (PL, PLP, PM and PMP) was based on the differences in height of the PL- and PLP-semicarbazone peaks in the presence and absence of glyoxylate. Vitamin B_6 values were expressed as micrograms of free base per gram of sample, permitting the addition of values for phosphorylated and nonphosphorylated vitamers. The limit of detection was *c*.20 μg/100 g for each vitamer in animal tissue samples, and 5 μg/100 ml for milk samples. This sensitivity was sufficient for the analysis of all tissues examined, but the quantification of certain milk samples was difficult because of the low vitamin B_6 content. For comparative microbiological (*S. carlsbergensis*) assay, the perchloric acid extracts of a limited number of samples were treated with acid phosphatase, and the individual PL, PN and PM fractions were collected manually by HPLC. Reasonable correlation was obtained between the results of the semicarbazone HPLC and microbiological procedures for the individual B_6 vitamers and total vitamin B_6, which supported the accuracy of the HPLC assay.

In a method proposed by Reitzer-Bergaentzlé, Marchioni and Hassel-mann (1993) for the determination of naturally occurring vitamin B_6 in yeast, wheat germ, unfortified breakfast cereal and muesli, the sample preparation entailed simultaneous dephosphorylation of B_6 vitamers and conversion of PM to PL by incubation with acid phosphatase and glyoxylic acid in the presence of ferrous ions. After filtration, the PL was reduced to PN by treatment with sodium borohydride in alkaline medium and the PN (representing total vitamin B_6) was determined using ion interaction HPLC with fluorimetric detection. The retention time of PN was 7 min under the selected conditions and the detection limit was 0.02 µg/g. In most of the foods studied, interfering substances prevented satisfactory separation and quantification of PN when dephos-phorylation was performed with hydrochloric acid hydrolysis. On the other hand, acid phosphatase hydrolysis led to a very good resolution of the PN peak in yeast, breakfast cereal and muesli. In wheat germ, how-ever, the separation of PN was not entirely satisfactory, but could be markedly improved by increasing the concentration of ion interaction agent from 0.5 mM to 0.7 mM. The proposed method has been subjected to a collaborative study (Bergaentzlé *et al.*, 1995) in which 12 participants analysed eight samples containing various amounts of vitamin B_6 (from 0.6 to 32.8 µg/g of PN). Repeatability CVs ranged from 3 to 18%, depend-ing on the vitamin B_6 concentration. Reproducibility CVs were around 30% for samples containing lowest concentrations of vitamin B_6 (< 3 µg/g of PN) and 12–13% when concentration exceeded 5 µg/g of PN, except for a sample of yeast for which the CV was 26%. The chromatographic method has been chosen as the official French method for vitamin B_6 determination in foods for nutritional purposes.

Van Schoonhoven *et al.* (1994) extracted samples of foods and feeds with 5% trichloroacetic acid and dephosphorylated B_6 vitamers by incu-bation with Takadiastase. PL, PN, PM and 4-deoxypyridoxine (internal standard) were separated by isocratic ion interaction HPLC within 35 min and detected by fluorimetry. To achieve a more sensitive and selective detection and to avoid observed interferences (e.g. in wheat and rice products), the pH of the mobile phase was elevated by mixing the eluent post-column with 1 M K_2HPO_4. An approximately 40% higher total vit-amin B_6 content was found with the HPLC method than with the classical microbiological method using *S. carlsbergensis*. It was demonstrated that the Takadiastase used for dephosphorylation contained β-glucosidase activity so the HPLC result represented total vitamin B_6 from both non-glycosylated and glycosylated forms.

Mahuren and Coburn (1990) accomplished the separation of all six B_6 vitamers, plus 4-pyridoxic acid, using a single silica-based strong cation exchange column and a ternary gradient elution programme. Post-column addition of pH 7.5 buffer and sodium bisulphite permitted the fluorimetric detection of all B$_6$ compounds at the same wavelength. The method was capable of measuring nanogram quantities of each B$_6$ compound in bovine, caprine and human milk. An internal standard (3,5-diaminobenzoic acid) was added to deproteinized milk samples immediately before injection. This compound, which eluted after the fourth peak of interest (PMP), could not be added to the sample at the start of the analysis, and hence served only to compensate for changes in the chromatographic conditions. Because 4-dPN eluted after the last vitamer (PM) and therefore would increase the running time, it was not used as an internal standard. Bisulphite was chosen in preference to semicarbazide as the derivatizing agent because it gave better detection sensitivity under the acidic conditions of the system. The pH 7.5 buffer was added to the bisulphite to facilitate the detection of PNP, which is not fluorescent under acidic conditions at the wavelengths selected. PNP, which has never been detected in biological samples, sometimes interfered with the quantitation of the first B$_6$ vitamer peak (PLP), and therefore it was not routinely included in the standards. The absence of a PNP peak might permit the use of 4'-deoxypyridoxine 5'-phosphate (4-dPNP) as an internal standard, as this analogue elutes between the PLP peak and the next peak of interest (4-pyridoxic acid) (Shephard, Louw and Labadarios, 1987). This research group used 4-dPNP as an internal standard to monitor sample preparation and chromatography in the analysis of human plasma using the HPLC technique of Coburn and Mahuren (1983).

Olds, Vanderslice and Brochetti (1993) developed an isocratic anion exchange HPLC method to determine individual B$_6$ vitamers in raw and fried chicken after extraction with 5% sulphosalicylic acid containing 3-hydroxypyridine as internal standard and anion exchange chromatographic clean-up. Detection was by programmable fluorescence after post-column addition of buffered sodium bisulphite. The extraction procedure did not separate PN from glucose, but this would not be necessary if only nonplant foods were analysed.

Hamaker *et al.* (1985) reported the reversed-phase isocratic separation and quantification of PMP, PM, PLP, PL and PN in human milk using fluorimetric detection after post-column addition of pH 7.5 buffer and sodium bisulphite. Separation of the PL and PLP was shown to depend specifically on the presence of trichloroacetic acid, and not on the final pH of the sample extract. PLP, PL and PN were detectable to at least 30 pmol/ml milk, and PM and PMP were detectable to 5 pmol/ml milk. Correlation between the HPLC results and the *S. carlsbergensis* growth assay was significant at $r = 0.87$.

Ang, Cenciarelli and Eitenmiller (1988) proposed a rapid method for the determination of B$_6$ vitamers in raw and cooked chicken tissues using metaphosphoric acid as the extracting agent. For raw samples the quantification limits per 20 μl injection volume were 0.10, 0.05, 1.00 and 0.1 ng for PMP, PM, PLP and PL, respectively. The fluorescence detection settings were adjusted to give more sensitivity for measuring the lower amounts of PLP and PL in cooked samples; this reduced the quantification limits of these vitamers to *c*. 0.30 and 0.03 ng, respectively. Recoveries of added vitamers in cooked tissues were between 90% and 107.8%, except in a few cases where the values were only 76.9–79.9% for PL and PLP.

Gregory and Feldstein (1985) determined the major B$_6$ vitamers in foods using an ODS analytical column and linear gradient elution with a phosphate buffer mobile phase containing octane sulphonic acid as an ion interaction agent and an increasing percentage of organic modifier (isopropanol). The fluorescence of PLP was enhanced by post-column addition of buffered sodium bisulphite. Food samples were extracted with sulphosalicylic acid in the presence of 4-dPN as internal standard, and subjected to the preparative anion exchange chromatographic technique devised by Vanderslice *et al.* (1980). Low recoveries were obtained for PLP and PL in pork loin (78.7% and 86.7%, respectively), as also reported by Ang, Cenciarelli and Eitenmiller (1988) for cooked chicken tissues after extraction with metaphosphoric acid. It appears that the aldehydic B$_6$ vitamers undergo limited entrapment or binding by muscle protein, even in the presence of the deproteinizing agent. PNP was not evaluated in this study because of its minor significance as a naturally occurring form of vitamin B$_6$. The metabolite 4-pyridoxic acid could be separated and detected by the HPLC system, but could not be quantified, as it is retained by the anion exchange resin used in the preparative system. Bitsch and Möller (1989) used perchloric acid rather than sulphosalicylic acid in their extraction procedure, which permitted the quantification of 4-pyridoxic acid.

The inclusion of PN-glucoside in the chromatographic assay is desirable as the proportion of glycosylated vitamin B$_6$ in the sample is a determinant of net bioavailability. Gregory and Ink (1987) used the HPLC analytical system described by Gregory and Feldstein (1985) with a slightly modified gradient elution programme to determine PLP, PMP, PL, PN-glucoside, PN, 4-dPN (internal standard) and PM in plant-derived foods. A similar procedure using a mixture of ion interaction agents (octane and heptane sulphonic acids) was reported by Addo and Augustin (1988) for the analysis of potatoes (Figure 10.4).

Gregory and Sartain (1991) determined the amounts of free and glycosylated forms of vitamin B$_6$ in plant-derived foods by homogenization with 5% sulphosalicylic acid and purification by anion exchange chromatography. Portions of the purified extracts were then subjected to the following treatments.

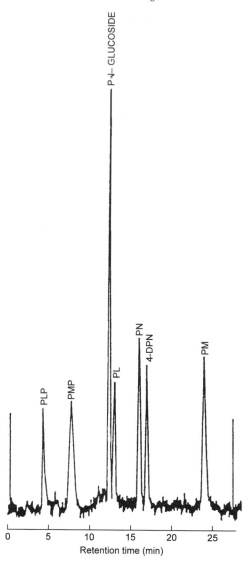

Figure 10.4 Ion interaction HPLC of B$_6$ vitamers extracted from potatoes. Conditions as given in Table 10.1. (Reprinted from Addo and Augustin, 1988, *J. Food Sci.*, **53**, 749–52. Copyright © Institute of Food Technologists.)

1. Direct analysis by ion interaction HPLC allowed the quantification of PN-glucoside, PMP, PL, PN and PM. The early-eluting PLP was not quantified in this chromatographic run, as it was subject to inter-

ference from a cluster of peaks which rendered its accurate measurement difficult or impossible.

2. Extracts treated simultaneously with acid phosphatase and β-glucosidase provided differential information regarding the content of PLP, PMP, PN-glucoside and PN oligosaccharides.
3. Extracts treated with 6 M KOH, followed by the combined enzymes, yielded differential information concerning saponifiable esterified PN-glucosides.

Alternatively, autoclaving for 2 h in the presence of 0.44 N HCl could be used to determine total vitamin B_6 (measured as PL, PN and PM). Gregory and Sartain pointed out that this treatment does not completely hydrolyse the glycosylated forms of vitamin B_6 in rice bran, and would lead to an underestimation of its vitamin B_6 value.

For the determination of free and glycosylated vitamin B_6 in wheat, Sampson *et al.* (1995) extracted samples using 5% metaphosphoric acid and quantified B_6 vitamers in a single binary-gradient ion interaction HPLC separation. The vitamers were detected fluorimetrically after post-column reaction with buffered sodium bisulphite.

An HPLC method for determining ε-pyridoxyllysine in infant formula products was reported by Grün *et al.* (1991).

10.6 BIOAVAILABILITY

10.6.1 Physiological aspects

Vitamin B_6 is present in foods mainly as the PN, PLP and PMP vitamers. In many fruits and vegetables, 30% or more of the total vitamin B_6 is present as PN-glucoside. The binding of PLP to protein through aldimine (Schiff base) and substituted aldamine linkages is reversibly dependent on pH, the vitamin–protein complexes being readily dissociated under normal gastric acid (low pH) conditions. The release of PLP from its association with protein is an important step in the subsequent absorption of vitamin B_6, as binding to protein inhibits the next step, hydrolysis of PLP by alkaline phosphatase (Middleton, 1986). It would appear, therefore, that the widespread practice of raising the postprandial gastric and upper small intestinal pH by the use of pharmaceutical antacids may impair vitamin B_6 absorption.

Intestinal absorption

Physiological amounts of PLP and PMP are largely hydrolysed by alkaline phosphatase in the intestinal lumen before absorption of free

PL and PM (Mehanso, Hamm and Henderson, 1979; Hamm, Mehanso and Henderson, 1979). When present in the lumen at nonphysiological levels which saturate the hydrolytic enzymes, substantial amounts of PLP and PMP are absorbed intact but at a slower rate than their non phosphorylated forms.

The absorption of PN, PL and PM takes place mainly in the jejunum and is a dynamic process involving several interrelated events. The vitamers cross the brush-border membrane by simple diffusion. In humans, PM is absorbed more slowly or metabolized differently, or both, than either PL or PN (Wozenski, Leklem and Miller, 1980). Within the enterocyte PN, PL and PM are converted to their corresponding phosphates by the catalytic action of cytoplasmic pyridoxal kinase, and transaminases interconvert PLP and PMP. The conversion of a particular vitamer to other forms by intracellular metabolism creates a concentration gradient across the brush border for that vitamer, thus enhancing its uptake by diffusion (Middleton, 1985). The phosphorylated vitamers formed in the cell are largely dephosphorylated by nonspecific phosphatases, thus permitting easy diffusion of B$_6$ compounds across the basolateral membrane. The major form of vitamin B$_6$ released to the portal circulation is the nonphosphorylated form of the vitamer predominant in the intestinal lumen.

In accordance with an absorption mechanism that is not carrier-mediated, intestinal uptake of vitamin B$_6$ is independent of the vitamin's dietary concentration (Heard and Annison, 1986).

Renal reabsorption

Little is known about the renal handling of vitamin B$_6$. Uptake of physiological concentrations of PN by isolated rat proximal tubule cells appears to take place by a carrier-mediated process that has substrate specificity and may be modulated by sodium–hydrogen exchange and/or pH gradient effects. Cellular uptake is followed by intracellular metabolic trapping catalysed by pyridoxal kinase (Bowman and McCormick, 1989).

Other metabolic considerations

Liver uptake of the B$_6$ vitamers is by passive diffusion, followed by metabolic trapping as phosphate esters. The PNP and PMP thus formed are oxidized to PLP by pyridoxine (pyridoxamine) 5'-phosphate oxidase. The newly formed PLP is contained in small rapidly mobilized pools that have a rapid rate of turnover and are not freely exchangeable with the endogenous coenzyme pools. A proportion of the newly formed PLP is bound to protein and released into the bloodstream in regulated amounts bound to plasma albumin. Much of the unbound PLP in the liver is hydrolysed to PL, which is also released and circulated bound to both albumin and haemoglobin in erythrocytes. The remaining unbound PLP is rapidly dephosphorylated and oxidized irreversibly to 4-pyridoxic acid, which is released into the plasma and excreted in the urine. The oxidation of PL to 4-pyridoxic acid can occur in many tissues through the action of an NAD-dependent aldehyde dehydrogenase. In the liver an additional enzyme, aldehyde oxidase, is active in converting PL to 4-pyridoxic acid. As the major end-product of vitamin B_6 metabolism, urinary 4-pyridoxic acid reflects the *in vivo* metabolic utilization of the vitamin.

Prior to tissue uptake, PLP is dissociated from the plasma albumin and dephosphorylated to facilitate entry into the cell. It is then rephosphorylated after cell entry. Interconversion between PLP and PMP occurs within the tissues. Some 80% of the total body pool of about 250 mg of the vitamin is in skeletal muscle, mostly in the form of PLP bound to glycogen phosphorylase. The liver contains a further 10% of the body pool and is the primary site of vitamin B_6 metabolism.

10.6.2 Dietary sources and their bioavailability

Dietary sources

Vitamin B_6 is present in all natural unprocessed foods, with yeast extract, wheat bran and liver containing particularly high concentrations. Other important sources include whole-grain cereals, nuts, pulses, lean meat, fish, kidney, potatoes and other vegetables. In cereal grains over 90% of the vitamin B_6 is found in the bran and germ (Polansky and Toepfer, 1969b), and 75–90% of the B_6 content of the whole grain is lost in the milling of wheat to low-extraction flour (Sauberlich, 1985). Thus, white bread is considerably lower in vitamin B_6 content than is whole wheat bread. Milk, eggs and fruits contain relatively low concentrations of the vitamin. Table 10.2 gives the vitamin B_6 content of selected foods.

In raw animal and fish tissue the major form of vitamin B_6 is PLP, which is reversibly bound to proteins as Schiff bases and substituted aldamines. PN and PNP are virtually absent in animal tissues, one exception being liver tissue, in which they are detectable at very low levels. Using a nonhydrolytic extraction procedure and HPLC, the vitamin B_6 content of whole pasteurized homogenized milk was found to comprise the following vitamers: PL (53%), PLP (23%), PMP (12%) and PM (12%); PN was not detected (Gregory and Feldstein, 1985). Siegel, Melnick and Oser (1943) estimated the free and total vitamin B_6 content of milk by assaying aliquots before and after acid hydrolysis; the difference indicated the amount of bound vitamin, which was found to be 14%

Table 10.2 Vitamin B_6 content of various foods (from Holland *et al.*, 1991)

Food	mg Vitamin B_6/100 g[a]
Cow's milk, whole, pasteurised	0.06
Egg, chicken, whole, raw	0.12
Wheat flour, wholemeal	0.50
Rice, brown, raw	N
Rice, white, easy cook, raw	0.31
Beef, lean only, raw, average	0.32
Lamb, lean only, raw, average	0.25
Pork, lean only, raw, average	0.45
Chicken, meat only, raw	0.42
Liver, ox, raw	0.83
Cod, raw, fillets	0.33
Potato, maincrop, old, average, raw	0.44
Soyabeans, dried, raw	0.38
Chick peas, whole, dried, raw	0.53
Lentils, green and brown, whole, dried, raw	0.93
Red kidney beans, dried, raw	0.40
Peas, raw	0.12
Brussels sprouts, raw	0.37
Apples, eating, raw	0.06
Orange juice, unsweetened	0.07
Peanuts, plain	0.59
Yeast extract (Marmite)	1.30

[a] Analytical methods: HPLC with fluorimetric detection (Brubacher, Müller-Mulot and Southgate, 1985) and microbiological assay using *S. carlsbergensis* (Bell, 1974).
N = The vitamin is present in significant quantities but there is no reliable information on the amount.

of the total. PNP does not occur to any measurable extent in natural products.

Plant tissue contains mostly PN, a proportion of which may be present as PN-glucoside or other conjugates. To date, PN-glucoside has not been found in animal products. No generalizations can be made as to one group of foods consistently having a high PN-glucoside content. Typical sources of PN-glucoside (expressed as a percentage of the total vitamin B_6 present) are bananas (5.5%), raw broccoli (35.1%), raw green beans (58.5%), raw carrots (70.1%) and orange juice (69.1%) (Gregory and Ink, 1987). PN-glucoside accounted for 10–15% of the total vitamin B_6 in the typical mixed diets used in a human study (Gregory *et al.*, 1991), but would be proportionally higher in vegetarian diets.

Bioavailability

The first indication that the bioavailability of vitamin B_6 in certain foods may be incomplete arose from observations that vitamin determinations by rat growth assay gave lower results than values obtained by micro-biological assay (Sarma, Snell and Elvehjem, 1947). Research gained further impetus after the occurrence of convulsions in infants during the 1950s was traced to the loss of vitamin activity in heat-sterilized unfortified infant formula. It is important in the following discussion to recognize the distinction between vitamin B_6 content and vitamin B_6 bioavailability, which should not be used interchangeably. Losses of vitamin B_6 content caused by thermal instability occur during food pro-cessing, but the remaining vitamin B_6 does not necessarily exhibit incom-plete bioavailability.

Although the six B_6 vitamers are considered to have approximately equivalent biological activity on the basis of their ultimate conversion to coenzymes, various studies have demonstrated different activities according to the experimental protocol and response criteria used (Gre-gory, 1997). For example, PL and PM were markedly less active than PN on the basis of erythrocyte aminotransferase activity in the rat, whereas plasma PLP values exhibited a fairly uniform response to these vitamers (Nguyen *et al.*, 1983). These experimentally determined differences in biological activity among the B_6 vitamers illustrate the need for careful selection of response criteria used for quantification of rat bioassays for biologically available vitamin B_6.

Factors affecting the bioavailability of vitamin B_6 in foods, experimen-tal approaches and interpretation of data have been discussed by Gre-gory and Ink (1985). Gregory (1997) also provided a critical assessment of methodology. The various response parameters used in rat bioassays each have their inherent problems. Growth assays are carried out with suboptimal amounts of vitamin B_6 in the diet and this may alter vitamin B_6-requiring metabolic pathways, resulting in a growth-retarding imbal-ance of metabolites. Plasma PLP concentrations have been found to correlate well with tissue vitamin B_6 stores and intake (Li and Lumeng, 1981) and would seem to be a better indicator of bioavailability. How-ever, with some test foods there is evidence of bias in rat bioassays using plasma PLP as an indicator. For example, beef and spinach diets pro-duced bioavailabilities significantly greater than 100% relative to stan-dard diets containing PN (Nguyen and Gregory, 1983). In addition, significantly higher levels of vitamin B_6 were found in faeces from spi-nach diets when compared with faeces from other diets. The dietary fibre present in spinach may stimulate the growth of colonic microorganisms which are capable of synthesizing vitamin B_6. Experiments in which coprophagy was prevented have shown that the rat assay based on plasma PLP still overestimates vitamin B_6 bioavailability when applied to spinach, cornmeal and potato (Gregory and Litherland, 1986). This finding indicates that the rat absorbs vitamin B_6 produced by colonic microflora, an ability that the rat is known to possess (Booth and Brain,

1962). The use of the chick as the test animal reduces the problem of coprophagy, but does not represent mammalian physiology.

The bioavailability of vitamin B$_6$ in foods is highly variable, owing largely to the presence of poorly utilized PN-glucoside in plant tissues. As expected, vitamin B$_6$ generally has a lower availability from plant-derived foods than from animal tissues (Nguyen and Gregory, 1983). In humans the vitamin B$_6$ from whole wheat bread and peanut butter was 75% and 63%, respectively, which was as available as that from tuna (Kabir, Leklem and Miller, 1983b). Urinary vitamin B$_6$ and 4-pyridoxic acid, faecal vitamin B$_6$ and plasma PLP were the four indices of vitamin B$_6$ bioavailability that were used and compared. The vitamin in soybeans was 6–7% less available than that in beef (Leklem, Shultz and Miller, 1980). Kies, Kan and Fox (1984) obtained experimental data indicating that vitamin B$_6$ provided by the bran fractions of wheat, rice and corn (maize) is unavailable to humans. Based on plasma PLP levels in male human subjects, the availability of the vitamin in an average American diet ranged from 61% to 81%, with a mean of 71% (Tarr, Tamura and Stokstad, 1981).

Effects of dietary fibre

Experimental studies using rats or chicks have shown that dietary fibre in purified or semi-purified form does not appear to be a major factor influencing the bioavailability of vitamin B$_6$ in animals. A wide variety of dietary fibre material (polysaccharides, lignin and wheat bran) did not bind or entrap B$_6$ vitamers *in vitro* (Nguyen *et al.*, 1981), and in a rat jejunal perfusion study cellulose, pectin and lignin did not adversely affect vitamin B$_6$ absorption (Nguyen, Gregory and Cerda, 1983). Cellulose, pectin and bran had little or no effect on the availability of vitamin B$_6$ in a chick bioassay (Nguyen, Gregory and Damron, 1981) and neither cellulose nor the indigestible component in wheat bran impaired the bioavailability in rats (Hudson, Betschart and Oace, 1988).

Little is known about the possible inhibitory effect of dietary fibre on the bioavailability of vitamin B$_6$ in humans. The availability of vitamin B$_6$ from whole wheat bread was only 5–10% less than that from white bread supplemented with vitamin B$_6$ (Leklem *et al.*, 1980), and the addition of wheat bran (15 g/day) to human diets resulted in only a minor decrease (maximum of 17%) in availability of the vitamin (Lindberg, Leklem and Miller, 1983). Supplementation of human diets with pectin (15 g supplement per day) had no effect on vitamin B$_6$ utilization, as measured by plasma PLP and urinary 4-pyridoxic acid (Miller, Shultz and Leklem, 1980). Although the pectin supplement stimulated the synthesis of vitamin B$_6$ by the intestinal microflora, the newly synthesized vitamin was apparently not absorbed by the host. An inhibitory effect of a component(s) of orange juice upon vitamin B$_6$ absorption has been shown using perfused segments of human jejunum (Nelson, Lane and Cerda, 1976). Absorption of the vitamin from orange juice was only about 50% of that from synthetic solutions. Further investigations (Nelson, Burgin and Cerda, 1977) revealed that vitamin B_6 in orange juice was bound to a small dialysable molecule which was heat stable and nonprotein in nature.

Since vegetarian diets are relatively high in fibre, such a diet may lead to a reduced bioavailability of vitamin B_6. However, two independent studies have shown no significant difference in vitamin B_6 status between vegetarian and nonvegetarian women (Shultz and Leklem, 1987; Löwik *et al.*, 1990). In these studies vitamin B_6 bioavailability was evaluated as it occurs in foods and after adaptive processes have resulted in steady-state conditions. Löwik *et al.* (1990) concluded from their results that dietary fibre (intake range: 5–64 g/day) does not have a negative impact on vitamin B_6 status. On the contrary, consumption of plant-derived foods would most likely exert a positive effect on vitamin B_6 intake and status because of their higher vitamin B_6 to protein ratio.

Effects of alcohol

The excessive consumption of alcohol meets most of the human energy needs and decreases food intake by as much as 50%. Alcoholic beverages that replace food are practically devoid of vitamin B_6 and the alcoholic is therefore likely to be consuming a diet that is deficient in the vitamin.

Chronic excessive alcohol ingestion can interfere with the normal processes of vitamin B_6 metabolism, thus leading to an increased requirement for the vitamin (Li, 1978). The conversion of intravenously administered pyridoxine to PLP in the plasma is impaired in alcoholic patients, suggesting that alcohol or its oxidation products may interfere directly with the metabolism of vitamin B_6. There is *in vitro* evidence that acetaldehyde facilitates the dissociation of PLP from its binding with protein, thereby making the PLP available for hydrolysis by membrane-bound alkaline phosphatase (Hoyumpa, 1986). Thus the generation of acetaldehyde associated with alcoholism accelerates the degradation of PLP, lowering plasma concentrations and also body stores. Ethanol may also stimulate the urinary excretion of nonphosphorylated B_6 vitamers (Hoyumpa, 1986).

Absorption of vitamin B_6 from food is significantly impaired in alcoholic patients with liver disease, although such patients are able to absorb synthetic vitamin B_6 normally. Liver disease may also impair the ability of the liver to synthesize PLP.

Glycosylated forms of vitamin B_6

The bioavailability of PN-glucoside in the rat is *c.*20–30% relative to PN (Trumbo, Gregory and Sartain, 1988). The conjugate is adequately absorbed by the rat without prior conversion to PN, although limited hydrolysis to PN by intestinal enzymes and/or microflora does occur (Trumbo and Gregory, 1988; Trumbo, Banks and Gregory, 1990). The absorption efficiency of PN-glucoside is about 50% that of PN in the rat (Ink Gregory and Saitain, 1986a). The incomplete bioavailability of the conjugate is due to a retarding effect on the metabolic utilization of co-ingested nonglycosylated forms of vitamin B$_6$ (Gilbert and Gregory, 1992a). PN-glucoside does not inhibit pyridoxal kinase and pyridoxine (pyridoxamine) 5'-phosphate oxidase, the two key enzymes in the metabolic utilization of nonglycosylated forms of vitamin B$_6$ (Gilbert and Gregory, 1992b).

In pregnant rats, the availability to the foetus of vitamin B$_6$ derived from maternal PN-glucoside seems to be similar to that of vitamin B$_6$ derived from maternal PN (Cheng and Trumbo, 1993). This increased utilization of PN-glucoside may be due to an increased activity of glycosidases in response to hormonal differences during pregnancy.

Zhang, Gregory and McCormick (1993) investigated the uptake and metabolism of PN-glucoside by isolated rat liver cells (hepatocytes) and came to the following conclusions. The conjugate is actively transported through the hepatocyte plasma membrane using the same facilitated transport system as used by free PN. However, the transport of PN-glucoside is only about 20% that of equimolar PN, suggesting the existence of a permeability barrier for the conjugate. On entry into the hepatocyte, PN-glucoside undergoes hydrolysis by a broad-specificity β-glucosidase before phosphorylation of PN can occur. The hydrolysis step is the rate-limiting step in the cellular metabolism of PN-glucoside. Much less PLP is formed when PN-glucoside is used as substrate than when PN is used.

Gregory *et al.* (1991) determined the bioavailability of PN-glucoside in humans through the use of a stable-isotope method. The utilization of orally administered deuterated PN-glucoside was $58 \pm 13\%$ (mean $\pm$ SEM) relative to that of deuterated PN. Intravenously administered PN-glucoside underwent approximately half the metabolic utilization of oral PN-glucoside, which suggested a role of β-glucosidase(s) of the intestinal mucosa, microflora, or both, in the release of free PN from dietary PN-glucoside. Stable isotope methodology provided evidence that PN-glucoside weakly retards the metabolic utilization of nonglycosylated forms of vitamin B$_6$ in humans (Gilbert *et al.*, 1991). Despite the relatively high consumption of glycosylated vitamin B$_6$, vegetarian women did not demonstrate any significant difference in vitamin B$_6$ status compared with nonvegetarian women (Shultz and Leklem, 1987; Löwik *et al.*, 1990). In addition, the intake of glycosylated vitamin B$_6$ had little, if any, effect upon maternal plasma PLP concentration and maternal urinary excretion of total vitamin B$_6$ and 4-pyridoxic acid in lactating women (Andon *et al.*, 1989). These observations suggest that there may be little practical significance to the human consumption of glyco-sylated vitamin B_6.

Bills, Leklem and Miller (1987) investigated the relationship between the PN-glucoside content of foods and vitamin B_6 bioavailability to ascertain if the PN-glucoside content could be used as an *in vitro* index of vitamin B_6 bioavailability. HPLC measurements of urinary 4-pyridoxic acid excretion were used to assess vitamin B_6 bioavailability in human subjects. A strong inverse correlation between percentage PN-glucoside and bioavailability was found in six of the 10 foods examined (walnuts, bananas, tomato juice, spinach, orange juice and carrots) but not in others (wheat bran, shredded wheat, broccoli and cauliflower). These data indicate that the PN-glucoside content of foods is not a reliable index of vitamin B_6 bioavailability, as previously supposed by Kabir, Leklem and Miller (1983c).

10.6.3 Effects of food processing, storage and cooking

Unlike other vitamins, in particular thiamin, riboflavin and vitamin C, data on the factors affecting vitamin B_6 stability in foods are relatively rare. Both the quantity and form of vitamin B_6 are affected by the processes of heating, concentration and dehydration (Gregory and Kirk, 1981; Gregory and Ink, 1985). In dehydrated model food systems sub-jected to roasting at $180\,°C$ for 25 min (conditions similar to but more severe than those used for the toasting of breakfast cereals) the degrada-tion of PN, PM and PLP was found to be 50–70% (Gregory and Kirk, 1978a). There is, however, good stability of vitamin B_6 in wheat flour during bread making (Perera, Leklem and Miller, 1979).

Using a combination of intrinsic and extrinsic labelling in the rat, Ink, Gregory and Sartain (1986b) showed that thermal processing of liver and muscle caused a $c.30\%$ reduction in the levels of $[^3H]B_6$ vitamers. Inter-conversions between the aldehyde (PL and PLP) and amine (PM and PMP) B_6 vitamers via reversible interaction with proteins or carbonyl compounds occur during the processing or storage of meat and dairy products (Gregory, Ink and Sartain, 1986). For example, in hydrolysed extracts of raw pork loin the predominant vitamer is PL, whereas in fully cooked ham it is PM (Polansky and Toepfer, 1969a).

In a report on the influence of cooking (Bognár, 1993), stewing reduced the content of total vitamin B_6 by 56% and braising by 58% in beef and by 58% and 45%, respectively, in pork. Part of the vitamin loss was due to release of meat juice. With meat juice included, the percentage of vitamin retained was 80% and 63% in beef, and 68% and 62% in pork. Losses of the total vitamin B_6 in vegetables after boiling in water were between 16% and 61%; losses were lower after steaming (between 10% and 24%) due to less leaching.

Thermal processing had little effect on the concentration of PN-glucoside in alfalfa sprouts, demonstrating the stability of the glycosidic bond as well as the stability of the PN moiety of the conjugate (Lik, Gregory and Sartain, 1900u).

Ekanayake and Nelson (1990) investigated the effect of retort processing on the total vitamin B$_6$ content and vitamin B$_6$ bioavailability of rehydrated lima beans. About 20% of the PN content was lost during the prior blanching of the beans, probably due to leaching into the blanch water. There was no further loss of total vitamin B$_6$ in the heat processing step, although interconversion of vitamers was evident. The unusually large (47%) loss of vitamin B$_6$ for canned lima beans reported by Schroeder (1971) may be explained by the observations of Chin and Coryell (1982) that up to 50% of the vitamin B$_6$ content of canned vegetables is found in the brine. Therefore if the drained product is used for vitamin B$_6$ analysis, the estimated loss would be much higher. Although Ekanayake and Nelson (1990) showed that vitamin B$_6$ is destroyed only to a small extent during thermal processing, the availability of the vitamin, as determined by an *in vitro* enzyme digestion method, was reduced by almost 50%. A 50% loss of vitamin B$_6$ activity in canned lima beans relative to frozen lima beans was also reported by Richardson, Wilkes and Ritchey (1961) using a rat growth assay. These authors observed an increase in vitamin B$_6$ activity of canned lima beans on storage as opposed to a decrease in activity for beef liver, boned chicken, cabbage and green beans. Ekanayake and Nelson (1990) postulated that PLP and PL are held in a moderately acid labile 'bound form' in the processed lima beans and that these bound forms do not release PL and PLP on enzymatic digestion.

Tadera *et al.* (1986) found that PN was converted into an unidentified compound when incubated at 30 °C in the presence of plant food homogenates. The reaction was neither thermal decomposition nor photodecomposition. A compound with identical physicochemical characteristics was also formed by incubating PN with ascorbic acid in the dark and, after purification, was identified as 6-hydroxypyridoxine. This compound had neither vitamin B$_6$ activity nor antivitamin B$_6$ activity for *S. carlsbergensis*. It was later shown to be also inactive for the rat (Gregory, 1990). Tadera *et al.* (1986) suggested that hydroxylation of PN and the PN moiety of PN conjugates in the presence of food compounds, especially ascorbic acid, was responsible for the loss of vitamin B$_6$ in plant foods during food processing, storage and cooking. However, Gregory (1990) stated that this reaction does not appear to be a significant mechanism for the loss of biologically active and available vitamin B$_6$ in foods.

Losses of indigenous vitamin B$_6$ in foods during storage have been observed (Harding, Plough and Friedemann, 1959; Everson *et al.*, 1964; Richardson, Wilkes and Ritchey, 1961; Kirchgessner and Kösters, 1977).

Vedrina-Dragojević and Šebečić (1994) reported that the storage of various foods at $-18\,°C$ for 5 months resulted in a 19–60% decrease in total vitamin B_6. The loss was significantly greater in foods of animal origin (an average of 55%) than in plant-derived foods. Unlike the loss in indigenous vitamin B_6 during storage, the stability of PN added to fortify various products is high. Bunting (1965) reported the retention of 90–100% of the PN added to corn meal and macaroni following storage at $38\,°C$ and 50% relative humidity for 1 year. There is good stability of vitamin B_6 in fortified cereal products after storage (Anderson *et al.*, 1976; Cort *et al.*, 1976; Rubin, Emodi and Scialpi, 1977). However, only 18–44% of the vitamin B_6 in a commercial rice-based PN-fortified breakfast cereal was found to be biologically available using a rat bioassay (Gregory, 1980a).

Much of the research into the effects of food processing upon vitamin B_6 bioavailability has been directed to the heat-sterilization of canned evaporated milk by retort processing. The heat-sterilization of milk and unfortified infant formula resulted in losses of 36–67% of naturally occurring vitamin B_6 using *S. carlsbergensis* as the assay organism, the losses appearing progressively during the first 10 days after processing (Hassinen, Durbin and Bernhart, 1954). The reduction of vitamin activity was due mainly to the loss of the predominant B_6 vitamer, PL. In the same study, PM and PL added to milk were degraded to roughly the same extent as natural vitamin B_6, whereas added PN.HCl was not appreciably destroyed by autoclaving. Gregory and Hiner (1983) reported that PN.HCl is stable during retort processing, while PM.2HCl and PL.HCl are 2.5- to 3.5-fold less stable. Vitamin B_6 losses are not as great when processing spray-dried milk products and condensed milk as they are when heat sterilization is employed (Hassinen, Durbin and Bernhart, 1954). High-temperature short-time (HTST) pasteurization (2–3 s at $92\,°C$) and boiling of milk for 2–3 min gave only 3% loss, while ultra-HTST sterilization (injection of steam at $143\,°C$ into preheated milk for 3–4 s) caused negligible loss of the vitamin (Tannenbaum, Young and Archer, 1985).

The research effort has led to the general conclusion that the thermal processing of milk products and other animal-derived foods promotes the chemical reduction of protein-bound Schiff base forms of PL and PLP to peptide-linked ε-pyridoxyllysine (Figure 10.5) or its 5'-phosphate derivative. This reaction has been demonstrated to take place in a model food system during thermal processing (Gregory and Kirk, 1977), in evaporated milk, chicken liver and muscle that were autoclaved to simulate retort processing (Gregory, Ink and Sartain, 1986) and in a dehydrated model food system that was stored at $37\,°C$ and 0.6 water activity (a_w) for 128 days (Gregory and Kirk, 1978b).

In vitro studies have shown that ε-pyridoxyllysine undergoes enzymatic phosphorylation and oxidation to form PLP in a manner similar

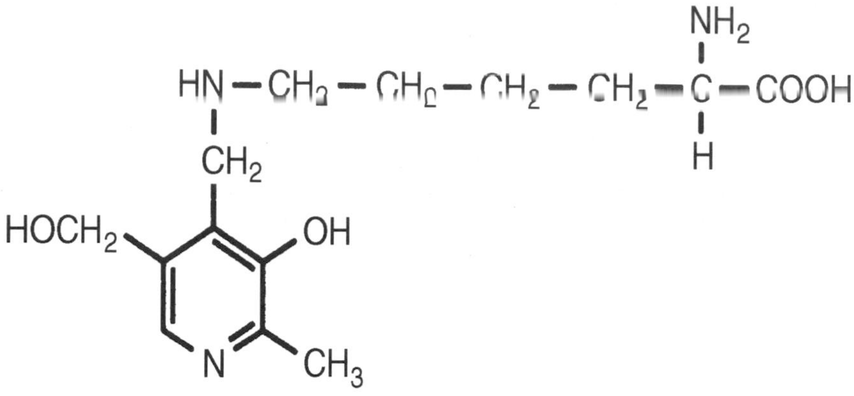

Figure 10.5 Structure of ε- pyridoxyllysine.

to that of free B$_6$ vitamers (Gregory, 1980b). These experiments provide a metabolic basis for the observed *c.*50% molar vitamin B$_6$ activity of this compound in rats (Gregory, 1980c). The effect of ε-pyridoxyllysine on the availability of vitamin B$_6$ in foods appears to depend on the total vitamin B$_6$ content of the diet and the ratio of the protein-bound vitamin to free B$_6$ vitamers. When fed to rats at low levels in vitamin B$_6$-deficient diets, ε-pyridoxyllysine demonstrated anti-vitamin B$_6$ activity, which could be counteracted by dietary supplementation with PN (Gregory, 1980c). This anti-vitamin B$_6$ activity is attributable, at least in part, to the competitive inhibition of pyridoxal kinase, which is the enzyme that catalyses the 5'-phosphorylation of PL (Gregory, 1980b). This antagonistic effect may have been responsible for the severe deficiency developed in infants fed unfortified, heat-sterilized, canned infant formulas (Coursin, 1954).

10.7 DIETARY INTAKE

It is well established that the human requirement for vitamin B$_6$ increases as the intake of protein increases (Baker *et al.*, 1964; Miller, Leklem and Shultz, 1985; Hansen, Leklem and Miller, 1996). This relationship can be explained by the increased activities of PLP-dependent enzymes which catalyse excess amino acids when dietary protein intake is high. There appears to be an age-dependent difference in the protein intake-related needs for vitamin B$_6$ in humans, whereby elderly subjects need less of the vitamin at a higher protein intake as compared with young adults (Pannemans, van den Berg and Westerterp, 1994). A dietary vitamin B$_6$ ratio of 16 µg/g protein appears to ensure acceptable nutritional status in adults of both sexes. The recommendations for adults in the United States and Australia are based on an allowance of 20 µg vitamin B$_6$/g protein, while those in the United Kingdom are based on an allowance of $15\,\mu g/g$ protein. The absolute RDA for vitamin B_6 in the United States is $2.0\,mg/$ day for males aged 15 years and upwards and $1.6\,mg/day$ for females aged 19 years and upwards. These allowances are adequate for the reported average protein intakes of $c.100\,g/day$ for men and $60\,g/day$ for women. The RDAs for women during pregnancy and lactation are increased to 2.2 and $2.1\,mg/day$, respectively. There are no increments for pregnancy or lactation in the UK recommendations. In the United States in 1985, the average vitamin B_6 intake was $1.87\,mg$ in adult males and $1.16\,mg$ in adult females (National Research Council, 1989).

Effects of high intake

Symptoms of neuropathy due to vitamin B_6 overdose have been observed in women receiving upwards of $50\,mg$ of the vitamin per day for 6 months or more (Dalton and Dalton, 1987). These observations suggest that chronic dosing with vitamin B_6 above $50\,mg/day$ is potentially dangerous. Acute toxicity of the vitamin, however, is low (McCormick, 1988).

Assessment of nutritional status

Various biochemical methods, such as plasma PLP concentrations, urinary excretion of 4-pyridoxic acid, measurement of aminotransferase activity and metabolic load tests have been used to assess vitamin B_6 status in humans (Bender, 1993; Leklem, 1994).

Metabolic load tests are based on an individual's ability to metabolize a test dose of a substrate whose catabolism is dependent on vitamin B_6. In the tryptophan load test, a $2\,g$ oral load of L-tryptophan is administered and the urinary excretion of the metabolites xanthurenic acid and kynurenic acid is determined. The sensitivity of the test to intakes of vitamin B_6 between 1.0 and $2.5\,mg$ (common intakes in adults) is not known, thus the test may only apply to situations in which the intake of the vitamin is low ($< 0.8\,mg/day$). In certain cases the test may give results suggesting a vitamin B_6 deficiency where no such deficiency exists. For example, during stress or illness the secretion of glucocorticoid hormones induces tryptophan dioxygenase, resulting in an increased formation of the metabolites being measured. The ability to metabolize a test dose of methionine is an alternative test, which does not seem to be as prone to misinterpretation as the tryptophan load test (Bender, 1993; Leklem, 1994).

10.8 SYNOPSIS

The widespread existence of glycosylated forms of vitamin B_6 in plant tissues, coupled with their known incomplete utilization by humans, accounts for the observed lower bioavailability of plant-derived foods compared with animal-derived foods. Although thermal destruction of B$_6$ vitamers can account for losses of vitamin B$_6$ content during food processing, the remaining vitamers exhibit nearly complete bioavailability. Only under conditions of severe vitamin B$_6$ deficiency would ε-pyridoxyllysine, formed during heat-sterilization of canned foods, normally exert a nutritionally significant antagonistic effect. Dietary fibre appears to have little or no influence on the bioavailability of vitamin B$_6$ under most dietary conditions.

Microbiological assays are currently the most widely used and accepted method for the determination of total vitamin B$_6$ in biological materials. The standard turbidimetric method using the yeast *S. carlsbergensis* underestimates the total vitamin B$_6$ content if the sample contains predominantly PM (e.g. a processed meat product), as the growth response of *S. carlsbergensis* to this vitamer is markedly less than that to PL or PN. This unequal response is of little concern in plant-derived foods or foods that are substantially fortified with PN.HCl. *K. apiculata* has been proposed as the assay organism in the standard turbidimetric and semi-automated radiometric–microbiological assay on the basis of an equivalent growth response to all three vitamers. However, this proposal has not found acceptance in certain other laboratories, in which *K. apiculata* was found to exhibit an even lower relative response to PM than that obtained with *S. carlsbergensis*.

Phosphorylated B$_6$ vitamers cannot be utilized for growth by the yeasts used as microbiological assay organisms, and it is standard practice to autoclave food samples in the presence of acid in order to liberate the PN, PL and PM. There is no entirely satisfactory single set of conditions that can quantitatively extract vitamin B$_6$ from both plant and animal products. The AOAC microbiological method specifies different extraction conditions for plant and animal products, but for composite foods and diets a compromise must be reached. Extraction of plant-derived foods by autoclaving in the presence of 0.44 N HCl for 2 h at 121 °C, in accordance with the AOAC method, will lead to an overestimation of vitamin B$_6$ bioavailability. This is because this rigorous hydrolysis treatment completely liberates PN from PN-glucoside, and this conjugate is only 58% available to humans (relative to PN).

HPLC methods have been developed for determining separately all of the B$_6$ vitamers plus PN-glucoside after an extraction procedure (deproteination) that preserves the integrity of the phosphorylated and glycosylated vitamers. Such HPLC methods may provide a better estimate of biologically available vitamin B$_6$ than would be obtained using typical microbiological assays. Other HPLC methods have employed enzyme digestion rather than acid hydrolysis in the extraction step in an attempt to simulate the human digestive process.

The absorption of free B_6 vitamers from the intestine takes place by simple diffusion and is unlimited. While vitamin B_6 in animal foods is well absorbed and bioavailable, a major part of the vitamin in plant foods is glycosylated and has a low bioavailability for humans. Further research is needed in order to clarify the contribution of plant foods to human vitamin B_6 nutriture.

REFERENCES

Addo, C. and Augustin, J. (1988) Changes in the vitamin B_6 content in potatoes during storage. *J. Food Sci.*, **53**, 749–52.

Anderson, R.H., Maxwell, D.L., Mulley, A.E. and Fritsch, C.W. (1976) Effects of processing and storage on micronutrients in breakfast cereals. *Food Technol.*, **30**(5), 110, 112–14.

Andon, M.B., Reynolds, R.D., Moser-Veillon, P.B. and Howard, M.P. (1989) Dietary intake of total and glycosylated vitamin B-6 and the vitamin B-6 nutritional status of unsupplemented lactating women and their infants. *Am. J. Clin. Nutr.*, **50**, 1050–8.

Ang, C.Y.W. (1979) Stability of three forms of vitamin B_6 to laboratory light conditions. *J. Ass. Off. Analyt. Chem.*, **62**, 1170–3.

Ang, C.Y.W., Cenciarelli, M. and Eitenmiller, R.R. (1988) A simple liquid chromatographic method for determination of B_6 vitamers in raw and cooked chicken. *J. Food Sci.*, **53**, 371–5.

AOAC (1990) Vitamin B_6 (pyridoxine, pyridoxal, pyridoxamine) in food extracts. Microbiological method. Final action 1975. In *AOAC Official Methods of Analysis*, 15th edn (ed. K. Helrich), Association of Official Analytical Chemists, Inc., Arlington, VA, 961.15.

Atkin, L. Schultz, A.S., Williams, W.L. and Frey, C.N. (1943) Yeast microbiological methods for determination of vitamins. Pyridoxine. *Ind. Engng Chem., Analyt. Edn*, **15**, 141–4.

Baker, E.M., Canham, J.E., Nunes, W.T. *et al.* (1964) Vitamin B_6 requirement for adult men. *Am. J. Clin. Nutr.*, **15**, 59–66.

Barton-Wright, E.C. (1971) The microbiological assay of the vitamin B_6 complex (pyridoxine, pyridoxal and pyridoxamine) with *Kloeckera brevis*. *Analyst, Lond.*, **96**, 314–18.

Bell, J.G. (1974) Microbiological assay of vitamins of the B group in foodstuffs. *Lab. Pract.*, **23**, 235–42, 252.

Bender, D.A. (1993) Vitamin B_6. Physiology. In *Encyclopaedia of Food Science, Food Technology and Nutrition*, Vol. 7 (eds R. Macrae, R.K. Robinson and M.J. Sadler), Academic Press, London, pp. 4795–4804.

Bergaentzlé, M., Arella, F., Bourguignon, J.B. and Hasselmann, C. (1995) Determination of vitamin B6 in foods by HPLC – a collaborative study. *Food Chem.*, **52**, 81–6.

Bills, N.D., Leklem, J.E. and Miller, L.T. (1987) Vitamin B-6 bioavailability in plant foods is inversely correlated with percent glycosylated vitamin B-6. *Fed. Proc.*, **46**, 1487 (abstr).

Bitsch, R. and Möller, J. (1989) Analysis of B_6 vitamers in foods using a modified high-performance liquid chromatographic method. *J. Chromat.*, **463**, 207–11.

Bliss, C.I. and György, P. (1967) Vitamin B_6. VII. Animal assays for vitamin B_6. In *The Vitamins. Chemistry, Physiology, Pathology, Methods*, 2nd edn, Vol. VII (eds P. György and W.N. Pearson), Academic Press, New York, pp. 205–8.

Bognár, A. (1993) Studies on the influence of cooking on the vitamin B$_6$ content of food. In *Bioavailability '93. Nutritional, Chemical and Food Processing Implications of Nutrient Availability*, Conference proceedings, Part 2 (ed. U. Schlemmer), Ettlingen, May 9–12, 1993, Bundesforschungsanstalt für Ernährung, pp. 346–51.

Booth, C.C. and Brain, M.C. (1962) The absorption of tritium-labelled pyridoxine hydrochloride in the rat. *J. Physiol.*, **164**, 282–94.

Bowman, B.B. and McCormick, D.B. (1989) Pyridoxine uptake by rat renal proximal tubular cells. *J. Nutr.*, **119**, 745–9.

Brin, M. (1978) Vitamin B$_6$: chemistry, absorption, metabolism, catabolism, and toxicity. In *Human Vitamin B$_6$ Requirements*, Food and Nutrition Board, National Research Council. National Academy of Sciences, Washington, DC, pp. 1–20.

Brubacher, G., Müller-Mulot, W. and Southgate, D.A.T. (1985) *Methods for the Determination of Vitamins in Food. Recommended by COST 91*. Elsevier Applied Science Publishers, London and New York.

Bunting, W.R. (1965) The stability of pyridoxine added to cereals. *Cereal Chem.*, **42**, 569–72.

Cheng, S. and Trumbo, P.R. (1993) Pyridoxine-5'-β-D-glucoside: metabolic utilization in rats during pregnancy and availability to the fetus. *J. Nutr.*, **123**, 1875–9.

Chin, H.B. and Coryell, P.B. (1982) Pantothenic acid and vitamin B-6 content of selected fruits and vegetables. Abs 238, presented at the 42nd Annual Meeting of the Institute of Food Technologists, Las Vegas, NV, June 22–25. (Cited by Ekanayake and Nelson, 1990.)

Coburn, S.P. and Mahuren, J.D. (1983) A versatile cation-exchange procedure for measuring the seven major forms of vitamin B$_6$ in biological samples. *Analyt. Biochem.*, **129**, 310–17.

Cort, W.M., Borenstein, B., Harley, J.H. *et al.*, (1976) Nutrient stability of fortified cereal products. *Food Technol.*, **30**(4), 52–60.

Coursin, D.B. (1954) Convulsive seizures in infants with pyridoxine-deficient diet. *J. Am. Med. Ass.*, **154**, 406–8.

Dakshinamurti, K. and Chauhan, M.S. (1981) Chemical analysis of pyridoxine vitamers. In *Methods in Vitamin B-6 Nutrition. Analysis and Status Assessment* (eds J.E. Leklem and R.D. Reynolds), Plenum Press, New York, pp. 99–122.

Dakshinamurti, K., Paulose, C.S. and Siow, Y.L. (1985) Neurobiology of pyridoxine. In *Vitamin B-6: its Role in Health and Disease* (eds R.D. Reynolds and J.E. Leklem), Alan R. Liss, Inc., New York, pp. 99–121.

Dalton, K. and Dalton, M.J.T. (1987) Characteristics of pyridoxine overdose neuropathy syndrome. *Acta Neurol. Scand.*, **76**, 8–11.

Daoud, H.N., Luh, B.S. and Miller, M.W. (1977) Effect of blanching, EDTA and NaHSO$_3$ on color and vitamin B$_6$ retention in canned garbanzo beans. *J. Food Sci.*, **42**, 375–8.

Ebadi, M. (1978) Vitamin B$_6$ and biogenic amines in brain metabolism. In *Human Vitamin B$_6$ Requirements*, Committee on Dietary Allowances, Food and Nutrition Board, National Research Council. National Academy of Sciences, Washington, DC, pp. 129–61.

Ekanayake, A. and Nelson, P.E. (1986) An *in vitro* method for estimating biologically available vitamin B$_6$ in processed foods. *Br. J. Nutr.*, **55**, 235–44.

Ekanayake, A. and Nelson, P.E. (1988) Applicability of an *in vitro* method for the estimation of vitamin B$_6$ biological availability. *J. Micronutr. Anal.*, **4**, 1–15.

Ekanayake, A. and Nelson, P.E. (1990) Effect of thermal processing on lima bean vitamin B-6 availability. *J. Food Sci.*, **55**, 154–7.

Everson, G.J., Chang, J., Leonard, S.L. *et al.* (1964) Aseptic canning of foods. III. Pyridoxine retention as influenced by processing method, storage time and temperature, and type of container. *Food Technol.*, **18**(1), 87–8.

Fujita, A., Fujita, D. and Fujino, K. (1955a) Fluorometric determination of vitamin B$_6$. 2. Determination of pyridoxamine. *J. Vitam.*, **1**, 275–8.

Fujita, A., Fujita, D. and Fujino, K. (1955b) Fluorometric determination of vitamin B$_6$. 3. Fractional determination of pyridoxal and 4-pyridoxic acid. *J. Vitam.*, **1**, 279–89.

Fujita, A., Matsuura, K. and Fujino, K. (1955) Fluorometric determination of vitamin B$_6$. 1. Determination of pyridoxine. *J. Vitam.*, **1**, 267–74.

Gare, L. (1968) The use of depleted cells as inocula in vitamin assays. *Analyst, Lond.*, **93**, 456–7.

Gilbert, J.A. and Gregory, J.F. III (1992a) Pyridoxine-5'-β-D-glucoside affects the metabolic utilization of pyridoxine in rats. *J. Nutr.*, **122**, 1029–35.

Gilbert, J.A. and Gregory, J.F. III (1992b) Pyridoxine-5'-β-D-glucoside does not inhibit pyridoxal kinase, pyridoxamine (pyridoxine) 5'-phosphate oxidase, or glycogen phosphorylase in vitro. *FASEB, J.* **6**, A1373 (abstr).

Gilbert, J.A., Gregory, J.F. III, Bailey, L.B. *et al.* (1991) Effects of pyridoxine-β-glucoside on the utilization of deuterium-labeled pyridoxine in the human. *FASEB, J.*, **5**, A586 (abstr).

Gregory, J.F. III (1980a) Bioavailability of vitamin B-6 in nonfat dry milk and a fortified rice breakfast cereal product. *J. Food Sci.*, **45**, 84–6, 114.

Gregory, J.F. III (1980b) Effects of ε-pyridoxyllysine and related compounds on liver and brain pyridoxal kinase and liver pyridoxamine (pyridoxine) 5'-phosphate oxidase *J. Biol. Chem.*, **255**, 2355–9.

Gregory, J.F. III (1980c) Effects of ε-pyridoxyllysine bound to dietary protein on the vitamin B-6 status of rats. *J. Nutr.*, **110**, 995–1005.

Gregory, J.F. III (1982) Relative activity of the nonphosphorylated B-6 vitamers for *Saccharomyces uvarum* and *Kloeckera brevis* in vitamin B-6 microbiological assay. *J. Nutr.*, **112**, 1643–7.

Gregory, J.F. III (1983) Response to Drs Guilarte and Tsan letter. *J. Nutr.*, **113**, 722–4.

Gregory, J.F. III (1988) Methods for determination of vitamin B6 in foods and other biological materials: a critical review. *J. Food Comp. Anal.*, **1**, 105–23.

Gregory, J.F. III (1989) Bioavailability of vitamin B-6 from plant foods. *Am. J. Clin. Nutr.*, **49**, 717.

Gregory, J.F. III (1990) The bioavailability of vitamin B$_6$. Recent findings. *Ann. NY Acad. Sci.*, **585**, 86–95.

Gregory, J.F. III (1997) Bioavailability of vitamin B-6. *Eur. J. Clin. Nutr.*, **51**, Suppl. 1, S43–8.

Gregory, J.F. III and Feldstein, D. (1985) Determination of vitamin B-6 in foods and other biological materials by paired-ion high-performance liquid chromatography. *J. Agric. Food Chem.*, **33**, 359–63.

Gregory, J.F. III and Hiner, M.E. (1983) Thermal stability of vitamin B6 compounds in liquid model food systems. *J. Food Sci.*, **48**, 1323–7, 1339.

Gregory, J.F. III and Ink, S.L. (1985) The bioavailability of vitamin B-6. In *Vitamin B-6: its Role in Health and Disease* (eds R.D. Reynolds and J.E. Leklem), Alan R. Liss, Inc., New York, pp. 3–23.

Gregory, J.F. III and Ink, S.L. (1987) Identification and quantification of pyridoxine-β-glucoside as a major form of vitamin B6 in plant-derived foods. *J. Agric. Food Chem.*, **35**, 76–82.

Gregory, J.F. III and Kirk, J.R. (1977) Interaction of pyridoxal and pyridoxal phosphate with peptides in a model food system during thermal processing. *J. Food Sci.*, **42**, 1554–7, 1561.

Gregory, J.F. III and Kirk, J.R. (1978a) Assessment of roasting effects on vitamin B$_6$ stability and bioavailability in dehydrated food systems. *J. Food Sci.*, **43**, 1585–9.

Gregory, J.F. III and Kirk, J.R. (1978b) Assessment of storage effects on vitamin B₆ stability and bioavailability in dehydrated food systems. *J. Food Sci.*, **43**, 1801–8, 1815.

Gregory, J.F. III and Kirk, J.R. (1981) The bioavailability of vitamin B-6 in foods. *Nutr. Rev.*, 09, 1 0.

Gregory, J.F. III and Litherland, S.A. (1986) Efficacy of the rat bioassay for the determination of biologically unavailable vitamin B₆. *J. Nutr.*, **116**, 87–97.

Gregory, J.F. III and Sartain, D.B. (1991) Improved chromatographic determination of free and glycosylated forms of vitamin B₆ in foods. *J. Agric. Food Chem.*, **39**, 899–905.

Gregory, J.F. III, Ink, S.L. and Sartain, D.B. (1986) Degradation and binding to food proteins of vitamin B-6 compounds during thermal processing. *J. Food Sci.*, **51**, 1345–51.

Gregory, J.F. III, Manley, D.B. and Kirk, J.R. (1981) Determination of vitamin B-6 in animal tissues by reverse-phase high-performance liquid chromatography. *J. Agric. Food Chem.*, **29**, 921–7.

Gregory, J.F. III, Trumbo, P.R., Bailey, L.B. *et al.* (1991) Bioavailability of pyridoxine-5′-β-D-glucoside determined in humans by stable-isotopic methods. *J. Nutr.*, **121**, 177–86.

Grün, I.U., Barbeau, W.E., Chrisley, B. and Driskell, J.A. (1991) Determination of vitamin B₆, available lysine, and ε-pyridoxyllysine in a new instant baby food product. *J. Agric. Food Chem.*, **39**, 102–8.

Guilarte, T.R. (1983) Radiometric microbiological assay of vitamin B₆: assay simplification and sensitivity study. *J. Ass. Off. Analyt. Chem.*, **66**, 58–61.

Guilarte, T.R. (1984) Effect of sodium chloride and potassium chloride on growth response of yeasts *Saccharomyces uvarum* and *Kloeckera brevis* to free vitamin B₆. *J. Ass. Off. Analyt Chem.*, **67**, 617–20.

Guilarte, T.R. (1986) Radiometric–microbiological assay of vitamin B₆ and derivatives. In *Vitamin B₆. Pyridoxal Phosphate. Chemical, Biochemical, and Medical Aspects*, part A (eds D. Dolphin, R. Poulson and O. Avramović), John Wiley & Sons, New York, pp. 595–627.

Guilarte, T.R. and Tsan, M.-F. (1983) Microbiological assays of total vitamin B-6 using the yeasts *Saccharomyces uvarum* and *Kloeckera brevis*. *J. Nutr.*, **113**, 721–2.

Guilarte, T., McIntyre, P.A. and Tsan, M.-F. (1980) Growth response of the yeasts *Saccharomyces uvarum* and *Kloeckera brevis* to the free biologically active forms of vitamin B-6. *J. Nutr.*, **110**, 954–8.

Guilarte, T., Shane, B. and McIntyre, P.A. (1981) Radiometric–microbiologic assay of vitamin B-6: application to food analysis. *J. Nutr.*, **111**, 1869–75.

Hamaker, B., Kirksey, A., Ekanayake, A. and Borschel, M. (1985) Analysis of B-6 vitamers in human milk by reverse-phase liquid chromatography. *Am. J. Clin. Nutr.*, **42**, 650–5.

Hamm, M.W., Mehanso, H. and Henderson, L.M. (1979) Transport and metabolism of pyridoxamine and pyridoxamine phosphate in the small intestine of the rat. *J. Nutr.*, **109**, 1552–9.

Hansen, C.M., Leklem, J.E. and Miller, L.T. (1996) Vitamin B-6 status of women with a constant intake of vitamin B-6 changes with three levels of dietary protein. *J. Nutr.*, **126**, 1891–1901.

Harding, R.S., Plough, I.C. and Friedemann, T.E. (1959) The effect of storage on the vitamin B₆ content of a packaged army ration, with a note on the human requirement for the vitamin. *J. Nutr.*, **68**, 323–31.

Hassinen, J.B., Durbin, G.T. and Bernhart, F.W. (1954) The vitamin B-6 content of milk products. *J. Nutr.*, **53**, 249–57.

Heard, G.S. and Annison, E.F. (1986) Gastrointestinal absorption of vitamin B-6 in the chicken (*Gallus domesticus*). *J. Nutr.*, **116**, 107–20.

Hennessy, D.J., Steinberg, A.M., Wilson, G.S. and Keaveney, W.P. (1960) Fluorometric determination of added pyridoxine in enriched white flour and in bread baked from it. *J. Ass. Off. Analyt. Chem.*, **43**, 765–8.

Holland, B., Welch, A.A., Unwin, I.D. *et al.* (1991) *McCance and Widdowson's The Composition of Foods*, 5th edn, Royal Society of Chemistry and Ministry of Agriculture, Fisheries and Foods.

Hoyumpa, A.M. (1986) Mechanisms of vitamin deficiencies in alcoholism. *Alcoholism Clin. Exp. Res.*, **10**, 573–81.

Hudson, C.A., Betschart, A.A. and Oace, S.M. (1988) Bioavailability of vitamin B-6 from rat diets containing wheat bran or cellulose. *J. Nutr.*, **118**, 65–71.

Ink, S.L., Gregory, J.F. III and Sartain, D.B. (1986a) Determination of pyridoxine-β-glucoside bioavailability using intrinsic and extrinsic labeling in the rat. *J. Agric. Food Chem.*, **34**, 857–62.

Ink, S.L., Gregory, J.F. III and Sartain, D.B. (1986b) Determination of vitamin B_6 bioavailability in animal tissues using intrinsic and extrinsic labeling in the rat. *J. Agric. Food Chem.*, **34**, 998–1004.

Kabir, H., Leklem, J.E. and Miller, L.T. (1983a) Measurement of glycosylated vitamin B-6 in foods. *J. Food Sci.*, **48**, 1422–5.

Kabir, H., Leklem, J.E. and Miller, L.T. (1983b) Comparative vitamin B-6 bioavailability from tuna, whole wheat bread and peanut butter in humans. *J. Nutr.*, **113**, 2412–20.

Kabir, H., Leklem, J.E. and Miller, L.T. (1983c) Relationship of the glycosylated vitamin B6 content of foods to vitamin B6 bioavailability in humans. *Nutr. Rep. Int.*, **28**, 709–16.

Kies, C., Kan, S. and Fox, H.M. (1984) Vitamin B6 availability from wheat, rice and corn brans for humans. *Nutr. Rep. Int.*, **30**, 483–91.

Kirchgessner, M. and Kösters, W.W. (1977) Effect of storage on the vitamin B_6 activity of foods. *Z. Lebensm. u.-Forsch.*, **164**, 15–16 (in German).

Leklem, J.E. (1994) Vitamin B_6. In *Modern Nutrition in Health and Disease*, 8th edn, Vol. 1 (eds M.E. Shils, J.A. Olson and M. Shike), Lea & Febiger, Philadelphia, pp. 383–94.

Leklem, J.E., Shultz, T.D. and Miller, L.T. (1980) Comparative bioavailability of vitamin B6 from soybeans and beef. *Fed. Proc.*, **39**, 558 (abstr).

Leklem, J.E., Miller, L.T., Perera, A.D. and Peffers, D.E. (1980) Bioavailability of vitamin B-6 from wheat bread in humans. *J. Nutr.*, **110**, 1819–28.

Li, T.-K. (1978) Factors influencing vitamin B_6 requirement in alcoholism. In *Human Vitamin B_6 Requirements*, Committee on Dietary Allowances, Food and Nutrition Board, National Research Council, National Academy of Sciences, Washington, DC, pp. 210–25.

Li, T.-K. and Lumeng, L. (1981) Plasma PLP as indicator of nutrition status: relationship to tissue vitamin B-6 content and hepatic metabolism. In *Methods in Vitamin B-6 Nutrition. Analysis and Status Assessment* (eds J.E. Leklem and R.D. Reynolds), Plenum Press, New York, pp. 289–96.

Lindberg, A.S., Leklem, J.E. and Miller, L.T. (1983) The effect of wheat bran on the bioavailability of vitamin B-6 in young men. *J. Nutr.*, **113**, 2578–86.

Löwik, M.R.H., Schrijver, J., van den Berg, H. *et al.* (1990) Effect of dietary fiber on the vitamin B_6 status among vegetarian and nonvegetarian elderly (Dutch Nutrition Surveillance System). *J. Am. College Nutr.*, **9**, 241–9.

McCormick, D.B. (1988) Vitamin B-6. In *Modern Nutrition in Health and Disease*, 7th edn (eds M.E. Shils and V.R. Young), Lea & Febiger, Philadelphia, pp. 376–82.

Mahuren, J.D. and Coburn, S.P. (1990) B₆ vitamers: cation exchange HPLC. *Meth. Nutr. Biochem.*, **1**, 659–63.

Mehanso, H., Hamm, M.W. and Henderson, L.M. (1979) Transport and metabolism of pyridoxal and pyridoxal phosphate in the small intestine of the rat. *J. Nutr.*, 109, 1542 51.

Middleton, H.M. III (1985) Uptake of pyridoxine by in vivo perfused segments of rat small intestine: a possible role for intracellular vitamin metabolism. *J. Nutr.*, **115**, 1079–88.

Middleton, H.M. III (1986) Intestinal hydrolysis of pyridoxal 5'-phosphate in vitro and in vivo in the rat. Effect of protein binding and pH. *Gastroenterology*, **91**, 343–50.

Miller, L.T., Leklem, J.E. and Shultz, T.D. (1985) The effect of dietary protein on the metabolism of vitamin B-6 in humans. *J. Nutr.*, **115**, 1663–72.

Miller, L.T., Shultz, T.D. and Leklem, J.E. (1980) Influence of citrus pectin on the bioavailability of vitamin B₆ in men. *Fed. Proc.*, **39**, 797.

Morino, Y. and Nagashima, F. (1986) Methods for isolation and purification of pyridoxal phosphate and derivatives. In *Vitamin B₆. Pyridoxal Phosphate. Chemical, Biochemical, and Medical Aspects*, Part A (eds D. Dolphin, R. Poulson and O. Avramović), John Wiley & Sons, New York, pp. 477–96.

Morris, J.G., Hughes, D.T.D. and Mulder, C. (1959) Observations on the assay of vitamin B₆ with *Saccharomyces carlsbergensis* 4228. *J. Gen. Microbiol.*, **20**, 566–75.

Morrison, L.A. and Driskell, J.A. (1985) Quantities of B₆ vitamers in human milk by high-performance liquid chromatography. *J. Chromat., Biomed. Appl.*, **337**, 249–58.

National Research Council (1980) Water-soluble vitamins. In *Recommended Dietary Allowances*, 9th revised edn, National Academy Press, Washington, DC, pp. 72–124.

National Research Council (1989) Water-soluble vitamins. In *Recommended Dietary Allowances*, 10th edn, National Academy Press, Washington, DC, pp. 115–73.

Nelson, E.W., Burgin, C.W. and Cerda, J.J. (1977) Characterization of food binding of vitamin B-6 in orange juice. *J. Nutr.*, **107**, 2128–34.

Nelson, E.W. Jr, Lane, H. and Cerda, J.J. (1976) Comparative human intestinal bioavailability of vitamin B-6 from a synthetic and a natural source. *J. Nutr.*, **106**, 1433–7.

Nguyen, L.B. and Gregory, J.F. III (1983) Effects of food composition on the bioavailability of vitamin B-6 in the rat. *J. Nutr.*, **113**, 1550–60.

Nguyen, L.B., Gregory, J.F. III and Cerda, J.J. (1983) Effect of dietary fiber on absorption of B-6 vitamers in a rat jejunal perfusion study. *Proc. Soc. Exp. Biol. Med.*, **173**, 568–73.

Nguyen, L.B., Gregory, J.F. III and Damron, B.L. (1981) Effects of selected polysaccharides on the bioavailability of pyridoxine in rats and chicks. *J. Nutr.*, **111**, 1403–10.

Nguyen, L.B., Gregory, J.F. III, Burgin, C.W. and Cerda, J.J. (1981) In vitro binding of vitamin B-6 by selected polysaccharides, lignin, and wheat bran. *J. Food Sci.*, **46**, 1860–2.

Nguyen, L.B., Hiner, M.E., Litherland, S.A. and Gregory, J.F. III (1983) Relative biological activity of nonphosphorylated vitamin B-6 compounds in the rat. *J. Agric. Food Chem.*, **31**, 1282–7.

Olds, S.J., Vanderslice, J.T. and Brochetti, D. (1993) Vitamin B₆ in raw and fried chicken by HPLC. *J. Food Sci.*, **58**, 505–507, 561.

Pannemans, D.L.E., van den Berg, H. and Westerterp, K.R. (1994) The influence of protein intake on vitamin B-6 metabolism differs in young and elderly humans. *J. Nutr.*, **124**, 1207–14.

Perera, A.D., Leklem, J.E. and Miller, L.T. (1979) Stability of vitamin B_6 during bread making and storage of bread and flour. *Cereal Chem.*, **56**, 577–80.

Polansky, M.M. (1981) Microbiological assay of vitamin B-6 in foods. In *Methods in Vitamin B-6 Nutrition. Analysis and Status Assessment* (eds J.E. Leklem and R.D. Reynolds), Plenum Press, New York, pp. 21–44.

Polansky, M.M. and Toepfer, E.W. (1969a) Vitamin B-6 components in some meats, fish, dairy products, and commercial infant formulas. *J. Agric. Food Chem.*, **17**, 1394–7.

Polansky, M.M. and Toepfer, E.W. (1969b) Nutrient composition of selected wheats and wheat products. IV. Vitamin B-6 components. *Cereal Chem.*, **46**, 664–74.

Polansky, M.M., Murphy, E.W. and Toepfer, E.W. (1964) Components of vitamin B_6 in grains and cereal products. *J. Ass. Off. Analyt. Chem.*, **47**, 750–3.

Polansky, M.M., Reynolds, R.D. and Vanderslice, J.T. (1985) Vitamin B_6. In *Methods of Vitamin Assay*, 4th edn (eds J. Augustin, B.P. Klein, D. Becker and P.B. Venugopal), John Wiley & Sons, New York, pp. 417–43.

Rabinowitz, J.C. and Snell, E.E. (1947a) Vitamin B_6 group. Extraction procedures for the microbiological determination of vitamin B_6. *Ind. Engng Chem., Analyt. Edn.*, **19**, 277–80.

Rabinowitz, J.C. and Snell, E.E. (1947b) The vitamin B_6 group. XII. Microbiological activity and natural occurrence of pyridoxamine phosphate. *J. Biol. Chem.*, **169**, 643–50.

Rabinowitz, J.C., Mondy, N.I. and Snell, E.E. (1948) The vitamin B_6 group. XIII. An improved procedure for determination of pyridoxal with *Lactobacillus casei*. *J. Biol. Chem.*, **175**, 147–53.

Reiber, H. (1972) Photochemical reactions of vitamin B_6 compounds, isolation and properties of products. *Biochim. Biophys. Acta*, **279**, 310–15.

Reitzer-Bergaentzlé, M., Marchioni, E. and Hasselmann, C. (1993) HPLC determination of vitamin B_6 in foods after pre-column derivatization of free and phosphorylated vitamers into pyridoxol. *Food Chem.*, **48**, 321–4.

Richardson, L.R., Wilkes, S. and Ritchey, S.J. (1961) Comparative vitamin B_6 activity of frozen, irradiated and heat-processed foods. *J. Nutr.*, **73**, 363–8.

Rubin, S.H. and Scheiner, J. (1946) The availability of vitamin B_6 in yeast to *Saccharomyces carlsbergensis*. *J. Biol. Chem.*, **162**, 389–90.

Rubin, S.H., Emodi, A. and Scialpi, L. (1977) Micronutrient additions to cereal grain products. *Cereal Chem.*, **54**, 895–904.

Rubin, S.H., Scheiner, J. and Hirschberg, E. (1947) The availability of vitamin B_6 in yeast and liver for growth of *Saccharomyces carlsbergensis*. *J. Biol. Chem.*, **167**, 599–611.

Saidi, B. and Warthesen, J.J. (1983) Influence of pH and light on the kinetics of vitamin B_6 degradation. *J. Agric. Food Chem.*, **31**, 876–80.

Sampson, D.A., Eoff, L.A., Yan, X.L. and Lorenz, K. (1995) Analysis of free and glycosylated vitamin B_6 in wheat by high-performance liquid chromatography. *Cereal Chem.*, **72**, 217–21.

Sarma, P.S., Snell, E.E. and Elvehjem, C.A. (1947) The bioassay of vitamin B_6 in natural materials. *J. Nutr.*, **33**, 121–8.

Sauberlich, H.E. (1985) Bioavailability of vitamins. *Prog. Food Nutr. Sci.*, **9**, 1–33.

Schroeder, H.A. (1971) Losses of vitamins and trace minerals resulting from processing and preservation of foods. *Am. J. Clin. Nutr.*, **24**, 562–73.

Šebečić, B. and Vedrina-Dragojević, I. (1992) Fluorometric method for determination of vitamin B_6 in soya bean. *Z. Lebensm. u.-Forsch.*, **194**, 144–7.

Shephard, G.S. and Labadarios, D. (1986) Degradation of vitamin B_6 standard solutions. *Clin. Chim. Acta*, **160**, 307–12.

Shephard, G.S., Louw, M.E.J. and Labadarios, D. (1987) Analysis of vitamin B₆ vitamers in plasma by cation-exchange high-performance liquid chromatography. *J. Chromat., Biomed. Appl.*, **416**, 138–43.

Shultz, T.D. and Leklem, J.E. (1987) Vitamin B-6 status and bioavailability in vegetarian women *Am. J. Clin. Nutr.*, **46**, 647–51.

Siegel, L., Melnick, D. and Oser, B.L. (1943) Bound pyridoxine (vitamin B₆) in biological materials. *J. Biol. Chem.*, **149**, 361–7.

Snell, E.E. (1958) Chemical structure in relation to biological activities of vitamin B₆. *Vitams Horm.*, **15**, 77–125.

Snell, E.E. (1981) Vitamin B-6 analysis: some historical aspects. In *Methods in Vitamin B-6 Nutrition. Analysis and Status Assessment* (eds J.E. Leklem and R.D. Reynolds), Plenum Press, New York, pp. 1–19.

Snell, E.E. (1986) Pyridoxal phosphate: history and nomenclature. In *Vitamin B₆ Pyridoxal Phosphate. Chemical, Biochemical, and Medical Aspects*, Part A (eds D. Dolphin, R. Poulson and O. Avramović), John Wiley & Sons, New York, pp. 1–12.

Strohecker, R. and Henning, H.M. (1966) *Vitamin Assay – Tested Methods*, Verlag Chemie, Weinheim.

Tadera, K. and Orite, K. (1991) Isolation and structure of a new vitamin B₆ conjugate in rice bran. *J. Food Sci.*, **56**, 268–9.

Tadera, K., Kaneko, T. and Yagi, F. (1986) Evidence for the occurrence and distribution of a new type of vitamin B₆ conjugate in plant foods. *Agric. Biol. Chem.*, **50**, 2933–4.

Tadera, K., Arima, M., Yoshino, S. *et al.* (1986) Conversion of pyridoxine into 6-hydroxypyridoxine by food components, especially ascorbic acid. *J. Nutr. Sci. Vitaminol.*, **32**, 267–77.

Tannenbaum, S.R., Young, V.R. and Archer, M.C. (1985) Vitamins and minerals. In *Food Chemistry*, 2nd edn (ed. O.R. Fennema), Marcel Dekker, Inc., New York, pp. 477–544.

Tarr, J.B., Tamura, T. and Stokstad, E.L.R. (1981) Availability of vitamin B-6 and pantothenate in an average American diet in man. *Am. J. Clin. Nutr.*, **34**, 1328–37.

Toepfer, E.W. and Polansky, M.M. (1964) Recent developments in the analysis for vitamin B₆ in foods. *Vitams. Horm.*, **22**, 825–32.

Toepfer, E.W. and Polansky, M.M. (1970) Microbiological assay of vitamin B₆ and its components. *J. Ass. Off. Analyt. Chem.*, **53**, 546–50.

Toepfer, E.W., Polansky, M.M., Richardson, L.R. and Wilkes, S. (1963) Comparison of vitamin B₆ values of selected food samples by bioassay and microbiological assay. *J. Agric. Food Chem.*, **11**, 523–5.

Trumbo, P.R. and Gregory, J.F. III (1988) Metabolic utilization of pyridoxine-β-glucoside in rats: influence of vitamin B-6 status and route of administration. *J. Nutr.*, **118**, 1336–42.

Trumbo, P.R., Banks, M.A. and Gregory, J.F. III (1990) Hydrolysis of pyridoxine-5'-β-D-glucoside by a broad-specificity β-glucosidase from mammalian tissues. *Proc. Soc. Exp. Biol. Med.*, **195**, 240–6.

Trumbo, P.R., Gregory, J.F. III and Sartain, D.B. (1988) Incomplete utilization of pyridoxine-β-glucoside as vitamin B-6 in the rat. *J. Nutr.*, **118**, 170–5.

van Schoonhoven, J., Schrijver, J., van den Berg, H. and Haenen, G.R.M.M. (1994) Reliable and sensitive high-performance liquid chromatographic method with fluorometric detection for the analysis of vitamin B-6 in foods and feeds. *J. Agric. Food Chem.*, **42**, 1475–80.

Vanderslice, J.T., Maire, C.E., Doherty, R.F. and Beecher, G.R. (1980) Sulfosalicylic acid as an extraction agent for vitamin B₆ in food. *J. Agric. Food Chem.*, **28**, 1145–9.

Vanderslice, J.T., Brownlee, S.G., Cortissoz, M.E. and Maire, C.E. (1985) Vitamin B₆ analysis; sample preparation, extraction procedures, and chromatographic separations. In *Modern Chromatographic Analysis of the Vitamins* (eds A.P. De Leenheer, W.E. Lambert, and M.G.M. De Ruyter), Marcel Dekker, Inc., New York, pp. 435–75.

Vedrina-Dragojević, I. and Šebečić, B. (1994) Effect of frozen storage on the degree of vitamin B₆ degradation in different foods. *Z. Lebensm. u.-Forsch.*, **198**, 44–6.

Wozenski, J.R., Leklem, J.E. and Miller, L.T. (1980) The metabolism of small doses of vitamin B-6 in men. *J. Nutr.*, **110**, 275–85.

Zhang, Z., Gregory, J.F. III and McCormick, D.B. (1993) Pyridoxine-5′-β-D-glucoside competitively inhibits uptake of vitamin B-6 into isolated rat liver cells. *J. Nutr.*, **123**, 85–9.

11

Pantothenic acid

11.1 INTRODUCTION

In 1933 R.J. Williams and his research team isolated from a variety of biological materials an acidic substance that acted as a growth factor for yeast. Williams named the substance pantothenic acid because of its apparently widespread occurrence (Greek *pantos*, everywhere). Pantothenic acid was recognized as a vitamin for animals in 1939, when Jukes and Wooley independently reported that pantothenic acid isolated in pure form from liver by R.J. Williams was identical with a chick antidermatitis factor first described by Norris and Ringrose in 1930. The biochemical role of pantothenic acid as a constituent of coenzyme A was identified by Lipmann in 1947.

The biological activity of pantothenic acid is attributable to its incorporation into the molecular structures of coenzyme A and acyl carrier protein. The steric configuration of the pantothenate structure is important for enzymatic recognition in biochemical reactions involving these coenzymes. As a component of coenzyme A, pantothenic acid is essential for numerous reactions involved in the release of energy from carbohydrates, fats and amino acids. In carbohydrate metabolism, the formation of acetyl-coenzyme A is necessary for introducing acyl groups into the tricarboxylic acid cycle. Other roles of acetyl coenzyme A include its requirement for the acetylation of amino sugars, which are constituents of various mucopolysaccharides of connective tissues, and also for the acetylation of choline to form the neurotransmitter acetylcholine. Succinyl-coenzyme A is a precursor for porphyrin and hence for haemoglobin and cytochromes. Acyl carrier protein plays a major role in the biosynthesis of fatty acids.

11.2 CHEMICAL STRUCTURE AND NOMENCLATURE

Structures of pantothenic acid and compounds containing a pantothenate moiety are shown in Figure 11.1. Pantothenic acid comprises a derivative of butyric acid (pantoic acid) joined by a peptide linkage to the amino acid β-alanine, and is systematically named as N-(2,4-dihydroxy-3,3-dimethyl-1-oxobutyl)-β-alanine ($C_9H_{17}O_5N$; MW = 219.23). The molecule is optically active but only the d(+)-enantiomorph occurs in nature. Synthetic pantothenic acid is a racemic mixture and, since only the d isomer is biologically active, this fact must be considered if the dl mixture is to be used therapeutically. Pantothenic acid is a pale yellow oil which is extremely hygroscopic, and so is unsuitable for commercial application. For human food supplements, calcium d-pantothenate $[(C_9H_{16}O_5N)_2Ca$; MW = 476.53] is used. In nature, pantothenic acid occurs only rarely in the free state, but it is very widely distributed as an integral part of the structures of coenzyme A and 4-phosphopantetheine. The latter serves as a covalently attached prosthetic group of acyl carrier protein.

The corresponding alcohol of pantothenic acid, pantothenol (referred to commercially as panthenol), is widely used as a source of pantothenate activity for pharmaceutical vitamin products, because it is more stable than the pantothenate salts, especially in liquid multivitamin products that must be slightly acid to preserve the thiamin content. Pantothenol does not occur naturally and itself has no pantothenate activity, but it is converted quantitatively to pantothenic acid *in vivo* and is equivalent to pantothenic acid in man (Bird and Thompson, 1967).

11.3 DEFICIENCY SYNDROMES

Because of the widespread distribution of pantothenic acid in foods, a dietary deficiency of the vitamin is virtually impossible, apart from circumstances of severe malnutrition. Although pantothenic acid deficiency can be produced easily in experimental animals, humans

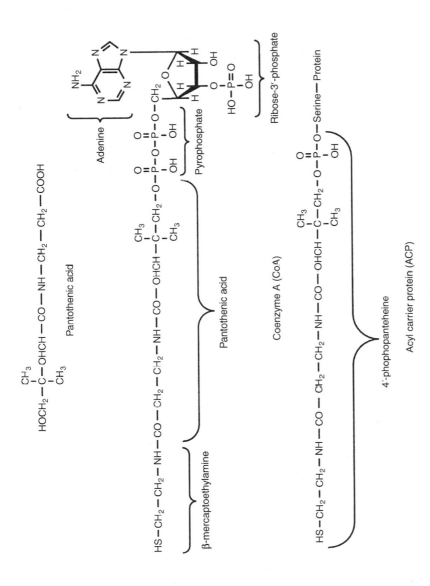

Figure 11.1 Structures of pantothenic acid and compounds containing a pantothenate moiety.

receiving purified diets devoid of the vitamin require long-term deprivation (> 12 weeks) to show symptoms, despite low urinary excretion of
the vitamin. Deficiency symptoms have, however, been induced in
humans by administering a metabolic antagonist φ-methylpantothenic
acid. In this compound the terminal hydroxymethyl group is replaced by
a methyl group which prevents phosphorylation of the analogue. When
volunteers were fed the antagonist, along with a diet low in pantothenic acid, the extensive list of symptoms reportedly reversed by subsequent administration of the vitamin included irritability, restlessness,
insomnia, the sensation of burning feet, muscle cramps, impaired muscular coordination, sensitivity to insulin, decreased antibody formation
leading to lowered resistance to infection, easy fatigue, mental depression, nausea, gastrointestinal disturbances and upper respiratory infections. The list probably reflects impaired cellular metabolism in many
tissues.

11.4 PHYSICOCHEMICAL PROPERTIES

Solubility and other properties

Calcium pantothenate is a colourless, odourless, bitter-tasting, moderately
hygroscopic microcrystalline powder which decomposes at *c*.200 °C.
The pH of an aqueous (1 in 20) solution of calcium pantothenate is
7.2–8.0; the pK_a value is 4.4. Pantothenic acid is readily soluble in water
and ethyl acetate and slightly soluble in diethyl ether; the calcium salt is
more soluble in water (40 g/100 ml), only slightly soluble in ethyl acetate
and insoluble in diethyl ether.

Spectroscopic properties

The pantothenic acid molecule does not contain a characteristic chromophore, and hence it exhibits only very weak absorbance at 204 nm, owing
to the presence of the carbonyl groups.

Stability

The stability of pantothenic acid and its calcium salt in aqueous solution
is highly dependent on the pH. In contrast to other B-vitamins, pantothenic acid becomes more stable as the pH of the solution increases.
Solutions of calcium pantothenate are most stable between pH 5 and 7
but, even so, are not stable to autoclaving, and therefore sterilization by
ultrafiltration is necessary. Below and above these pH values, solutions
of calcium pantothenate are thermolabile. Alkaline hydrolysis yields
pantoic acid and β-alanine, whereas acid hydrolysis yields the γ-lactone

of pantoic acid (Tannenbaum, Young and Archer, 1985). Pantothenic acid is unaffected by atmospheric oxygen and light.

11.5 ANALYSIS

11.5.1 Scope of analytical techniques

The majority of naturally occurring pantothenic acid is in the form of coenzyme A and is routinely determined by microbiological assay after suitable extraction. Biospecific methods of analysis include the radio-immunoassay and the enzyme-linked immunosorbent assay (ELISA). Pantothenate can be assayed biologically in the curative chick growth test.

No IU of pantothenic acid activity has been defined. Analytical results are generally expressed in weight units (mg) of pantothenic acid, using calcium pantothenate as the standard material: 1 mg of pantothenic acid is equivalent to 1.087 mg of calcium pantothenate.

11.5.2 Extraction techniques

Before pantothenic acid can be determined by methods other than an animal bioassay, it is necessary to liberate the vitamin from its bound forms, chiefly coenzyme A. Neither acid nor alkaline hydrolysis can be used, since the pantothenic acid is degraded by such treatments. The only practicable alternative is enzymatic hydrolysis, and this was successfully accomplished through the simultaneous action of intestinal phosphatase and an avian liver enzyme (Novelli, Kaplan and Lipmann, 1949). This double enzyme combination liberates practically all of the pantothenic acid from coenzyme A, but it does not release the vitamin from acyl carrier protein (Wyse *et al.*, 1985). The phosphatase splits the coenzyme A molecule between the phosphate-containing moiety and pantetheine, while the liver enzyme breaks the link in pantetheine between the pantothenic acid and β-mercaptoethylamine moieties.

11.5.3 Microbiological assays

The usual assay organism for the microbiological determination of pantothenic acid in foods is *L. plantarum*. The basal medium has the same composition as that used for the determination of niacin, except that pantothenic acid instead of nicotinic acid is the limiting factor. Fatty acids are stimulatory in the presence of suboptimal amounts of pantothenic acid (Skeggs and Wright, 1944), so a preliminary ether extraction step may be necessary.

Both intestinal phosophatase and avian liver enzyme preparations required to liberate bound pantothenic acid are available commercially as powdered extracts. Liver enzyme preparations contain a relatively high amount of coenzyme A, which is converted to pantothenate during the incubation period, thus creating an unacceptably high blank value. Pigeon or chicken acetone-dried liver powder obtained from Sigma Chemicals can be purified quite simply by treatment with Dowex 1-X4 anion exchange resin, as reported by Novelli and Schmetz (1951) and modified by Toepfer, Zook and Richardson (1954), Clegg (1958) and Bell (1974). Purification of the enzyme reduces the blank value to a very low level without appreciable loss of coenzyme A-splitting activity. Intestinal phosphatase preparations contain negligible amounts of coenzyme A and do not require purification.

Toepfer, Zook and Richardson (1954) used the phosphatase–liver enzyme treatment in a standardized microbiological assay and obtained pantothenic acid values for whole egg powder, kale, peanuts, pig liver and brewer's yeast that did not differ significantly from values obtained on the same samples using a rat bioassay. The microbiological results for carrots, however, were significantly lower than the bioassay results.

A collaborative study (Toepfer, 1957) showed that using Dowex-treated pigeon liver enzyme or a commercially purified hog kidney enzyme in combination with intestinal phosphatase produced similar results for the analysis of alfalfa leaf meal, whole egg powder and dried brewer's yeast. However, there was considerable variation among the data reported by collaborators. A loss of activity of these enzyme preparations showed a need for establishing the activity of the enzymes before they could be relied upon for use in an official assay procedure. Measurement of enzyme activity requires a stable standard of bound pantothenic acid, but such a standard has yet to be found. Coenzyme A did not prove satisfactory because only minute quantities of enzymes could release most of the bound pantothenate (Toepfer, 1960). The AOAC microbiological method (AOAC, 1990) is limited to the determination of free pantothenate in vitamin preparations, and omits enzymatic hydrolysis.

In the procedure employed by Bell (1974), the test material is homogenized with 10 ml of 0.2 M tris-(hydroxymethyl)-methylamine buffer (pH 8.0) at 70 °C, and then autoclaved for 15 min at 121 °C. After mixing and cooling to room temperature, 1.0 ml of 2% w/v alkaline phosphatase (Sigma) in 'tris' buffer and 0.5 ml of Dowex-treated pigeon liver enzyme solution are added, and the mixture is incubated at 37 °C overnight. On the following day, the extracts are steamed in an autoclave for 5 min to destroy any remaining enzyme activity, cooled to room temperature and diluted to a suitable volume with 'tris' buffer. The extracts are centrifuged at 15 000 × g for 20 min (additional step used by Finglas *et al.*, 1988), and then filtered through Whatman No. 42 (or, if non-fatty, through

Whatman No. 541) filter paper into polythene bottles for storage at −20 °C pending analysis. For conventional assay procedures using double-strength media, the filtrates are adjusted to pH 6.8 with NaOH solution and diluted to place them in the range of the standard curve. A reagent blank is taken through the same procedure.

Walsh, Wyse and Hansen (1979) omitted the slow and troublesome filtration step, and used dialysis to isolate the liberated pantothenate from the large amount of protein in the food digest.

11.5.4 Radioimmunoassay

Walsh, Wyse and Hansen (1979) compared a radioimmunoassay method with the microbiological (*L. plantarum*) method for the determination of pantothenic acid in 75 processed and cooked foods. The pantothenic acid was released from aqueous sample homogenates by autoclaving at 121 °C for 10 min, followed by incubation with phosphatase–liver enzyme, and the protein was removed by dialysis. Antibody was prepared by injecting rabbits with a pantothenic acid–bovine serum albumin (PA–BSA) conjugate, and the resulting antiserum was diluted 100-fold with a solution of rabbit albumin. Each assay tube contained 0.5 ml of the diluted antiserum, 0.5 ml of standard or sample, and 50 μl of [^{3}H]sodium d-pantothenate. After incubation at room temperature for 15 min, neutral saturated ammonium sulphate was added to achieve a 50% saturation, and the suspension was centrifuged. The precipitate was washed with 0.5 ml of 50%-saturated ammonium sulphate and recentrifuged. The washed precipitate, containing antibody-bound pantothenic acid, was dissolved in 0.5 ml of tissue solubilizer and transferred quantitatively to vials containing 12 ml of scintillation fluid. The radioactivity results (cpm) were read on a 5–150 ng (per 0.5 ml) standard curve. An enzyme blank value was subtracted from each sample value. Although the results from the radioimmunoassay and microbiological assay were highly correlated ($r = 0.94$), the microbiological assay produced a higher average result for the food types meats, fruits and vegetables, and breads and cereals. It was postulated that bacterial enzymes in the assay organism promote further breakdown of bound pantothenic acid, or non-enzymatic breakdown occurs during the long microbiological incubation period. The results of the individual foods analysed by the radioimmunoassay method have been reported by Walsh, Wyse and Hansen (1981).

11.5.5 Enzyme-linked immunosorbent assay (ELISA)

Morris *et al.* (1988) developed an indirect two-site noncompetitive ELISA by raising polyclonal antibodies in rabbits against pantothenic acid. The PA–BSA immunogen was prepared by reacting the primary alcohol group of the PA molecule with bromoacetyl bromide to form bromoacetyl pantothenate which, in turn, was reacted with denatured reduced BSA. The immunogen was purified by extensive dialysis and column chromatography on Sephadex G 25. The protein conjugate for plate coating was pantothenic acid–keyhole limpet haemocyanin (PA–KLH) produced by the bromoacetyl procedure, as for the immunogen. Anti-rabbit immunoglobulin–alkaline phosphatase conjugate was used as the enzyme-labelled second antibody. The ELISA system was highly specific for pantothenic acid, and did not recognize coenzyme A, panthenol or pantheneine. The lower limit of detection was 0.5 ng pantothenic acid per well.

The validation and application of the ELISA system for the analysis of six foods representing major sources of pantothenic acid in the UK diet was reported by Finglas *et al.* (1988). Sample preparation entailed autoclaving at 121 °C for 15 min, homogenization, and overnight incubation with phosphatase–pigeon liver enzyme. The following day, the sample hydrolysates were autoclaved at 121 °C for 5 min to destroy any remaining enzyme activity. The ELISA values obtained for the six foods compared favourably ($r = 0.999$) with values obtained by the microbiological method of Bell (1974) using *L. plantarum*.

11.6 BIOAVAILABILITY

11.6.1 Physiological aspects

Intestinal absorption

Ingested coenzyme A, the major dietary form of pantothenic acid, is hydrolysed in the intestinal lumen to pantetheine by the nonspecific action of pyrophosphatases and phosphatase (Hoffmann–La Roche Inc., 1992). Pantetheine is then split into pantothenic acid and β-mercaptoethylamine by the action of pantotheinase secreted from the intestinal mucosa into the lumen (Shibata, Gross and Henderson, 1983). Absorption of the liberated pantothenic acid takes place mainly in the jejunum. Earlier reports that absorption occurs by simple diffusion resulted from the use of unphysiologically high pantothenic acid concentrations. At physiological intakes pantothenic acid traverses the brush-border membrane by a sodium-dependent secondary active transport process similar to that described for the uptake of glucose. Within the alkaline medium of the intestinal chyme, the vitamin exists primarily as the pantothenate anion. As the transport process does not respond to an electrical gradient, the coupling ratio of Na^+ to pantothenate$^-$ is assumed to be $1:1$ (Fenstermacher and Rose, 1986). The process by which pantothenic acid exits the enterocyte at the basolateral membrane has not been characterized.

Unlike other water-soluble vitamins which are absorbed by specific carrier-mediated mechanisms, the absorption of pantothenic acid is not regulated by its level of dietary intake (Stein and Diamond, 1989). Clear-cut pantothenic acid deficiency symptoms in humans are rarely found in practice and the vitamin is nontoxic at high doses. These factors could explain why a regulated absorption mechanism has not evolved for this vitamin.

Renal reabsorption

Reabsorption of pantothenic acid in the brush-border membrane of the proximal convoluted tubule takes place by an electrogenic sodium-dependent secondary active transport system with a $2:1$ Na^+/ pantothenate$^-$ stoichiometry (Barbarat and Podevin, 1986).

Other metabolic considerations

After absorption, free pantothenic acid is conveyed to the body tissues in the plasma from which it is taken up by most cells, including the erythrocytes. Uptake by the liver (Smith and Milner, 1985) and heart (Lopaschuk, Michalak and Tsang, 1987) takes place by secondary active transport using the sodium electrochemical gradient.

Altered metabolism of CoA has been observed in starvation, alcoholism, diabetes and diverse disease states. Hormones such as glucocorticoids, insulin and glucagon also affect tissue CoA levels. It is not known whether the altered CoA metabolism can be implicated in the manifestations of a disease (Tahiliani and Beinlich, 1991).

Reibel *et al.* (1981) studied the effects of fasting and diabetes on pantothenic acid (PA) metabolism in rats. Both fasting and diabetes resulted in accelerated rates of [^{14}C]PA uptake, higher tissue concentrations of PA, increased incorporation of [^{14}C]PA into CoA, and elevated tissue concentrations of [^{14}C]PA in the liver. In cardiac muscle, PA uptake and PA levels in fasting and diabetic animals were lowered and yet incorporation of [^{14}C]PA into CoA and CoA levels were increased. Thus CoA synthesis is not governed by PA availability in the heart. Uptake of [^{14}C]PA by skeletal muscle was also reduced in diabetic animals. The data reported suggest that pantothenic acid uptake by tissues is under metabolic or hormonal control. The authors postulated that control of uptake provides a means whereby the large body pool of pantothenic acid present in muscle can be diverted to the liver during hormonal or dietary imbalance. This regulatory mechanism would ensure that the liver, which normally has low endogenous levels of pantothenic acid, has a readily available supply of the vitamin for CoA synthesis.

Free pantothenic acid is excreted in human urine, mainly as a metabolite of CoA. There is a positive correlation between urinary excretion and dietary intake of the vitamin in adults (Tahiliani and Beinlich, 1991). In rats, both fasting and diabetes resulted in decreased urinary pantothenic acid excretion (Reibel *et al.*, 1981), thus conserving the vitamin under these conditions.

11.6.2 Dietary sources and their bioavailability

Dietary sources

Pantothenic acid is widely distributed in foods of animal and plant origin. Coenzyme A is the major pantothenic acid-containing compound present in animal tissues, but small amounts of other bound forms (phosphopantothenic acid, pantetheine and phosphopantetheine), in addition to free pantothenic acid, have also been detected (Brown, 1959). In human breast milk the vitamin is present entirely as free pantothenic acid (Song, 1993). Significant amounts of bound pantothenic acid other than coenzyme A have been found in milk (Voigt and Eitenmiller, 1978).

Pantothenic acid is particularly abundant in animal sources, legumes, nuts and whole grain cereals. Rich food sources (100–200 µg/g dry weight) include egg yolk, kidney, liver and yeast (McCormick, 1988). Lean meat, milk, potatoes and green leafy vegetables contain lesser amounts, but are important food sources because of the quantities consumed. Especially high levels are found in royal jelly from the queen bee. Relatively low amounts of the vitamin are found in many highly processed foods (Walsh, Wyse and Hansen, 1981).

Bioavailability

Biosynthesis of pantothenic acid by microflora in the human intestine has been suspected, but the amounts produced and the nutritional availability of the vitamin from this source are unknown (National Research Council, 1989)

Little information is at hand regarding the nutritional availability to humans of pantothenic acid in food commodities. Based on the urinary excretion of pantothenic acid, the availability for male human subjects ingesting the 'average American diet' ranged from 40% to 61%, with a mean of 50% (Tarr, Tamura and Stokstad, 1981).

Effects of alcohol

A review of studies investigating pantothenic acid status in alcoholics indicates a lowering of urinary excretion and blood levels (Bonjour, 1980). Ethanol administered to rats *in vivo* or added to cultured hepatocytes *in vitro* inhibited conversion of [^{14}C]PA to [^{14}C]CoA by inhibition of pantothenate kinase (Iannucci *et al.*, 1982; Israel and Smith, 1987).

11.6.3 Effects of food processing and cooking

Pantothenic acid has good stability in most foods during processing, but is susceptible to leaching in the blanching of vegetables and home cooking. The roasting of meat causes degradation of < 10%, but the meat drippings contain 20–25% of the initial vitamin content (Hoffmann–La Roche Inc., 1992).

It has been reported by Schroeder (1971) that the canning and freezing of foods incurs large losses of pantothenic acid. In canned foods of animal origin, losses ranged from 20% to 35%, and in vegetable foods from 46% to 78%. Freezing losses in animal foods ranged from 21% to 70%, and in vegetable foods from 37% to 57%. Grains lost 37–74% of the original pantothenic acid content during conversion to various cereal products, and meats lost 50–75% during conversion to comminuted products. These processing losses were calculated from data on corresponding raw materials rather than on identical materials before and after processing, and have been strongly criticized (Orr and Watt, 1972). However, later studies have also indicated large losses of pantothenic acid in many highly processed foods, including products made from refined grains, fruit products and fat- or cereal-extended meats and fish (Walsh, Wyse and Hansen, 1981). Losses of pantothenic acid during milk pasteurization, sterilization and drying are usually less than 10% (Archer and Tannenbaum, 1979).

11.7 DIETARY INTAKE

The typical intake of pantothenic acid in the United States has been reported to range from 5 to 10 mg/day, with an average of *c.*6 mg/day (National Research Council, 1989). There is insufficient evidence to establish an RDA for the vitamin. However, the Food and Nutrition Board of the US National Academy of Sciences suggested that a safe and adequate intake for adults and adolescents is 4–7 mg/day. The Board suggested lower intakes for younger age groups, ranging from 2–3 mg/day for infants up to 3 years to 4–5 mg/day for children of 7–10 years.

Tarr, Tamura and Stokstad (1981) reported that the average American diet contained 5–8 mg of available pantothenic acid. If the bioavailability of the vitamin from food is 50% as suggested by these authors, the provisional requirements may be more than currently accepted.

Effects of high intakes

There is no evidence that therapeutic doses (10–100 mg range) of pantothenate cause any harmful effects. Even with oral amounts as high as
10 20 g of the calcium salt, the only reported problem was occasional
diarrhoea (McCormick, 1988).

Assessment of nutritional status

There are no satisfactory functional tests of pantothenic acid nutritional
status for humans and serum or plasma concentrations of the vitamin are
not applicable. Since urinary output of pantothenic acid is directly proportional to dietary intake, urinary excretion provides a means of assessing status. Urinary excretion is normally within the range 1–15 mg/day
(mean 4 mg/day) and an output of < 1 mg/day is considered to be
abnormally low. Whole blood concentrations normally range between
103 and 183 µg/100 ml and a concentration of < 100 µg/100 ml signifies
inadequate intake of pantothenic acid (McCormick, 1988). Measurement
of pantothenic acid concentration in urine and blood is carried out by
microbiological assay or radioimmunoassay.

11.8 SYNOPSIS

Pantothenic acid is widely distributed in foods and a dietary deficiency of
this vitamin is extremely unlikely, apart from circumstances of severe
malnutrition.

The inclusion of knowledge of the bioavailability of dietary pantothenic
acid to humans, but one study has shown that the availability for male
subjects ingesting the 'average American diet' is 50% on average. Large
losses of pantothenic acid have been reported to take place in many
highly processed foods.

The 'free' (no enzyme treatment) or 'total' (after enzyme treatment)
pantothenic acid content of a food is traditionally determined by microbiological assay with *L. plantarum*. Either the radioimmunoassay or the
ELISA can substitute for the microbiological assay in the determination
of total pantothenic acid, as shown by the high correlation obtained for
the analysis of foods using the same extract. The inherent practical
advantages of the ELISA over the radioimmunoassay, and the better
agreement with microbiological results, suggest that the ELISA could
be the method of choice for determining pantothenic acid. The double
enzyme extraction procedure is the time-limiting step in the batchwise
analysis of large numbers of samples, and the need to establish the
activity of the enzymes is an additional problem.

REFERENCES

AOAC (1990) Pantothenic acid in vitamin preparations. Microbiological methods. Final action 1960. In *AOAC Official Methods of Analysis*, 15th edn (ed. K. Helrich), Association of Official Analytical Chemists, Inc., Arlington, VA, 954.74.

Archer, M.C. and Tannenbaum, S.R. (1979) Vitamins. In *Nutritional and Safety Aspects of Food Processing* (ed. S.R. Tannenbaum), Marcel Dekker, Inc., New York, pp. 47–95.

Barbarat, B. and Podevin, R.-A. (1986) Pantothenate-sodium cotransport in renal brush-border membranes. *J. Biol. Chem.*, **261**, 14455–60.

Bell, J.G. (1974) Microbiological assay of vitamins of the B group in foodstuffs. *Lab. Pract.*, **23**, 235–42, 252.

Bird, O.D. and Thompson, R.Q. (1967) Pantothenic acid. I. Introduction. In *The Vitamins. Chemistry, Physiology, Pathology, Methods*, 2nd edn, Vol. VII (eds P. György and W.N. Pearson), Academic Press, New York, pp. 209–19.

Bonjour, J.-P. (1980) Vitamins and alcoholism. V. Riboflavin, VI. Niacin, VII. Pantothenic acid, and VIII. Biotin. *Int. J. Vitam. Nutr. Res.*, **50**, 425–40.

Brown, G.M. (1959) Assay and distribution of bound forms of pantothenic acid. *J. Biol. Chem.*, **234**, 379–82.

Clegg, K.M. (1958) The microbiological determination of pantothenic acid in wheaten flour. *J. Sci. Food Agric.*, **9**, 366–70.

Fenstermacher, D.K. and Rose, R.C. (1986) Absorption of pantothenic acid in rat and chick intestine. *Am. J. Physiol.*, **250**, G155–60.

Finglas, P.M., Faulks, R.M., Morris, H.C. *et al.* (1988) The development of an enzyme-linked immunosorbent assay (ELISA) for the analysis of pantothenic acid and analogues, part II – determination of pantothenic acid in foods. *J. Micronutr. Anal.*, **4**, 47–59.

Hoffmann–La Roche Inc. (1992) Vitamins, part XVI: pantothenic acid. In *Encyclopedia of Food Science and Technology*, Vol. 4 (ed. Y.H. Hui), John Wiley & Sons, Inc., New York, pp. 2783–7.

Iannucci, J., Milner, R., Arbizo, M.V. and Smith, C.M. (1982) The effect of ethanol and acetaldehyde on [^{14}C]pantothenate incorporation into CoA in cultured rat liver parenchymal cells. *Arch. Biochem. Biophys.*, **217**, 15–29.

Israel, B.C. and Smith, C.M. (1987) Effects of acute and chronic ethanol ingestion on pantothenate and CoA status of rats. *J. Nutr.*, **117**, 443–51.

Lopaschuk, G.D., Michalak, M. and Tsang, H. (1987) Regulation of pantothenic acid transport in the heart. *J. Biol. Chem.*, **262**, 3615–9.

McCormick, D.B. (1988) Pantothenic acid. In *Modern Nutrition in Health and Disease*, 7th edn (eds M.E. Shils and V.R. Young), Lea & Febiger, Philadelphia, pp. 383–7.

Morris, H.C., Finglas, P.M., Faulks, R.M. and Morgan, M.R.A. (1988) The development of an enzyme-linked immunosorbent assay (ELISA) for the analysis of pantothenic acid and analogues. Part 1 – Production of antibodies and establishment of ELISA systems. *J. Micronutr. Anal.*, **4**, 33–45.

National Research Council (1989) Water-soluble vitamins. In *Recommended Dietary Allowances*, 10th edn, National Academy Press, Washington, DC, pp. 115–73.

Novelli, G.D. and Schmetz, F.J. Jr (1951) An improved method for the determination of pantothenic acid in tissues. *J. Biol. Chem.*, **192**, 181–5.

Novelli, G.D., Kaplan, N.O. and Lipmann, F. (1949) The liberation of pantothenic acid from coenzyme A. *J. Biol. Chem.*, **177**, 97–107.

Orr, M.L. and Watt, B.K. (1972) Losses of vitamins and trace minerals resulting from processing and preservation of foods. *Am. J. Clin. Nutr.*, **25**, 647–8.

Reibel, D.K., Wyse, B.W., Berkich, D.A. *et al.* (1981) Effects of diabetes and fasting on pantothenic acid metabolism in rats. *Am. J. Physiol.*, **240**, E597–E601.

Schroeder, H.A. (1971) Losses of vitamins and trace minerals resulting from processing and preservation of foods. *Am. J. Clin. Nutr.*, **24**, 562–73.

Thibrala, K., Gross, C.J. and Henderson, L.M. (1983) Hydrolysis and absorption of pantothenate and its coenzymes in the rat small intestine. *J. Nutr.*, **113**, 2107–15.

Skeggs, H.R. and Wright, L.D. (1944) The use of *Lacotobacillus arabinosus* in the microbiological determination of pantothenic acid. *J. Biol. Chem.*, **156**, 21–6.

Smith, C.M. and Milner, R.E. (1985) The mechanism of pantothenate transport by rat liver parenchymal cells in primary culture. *J. Biol. Chem.*, **260**, 4823–31.

Song, W.O. (1993) Pantothenic acid. Physiology. In *Encyclopaedia of Food Science, Food Technology and Nutrition*, Vol. 6 (eds R. Macrae, R.K. Robinson and M.J. Sadler), Academic Press, London, pp. 3402–3408.

Stein, E.D. and Diamond, J.M. (1989) Do dietary levels of pantothenic acid regulate its intestinal uptake in mice? *J. Nutr.*, **119**, 1973–83.

Tahiliani, A.G. and Beinlich, C.J. (1991) Pantothenic acid in health and disease. *Vitams Horm.*, **46**, 165–228.

Tannenbaum, S.R., Young, V.R. and Archer, M.C. (1985) Vitamins and minerals. In *Food Chemistry*, 2nd edn (ed. O.R. Fennema), Marcel Dekker, Inc., New York, pp. 477–544.

Tarr, J.B., Tamura, T. and Stokstad, E.L.R. (1981) Availability of vitamin B-6 and pantothenate in an average American diet in man. *Am. J. Clin. Nutr.*, **34**, 1328–37.

Toepfer, E.W. (1957) Report on pantothenic acid. Assay for total pantothenic acid: 1956. Collaborative study. *J. Ass. Off. Agric. Chem.*, **40**, 853–5.

Toepfer, E.W. (1960) Microbiological assay for total pantothenic acid. *J. Ass. Off. Analyt. Chem.*, **43**, 28–9.

Toepfer, E.W., Zook, E.G. and Richardson, L.R. (1954) Microbiological procedure for the assay of pantothenic acid in foods: results compared with those by bioassay. *J. Ass. Off. Agric. Chem.*, **37**, 182–90.

Voigt, M.N. and Eitenmiller, R.R. (1978) Comparative review of the thiochrome, microbial and protozoan analyses of B-vitamins. *J. Food Protection*, **41**, 730–8.

Walsh, J.H., Wyse, B.W. and Hansen, R.G. (1979) A comparison of microbiological and radioimmunoassay methods for the determination of pantothenic acid in foods. *J. Food Biochem.*, **3**, 175–89.

Walsh, J.H., Wyse, B.W. and Hansen, R.G. (1981) Pantothenic acid content of 75 processed and cooked foods. *J. Am. Diet. Ass.*, **78**, 140–4.

Wyse, B.W., Song, W.O., Walsh, J.H. and Hansen, R.G. (1985) Pantothenic acid. In *Methods of Vitamin Assay*, 4th edn (eds J. Augustin, B.P. Klein, D. Becker and P.B. Venugopal), John Wiley & Sons, New York, pp. 399–416.

12

Biotin

12.1 INTRODUCTION

Biotin was discovered as a result of several independent lines of investigation. In 1916 Bateman reported that the inclusion of large proportions
of raw egg white in experimental diets produced a characteristic 'spec-
tacle-eye' hair loss and dermatitis in rats. The causative factor later
proved to be avidin, a glycoprotein component of egg white, which
binds biotin and renders it biologically unavailable to the rat. Boas in
1927 and later György in 1931 found a curative factor in liver against the
egg white injury. In 1936 Kögl and Tönnis isolated from egg yolk a
crystalline substance which acted as a growth factor for yeast, and
which they named biotin. Independent research conducted by György
and du Vigneaud led to the conclusion in 1940 that the above curative
and growth factors were chemically identical to each other. The structure
of biotin was established in 1942 by du Vigneaud's group.

Biotin functions as the prosthetic group in carboxylase enzymes, which
are capable of taking up bicarbonate ions as the carboxyl donor and
transferring them to specific substrates. The biotin-dependent enzymes

catalyse many metabolic reactions including pathways for fatty acid biosynthesis, gluconeogenesis, and catabolism of several branched-chain amino acids and odd-carbon-chain fatty acids.

12.2 CHEMICAL STRUCTURE AND NOMENCLATURE

The biotin molecule is a fusion of an imidazolidone ring with a tetrahydrothiophene ring bearing a pentanoic acid side-chain and named hexahydro-2-oxo-1*H*-thienol(3,4-d)imidazole-4-pentanoic acid ($C_{10}H_{16}O_3$ N_2S; MW = 244.31). The molecule contains three asymmetric carbon atoms, and hence eight stereoisomers are possible. Of these, only the dextrorotatory d-biotin occurs in nature and possesses vitamin activity. Biotin synthesized industrially by the Hoffmann–La Roche process is also in the d-form (Achuta Murthy and Mistry, 1977). The stereochemical structure of d-biotin (Figure 12.1) reveals that the two rings are fused in the *cis* configuration and the aliphatic side chain is *cis* with respect to the imidazolidone ring.

In animal and plant tissues only a small proportion of the biotin present occurs in the free state. The majority is covalently bound to the protein structure (apoenzyme) of biotin-dependent enzymes via an amide bond between the carboxyl group of biotin and the ε-amino group of a lysine residue (Zapsalis and Beck, 1985). Proteolysis of the enzyme liberates a natural water-soluble fragment called biocytin (ε-*N*-biotinyl-L-lysine) (Figure 12.2), which is biologically active.

12.3 DEFICIENCY SYNDROMES

The human requirements for biotin are normally met through the combined dietary supply and endogenous microbial synthesis in the gut. A primary biotin deficiency state is therefore extremely rare, especially where a well-balanced diet prevails. Deficiency states have been induced in adult volunteers by feeding low-biotin diets containing a high proportion of raw egg white. An initial scaly dermatitis was followed by nonspecific symptoms that included extreme lassitude, anorexia,

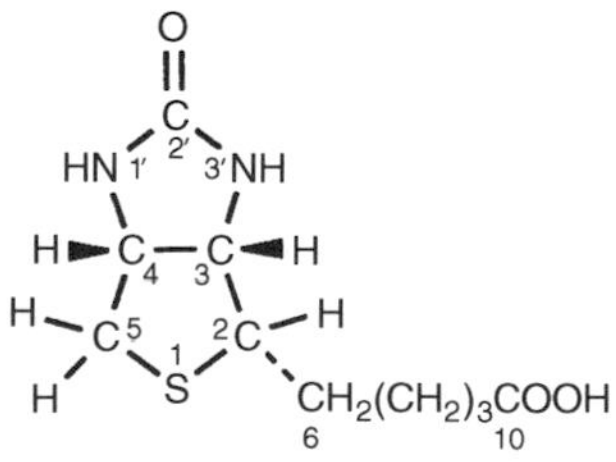

Figure 12.1 The absolute stereochemistry of d-biotin.

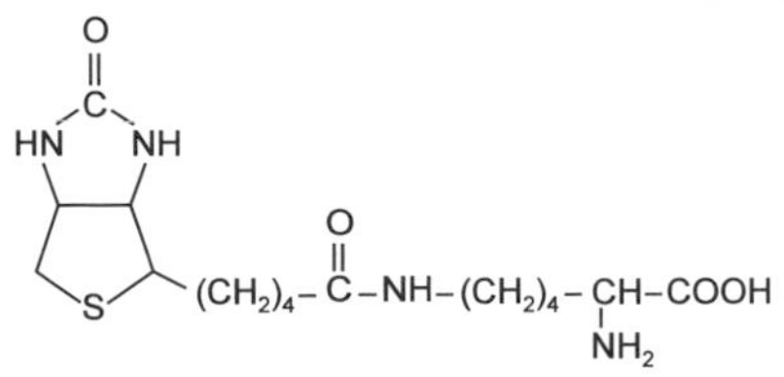

Figure 12.2 Structure of biocytin.

muscular pains, hyperaesthesia and localized paraesthesia. All of these symptoms responded to injections of 150–300 μg of biotin per day. Seborrhoeic dermatitis of the scalp and a more generalized dermatitis known as Leiner's disease have been reported in breast-fed infants when the mother is malnourished. These symptoms are relieved when biotin is administered to the mother.

12.4 PHYSICOCHEMICAL PROPERTIES

Solubility and other properties

Synthetic d-biotin crystallizes as fine colourless needles which melt at 232 °C with decomposition. In its free acid form biotin is only very sparingly soluble in water at 25 °C (20 mg/100 ml) and in 95% ethanol (80 mg/100 ml) but is more soluble in hot water. It is soluble in dilute alkali, sparingly soluble in dilute acid, and practically insoluble in fat solvents, including chloroform, diethyl ether and petroleum ether. The pH of a 0.01% aqueous solution of biotin is 4.5; the pK_a value is 4.51. Biotin salts are significantly more water-soluble than the acid form of biotin.

Spectroscopic properties

The biotin molecule does not contain a characteristic chromophore, apart from the carbonyl groups; hence it exhibits only very weak absorbance at 204 nm.

Stability

Dry crystalline biotin is stable to heat and atmospheric oxygen but is gradually destroyed by UV radiation. Aqueous solutions within the pH range 4–9 are stable up to 100 °C, but solutions that are more acid or alkaline are unstable to heating. The sulphur atom can be readily oxidized with various reagents to yield a mixture of isomeric d- and l-biotin sulphoxides; stronger oxidants form biotin sulphone. These

oxidation products are essentially inactive for humans, although the
d-sulphoxide can be slowly reduced metabolically to biotin (McCormick,
1988). Bound forms of biotin may be freed by strong acid hydrolysis at
elevated temperature or with a proteolytic enzyme (Lampen, Bahler and
Peterson, 1942). Aqueous solutions of biotin are very susceptible to
mould growth.

12.5 ANALYSIS

12.5.1 Scope of analytical techniques

Biotin occurs covalently bound to proteins as well as in the free state. In
addition to the conventional microbiological assay for determining biotin, biospecific methods have been developed utilizing radiometric protein-binding assays (radioassays) and non-isotopic protein-binding
assays. Biotin can be assayed biologically by the egg white injury test in
rats.

No IU of biotin activity has been defined and analytical results are
expressed in weight units (μg) of pure d-biotin.

12.5.2 Extraction techniques

Bound forms of biotin, including biocytin, cannot be utilized by *L. plantarum*, the organism usually employed in microbiological assays, and
strong acid hydrolysis at elevated temperature is required to liberate
biotin completely from natural materials (Thompson, Eakin and Williams, 1941). Animal tissues require more stringent hydrolysis conditions
than do plant tissues, because the latter contain a higher proportion of
free water-extractable biotin (Lampen, Bahler and Peterson, 1942).
Experimental studies with meat and meat products (Schweigert *et al.*,
1943) and feedstuffs of animal origin (Scheiner and De Ritter, 1975)
showed that maximum liberation of biotin in animal-derived products
is obtained by autoclaving with $6 N$ H_2SO_4 for 2 h at 121 °C. This procedure promotes losses of biotin in plant materials, which are extracted
most efficiently by autoclaving with $4 N$ H_2SO_4 for 1 h at 121 °C (Lampen,
Bahler and Peterson, 1942) or with $2 N$ H_2SO_4 for 2 h at 121 °C (Scheiner
and De Ritter, 1975). Because of the differences in extractability between
animal and plant tissues, a single extraction procedure to cover all food
commodities must be a compromise, and no such procedure has been
universally adopted. Representative methods for extracting foods of any
type entail autoclaving with $2 N$ H_2SO_4 at 121 °C for 2 h (Lampen, Bahler
and Peterson, 1942; Strohecker and Henning, 1966) or $3 N$ H_2SO_4 at
121 °C for 1 h (Shull, Hutchings and Peterson, 1942) or 30 min (Barton-
Wright, 1952; Bell, 1974).

Hydrolysis with 6 N H_2SO_4 destroys the synthetic sodium salt of biotin added to feed premixes. A suggested procedure for extracting feed premixes with biotin potencies up to 1 g/lb entailed the addition of 50 ml of 0.1 N NaOH and 250 ml of water to 5 g of sample, shaking vigorously, and then standing for 30 min at room temperature with occasional swirling (Scheiner, 1966).

Sulphuric acid, rather than hydrochloric acid, is invariably used for sample hydrolysis, as the biotin content of dilute (30 ng/ml) solutions is almost completely destroyed by autoclaving with 2 N HCl (Axelrod and Hofmann, 1950). Evidence from differential microbiological assay points to the oxidation of biotin to a mixture of its sulphoxide and sulphone derivatives, possibly caused by trace impurities (e.g. chlorine) in the acid. This loss of vitamin activity does not necessarily occur when autoclaving actual food samples, as many natural products are capable of preventing this oxidation (Langer and György, 1968). Finglas, Faulks and Morgan (1986) reported no loss of biotin from liver using 3 N HCl.

An alternative extraction procedure to acid hydrolysis is digestion of the sample with the proteolytic enzyme papain (György, 1967). Bitsch, Salz and Hötzel (1989a) compared papain digestion (after treatment with liquid nitrogen and homogenization) with conventional acid hydrolysis for its efficiency in releasing the biotin from a variety of food samples (wheat flour, oat flakes, whole milk and pork liver). No significant differences in biotin values were obtained between extraction methods using either a microbiological assay with *L. plantarum* or a radiometric protein-binding assay. The results indicated that enzymatic hydrolysis with papain, after treatment of the sample with liquid nitrogen and homogenization, is capable of quantitatively releasing protein-bound biotin from both plant and animal tissues. Similar experiments on human plasma within the same study revealed slightly enhanced biotin values obtained with the microbiological assay in relation to the protein-binding assay. In plasma, biotin exists mainly as protein-bound biocytin which, when released from the plasma proteins, is bound to the protein (avidin) used in the protein-binding assay. If papain digestion is ineffective in hydrolysing biocytin, as reported by Wright *et al.* (1952), one would expect the microbiological results to be low in relation to the protein-binding assay results, since *L. plantarum* responds only to free biotin. These results for plasma therefore suggest that biotin was, in fact, released from biocytin by papain digestion under the conditions used.

12.5.3 Microbiological assay

The assay organism *L. plantarum* shows a high specificity towards free d-biotin and does not respond to biocytin or other bound forms. Biotin-d-sulphoxide elicits an equal growth response to that of d-biotin, while biotin sulphone is inhibitory (Langer and György, 1968). However, these oxidation products are not likely to be present in food samples in sufficient concentration to cause significant interference. *L. plantarum* is stimulated by non esterified unsaturated fatty acids and other lipids when biotin is present in suboptimal amounts (Williams, 1945; Williams and Fieger, 1945; Broquist and Snell, 1951). The high sensitivity of the biotin assay (0.1–1.0 ng per assay tube; Dakshinamurti *et al.*, 1974) necessitates a high sample dilution, and this will reduce this growth stimulation. The high sample dilution has the added advantage of lowering the initially high concentration of salt produced on neutralization of the acid extracts. The basal medium used for nicotinic acid and pantothenic acid can be used for biotin assays with the exclusion of the relevant vitamin (in this case biotin) from the formulation. The use of the same organism (*L. plantarum*) for assaying the three vitamins also eliminates the need for preparing separate stock cultures.

It has been recommended to remove lipid material by adjusting the pH of the acid hydrolysate to 4.5 and filtering through paper (Skeggs, 1963; Strohecker and Henning, 1966). [Note: Guilarte (1985) reported a possible loss of biotin during filtration.] This procedure might not be adequate for samples with a high fat content (e.g. wheat germ, meat, eggs), which should be defatted by a preliminary Soxhlet extraction with light petroleum (Barton-Wright, 1952).

12.5.4 Radioassay

Radioassay techniques for determining biotin are based on the high affinity of the glycoprotein avidin for the functional ureido group of the biotin molecule. Early methods used [^{14}C]biotin, which has a specific radioactivity of 45 mCi/mmol, but a higher sensitivity can be obtained using [^{3}H]biotin of specific activity 2.5 Ci/mmol (Hood, 1979). In a procedure described by Hood (1975), samples of pelleted poultry feeds and wheat were autoclaved with 2 N H_2SO_4 for 1 h at 121 °C, and then neutralized with 20% NaOH. The filtered extracts were incubated with [^{14}C]biotin and the avidin–biotin complex was precipitated with 2% zinc sulphate solution. The method was reported to be capable of measuring biotin levels down to 5 µg/kg of biological material and was more than adequate for analysing wheat and poultry feeds, which contained from 68 to 341 µg/kg. Results obtained by radioassay and microbiological (*L. plantarum*) assay were similar for poultry feeds, but the radioassay values for two wheat samples were approximately 20% and 55% higher than the microbiological assay values.

Bitsch, Salz and Hötzel (1986, 1989a) used [^{3}H]biotin as the tracer and removed nonbound biotin by adsorption on dextran-coated charcoal. Values for biotin concentration obtained by this method for meat, offal, cereal products, milk and vegetables generally agreed well with data from food composition tables, but radioassay values for cabbage and bananas were higher than literature values.

12.6 BIOAVAILABILITY

12.6.1 Physiological aspects

Biotin in foods exists as the free vitamin and as protein-bound forms in variable proportions. Because the amide bond linking biotin and lysine is not hydrolysed by gastrointestinal proteases, proteolytic digestion of protein-bound biotin releases not biotin, but biocytin and biotin-containing short peptides (biotinyl peptides). Biotinidase, which is present in pancreatic juice and in intestinal mucosa, is capable of hydrolysing biocytin to yield free biotin. The release of biotin might occur during the luminal phase of proteolysis by the action of pancreatic biotinidase or in the intestinal mucosa by the action of mucosal biotinidase. Biocytin may also be absorbed intact and be acted upon by biotinidase present in plasma. Biotinyl peptides can also be absorbed directly.

Avidin, the glycoprotein in egg white, uniquely binds biotin, both in the free form and as the prosthetic group of enzymes. The complex is stable over a wide pH range and is not dissociated in the gastrointestinal tract. When eggs are cooked, the avidin is denatured and the biotin is liberated.

Biotin is conserved very efficiently in the body owing to reabsorption by the kidney and the protein binding of plasma biotin, which reduces filtration at the glomerulus.

Intestinal absorption

Transport of biotin across the brush-border membrane of the enterocyte takes place by a carrier-mediated process at physiological concentrations and by simple diffusion at higher concentrations (Bowman, Selhub and Rosenberg, 1986; Dakshinamurti, Chauhan and Ebrahim, 1987; Said and Redha, 1987). Transport occurs without metabolic alteration to the biotin molecule. In the human intestine, the carrier-mediated system is sodium- and energy-dependent, and is predominant in the jejunum (Said, Redha and Nylander, 1987). The system resembles that for D-glucose in that a sodium gradient maintained by the basolateral sodium pump provides the energy required for the uphill movement of biotin against its concentration gradient. However, unlike the electrogenic transport of glucose, the transport of biotin by the carrier-mediated system is electroneutral. Biotin at physiological pH exists mostly in the anionic form and therefore the stoichiometric coupling ratio of biotin and sodium transport is probably 1 : 1. Transport in both the jejunum and the ileum of the rat is appropriately increased and suppressed by biotin deficiency and biotin excess, respectively, due to changes in the number of transport carriers (Said, Mock and Collins, 1989)

The rate of biotin transport was shown to increase with a decrease in pH of the incubation medium from 8 to 5.5 (Said, Redha and Nylander, 1988a). This effect presumably occurs through an increase in the transport of biotin by the nonmediated process and is due to an increase in the percentage of the easily diffusible unionized form of the vitamin under acidic conditions.

The exit of biotin from the enterocyte, i.e transport across the basolateral membrane, is a carrier-mediated process which has a higher affinity for the substrate than the transport system at the brush-border membrane. The basolateral transport process is independent of sodium, electrogenic in nature, and not capable of accumulating biotin against a concentration gradient (Said, Redha and Nylander, 1988b).

There are conflicting data regarding the intestinal transport of biocytin. Dakshinamurti, Chauhan and Ebrahim (1987) concluded that the transport is carrier-mediated and more efficient than the uptake of free biotin. Said *et al.* (1993), on the other hand, found that the transport of biocytin occurs by a nonmediated process that is independent of sodium, pH, energy and temperature, and is significantly less efficient than that of free biotin. The mechanism for the transport of biotinyl peptides is unknown; it might involve a specific biotin transporter or a nonspecific pathway for peptide absorption (Mock, 1990).

Renal reabsorption

Free biotin is transported across the brush-border membrane of the renal proximal tubular cell by an electrogenic, sodium–biotin coupled process with a 2 : 1 Na^+/biotin stoichiometry. Intracellular biotin diffuses passively across the basolateral membrane into the peritubular fluid to complete the process of transepithelial absorption (Podevin and Barbarat, 1986). The mediated transport system of the renal brush-border membrane has an affinity for biocytin as well as for biotin (Spencer and Roth, 1988). This property contributes to the conservation of biotin, since biotinidase is not enriched in the renal brush-border membrane.

Other metabolic considerations

There is controversy concerning the transport of biotin to the liver and peripheral tissues (Mock, 1990), but it is possible that the enzyme biotinidase is the major carrier of the vitamin in plasma. Chauhan and Dakshinamurti (1988) demonstrated that biotinidase is the only protein in human serum that specifically binds biotin. Moreover, biotin was found to bind at least 100-fold more tightly than the enzyme's substrate, biocytin.

The liver extracts the majority of the newly absorbed biotin from the blood and is the major site of biotin utilization and metabolism. In the human, biotin enters the liver by transport across the basolateral membrane of the hepatocyte in a similar manner to the brush-border transport in the intestine, i.e. a carrier-mediated, Na^+ gradient-dependent mechanism that is electroneutral in nature (Said *et al.*, 1992).

Biotin is excreted in the urine primarily unchanged, with only a small amount as metabolites (Dakshinamurti, 1994).

12.6.2 Dietary sources and their bioavailability

Dietary sources

Biotin is present in all natural foodstuffs, but the content of even the richest sources is very low when compared with the content of most other water-soluble vitamins. Typical values of some rich sources of biotin include ox liver (33 µg/100 g), whole eggs (20 µg/100 g), dried soya beans (65 µg/100 g) and peanuts (72 µg/100 g) (Holland *et al.*, 1991). Other good sources include yeast, wheat bran, oatmeal and some vegetables. Muscle meats, fish, dairy products and cereals contain smaller amounts, but are important contributors to the dietary intake. Most of the biotin content of animal products, nuts, cereals and yeast is in a protein-bound form. A higher percentage of free water-extractable biotin occurs in vegetables, green plants, fruit, milk and rice bran (Lampen, Bahler and Peterson, 1942). Most, if not all, of the biotin content of human milk is in the free form and thus is likely to be completely available to the infant (Heard, Redmond and Wolf, 1987).

Biotin is not commonly used in fortified foods, apart from infant formulas.

Bioavailability

The microflora in the human colon (more specifically in the caecum) can synthesize large amounts of biotin in the free (unbound) form. This is inferred by the faecal content of biotin exceeding dietary intake three-to six-fold. Balance studies have shown that urinary content often exceeds dietary intake, suggesting that biotin synthesized by colonic microflora can be absorbed and utilized to some extent by the host. Further indirect evidence of absorption is the observation that when biotin is instilled directly into the lumen of the colon, the concentration of plasma biotin increases.

Table 12.1 Bioavailability data for various animal feed ingredients based on activity of blood pyruvate carboxylase (from Whitehead, Armstrong and Waddington, 1982)

Ingredient	Estimated biotin bioavailability (%)	Estimated available biotin (µg/kg)
Cereals		
Barley	11	12
Maize	133	65
Wheat	5	4
Oilseed meals		
Rapeseed meal	62	574
Safflower meal	32	385
Sunflower meal	35	415
Soybean meal	108	278

The bioavailability to the human host of enterically synthesized biotin is unknown, but certain observations indicate that this nondietary source may not contribute significantly to the biotin nutriture of the host. The transport of biotin is minimal in the colon and the vitamin is absorbed to a greater extent when given orally than when instilled directly into the colon. Also, urinary excretion varies with dietary intake whereas faecal excretion is independent of dietary intake (Dakshinamurti, 1994).

The bioavailability of biotin in various foods and the effect of protein binding on bioavailability have not been studied directly in human subjects. Studies of animal feed ingredients using growth assays have shown that biotin in various cereals (but not maize) is largely unavailable to the chicken (Frigg, 1984), turkey (Misir and Blair, 1988a) and weanling pig (Misir and Blair, 1988b). These observations have been confirmed using a chick bioassay based on the activity of blood pyruvate carboxylase (Whitehead, Armstrong and Waddington, 1982). Measurements of this enzyme in the blood of chicks give a specific indication of the relative biotin contents of their diets, provided the dietary contents of protein and fat are similar. Data from this study (Table 12.1) show the low biotin availability in barley and wheat. In contrast, the biotin content of maize is completely available and was repeatedly found to be higher than the amount determined microbiologically. Biotin bioavailability varies widely in the oil seed meals tested. In soybean meal the vitamin is fully available. Bioavailability in sunflower- and safflower-seed meals were comparatively low but, because of the high total contents, these meals are still a good source of biotin. Rapeseed meal was the richest source of available biotin among the ingredients tested.

Bryden, Mollah and Gill (1991) determined the digestibility of biotin by analysing feed residues in the ileum, prior to excretion, rather than in the excreted faecal material. The method of ileal digestibility largely eliminated the difficulties in interpreting values obtained from excreta due to microbial activity in the hindgut. The low digestibility values obtained for wheat suggested that < 12% of the biotin is available from this cereal compared with nearly complete availability from maize. It was concluded that low biotin availability in wheat results from the inability of birds to liberate biotin from bound forms in a form suitable for absorption.

12.6.3 Effects of food processing, cooking and storage

In vegetables, in which a significant proportion of the biotin present is not bound, leaching of the vitamin may occur during the washing and blanching steps of commercial processing. The canning of carrots, maize, mushrooms, green peas, spinach and tomatoes resulted in biotin losses of 40–78% in contrast to asparagus with 0% loss (Lund, 1988). Evaporated milk and milk powder incur losses of 10–15% during processing (Hoffmann–La Roche Inc., 1992). In canned infant foods an approximately 15% reduction in biotin content after a 6-month storage period has been observed (Bonjour, 1984). The cooking of foods can convert a significant portion of biotin to the sulphoxide and sulphone oxidation products, with consequent loss of vitamin activity (Zapsalis and Beck, 1985).

12.7 DIETARY INTAKE

The average biotin intake in the United Kingdom has been estimated to be 39 µg/day (range 15–70 µg) for men and 26 µg/day (range 10 – 58 µg) for women (Department of Health, 1991). There is no evidence of biotin deficiency at these levels of intake. In adult patients receiving long-term total parenteral nutrition, daily administration of 60 µg prevented signs of biotin deficiency (Dakshinamurti, 1994).

In view of the incomplete knowledge of the bioavailability of biotin in foods and of the uncertain contribution of intestinal synthesis to the total intake, no RDA has been calculated. However, a range of 30 to 100 µg/day has been provisionally recommended for adults (National Research Council, 1989).

Effects of high intakes

Biotin toxicity in humans has not been reported. It is presumably low, since infants have tolerated injections of 10 mg daily for 6 months with no adverse effects (Miller and Hayes, 1982).

Assessment of nutritional status

This is evaluated by measuring biotin concentrations in whole blood or urine. However, such data as are available among healthy adults cover a broad range and are, therefore, of limited use (Bonjour, 1991). Knowledge of the body stores of biotin is an important aspect for the evaluation of blood and urine levels as indicators of biotin status. Biokinetic studies, as carried out for example by Bitsch, Salz and Hötzel (1989b), may be helpful in this context.

12.8 SYNOPSIS

There is little knowledge of the bioavailability of dietary biotin to the human, and the nutritional significance of biotin synthesized by intestinal bacteria is uncertain. Animal studies have shown that the bioavailability can vary from one extreme to the other (as in maize and wheat), depending on whether or not the vitamin occurs in an indigestible bound form. Thus it is important to know the form in which biotin occurs, as well as the amount in a foodstuff or diet.

The microbiological assay using *L. plantarum* is the standard method for determining biotin in foods, but the nutritional significance of the results obtained depends largely on the extraction procedure employed. Maximum liberation of bound biotin from animal tissues requires autoclaving the sample with $6 N H_2SO_4$ for 2 h at 121 °C, but these conditions promote losses of biotin in plant tissues, and a somewhat milder acid hydrolysis must be used for plant foods and food composites. Treatment of samples with liquid nitrogen, followed by digestion with papain, is capable of quantitatively releasing bound biotin (including biotin from biocytin) from both plant and animal tissues.

REFERENCES

Achuta Murthy, P.N. and Mistry, S.P. (1977) Biotin. *Prog. Food Nutr. Sci.*, **2**, 405–55.
Axelrod, A.E. and Hofmann, K. (1950) The inactivation of biotin by hydrochloric acid. *J. Biol. Chem.*, **187**, 23–8.
Barton-Wright, E.C. (1952) *The Microbiological Assay of the Vitamin B-Complex and Amino Acids*, Sir Isaac Pitman & Sons Ltd, London, pp. 63–7.
Bell, J.G. (1974) Microbiological assay of vitamins of the B group in foodstuffs. *Lab. Pract.*, **23**, 235–42, 252.
Bitsch, R., Salz, I. and Hötzel, D. (1986) Determination of biotin in foods by a protein binding assay. *Deutsche Lebensmittel-Rundschau*, **82**, 80–3.
Bitsch, R., Salz, I. and Hötzel, D. (1989a) Biotin assessment in foods and body fluids by a protein binding assay (PBA). *Int. J. Vitam. Nutr. Res.*, **59**, 59–64.
Bitsch, R., Salz, I. and Hötzel, D. (1989b) Studies on bioavailability of oral biotin doses for humans. *Int. J. Vitam. Nutr. Res.*, **59**, 65–71.
Bonjour, J.-P. (1984) Biotin. In *Handbook of Vitamins. Nutritional, Biochemical, and Clinical Aspects* (ed. L.J. Machlin), Marcel Dekker, Inc., New York, pp. 403–35.

Bonjour, J.-P. (1991) Biotin. In *Handbook of Vitamins*, 2nd edn (ed. L.J. Machlin), Marcel Dekker, Inc., New York, pp. 393–427.

Bowman, B.B., Selhub, J. and Rosenberg, I.H. (1986) Intestinal absorption of biotin in the rat. *J. Nutr.*, **116**, 1266–71.

Broquist, H.P. and Snell, E.E. (1951) Biotin and bacterial growth. I. Relation to aspartate, oleate, and carbon dioxide. *J. Biol. Chem.*, **188**, 431–44.

Bryden, W.L., Mollah, Y. and Gill, R.J. (1991) Bioavailability of biotin in wheat. *J. Sci. Food Agric.*, **55**, 269–75.

Chauhan, J. and Dakshinamurti, K. (1988) Role of human serum biotinase as biotin-binding protein. *Biochem. J.*, **256**, 265–70.

Dakshinamurti, K. (1994) Biotin. In *Modern Nutrition in Health and Disease*, 8th edn, Vol. 1 (eds M.E. Shils, J.A. Olson and M. Shike), Lea & Febiger, Philadelphia, pp. 426–31.

Dakshinamurti, K., Chauhan, J. and Ebrahim, H. (1987) Intestinal absorption of biotin and biocytin in the rat. *Bioscience Rep.*, **7**, 667–73.

Dakshinamurti, K., Landman, A.D., Ramamurti, L. and Constable, R.J. (1974) Isotope dilution assay for biotin. *Analyt. Biochem.*, **61**, 225–31.

Department of Health (1991) *Dietary Reference Values for Food Energy and Nutrients for the United Kingdom*. Report on Health and Social Subjects No. 41. H.M. Stationary Office, London.

Finglas, P.M., Faulks, R.M. and Morgan, M.R.A. (1986) The analysis of biotin in liver using a protein-binding assay. *J. Micronutr. Anal.*, **2**, 247–57.

Frigg, M. (1984) Available biotin content of various feed ingredients. *Poultry Sci.*, **63**, 750–3.

Guilarte, T.R. (1985) Analysis of biotin levels in selected foods using a radiometric–microbiological method. *Nutr. Rep. Int.*, **32**, 837–45.

György, P. (1967) Biotin. In *The Vitamins. Chemistry, Physiology, Pathology, Methods*, 2nd edn, Vol. VII (eds P. György and W.N. Pearson), Academic Press, New York, pp. 303–12.

Heard, G.S., Redmond, J.B. and Wolf, B. (1987) Distribution and bioavailability of biotin in human milk. *Fed. Proc.*, **46**, 897 (abstr).

Hoffmann–La Roche Inc. (1992) Vitamins, part XIII: biotin. In *Encyclopedia of Food Science and Technology*, Vol. 4 (ed. Y.H. Hui), John Wiley & Sons, Inc., New York, pp. 2764–70.

Holland, B., Welch, A.A., Unwin, I.D. *et al.* (1991) *McCance and Widdowson's The Composition of Foods*, 5th edn, Royal Society of Chemistry and Ministry of Agriculture, Fisheries and Food.

Hood, R.L. (1975) A radiochemical assay for biotin in biological materials. *J. Sci. Food Agric.*, **26**, 1847–52.

Hood, R.L. (1979) Isotopic dilution assay for biotin: use of [^{14}C]biotin. *Methods Enzymol.*, **62D**, 279–83.

Lampen, J.O., Bahler, G.P. and Peterson, W.H. (1942) The occurrence of free and bound biotin. *J. Nutr.*, **23**, 11–21.

Langer, B.W. Jr and György, P. (1968) Biotin. VIII. Active compounds and antagonists. In *The Vitamins. Chemistry, Physiology, Pathology, Methods*, 2nd edn, Vol. II (eds W.H. Sebrell, Jr and R.S. Harris), Academic Press, New York, pp. 294–322.

Lund, D. (1988) Effects of heat processing on nutrients. In *Nutritional Evaluation of Food Processing*, 3rd edn (eds E. Karmas and R.S. Harris), Van Nostrand Reinhold Company, New York, pp. 319–54.

McCormick, D.B. (1988) Biotin. In *Modern Nutrition in Health and Disease*, 7th edn (eds M.E. Shils and V.R. Young), Lea & Febiger, Philadelphia, pp. 436–9.

Miller, D.R. and Hayes, K.C. (1982) Vitamin excess and toxicity. In *Nutritional Toxicology*, Vol. I (ed. J.N. Hathcock), Academic Press, New York, pp. 81–133.

Misir, R. and Blair, R. (1988a) Biotin bioavailability of protein supplements and cereal grains for starting turkey poults. *Poultry Sci.*, **67**, 1274–80.

Misir, R. and Blair, R. (1988b) Biotin bioavailability from protein supplements and cereal grains for weanling pigs. *Can. J. Anim. Sci.*, **68**, 523–32.

Mock, D.M. (1990) Biotin. In *Present Knowledge in Nutrition* 6th edn (ed. M.L. Brown), International Life Sciences Institute, Nutrition Foundation, Washington, DC, pp. 189–207.

National Research Council (1989) Water-soluble vitamins. In *Recommended Dietary Allowances*, 10th edn, National Academy Press, Washington, DC, pp. 115–73.

Podevin, R.-A. and Barbarat, B. (1986) Biotin uptake mechanisms in brush-border and basolateral membrane vesicles isolated from rabbit kidney cortex. *Biochim. Biophys. Acta*, **856**, 471–81.

Said, H.M. and Redha, R. (1987) A carrier-mediated system for transport of biotin in rat intestine in vitro. *Am. J. Physiol.*, **252**, G52–5.

Said, H.M., Mock, D.M. and Collins, J.C. (1989) Regulation of intestinal biotin transport in the rat: effect of biotin deficiency and supplementation. *Am. J. Physiol.*, **256**, G306–11.

Said, H.M., Redha, R. and Nylander, W. (1987) A carrier-mediated, Na$^+$ gradient-dependent transport for biotin in human intestinal brush-border membrane vesicles. *Am. J. Physiol.*, **253**, G631–6.

Said, H.M., Redha, R. and Nylander, W. (1988a) Biotin transport in the human intestine: site of maximum transport and effect of pH. *Gastroenterology*, **95**, 1312–7.

Said, H.M., Redha, R. and Nylander, W. (1988b) Biotin transport in basolateral membrane vesicles of human intestine. *Gastroenterology*, **94**, 1157–63.

Said, H.M., Hoefs, J., Mohammadkhani, R. and Horne, D.W. (1992) Biotin transport in human liver basolateral membrane vesicles: a carrier-mediated, Na$^+$ gradient-dependent process. *Gastroenterology*, **102**, 2120–5.

Said, H.M., Thuy, L.P., Sweetman, L. and Schatzman, B. (1993) Transport of the biotin dietary derivative biocytin (*N*-biotinyl-L-lysine) in rat small intestine. *Gastroenterology*, **104**, 75–80.

Scheiner, J. (1966) Extraction of added biotin from animal feed premixes. *J. Ass. Off. Analyt. Chem.*, **49**, 882–3.

Scheiner, J. and De Ritter, E. (1975) Biotin content of feedstuffs. *J. Agric. Food Chem.*, **23**, 1157–62.

Schweigert, B.S., Nielsen, E., McIntire, J.M. and Elvehjem, C.A. (1943) Biotin content of meat and meat products. *J. Nutr.*, **26**, 65–71.

Shull, G.M., Hutchings, B.L. and Peterson, W.H. (1942) A microbiological assay for biotin. *J. Biol. Chem.*, **142**, 913–20.

Skeggs, H.R. (1963) Biotin. In *Analytical Microbiology* (ed. F. Kavanagh), Academic Press, New York, pp. 421–30.

Spencer, P.D. and Roth, K.S. (1988) On the uptake of biotin by the rat renal tubule. *Biochem. Med. Metabolic Biol.*, **40**, 95–100.

Strohecker, R. and Henning, H.M. (1966) *Vitamin Assay. Tested Methods*, Verlag Chemie, Weinheim.

Thompson, R.C., Eakin, R.E. and Williams, R.J. (1941) The extraction of biotin from tissues. *Science*, **94**, 589–90.

Whitehead, C.C., Armstrong, J.A. and Waddington, D. (1982) The determination of the availability to chicks of biotin in feed ingredients by a bioassay based on the response of blood pyruvate carboxylase (*EC* 6.4.1.1) activity. *Br. J. Nutr.*, **48**, 81–8.

Williams, V.R. (1945) Growth stimulants in the *Lactobacillus arabinosus* biotin assay. *J. Biol. Chem.*, **159**, 237–8.

Williams, V.R. and Fieger, E.A. (1945) Growth stimulants for microbiological biotin assay. *Ind. Engng Chem., Analyt. Edn*, **17**, 127–30.

Wright, L.D., Cresson, E.L., Leibert, K.V. and Skeggs, H.R. (1952) Biological studies of biocytin. *J. Am. Chem. Soc.*, **74**, 2004–6.

Zapsalis, C. and Beck, R.A. (1985) Vitamins and vitamin-like substances. In *Food Chemistry and Nutritional Biochemistry*, John Wiley & Sons, New York, pp. 189–312.

13

Folate

13.1 INTRODUCTION

In 1931, a research group led by Lucy Wills showed that an autolysed yeast preparation (Marmite), which was therapeutically ineffective against the pernicious anaemia caused by vitamin B_{12} deficiency, was effective against nutritional megaloblastic anaemia in pregnant women. These researchers induced a similar anaemia in monkeys which then responded to crude liver extracts. Other substances that cured specific deficiency anaemias in monkeys and chicks were isolated from yeast by different research groups and assigned the names 'vitamin M' and 'vitamin B_c'. Another substance isolated from liver was shown to be essential to the growth of *Lactobacillus casei* and was therefore called the '*L. casei* factor'. In 1941, Mitchell and co-workers processed four tons of spinach leaves to obtain a purified substance with acidic properties which was an active growth factor for rats and *L. casei*. They named the factor 'folic acid' (from *folium*, the Latin word for leaf). Eventually, all of the above substances proved to be the same when

Angier's group in 1946 accomplished the synthesis and chemical structure of folic acid.

Folate plays an essential role in the mobilization of one-carbon units in intermediary metabolism. Such reactions are essential for the synthesis of the purines (adenine and guanine) and one of the pyrimidines (thymine), which are the base constituents of DNA. Folate is also directly involved in the metabolism of certain amino acids. Reactions include the *de novo* biosynthesis of methionine (an important methyl donor) and glycine–serine interconversions.

13.2 CHEMICAL STRUCTURE AND NOMENCLATURE

The term folate is used as the generic descriptor for all derivatives of pteroic acid that demonstrate vitamin activity in humans. The structure of the parent folate compound, folic acid, comprises a bicyclic pterin moiety joined by a methylene bridge to *p*-aminobenzoic acid, which in turn is coupled via an α-peptide bond to a single molecule of L-glutamic acid (I in Figure 13.1). Folic acid can be systematically named as *N*-[(6-pteridinyl)methyl]-*p*-aminobenzoyl-L-glutamic acid ($C_{19}H_{19}N_7O_6$; MW = 441.42). The conformation and absolute configuration of the folates have been described by Temple and Montgomery (1984).

(Note: In the present context, the term 'folic acid' refers specifically to pteroylmonoglutamic acid which, with reference to the pteroic acid and glutamate moieties, can be abbreviated to PteGlu. 'Folate' is a nonspecific term referring to any folate compound with vitamin activity. 'Folacin' is a non-approved term synonymous with 'folate'.)

Folic acid is not a natural physiological form of the vitamin. In nature, the pteridine ring is reduced to give either the 7,8-dihydrofolate (DHF) or 5,6,7,8-tetrahydrofolate (THF) (Figure 13.1). These reduced forms can be substituted with a covalently bonded one-carbon adduct attached to nitrogen positions 5 or 10 or bridged across both positions. The following substituted forms of THF are important intermediates in folate metabolism: 10-formyl-, 5-methyl-, 5-formimino-, 5,10-methylene- and 5,10-methenyl-THF (Figure 13.1).

An important structural aspect of the 5,6,7,8-tetrahydrofolates is the stereochemical orientation at the C-6 asymmetric carbon of the pteridine ring (Gregory, 1989). Of the two 6S and 6R stereoisomers (formerly called 6*l* and 6*d*) only the 6S is biologically active and occurs in nature. Methods of chemical synthesis of tetrahydrofolates, whether by catalytic hydrogenation or chemical reduction, yield a racemic product (i.e. a mixture of both stereoisomers).

All folate compounds exist predominantly as polyglutamates containing typically from five to seven glutamate residues in γ-peptide linkage. The γ-peptide bond is almost unique in nature, its only other known

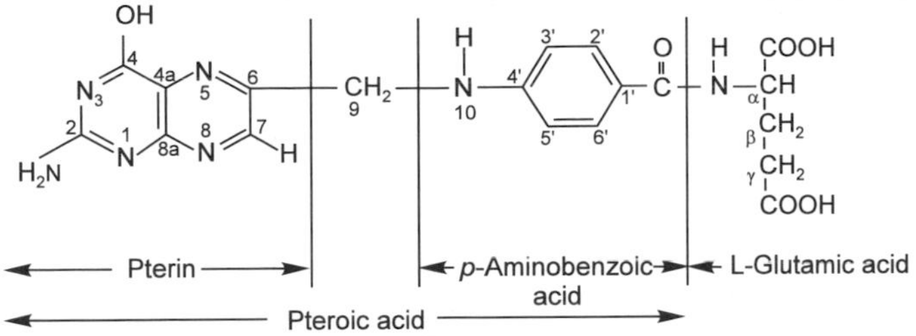

Pterin

p-Aminobenzoic acid

L-Glutamic acid

Pteroic acid

I

Pteroylmonoglutamic acid (folic acid)

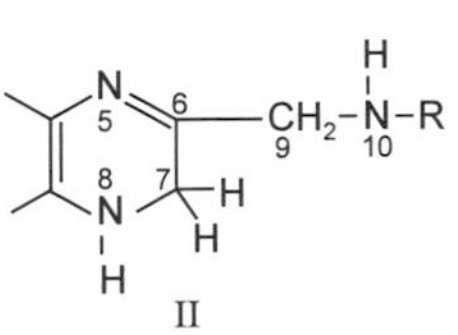

II

7,8-Dihydropteroylmonoglutamic acid (DHF)

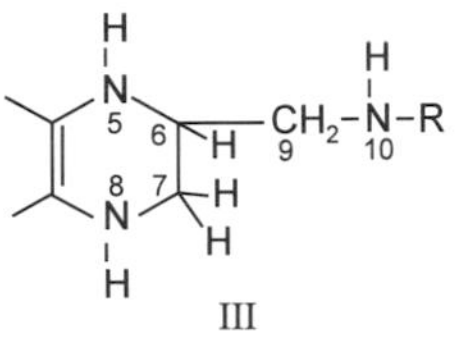

III

5,6,7,8-Tetrahydropteroylmonoglutamic acid (THF)

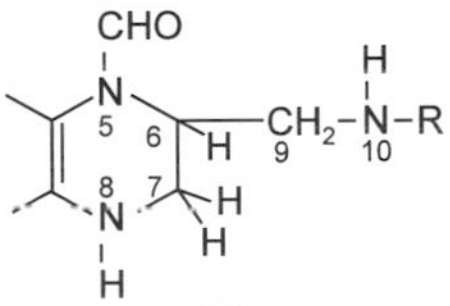

IV

5-Formyl-THF (5-CHO-THF)

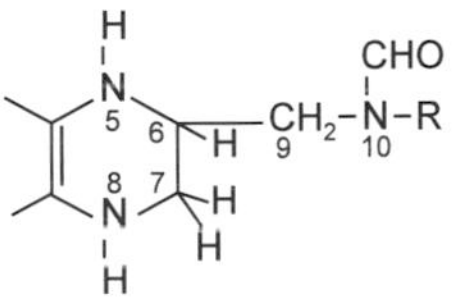

V

10-Formyl-THF (10-CHO-THF)

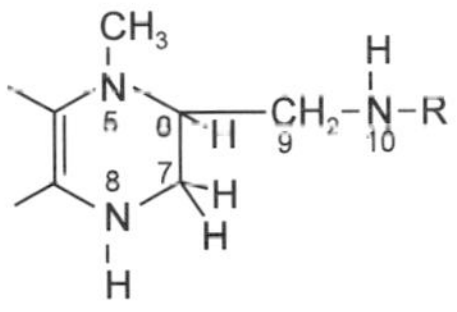

VI

5-Methyl-THF (5-CH₃-THF)

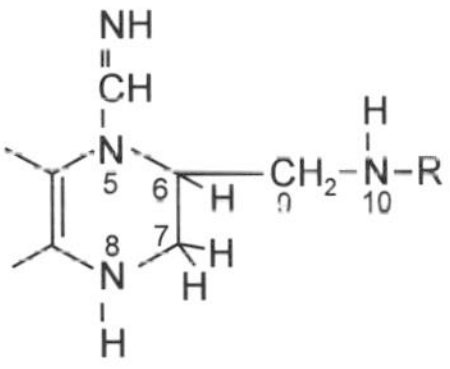

VII

5-Formimino-THF (5-CH=NH-THF)

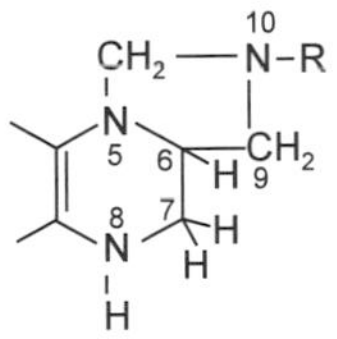

VIII

5,10-Methylene-THF (5,10-CH₂-THF)

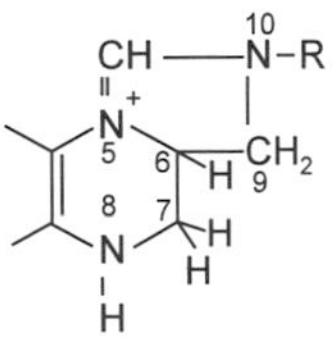

IX

5,10-Methenyl-THF (5,10=CH-THF)

Figure 13.1 Structures of folate compounds.

occurrence being in peptides synthesized by two *Bacillus* species (Mason and Rosenberg, 1994). Folate conjugates are abbreviated to PteGlu$_n$ derivatives, where n is the number of glutamate residues; for example, 5-CH$_3$-H$_4$PteGlu$_2$ refers to triglutamyl 5 methyl tetrahydrofolic acid. Assuming that the polyglutamyl side-chain extends to no more than seven residues, the theoretical number of folates approaches 150 (Baugh and Krumdieck, 1971).

13.3 DEFICIENCY SYNDROMES

A deficiency in folate leads to a lack of adequate DNA replication and consequent impaired cell division, especially in the haemopoietic tissue of the bone marrow and the epithelial cells of the gastrointestinal tract. In the bone marrow the erythroblasts fail to divide properly and become enlarged, and the circulating red blood cells are macrocytic and fewer in number than normal. This condition, megaloblastic anaemia, is particularly common in pregnancy, during which there is an increased metabolic demand for folate. The effect of folate deficiency upon cell renewal of the intestinal mucosa causes gastrointestinal disturbances and also has adverse consequences upon overall nutritional status. Experimental folate deficiency is difficult to produce under normal circumstances, but the study of patients suffering from malabsorption problems such as tropical sprue (atrophy of the jejenum) or the use of folate antagonists has yielded much clinical information.

A poor folate status in early pregnancy appears to be related to the occurrence of neural tube defects in the foetus. These defects – anencephaly, spina bifida and encephalocoele – comprise a group of congenital malformations of the skull and spinal column resulting from the failure of the neural tube to close completely during the third or fourth week postconception (Picciano, Green and O'Connor, 1994). The aetiology is unknown but is believed to be multifactorial, involving both genetic and environmental determinants (Anon., 1992). In 1981 two studies reported a reduction in the incidence of neural tube defects when either folate or multivitamins including folate were given before and after conception to women who had previously given birth to an afflicted child (Smithells *et al.*, 1981; Laurence *et al.*, 1981). These reports were confirmed in later independent studies (Milunsky *et al.*, 1989; MRC Vitamin Study Research Group, 1991). Earlier studies had shown that women giving birth to afflicted infants usually had normal blood levels of folate; thus neural tube defects do not appear to be the result of maternal folate deficiency.

Interrelationship with vitamin B$_{12}$

A close synergistic relationship exists between folate and vitamin B$_{12}$. Adequate vitamin B$_{12}$ must be present for the activity of

$$\text{5,10-CH}_2\text{-THF} \longrightarrow \text{5-CH}_3\text{-THF} + \text{homocysteine} \xrightarrow[\text{B}_{12}]{\text{methionine synthetase}} \text{THF} + \text{methionine}$$

Figure 13.2 The involvement of vitamin B_{12} as a coenzyme in the demethylation of 5-methyl-THF.

methionine synthetase, the enzyme that removes the methyl group from 5-methyl-THF for the formation of methionine from homocysteine. The inability to synthesize methionine from homocysteine in the absence of vitamin B_{12} means that THF cannot be regenerated from the demethylation of 5-methyl-THF. The folate thus becomes trapped in the form of 5-methyl-THF because the formation of this derivative by reduction of 5,10-methylene-THF is thermodynamically irreversible (Figure 13.2). This situation could lead to the inability to form the other THF derivatives that are necessary for purine and pyrimidine synthesis. The consequent lack of DNA synthesis causes many haemo-poietic cells to die in the bone marrow. In this event a megaloblastic anaemia that is clinically indistinguishable from that induced by folate deficiency results. When this type of anaemia is due to deficiency of vitamin B_{12} it is called pernicious or Addisonian anaemia (Herbert and Das, 1994).

13.4 PHYSICOCHEMICAL PROPERTIES

Solubility and ionic characteristics

Folic acid is synthesized commercially for use in food fortification as either the free acid or the disodium salt. The free acid is practically insoluble in cold water and sparingly soluble in boiling water (200 µg/ml). It is soluble and stable in dilute alkaline solution and it dissolves in warm dilute hydrochloric acid, but with degradation that increases with increasing temperature and acid strength. There is slight solubility in methanol, appreciably less solubility in ethanol, and no solubility in acetone, diethyl ether, chloroform or benzene. The disodium salt is soluble in water (15 mg/ml at 0 °C).

Folates are ionogenic and amphoteric molecules. Ionogenic groups of particular significance in the range of pH values relevant to foods and biological systems are the N-5 positions of THF ($pK_a = 4.8$) and the glutamate carboxyl groups ($\gamma\ pK_a = 4.8; \alpha\ pK_a = 3.5$). Polyglut-amyl folates, because of the free α-carboxyl groups situated on each glutamate residue, exhibit greater ionic character than the monoglut-amyl forms when dissociated under neutral to alkaline conditions (Gregory, 1989).

Spectroscopic properties

Absorption

In a buffered medium at pH 7, folic acid and derivatives of THF exhibit absorption spectra with maxima at around 290 nm (except for 5,10-methenyl-THF with a λ_{max} at 352 nm) and ϵ values ranging between 18 200 and 35 400 (Temple and Montgomery, 1984).

Fluorescence

The naturally occurring folates and several pterin derivatives exhibit native fluorescence as reported by Uyeda and Rabinowitz (1963). The fluorescence intensity of folic acid increases in media above or below pH 6, there being no fluorescence at pH 6. The excitation and emission maxima of folic acid at the pH of optimum fluorescence (pH 9) are 313 nm and 450–460 nm, respectively.

Stability

The limited information on the kinetics of folate degradation with regard to food processing has been reviewed by Gregory (1989) and Hawkes and Villota (1989). The folate vitamers differ widely with respect to their susceptibility to oxidative degradation, their thermal stability, and the pH dependence of their stability. Folate activity is gradually destroyed by exposure to sunlight, especially in the presence of riboflavin, to yield *p*-aminobenzolglutamic acid (PABG) and pterin-6-carboxaldehyde (Tannenbaum, Young and Archer, 1985). The length of the glutamyl side chain has little or no influence on the stability properties of the folate compounds (Gregory, 1985a).

The most stable of the various folates at ambient and elevated temperatures is the parent compound, folic acid. In aqueous solution folic acid is stable at 100 °C for 10 h in a pH range 5.0–12.0 when protected from light, but becomes increasingly unstable as the pH decreases below 5.0 (Paine-Wilson and Chen, 1979). Alkaline hydrolysis under aerobic conditions promotes oxidative cleavage of the folic acid molecule to yield PABG and pterin-6-carboxylic acid, whereas acid hydrolysis under aerobic conditions yields 6-methylpterin (Tannenbaum, Young and Archer, 1985). The stability of folic acid (and indeed all folates) is greatly enhanced in the presence of an antioxidant such as ascorbic acid. Polyglutamyl derivatives of folic acid can be hydrolysed by alkali in the absence of air to yield folic and glutamic acids (Tannenbaum, Young and Archer, 1985).

The unsubstituted reduced structure THF is extremely susceptible to oxidative cleavage. DHF is included with the PABG and pterin breakdown products at pH 10 but not at pH 4 or 7 (Reed and Archer, 1980).

Trace metals, particularly iron (III) and copper (II) ions, catalyse the oxidation of THF (Hawkes and Villota, 1989).

The presence of substituent groups in the N-5 position greatly increases the oxidative stability of the reduced folates relative to that of THF. 5-Methyl-THF, an important dietary folate in its polyglutamate form, exhibits a half-life of 21 min at 100 °C in aqueous solution compared with 2 min for THF (Chen and Cooper, 1979). Temperature rather than light is the predominant factor influencing the stability of 5-methyl-THF. The rate of 5-methyl-THF loss varies dramatically with pH. At pH 9.0 in the absence of antioxidant it is very unstable at 25 °C, whilst at pH 7.3 and 3.5 the stability is much greater, with the latter two pH values producing very similar rates of loss. In the presence of an antioxidant (dithiothreitol) at 25 °C, 5-methyl-THF is relatively stable at pH 7.3 and 9.0, but at pH 3.5 the antioxidant has little or no protective effect (Lucock *et al.*, 1993).

Oxidation of 5-methyl-THF under mild conditions at or near neutral pH yields 5-methyl-5,6-DHF (Donaldson and Keresztesy, 1962). The latter compound is rapidly reduced back to 5-methyl-THF by ascorbate and mercaptoethanol, which are commonly used as antioxidants in folate analysis, and therefore the specific presence of 5-methyl-5,6-DHF would not be detected in most chromatographic methods for determining folates. In strongly acidic media 5-methyl-5,6-DHF undergoes C-9–N-10 bond cleavage (Maruyama, Shiota and Krumdieck, 1978; Lewis and Rowe, 1979), while in mildly acidic media it undergoes rearrangement of the pteridine ring system (Gregory, 1985a). In both cases there is a consequent loss of folate activity (Gregory, 1985a).

10-Formyl-THF is readily oxidized by air to 10-formylfolic acid (Maruyama, Shiota and Krumdieck, 1978; Lewis and Rowe, 1979) with no loss of biological activity (Gregory *et al.*, 1984). The stability of 10-formylfolic acid is comparable to that of folic acid (Gregory, 1985a). Under anaerobic conditions, 10-formyl-THF undergoes isomerization to 5-formyl-THF at neutral pH after prolonged standing and especially at elevated temperature (Robinson, 1971).

5-Formyl-THF exhibits equal thermal stability to folic acid at neutral pH, but under acidic conditions, and especially at high temperatures, it loses a molecule of water to form 5,10-methenyl-THF (Paine-Wilson and Chen, 1979). The latter compound is stable to atmospheric oxidation in acid solution, but it is hydrolysed to 10-formyl-THF in neutral and slightly alkaline solutions (Stokstad and Koch, 1967).

The milk protein casein, iron(II) and ascorbate, which are capable of lowering the concentration of dissolved oxygen, have all been shown to increase the thermal stability of folic acid and 5-methyl-THF (Day and Gregory, 1983). Other reducing agents that occur in foods, such as thiols and cysteine, would also retard folate oxidation.

Lucock *et al.* (1994) reported that at pH 6.4, 5-methyl-THF is pro-
foundly unstable in the presence of 0.1 M ZnCl$_2$. Instability to a lesser
degree was also observed in the presence of other metal cations, the order
of the effect being Zn^{2+} > Ca^{2+} ~ K$^+$ > Mg^{2+} or Na$^+$. This effect was
negated in the presence of reduced glutathione, suggesting that the loss
is due to oxidative degradation. The oxidative process seems to depend
on the ionic state of 5-methyl-THF. At pH 6.4, 5-methyl-THF exists in its
anionic form, which renders it more labile in the presence of metal
cations through the formation of a complex. Since the stability of 5-
methyl-THF increased at pH 3.5 in the presence of the same cations, the
protonated free acid is probably less available for complex formation and
consequently is more stable.

Sulphurous acid and nitrite, two chemicals that are used in food
processing, cause a loss of folate activity in aqueous systems. Reactions
between folates and sulphurous acid promote C-9–N-10 bond cleavage
(Tannenbaum, Young and Archer, 1985). Nitrite ions react with folic acid
to yield exclusively 10-nitrosofolic acid and with 5-formyl-THF to yield
the 10-nitroso derivative; interaction with THF and 5-methyl-THF yields
PABG and several pterin products (Reed and Archer, 1979).

13.5 ANALYSIS

13.5.1 Scope of analytical techniques

Folate compounds constitute a large family of pteroylpolyglutamates
which differ in the reduction state of the pteridine nucleus, the nature
of the single carbon substituent at the N-5 and/or N-10 positions, and the
number of glutamate residues. The parent compound, folic acid, is relatively stable and is used in food fortification. Although up to 150 different
folate structures are theoretically possible, fewer than 50 principal folates
probably exist in most animal and plant tissues (Gregory, 1989). Naturally occurring folates exist in protein-bound form, the predominant
vitamers being polyglutamyl forms of tetrahydrofolic acid (THF),
5-methyl-THF and 10-formyl-THF (Gregory, 1984). THF is extremely
susceptible to oxidation but the presence of substituent groups in the
N-5 position greatly increases the oxidative stability.

The muliplicity and diversity of natural folates, their instability and
their existence at low concentration in biological tissues pose formidable
obstacles to the separation, identification and quantification of these
compounds. The study of folates is further complicated by their association with enzymes capable of modifying or degrading them during
sample preparation. Attempts to determine intact pteroylpolyglutamates
in biological tissues have involved fractionation of tissue extracts by
column chromatography and comparison of elution profiles with those

of authentic folate standards. The effluent fractions are then further characterized and quantified by differential microbiological assay both before and after treatment with conjugase (Krumdieck, Tamura and Eto, 1983).

Methods for determining intact folates are powerful analytical tools, but they cannot possibly yield complete and accurate qualitative and quantitative information about all of the folates present in a biological sample. The methods also suffer from being slow and extremely laborious. The simplest approach for determining the chain length of pteroylpolyglutamates would be to cleave the C-9–N-10 bond and separate the *p*-aminobenzoylpolyglutamates (PABGlu$_n$) by ion exchange or gel chromatography. Converting the folates to a common chemical form differing only in polyglutamyl chain length would reduce the number of potential compounds to a level readily manageable by column chromatographic techniques, whilst also eliminating the need for enzyme treatment and microbiological assay. Methods using this simplified approach have been published, but the direct oxidation or reductive procedures originally employed have since been found to be inconsistent (Krumdieck, Tamura and Eto, 1983). A quantitative conversion of all naturally occurring folates to PABGlu$_n$ has been achieved by prior conversion of 5- and 10-formyl-THFs to 5-methyl-THFs, followed by standard Zn/HCl reductive cleavage. The PABGlu$_n$ fragments were converted to azo dyes by coupling their diazonium salts with naphthylethylene diamine, and the dye compounds were then separated according to glutamate chain length by polyacrylamide gel chromatography (Brody, Shane and Stokstad, 1979). Several variations of the cleavage procedure have been proposed, including a differential procedure by which the chain lengths of specific pools of folates are ascertained. More recently, HPLC has been applied to the separation of the azo dye derivatives of PABGlu$_n$ (Krumdieck, Tamura and Eto, 1983).

For most food analysis purposes, the determination of the chain length of folate polyglutamates is not required, and the folates are determined as their monoglutamyl forms after enzymatic deconjugation. The microbiological assay of total folate activity is the most widely used and accepted procedure. The radioassay and non-isotopic protein-binding assay represent current biospecific techniques. HPLC assays permit the determination of the major folate vitamers that occur in foods. Folate activity can be assayed biologically in chick or rat growth tests.

No IU for folate activity has been defined, and analytical results are expressed in weight units (μg or mg) of pure crystalline folic acid. The various monoglutamyl folates being totally interconvertible in mammalian metabolism, are considered to have equal biological activity in humans on a molar basis.

13.5.2 Extraction techniques

At one time the folate in food was classified into two main groups according to its availability to *L. casei*. 'Free' folate referred to the mono-, di- and triglutamates that were assayed by *L. casei* without prior deconjugation with poly-γ-glutamyl hydrolase ('conjugase'), whilst 'total' folate was assayed by *L. casei* after treatment with conjugase. It has since been shown that the action of endogenous conjugases present in vegetable samples can cause an increase in the apparent 'free' folate during extraction of the sample (Leichter, Landymore and Krumdieck, 1979). Thus, 'free' (i.e. short-chain) folate is only a valid concept if care is taken to inactivate conjugases before homogenization (Gregory, 1989).

For the determination of monoglutamyl folate, it is first necessary to liberate folate from its association with proteins and polysaccharides, whilst simultaneously denaturing folate-binding proteins and enzymes that may catalyse folate degradation or interconversion.

Two distinct approaches for extracting folate from food matrices have evolved from the many published research studies, both of which merit discussion. The first approach is a two-step procedure involving a thermal treatment to denature proteins, followed by enzymatic deconjugation with conjugase. The second approach is entirely enzymatic and utilizes a proteolytic enzyme to liberate protein-bound folate, α-amylase to break down starch, and deconjugation with conjugase.

Thermal treatment and enzymatic deconjugation

The inclusion of an antioxidant (usually ascorbic acid) is essential in preventing the destruction of labile folates during heat treatment (Malin, 1975). Wilson and Horne (1983) reported that at neutral pH the presence of ascorbate actually induced thermal degradation and interconversion of folates during extraction at 100 °C. This was attributed, at least in part, to the action of formaldehyde produced by the degradation of ascorbate. It was later reported (Wilson and Horne, 1984) that the use of a pH 7.85 extraction buffer containing ascorbate (101 mM) and 2-mercaptoethanol (200 mM) eliminated these problems, probably through removal of the formaldehyde by reaction with the thiol compound. Little or no conversion of 10-formyl-THF to 5-formyl-THF occurred using the pH 7.85 buffer containing ascorbate and 2-mercaptoethanol.

Gregory *et al.* (1990) compared the merits of three extraction procedures employing: (1) acetate buffer (pH 4.9) containing 50 mM ascorbate (Gregory, Sartain and Day, 1984); (2) phosphate buffer (pH 7.0) containing 50 mM ascorbate; and (3) mixed HEPES/CHES buffers (pH 7.85) containing 101 mM ascorbate and 200 mM 2-mercaptoethanol (Wilson and Horne, 1984). In a preliminary experiment, the pH of the buffer (pH 4.9 versus 7.0) did not consistently affect the recovery of radiolabelled mono-

and polyglutamyl folates added to various food homogenates. However, radiolabelled folates from the livers of rats previously dosed with tritium-labelled folate were more effectively extracted at pH 7.0 and 7.85 than at 4.9. Microbiological assay of total folate in beef liver and green peas extracted with pH 4.9, 7.0 or 7.85 buffers indicated the superiority of the pH 7.85 buffer. In all experiments, the completeness of extraction was improved when the residue from the first extraction was suspended in a new portion of buffer and recentrifuged. The results obtained supported the superiority of the Wilson and Horne extraction buffer, and indicated the need for a second extraction in the analysis of many foods, regardless of the extraction method used. The improved efficiency of extraction at the higher pH might be at least partly explained by the increased ionization of the polyglutamyl folate molecule, which would tend to promote dissociation from anionic sites of the sample matrix.

Conjugases from three biological sources have been used in folate deconjugation, each differing in its pH optimum and mode of action. Conjugase from chicken pancreas exhibits a neutral pH optimum and yields a folate diglutamate end-product; hog kidney conjugase exhibits an acidic pH optimum (pH 4.5–4.9) and yields a monoglutamate end-product; and human plasma conjugase also exhibits a pH optimum of 4.5 and yields a monoglutamate (Gregory, 1989). In a detailed study of the properties and efficacy of hog kidney conjugase, Engelhardt and Gregory (1990) noted that a variety of foods caused detectable inhibition of enzyme activity. However, the use of hog kidney conjugase was judged to be appropriate for the various assay methods, provided that the proper combination of enzyme concentration and incubation time is determined for each type of sample to be assayed, and for each batch of enzyme preparation.

Enzymatic digestion and deconjugation

Martin *et al.* (1990) published a triple enzyme digestion procedure using chicken pancreas conjugase, α-amylase and protease in the microbiological determination of total folate in foods. Ground food samples were homogenized in phosphate buffer (pH 6.8) containing 56.8 mM ascorbate, and diluted aliquots were placed in a boiling water bath for 5 min to gelatinize starch before α-amylase treatment. After cooling, conjugase and α-amylase were added, plus two drops of toluene to layer the tubes, and the capped tubes were incubated at 37 °C for 4 h. Protease solution and two drops of toluene were then added, and the capped tubes were incubated overnight at 37 °C. It was necessary to add the protease after the action of conjugase and α-amylase to prevent it from deactivating these enzymes. After the second incubation period all extracts were steamed in an autoclave for 5 min at 100 °C to remove the toluene and inactivate the enzymes.

The triple enzyme treatment increased measurable folate from several foods when compared with levels found after digestion with chicken pancreas enzyme *per se*. Both protease and α-amylase appeared to be necessary for the complete liberation of bound folate, since increased folate values were also observed using just protease or α-amylase in addition to chicken pancreas conjugase (De Souza and Eitenmiller, 1990). The role of α-amylase in releasing folate bound to starch by physical adsorption was reported by Černá and Káš (1983). According to Pedersen (1988), specific amylase addition is of little benefit if chicken pancreas conjugase is used, as the latter was shown to contain equally high amounts of amylase activity as did α-amylase or amyloglucosidase sources.

13.5.3 Microbiological assays

For routine food analysis purposes, the monoglutamate folate activity is measured using *L. casei* after treatment with hog kidney conjugase. The basic premise using this approach is that all active monoglutamyl folates have identical equimolar growth-support activities for *L. casei*, under the conditions of the assay. This assumption is, however, a subject of controversy. While some investigators (O'Broin *et al.*, 1975; Shane, Tamura and Stokstad, 1980) reported that various monoglutamates gave essentially the same response, others (Bird, McGlohon and Vaitkus, 1969; Ruddick, Vanderstoep and Richards, 1978; Reingold and Picciano, 1982; Phillips and Wright, 1982; Goli and Vanderslice, 1989) have demonstrated different responses.

Phillips and Wright (1982) observed a concentration-dependent reduced response of *L. casei* to 5-methyl-THF in sample extracts relative to folic acid and 5-formyl-THF standards using the assay procedure of Bell (1974). The resultant underestimation of total folate caused by this positive drift effect could be rectified by increasing the buffering capacity of the medium, and using an initial pH of 6.2 instead of 6.8 (Phillips and Wright, 1982, 1983; Wright and Phillips, 1985). These investigators further modified the Bell (1974) method by adding the standards later by aseptic addition, instead of autoclaving them, and by increasing the ascorbate concentration from 0.25 to 1 g/litre in the standard solutions and in the incubation medium. These changes were made to ensure that the folate did not deteriorate due to heat or by atmospheric oxidation. It was also important to conduct the assay within the linear response range of 0–1 ng/ml.

13.5.4 Radioassay

Waxman, Schreiber and Herbert (1971) developed a radioassay for measuring folate levels in blood serum using a folate-binding protein (FBP) isolated from milk. Subsequently, many variations of this technique have been applied to the measurement of folate levels in serum, plasma or red blood cells, and several radioassay kits are commercially available for such analyses.

Sources of FBP used in radioassays have included non-fat dry milk, skim milk and whey protein concentrate (Gregory, 1985b), as well as crystalline bovine β-lactoglobulin (Waxman and Schreiber, 1980). At the physiological pH range of 7.3–7.6, milk FBP shows a greater affinity for binding folic acid than it does for 5-methyl-THF, whereas at pH 9.3 milk FBP exhibits a similar binding capacity for these two folates (Gregory, 1989). The presence of at least one glutamate residue is required for binding to take place, as shown by the nonbinding of pteroic acid (Ghitis, Mandelbaum-Shavit and Grossowicz, 1969). Pterin-6-carboxylic acid and p-aminobenzoylglutamic acid exhibit little or no affinity for FBP, indicating that these folate degradation products would not significantly interfere with the accuracy of the radioassay (Gregory, Day and Ristow, 1982).

Radioactive folic acid, labelled with tritium in the 3′-, 5′-, 7- and 9-positions, is commercially available in high specific activity (43 Ci/mmol) (Gregory, 1985b). The use of γ emitters such as ^{125}I- and ^{75}Se-labelled folic acid simplifies the assay procedure by eliminating liquid scintillation counting (Waxman and Schreiber, 1980).

In contrast to blood plasma or serum, in which monoglutamyl 5-methyl-THF is virtually the sole folate present (Herbert, Larrabee and Buchanan, 1962), the naturally occurring folates in foodstuffs comprise a variety of polyglutamyl forms. In applying the pH 9.3 radioassay to foods, the extraction procedure should avoid or minimize the thermal conversion of 10-formyl-THF to 5-formyl-THF, because FBP does not exhibit significant affinity for 5-formyl-THF (Gregory, Day and Ristow, 1982). An initial heating step is, however, essential when analysing milk and other dairy products in order to denature indigenous FBP. Deconjugation of folates to monoglutamyl forms is obligatory, because of the dependency of the binding affinity on polyglutamyl chain length. Shane, Tamura and Stokstad (1980) reported that the molar response of different folate compounds in radioassay procedures varied considerably, depending on the stereochemical form, the reduction state of the pteridine nucleus, the nature of the one-carbon substituent and the number of glutamate residues. This observation appears to rule out the application of radioassays for accurately determining the naturally occurring folates in foods.

Several studies have been conducted in which the results of the pH 9.3 radioassay have been compared with those of the *L. casei* assay for the determination of total folate in foods (Ruddick, Vanderstoep and Richards, 1978; Graham, Roe and Ostertag, 1980; Klein and Kuo, 1981; Gregory, Day and Ristow, 1982; Österdahl and Johansson, 1989).

Ruddick, Vanderstoep and Richards (1978) concluded that the radioassay is suitable only for measuring the particular folate compound used to construct the standard curve for the assay.

13.5.5 High-performance liquid chromatography

Selected HPLC methods for determining folic acid and the principal naturally occurring folates in foods are summarized in Table 13.1.

To determine the added folic acid in fortified breakfast cereals and infant formula products, Day and Gregory (1981) extracted the folic acid by homogenization or dilution in a pH 7.0 phosphate buffer containing ascorbate as an antioxidant. Cereal extracts were centrifuged directly, while formula extracts were adjusted to pH 4.5 to precipitate the proteins, and then centrifuged. The extracts were analysed by isocratic reversed-phase HPLC using a C_{18} column and a phenyl column connected in series. Fluorimetric detection was achieved using in-line post-column chemistry to convert the folic acid to a highly fluorescent pterin compound.

Gregory, Sartain and Day (1984) subjected food samples to heat treatment and enzymatic deconjugation, followed by purification by anion exchange chromatography on DEAE-Sephadex A-25 columns. Extraction of food samples at the mildly acidic pH of 4.9 was favourable to the stability of ascorbic acid and also to the subsequent enzymatic deconjugation step. A prolonged heat treatment (100 °C for 1 h) was employed intentionally to convert 10-formyl-THF to 5-formyl-THF, as a means of simplifying the quantification of formyl tetrahydrofolates by HPLC. No degradation of 5-methyl-THF occurred during the heat treatment, while less than 10% loss of THF was observed. Chromatographic retention is governed by polyglutamyl chain length, and therefore hog kidney conjugase was used to promote deconjugation to monoglutamyl folates. Analysis was performed by single-column reversed-phase HPLC using a linear gradient elution programme. Determination of THF and its substituted derivatives was accomplished by monitoring the fluorescence produced by excitement at 290–295 nm in the acidic environment of the mobile phase. A fluorimetric detector which is capable of extremely high sensitivity is required for this purpose. Because of the much greater fluorescence of THF and 5-methyl-THF compared with 5-formyl-THF, a change in excitation and emission slit widths was necessary before elution of 5-formyl-THF to maintain adequate detector response. With the aid of a second fluorescence detector, folic acid, DHF and also THF could be measured after post-column derivatization to pterin compounds. The suitability of the method was supported by recovery data and fluorescence spectral studies, and by its application to a wide range of biological materials.

Vahteristo *et al.* (1996) employed rapid heat extraction by microwave heating and an enzymatic deconjugation with hog kidney and chicken pancreas conjugases for the determination of folate vitamers in foods by reversed-phase HPLC. The extracts were purified with the use of strong anion exchange cartridges before injection. The combined use of ascorbic acid and mercaptoethanol in the extraction step markedly improved the stability of THF. The necessity of flushing nitrogen through the test solution was also observed, at least if only ascorbic acid was used as an antioxidant. The possible presence of 10-formyl-THF in foods was not monitored. Most probably, the heat treatment employed caused only partial conversion of this vitamer to 5-formyl-THF. The gradient elution used allowed good separation of the four vitamers studied (THF, 5-methyl-THF, 5-formyl-THF and folic acid) and native fluorescence of reduced folates at low pH was used for detection. Folic acid was detected at its absorbance maximum of 290 nm.

Holt, Wehling and Zeece (1988) accomplished the separation of THF, 5-formyl-THF, DHF, folic acid and 5-methyl-THF using ion interaction HPLC with tetrabutylammonium phosphate and a mobile phase pH of 6.8. The use of post-column acidification to allow fluorimetric detection of reduced folates would extend column life, as acidic mobile phases gradually erode silica-based column packings.

A method developed by White (1990) for the determination of 5-methyl-THF in citrus juice involved centrifugation, adjustment to pH 5.0 and treatment with conjugase, followed by solid-phase extraction using a phenyl-bonded phase with the aid of an ion interaction agent (tetrabutylammonium phosphate). At pH 5.0, 5-methyl-THF is essentially anionic, and the use of an ion interaction agent improved its retention on the phenyl-bonded phase. HPLC was performed using a C_{18} column, a methanol/buffer mobile phase of pH 5.5, and amperometric detection. The detection limit was in the low nanogram range. Ascorbic acid, which is present in natural orange juice at a concentration roughly 1000 times that of 5-methyl-THF, is easily oxidized at the detector operating potential of +0.2 V, and the detector response was found to be about equal for the two compounds. Clean-up by solid-phase extraction was capable of lowering the ascorbate to a residual concentration of only about 10 times that of 5-methyl-THF. Without solid-phase extraction, the ascorbate, although not retained on the HPLC column, caused severe tailing, which swamped the detector signal and obscured the 5-methyl-THF peak. The standard was prepared to contain ascorbate at 10 times the concentration of 5-methyl-THF. In an automated HPLC method for the determination of 5-methyl-THF in citrus juices (White, Lee and Krüger, 1991), sample clean-up took place on a C_{18} pre-column, followed by backflush of the analyte to the C_{18} analytical column.

Table 13.1 HPLC methods used for the determination of folate compounds

Food	Sample preparation	Column	Mobile phase	Compounds separated	Detection	Reference
Reversed-phase chromatography						
Fortified milk- and soy-based infant formula	Disperse powders in water plus 0.05 M phosphate–citrate buffer (pH 8.0) containing ascorbate. Incubate with papain at 40 °C overnight, then centrifuge and filter. Resuspend pellet in 0.005 M phosphate–citrate buffer (pH 8.0) and rinse filter with the diluted buffer. *Clean-up and concentration*: pass extract through a DEAE cellulose column, wash column bed with 0.05 M phosphate buffer (pH 7.0) and elute folate with 0.1 M phosphate buffer (pH 7.0) containing 0.5 M NaCl	Spherisorb ODS 10 μm 250 × 4.6 mm i.d.	30 min linear gradient from 2% to 30% MeCN in 0.1 M acetate buffer (pH 4.0)	Folic acid	UV 280 nm	Hoppner and Lampi (1982)
Fortified commercial diets	Homogenize by mechanically shaking with 0.01 M phosphate buffer (pH 7.4). Filter through glass fibre filter and pass a 10 ml portion of the filtrate through a preconditioned anion exchange solid-phase extraction cartridge. Wash the column bed with water and then elute the folic acid with a 10% NaCl–0.1 M sodium acetate solution. Dilute the eluate with the NaCl–sodium acetate solution	μBondapak C$_{18}$ 10 μm 300 × 3.9 mm i.d.	0.1 M sodium acetate buffer (pH 5.7)/ MeCN (94:6)	Folic acid	UV 365 nm	Schieffer, Wheeler and Cimino 1984
Fortified breakfast cereal, fortified liquid infant formula	Cereal: homogenize in pH 7.0 phosphate–ascorbate buffer, centrifuge, filter. Formula: dilute with pH 7.0 phosphate–ascorbate buffer, adjust to pH 4.5, centrifuge, filter	Ultrasphere ODS 5 μm 250 × 4.6 mm i.d. coupled with μBondapak Phenyl 10 μm 300 × 3.9 mm i.d.	33 mM phosphate buffer (pH 2.3) containing 9.5% MeCN	Folic acid	Fluorescence (ex. filter 365 nm max, em. filter > 415 nm) after post-column oxidation of folic acid to pterin fragment with calcium hypochlorite reagent	Day and Gregory (1981)

Sample	Sample preparation	Column conditions	Mobile phase	Compounds	Detection	Reference
Milk	Place milk sample in centrifuge tube and add sodium ascorbate to yield 1%. Place tube in a boiling water bath for 15 min and then in an ice bath until it reaches room temperature. Add human plasma conjugase containing 2-mercaptoethanol then add extraction buffer (0.1 M phosphate buffer, pH 6.0, containing sodium ascorbate). Incubate at 37°C for 1 h (protected from light). Centrifuge and load supernatant on to a pretreated strong anion exchange solid-phase extraction cartridge. Discard the initial column effluent and wash column bed twice with water. Elute the folates with 10% aqueous NaCl solution containing freshly added 1% sodium ascorbate, filter (0.45 µm)	LiChrospher 100 RP-18 5 mm; 250 × 4 mm i.d.; $T = 27°C$	8% MeCN in acetic acid (pH 2.3)	THF, 5-CH$_3$-THF, 5-CHO-THF	Fluorescence ex. 310 nm em. 352 nm	Wigertz and Jägerstad (1995)
Infant formulas and liquid medical nutritional diets fortified with folic acid	Dilute sample with water and sterilize by autoclaving at 100 °C for 5 min. Incubate with buffered (pH 8.4) papain + bacterial protease enzyme solution at 4°C for 1 h with continuous shaking, then filter	Automated in-line column system using column switching; *Preliminary separation:* Bio Series SAX 80 × 6.2 mm i.d. $T = 25°C$; *Analytical separation:* Zorbax RX C$_8$ (octyl) 250 × 4.6 mm i.d. $T = 25°C$	Solvent A: 0.02 M sodium acetate + 0.02 M Na$_2$SO$_4$, pH 5.3; Solvent B: Solvent A/MeCN (64 + 36); *Preliminary separation:* use Solvent A; *Analytical separation:* MeCN gradient	Folic acid	UV 345 nm	Jacoby and Henry (1992)
Cabbage, cow's milk, calf liver, orange juice, whole-wheat flour	*Extraction and deconjugation:* blend samples with 0.05 M acetate buffer (pH 4.9) containing 1% ascorbic acid. Heat at 100°C for 1 h in sealed N$_2$-flushed tubes, then cool on ice and centrifuge at 2°C. Incubate supernatant with hog kidney conjugase at 37°C for 1 h in sealed N$_2$-flushed tubes	µBondapak Phenyl 10 µm; 300 × 3.9 mm i.d.	15 min linear gradient from 7.2 to 11.3% MeCN in 33 mM phosphate buffer (pH 2.3)	THF, 5-CH$_3$-THF, DHF, 5-CHO-THF, folic acid	Fluorescence (ex. 295 nm, em. 365 nm) for THF, 5-CH$_3$-THF and 5-CHO-THF Fluorescence (2nd detector) (ex. filter 365 nm max, em. filter > 415 nm)	Gregory, Sartain and Day (1984)

Table 13.1 *Continued*

Food	Sample preparation	Column	Mobile phase	Compounds separated	Detection	Reference
	Clean-up: adjust pH of deconjugated extract to 7.0 and apply to DEAE-Sephadex A-25 (Pharmacia) anion exchange column. Wash column bed with 0.01 M phosphate buffer (pH 7.0) containing ascorbate, then with 0.2M phosphate buffer (pH 7.0) containing ascorbate and 1 M sodium sulphate. Elute monoglutamyl folate with additional phosphate/ascorbate/sulphate solution				after post-column oxidation of DHF, THF and folic acid to pterin fragments with calcium hypochlorite reagent	
Various	*Extraction and deconjugation:* homogenize samples with extraction buffer (75 mM K_2HPO_4 containing 52 mM ascorbic acid and 0.1% w/v 2-mercaptoethanol, adjusted to pH 6.0 with phosphoric acid) plus 2-octanol to reduce foaming, and under N_2 flow. Divide homogenate into two equal portions in 50 ml tubes (duplicate analysis). Flush tube with N_2, heat for 1 min in a microwave oven at 75% power and shake once. Cap tube tightly and heat on a boiling water bath for 10 min with occasional shaking. Cool extracts rapidly in ice and centrifuge at 2–4 °C. Redissolve residue in extraction buffer and recentrifuge. Combine supernatants and dilute with extraction buffer. Adjust pH of the extract (3–5 ml) to 4.9 with acetic acid and add hog kidney and chicken pancreas conjugases. Flush mixture with N_2, cap tubes and incubate at 37 °C for 2 h. Heat on a boiling water bath for 5 min to inactivate the enzymes and then cool in ice.	Hypersil ODS 3 μm 150 × 4.6 mm i.d. T = 30 °C	Gradient elution with MeCN and 30 mM phosphate buffer (pH 2.2)	THF, 5-CH_3-THF, 5-CHO-THF, folic acid (main folates present in foods) 10-CHO-folic acid (oxidation product of 10-CHO-THF) could also be determined	Fluorescence (programmable) ex. 290 nm, em. 356 nm for THF, 5-CH_3-THF and 5-CHO-THF ex. 360 nm, em. 460 nm for 10-CHO-folic acid UV 290 nm for folic acid and PteGlu$_3$	Vahteristo *et al.* (1996)

Sample	Clean-up and concentration	HPLC column	Mobile phase	Compounds	Detection	Reference
(continued)	*Clean-up and concentration*: dilute test extract with water and add 2-mercaptoethanol before applying the extract on an activated and preconditioned strong anion exchange solid-phase extraction cartridge. Wash cartridge twice with conditioning buffer and then elute the folate compounds with 0.1 M sodium acetate containing 10% NaCl and 1% ascorbic acid					
Liver (pig, beef, chicken), kidney (beef)	As above, except that deconjugation is performed using hog kidney conjugase only	As above	As above	As above and also 5,10-methenyl-THF	As above ex. 360 nm, em. 460 nm for 10-CHO-folic acid and 5,10-methenyl-THF	Vahteristo, Ollilainen and Varo (1996)
Orange juice	Add β-hydroxyethyltheophylline (β-HET) to orange juice sample as an internal standard. Transfer to preconditioned reversed-phase phenyl bonded-silica solid-phase extraction cartridge (Bond-Elut Phenyl, 1 ml capacity) and allow to equilibrate. Wash column bed with 0.1 M, pH 3.4 citrate buffer then elute 5-CH$_3$-THF and β-HET with MeOH into a centrifuge tube containing DL-dithiothreitol (DTT) as antioxidant. Evaporate to dryness in a centrifugal evaporator and reconstitute sample in HPLC mobile phase	Nova-PAK phenyl 4 µm 75 × 3.9 mm i.d.	15% MeOH in 0.05 M phosphate buffer (pH 3.5)	5-CH$_3$-THF, β-HET (internal standard)	*Internal standard* UV 254 nm *5-CH$_3$-THF* Amperometric: glassy carbon electrode, +0.35 V vs Ag/AgCl	Lucock, Hartley and Smithells (1989)

Table 13.1 Continued

Food	Sample preparation	Column	Mobile phase	Compounds separated	Detection	Reference
Orange juice, grapefruit juice	*Deconjugation*: centrifuge and adjust pH to 5.0. Incubate with hog kidney conjugase at 37°C for 1h. Cool on ice and centrifuge at 2°C *Clean-up*: pass supernatant through phenyl solid-phase extraction cartridge, wash cartridge with 0.05 M phosphate–acetate buffer (pH 5) containing 0.005 M tetrabutylammonium phosphate (TBAP), followed by the buffer/TBAP containing 10% MeOH. Elute folate with the buffer containing 30% MeOH without TBAP	Zorbax ODS 250 × 4.6 mm i.d.	25% MeOH in phosphate–acetate buffer (pH 5.5)	5-CH₃-THF	Amperometric: glassy carbon electrode, +0.2 V vs Ag/AgCl	White (1990)
			Ion interaction chromatography			
Orange juice, grapefruit juice	*Deconjugation*: as for White (1990) (see above) except longer (90 min) incubation with conjugase *Clean-up*: direct injection of deconjugated sample onto a Nova-PAK C₁₈ 4 µm pre-column (75 × 3.9 mm i.d.) followed by automatic backflush of the analyte to the analytical column	Zorbax ODS 250 × 4.6 mm i.d.	A. 10% MeOH in phosphate–acetate buffer (pH 5.0) containing 5 mM tetrabutyl-ammonium phosphate for pre-column elution B. Similar to A but containing 30% MeOH for backflushing and analysis	5-CH₃-THF	As above	White, Lee and Krüger (1991)

Food	Extraction procedure	Column	Mobile phase	Analyte	Detection	Reference
Breakfast cereals and beverages enriched with folic acid	Digest homogenized sample with Claradiastase and trypsin for 3 h at 40 °C, then centrifuge and filter. *Clean-up:* DEAE-Sephadex ion exchange column chromatography followed by C$_{18}$ solid-phase extraction. Analyse directly using System 1 or proceed as follows for System 2. Adjust pH of purified sample extract to 4.0–4.5 and derivatize by addition of FMOC-Cl	*Systems 1 and 2* Nucleosil C$_{18}$ 5 μm	*System 1* 0.5% tetrabutyl-ammonium-hydrogen sulphate, pH 8.5/MeCN, 85 + 15 *System 2* 0.5% tetrabutyl-ammonium hydrogen sulphate/MeOH, 70 + 30	Folic acid	*System 1* UV 300 and 340 nm *System 2* Fluorescence: ex. 360 nm em. 450 nm	Gauch, Leuenberger and Müller (1993)
Milk, dairy products	*Extraction and deconjugation:* homogenize solid products, adjust to pH 4.5 with acetic acid and centrifuge. Add phosphate buffer (pH 4.5) containing 10% ascorbate and 1 M 2-mercaptoethanol. Incubate with hog kidney conjugase at 37 °C for 2 h and centrifuge	C$_{18}$ Microsorb 3 μm (Rainin) 100 × 4.6 mm i.d.	MeOH/phosphate buffer (pH 6.8) containing tetrabutyl-ammonium phosphate at 50 ml/l delivered by high-pressure mixing of MeOH/buffer (50:50) and 100% aqueous buffer	THF, 5-CHO-THF, DHF, folic acid, 5-CH$_3$-THF	Fluorescence (ex. 238 nm, em. filter > 340 nm) after post-column pH adjustment of the eluent with 4.25% (v/v) phosphoric acid for the reduced forms of the vitamin. Folic acid requires oxidation with buffered hypochlorite (pH 3.0)	Holt, Wehling and Zeece (1988)

Abbreviations: see footnote to Table 2.4, except THF is tetrahydrofolate.

Lucock, Hartley and Smithells (1989) determined endogenous 5-methyl-THF in plasma using reversed-phase HPLC on a phenyl column and amperometric detection. The use of dithiothreitol as an antioxidant overcame problems of interference experienced with ascorbic acid. The method described could also be applied to the analysis of orange juice (M.D. Lucock, personal communication). An internal standard, β-hydroxyethyltheophylline, was added to the plasma or orange juice sample at the start of the analytical procedure and finally measured by its absorbance at 254 nm. Lucock *et al.* (1995a) discussed the full potential of HPLC for the determination of folates in foods and biological tissues, and provided physical data pertaining to absorbance, fluorimetric and electrochemical detection methods.

13.6 BIOAVAILABILITY

The bioavailability of dietary folate can be affected by: (1) dietary factors and physiological conditions influencing the rate or extent of intestinal deconjugation of polyglutamyl folate; (2) intestinal absorption; (3) entry into folate metabolism and enterohepatic circulation; and (4) the rate of urinary excretion.

13.6.1 Physiological aspects

Folate homeostasis

The liver plays a major role in maintaining folate homeostasis because of its mass, relatively rapid folate turnover, and the large folate flux through the enterohepatic circulation (Steinberg, 1984). It is fundamental that folate monoglutamates are the circulatory and membrane-transportable forms of the vitamin, whereas polyglutamates are the intracellular biochemical and storage forms. Folate homeostasis depends on the involvement of various folate-binding proteins as reviewed by Wagner (1985) and Henderson (1990).

The maintenance of a normal level of plasma folate depends on regular increments of exogenous folate from the diet. The enterohepatic circulation of folate evens out the intermittent nature of intake of dietary folate. In situations of dietary folate deficiency, the liver does not respond by releasing its folate stores. Rather, the less metabolically active tissues mobilize their folate stores and return monoglutamyl folate to the liver, which then redistributes the folate to the tissues that most require it (e.g. the actively proliferating cells of the placenta and foetus). There is evidence that some of the folate returning to the liver is in the form of nonsubstituted (and possibly oxidized) folate, which is subsequently remethylated in the liver. The preferential return of nonmethylated folate to the liver might be accomplished by a specific plasma folate-binding protein.

The kidney plays its part in conserving body folate by actively re-absorbing folate from the glomerular filtrate. In addition, a pathway exists that is capable of salvaging folate released from senescent red blood cells (Steinberg, 1984).

Deconjugation of dietary folate

Before the polyglutamate forms of folate in foods can be absorbed, they must be hydrolysed to monoglutamate forms. The presence of conjugase (a trivial name representing a group of poly-γ-glutamyl hydrolases) in many raw foods of both plant and animal origin results in a high proportion of the dietary folate being already monoglutamyl when presented to the intestinal mucosa (Gregory, 1989). The buffering action of the food matrix would tend to protect dietary folates against the acidic gastric environment. Ingested polyglutamates, mostly in the reduced tetrahydro form, are hydrolysed by conjugase in the mucosal cell to monoglutamates. As much as 50–75% of dietary polyglutamyl folate can be absorbed after deconjugation (Butterworth, Baugh and Krumdieck, 1969).

None of the known proteases in saliva, gastric juice or pancreatic secretions are capable of splitting the γ-peptide bonds in the polyglutamyl side chain. Two folate conjugase activities have been found in human jejunal tissue fractions. One, a brush-border exopeptidase, has a pH optimum of 6.7–7.0 and is activated by Zn^{2+}. The other, an intracellular endopeptidase of mainly lysosomal origin, has a pH optimum of 4.5 and no defined metal requirement. The brush-border conjugase splits off terminal glutamate residues one at a time and is thought to be the principal enzyme in the hydrolysis of polglutamyl folate. The intracellular conjugase may play no role in the digestion of dietary folate, being, instead, concerned with folate metabolism within the enterocyte (Halsted, 1990).

In contrast to pigs and humans, rats and many other species exhibit little or no brush-border conjugase activity in jejunal mucosa, but the pancreatic juice of the rat has a high conjugase activity. Significant conjugase activity has also been reported in the pancreatic juice of pigs and humans (Gregory, 1995). Bhandari *et al.* (1990) investigated the properties of pancreatic conjugase in the pig and found the enzyme to be Zn^{2+}-dependent with maximum activity at pH 4.0–4.5. Feeding stimulated secretion of pancreatic juice, including conjugase activity. The volume of the postprandial secretion, activity under mildly acidic conditions similar to those known to exist in the upper small intestine, and stability in the presence of other digestive enzymes suggested that pancreatic conjugase may initiate the deconjugation of dietary folate in the pig and potentially in humans prior to the action of jejunal brush-border conjugase. Chandler *et al.* (1991) calculated that conjugase activity in porcine pancreatic juice was minor relative to the activity of the jejunal brush-border conjugase.

Absorption of folate in the small intestine

In the human, the entire small intestine is capable of absorbing mono-glutamyl folate. Absorption is somewhat greater in the proximal than in the distal jejunum which, in turn, is much greater than in the ileum. Folate transport across the brush-border membrane of the enterocyte takes place by two processes (Strum, 1979). At physiological concentrations of luminal folate (10^{-5} M or less), transport occurs primarily by a saturable process. At higher concentrations, transport occurs by a non-saturable process, presumably simple diffusion. The saturable component will be discussed here.

Transport of folate is a carrier-mediated system that is markedly influenced by the luminal pH and inhibited competitively by sulphasalazine (Mason and Rosenberg, 1994). *In vitro* studies using everted rat jejunal rings showed that absorption was maximal at pH 6.3 and fell off sharply between pH 6.3 and 7.6 (Russell *et al.*, 1979). Schron, Washington and Blitzer (1985, 1988) proposed, from experimental evidence, that folate uptake takes place by an anion-exchange mechanism driven by the energy provided by the transmembrane pH gradient. Folate exists primarily as an anion at the pH of the lumen and is exchanged specifically for a hydroxyl anion.

The presence of several folate-binding proteins on the intestinal brush-border membrane has been implied (Leslie and Rowe, 1972; Reisenauer, Chandler and Halsted, 1986) and two distinct binding activities have been demonstrated (Shoda *et al.*, 1990). However, none of these studies determined whether the observed binding was associated with folate uptake by the enterocyte. Reisenauer (1990) used the technique of affinity labelling to identify a single folate-binding protein that was associated with folate transport in pig intestinal brush-border membrane vesicles. Labelling the vesicles with the affinity labelling reagent inhibited folic acid transport by 60% and binding by 80%. The addition of excess folate during the labelling step protected the transport system against inactivation, indicating specificity of labelling. These results suggest that the folate-binding protein identified is a component of the intestinal transport system that mediates folate uptake by the enterocyte.

Folic acid and reduced monoglutamyl folates appear to share the same transport system and are transported with similar avidity (Mason and Rosenberg, 1994). The lack of hierarchy with respect to various reduced and nonreduced forms of folate seems to be unique among epithelial folate transport systems. Despite this unique characteristic, intestinal folate transport is highly specific. For example, stereospecificity is clearly present and pterin is not inhibitory (Schron, Washington and Blitzer, 1988).

Dietary folate is composed mainly of reduced derivatives of folic acid. When these are ingested most is methylated during absorption, so that what enters the portal circulation is largely 5-methyl-THF (Cooper, 1977). The conversion of DHF, THF, 5-formyl-THF and 10-formyl-THF to 5-methyl-THF by the human intestine was demonstrated by Chanarin and Perry (1969). Strum *et al.* (1971) showed that 5-methyl-THF is not altered during passage across the enterocyte.

Physiological doses of folic acid undergo reduction and methylation during passage across the enterocyte, resulting in the appearance of 5-methyl-THF in the portal circulation (Rosenberg, 1977). At higher doses the percentage conversion of folic acid to 5-methyl-THF is diminished, thus indicating saturation of the reduction and methylation process (Strum, 1979). Olinger *et al.* (1973) demonstrated slow reduction of folic acid by dihydrofolate reductase in rat intestinal mucosa *in vitro*. The jejunum, because of its higher concentration of dihydrofolate reductase as compared with the ileum, was more efficient in carrying out this reduction. The combined reduction and methylation process was shown to be maximal in the jejunum, but occurred throughout the small intestine (Strum, 1979); thus it may be inferred that dihydrofolate reductase is the rate-limiting enzyme in the metabolic processing. The metabolic conversion, like the intestinal uptake process, is pH-dependent, being extensive at pH 6.0 and negligible at pH 7.5 (Strum, 1979). The pH effect may reflect in part that dihydrofolate reductase has an acidic pH optimum. The abolition of folate reduction by methotrexate does not significantly interfere with the transport process (Selhub, Brin and Grossowicz, 1973), showing that the reduction step is not a prerequisite for absorption.

When folic acid is ingested in amounts that saturate the intestinal metabolic processing, it is transported through the enterocyte and into the portal circulation without modification. However, the metabolized folic acid (5-methyl-THF) is transported more rapidly through the intestinal wall than is unmodified folic acid (Darcy-Vrillon, Selhub and Rosenberg, 1988).

The exit of folate from the enterocyte into the lamina propria of the villus has been examined using basolateral membrane vesicles of rat intestine (Said and Redha, 1987). The transport system is similar in several respects to the brush-border transporter, being carrier-mediated and sensitive to the effect of anion exchange inhibition. In addition, the exit mechanism is electroneutral and sodium-independent and has a higher affinity for the substrate than has the system at the brush-border membrane.

The brush-border uptake of folate is increased in the presence of sodium ions due to an effect upon the nonsaturable component of folate transport (Zimmerman, Selhub and Rosenberg, 1986).

Absorption of folate in the colon

Direct evidence for folate absorption in the colon has been provided by studies utilizing organ cultures of human colonic biopsies (Zimmerman, 1990). Folate uptake exhibited marked pH dependence as seen in the small intestine, but concentrative transport could not be demonstrated. The data indicated that folate absorption in the colon is a process of facilitated diffusion through a low-affinity carrier. The capacity of the colonic transport system was not as great as that observed in the jejunum.

Intestinal absorption of milk folate

Human milk and the milk of other mammals contain soluble proteins that avidly and specifically bind the folate present in the milk. Cow's milk FBP exists as an aggregating system at pH values above 6.0 in which a polymer composed of at least 16 monomers is reversibly formed (Svendsen *et al.*, 1979; Pedersen *et al.*, 1980). Concentration and pH effects on the equilibrium constant of this reversible system explain the variations in the molecular weight of FBP reported in the literature. FBP has both physicochemical and immunological species-specificity among mammals (Iwai, Tani and Fushiki, 1983). The binding of folate to FBP in milk is tight, but not covalent (Ford, Salter and Scott, 1969). FBP is present in excessive amounts, so that the milk has the capacity to bind added folate (Ford, 1974). The folate in bovine milk occurs predominantly as 5-methyl-THF, of which about 60% is the monoglutamate (Wagner, 1985). In unprocessed and pasteurized milk the folate is bound to FBP, but in UHT-processed milk and yoghurt the folate exists in the free (unbound) form (Wigertz *et al.*, 1996).

Bovine milk is an important dietary source of folate for humans and therefore the bioavailability of bound milk folate merits investigation. Little is known about the manner of absorption of bound milk folate. Tani, Fushiki and Iwai (1983) showed that, *in vitro*, folic acid was entirely dissociated from FBP at pH 4.5 and below, but at pH 7.4 dissociation was negligible. The dissociation was completely reversible and pepsin did not impair the binding activity of the protein. These observations indicate that, *in vivo*, the bovine milk FBP is not inactivated under the acidic conditions in the stomach.

Tani, Fushiki and Iwai (1983) studied the influence of FBP from bovine milk on the absorption of folic acid (PteGlu) in the gastrointestinal tract of the rat. When bound [^{14}C]PteGlu was administered intragastrically to rats by stomach tube, a considerable amount of PteGlu was released from FBP in the stomach and it recombined with FBP in the jejunum. It appears that when the ingested material is transferred from the stomach to the duodenum, it is not yet fully neutralized, allowing a small part of the unbound PteGlu to be absorbed before being recombined with FBP as the material is neutralized in the jejunum. Compared with free PteGlu, bound PteGlu was more gradually absorbed in the small intestine, but finally the total amount of bound PteGlu absorbed was almost the same as that of free PteGlu. Free PteGlu was rapidly absorbed in the jejunum, whereas bound PteGlu was only slightly absorbed in this region. On the other hand, the absorption rates of the two forms of PteGlu were almost similar to each other in the ileum. These results suggest that PteGlu bound to FBP is absorbed by a manner different from that of free PteGlu in the adult rat. A slower rate of folate absorption was also demonstrated in adult rats by Said, Horne and Wagner (1986) using human milk FBP and 5-methyl-THF as the substrate.

Milk has evolved for the nutritional support of the suckling animal and absorption studies in neonates might shed more light on the role of FBP in milk than studies in adults. Mason and Selhub (1988) demonstrated that, with respect to preferential jejunal uptake and sulphasalazine inhibition, free folic acid absorption in the neonatal rat is similar to that in the adult. However, the characteristics of FBP-bound folic acid absorption are very different from those of free folic acid. The uptake of folate bound to FBP from rat milk occurs preferentially in the ileum of the suckling rat as opposed to the jejunum. The absorption rate was slower than that for free folic acid and sulphasalazine was ineffective. In the suckling animal, a variety of pancreatic and intestinal absorptive functions are not fully developed, a situation which could allow the FBP to reach the ileum in an active form. This situation was demonstrated by Salter and Mowlem (1983), who showed that a proportion of goat's milk FBP administered orally to neonatal goats survived along the length of the small intestine. Protease inhibitors inherent to colostrum may assist the passage of bound folate into the small intestine (Laskowski and Laskowski, 1951). Additional evidence for the role of FBP in the absorption of milk folate by neonates was provided by Salter and Blakeborough (1988), who reported that the addition of goat's milk FBP enhanced the transport of 5-methyl-THF in brush-border membrane vesicles isolated from the small intestine of neonatal goats.

Tani and Iwai (1984) reported that the incorporation of radiolabelled folic acid into folate-requiring intestinal bacteria was considerably diminished when it was bound to FBP. Also, the presence of goat's colostrum considerably reduced the uptake of folic acid by folate-requiring bacteria (Ford, 1974). These observations suggest that milk FBP prevents the uptake of bound folate by intestinal bacteria, thereby indirectly promoting folate absorption by the infant.

Mason and Selhub (1988) postulated that in neonates the absorption of bound folate occurs by endocytotic (or similar process) of the intact folate protein complex. Subsequent proteolytic digestion inside the enterocyte would then release the bound folate for export into the portal circulation. Suckling animals are known to absorb proteins by a process that decreases markedly at the time of weaning and is more active in the ileum than in the jejunum (Henderson, 1990).

Plasma transport

Folate circulates in the plasma as reduced monoglutamates, primarily 5-methyl-THF. Any polyglutamyl folate released into plasma from tissues is hydrolysed to monoglutamate by plasma conjugase. Folate circulates as free folate and as protein-bound folate in approximately equal proportions (Herbert and Das, 1994). Most of the binding is a low-affinity, nonspecific binding to a variety of proteins, including albumin. Plasma also contains low concentrations of a high-affinity, specific folate-binding protein which has a greater capacity and a much stronger affinity for folic acid than for 5-methyl-THF. Folate bound to this protein is very efficiently transported to the liver, but not efficiently to other tissues.

Enterohepatic circulation

The kinetics of the folate enterohepatic circulation in the rat have been studied by Steinberg, Campbell and Hillman (1979). After liver uptake, 5-methyl-THF, accompanied by larger amounts of nonmethylated folates (Shin *et al.*, 1995), is rapidly and quantitatively secreted into bile. In contrast, folic acid is either methylated and transported into bile or incorporated into a hepatic polyglutamate pool. Bile folate is then discharged into the small intestine for reabsorption and distribution to the liver and extrahepatic tissues, thus completing the cycle. This recirculation process may account for as much as 50% of the folate that ultimately reaches the tissues.

The importance of the enterohepatic circulation to folate homeostasis was demonstrated by biliary diversion, which resulted in a fall of the serum folate level to 30% of normal within 6 h. This is a much more dramatic drop than that seen with folate-free diets alone. Eventually, the serum folate level stabilizes, despite continuing losses in the bile. The quantitative significance of the enterohepatic circulation of folate and the efficiency of reabsorption have not been determined in humans.

Hepatic uptake of 5-methyltetrahydrofolate

Approximately 10–20% of absorbed folate, mainly in the form of 5-methyl-THF, is taken up by the liver on the first pass, the remainder being distributed to other tissues (Steinberg, 1984).

Uptake of 5-methyl-THF by mammalian hepatocytes is an active transport process that is not energized by Na^+ cotransport (Horne, 1990; Horne, Reed and Said, 1992) and is different from that of folic acid. The initial rate of 5-methyl-THF uptake can be divided into at least two components (Horne, Briggs and Wagner, 1978). At low extracellular concentrations (below $5\,\mu M$) uptake of 5-methyl-THF is effected by a saturable system with a high affinity for the substrate, whereas at higher concentrations (up to $20\,\mu M$) uptake is a linear function of concentration (nonsaturable) and the system has a low affinity for the substrate. Uptake of 5-methyl-THF by the primary saturable component is inhibited by the folate antagonist methotrexate and by 5-formyl-THF, whereas the secondary, nonsaturable component is unaffected by high concentrations of these compounds. The difference in 5-methyl-THF and methotrexate transport by the hepatic saturable system is in contrast to intestinal cells and most other mammalian cells, whose saturable transport systems are similar for the uptake of reduced folates and methotrexate.

Acute exposure of isolated rat hepatocytes to ethanol increased the uptake of 5-methyl-THF in a concentration-dependent and saturable manner. The stimulatory effect was abolished by the addition of pyrazole, an inhibitor of alcohol dehydrogenase, showing that the effect must be a consequence of ethanol metabolism and not due to ethanol itself. The phenomenon was not caused by acetaldehyde, the oxidation product of ethanol, but rather to the altered cellular redox state, i.e. a shift in the $NADH/NAD^+$ ratio in favour of NADH. Uptake of folic acid and methotrexate was inhibited by ethanol, consistent with the concept of a separate transport mechanism for the hepatic uptake of 5-methyl-THF (Horne, Briggs and Wagner, 1979).

Hepatic uptake of 5-methyl-THF by both saturable (Horne, 1990) and nonsaturable (Horne, 1986) transport systems appears to involve cotransport with hydrogen ions. Horne *et al.* (1993) provided the following evidence to support the existence of this transport mechanism in saturable hepatic uptake using basolateral membrane vesicles isolated from human liver. This technique obviates the complications associated with the hepatocyte being polar (i.e. having both basolateral and bile canaliculi membrane faces), as well as those represented by intracellular binding, metabolism, and uptake of substrates by intracellular organelles. Firstly, transport was examined as a function of osmolarity in order to differentiate between transport of the substrate into the intravesicular space and binding to the vesicle membrane. The results of this experiment showed that the majority of 5-methyl-THF uptake was contributed by transport (only *c.*20% could be accounted for by binding). Next, transport was examined as a function of time at different intra- and extravesicular pH. Transport of 5-methyl-THF under the influence of an imposed transmembrane pH gradient ($pH_{extra} = 5.0, pH_{intra} = 7.5$) displayed a transient overshoot; uptake at 60 s was 4.2 times higher than at equilibrium (60 min). Transport was saturable in the presence, but not in the absence, of this pH gradient, suggesting that protonation of the carrier is necessary for interaction with the substrate.

Cotransport of 5-methyl-THF with H^+ requires that the hepatocyte must have a mechanism for maintaining a gradient of H^+ across the basolateral membrane, but no such gradient has been demonstrated. Horne (1990) speculated that H^+ extruded via plasma membrane oxidation of NADH and/or Na^+–H^+ exchange could be conducted along the membrane and interact with the carrier, thereby generating a 'localized' proton gradient.

It was further demonstrated (Horne, 1990; Horne, Reed and Said, 1992; Horne *et al.*, 1993) that hepatic uptake of 5-methyl-THF is an electroneutral process with a probable stoichiometry of 1 : 1 for the interaction of protons with the substrate.

Metabolism and storage

Monoglutamyl 5-methyl-THF is taken up by tissue cells, converted into polyglutamate, and then further metabolized if necessary to other one-carbon derivatives according to the biochemical requirements of the cell or for storage. Conversion of 5-methyl-THF into a polyglutamate requires vitamin B_{12}-dependent methionine synthetase and folate polyglutamate synthetase. Methionine synthetase removes the methyl group from the folate, leaving THF, which can now have glutamate residues (usually five to eight) added to it. Polyglutamyl folates of chain length three and above cannot cross mammalian cell membranes; therefore, adding glutamate residues traps the folate within the cell.

Any folic acid that might have been absorbed and released into the portal circulation without modification accumulates in the liver. There is a short delay before systemic plasma folate concentrations begin to rise, the increase consisting of both absorbed exogenous folic acid and endogenous 5-methyl-THF displaced from the liver (Whitehead, 1986).

The liver has the highest concentration of folate and represents *c.*50% of the folate body store. The liver needs a supply of nonmethylated folate from other tissues in order to maintain an intracellular polyglutamate folate pool, as 5-methyl-THF is a poor substrate for the formation of polyglutamate. Release of stored folate from cells of any tissue requires hydrolysis of the polyglutamates to monoglutamates by intracellular conjugase. Folate coenzymes vary in their number of glutamate residues, and therefore both methionine synthetase and conjugase have a metabolic role as well as controlling folate storage.

Folate catabolism increases dramatically during pregnancy, thereby increasing the folate requirement and the risk of gestational folate deficiency (McPartlin *et al.*, 1993).

Renal reabsorption

The reabsorption of circulating 5-methyl-folate by the renal proximal tubule fulfils an important role in folate homeostasis by maintaining plasma concentrations of this essential vitamin. The reabsorption mechanism appears to be receptor-mediated endocytosis involving a high-affinity, high-specificity folate receptor whose expression is regulated by the folate content of the epithelial cell (Kamen and Capdevila, 1986). It has been postulated (Hjelle *et al.*, 1991; Birn, Selhub and Christensen, 1993) that folate in the glomerular filtrate binds to the receptor in the brush-border membrane and the protein–folate complex is subsequently internalized. After dissociation in the endocytic vacuoles, folate is transported across the basolateral membrane for release into the renal vascular circulation, while the receptor migrates back into the brush-border membrane.

Salvage from senescent red blood cells

Erythrocytes have significant folate stores by virtue of both large cell mass and high folate content; they also have a limited life span. Folate can be salvaged from erythrocytes that die and release their intracellular contents. After erythrocytes are cleared by the reticuloendothelial system, the folate is recovered, released for transport to the liver, and rapidly appears in the bile for subsequent distribution to the tissues (Steinberg, 1984).

13.6.2 Dietary sources and their bioavailability

Dietary sources

Polyglutamyl folate is an essential biochemical constituent of living cells, and most foods contribute some folate. In the United States, dried beans, eggs, greens, orange juice, sweet corn, peas and peanut products are good sources of folate that are inexpensive and available all the year round. The folate content of a selection of foods is shown in Table 13.2. Although liver, all types of fortified breakfast cereals, cooked dried beans, asparagus, spinach, broccoli and avocado provide the highest

Table 13.2 Folate content of selected foods in the US diet according to NHANES II[a] data, 1976–80 (from Subar, Block and James, 1989)

Food	µg Folate/usual serving	µg Folate/100 g
Liver	383	428
Cold cereals (not bran or superfortified)	112	275
Pinto, navy and other dried beans (cooked)	84	100
Spinach	70	128
Broccoli	53	65
Orange juice	43	41
Eggs	31	43
Beef stew	26	10
Pizza	24	16
Tomatoes	21	10
Green salad	20	48
Bananas	19	19

[a] National health and nutrition survey.

amount of folate per average serving, several of these foods do not rank highly in terms of actual dietary folate intake in the United States because of their low rate of consumption. Table 13.3 presents data on the major contributors of folate to the US diet and the number of persons per 10 000 consuming the food on any given day. Orange juice is the biggest source, contributing 9.70% to dietary folate intake.

Approximately 75% of the folate in mixed American diets is present in the form of polyglutamates (Sauberlich *et al.*, 1987). The predominant folate vitamers in animal tissues are polyglutamyl forms of THF, 5-methyl-THF and 10-formyl-THF, while plant tissues contain mainly polyglutamyl 5-methyl-THF (Gregory, Sartain and Day, 1984). Folic acid does not occur naturally to a significant extent, but it is often found in small quantities as an oxidation product of THF in foods stored under conditions that permit exposure to oxygen (Gregory, 1984). 5-Formyl-THF is a minor vitamer in most tissues, but thermal processing can promote its formation through isomerization of 10-formyl-THF (Gregory, 1989). 5-Methyl-5,6-DHF and, to a lesser extent, 10-formylfolic acid may be present in processed foods as oxidation products of 5-methyl-THF and 10-formyl-THF, respectively.

The folates generally exist in nature bound to proteins (Baugh and Krumdieck, 1971) and they are also bound to storage polysaccharides (various types of starch and glycogen) in foods (Černá and Káš, 1983).

Nonenzymatic degradation and salvage of dietary folate

An important factor influencing folate bioavailability in humans is the facile oxidation of 5-methyl-THF to 5-methyl-5,6-DHF. Ratanasthien *et al.*

Table 13.3 Major contributors of folate in the US diet according to NHANES II[a] data, 1976–80 (from Subar, Block and James, 1989)

Ranking	Description	Percentage of total folate (%)	Cumulative percentage of folate (%)	Persons per 10 000 population consuming the indicated food
1	Orange juice	9.70	9.70	2296
2	White bread, rolls, crackers	8.61	18.30	7724
3	Pinto, navy and other dried beans (cooked)	7.08	25.30	1074
4	Green salad	6.85	32.24	4083
5	Cold cereals (not bran or superfortified)	4.96	37.20	1182
6	Eggs	4.63	41.83	3141
7	Alcoholic beverages	3.85	45.68	2689
8	Coffee, tea	3.40	49.08	8007
9	Liver	3.07	52.15	174
10	Superfortified cereals	3.06	55.21	214
11	Whole milk, whole-milk beverages	2.90	58.11	4166
12	Bran and granola cereals	2.48	60.59	669

[a] National health and nutrition survey.

(1977) have stated that the latter compound may constitute as much as 50% of the total folate in processed foods.

The physiological significance of the presence of 5-methyl-5,6-DHF in the human diet has been studied by Lucock *et al.* (1995b). In the mildly acidic postprandial gastric environment, 5-methyl-5,6-DHF is rapidly degraded, while 5-methyl-THF is relatively stable. Fortunately, where gastric histology is normal, endogenous ascorbic acid present in gastric juice can salvage labile 5-methyl-5,6-DHF by reducing it back to 5-methyl-THF. This appears to be a true physiological process as 5-methyl-5,6-DHF exhibits nearly complete folate activity when given orally to the rat and chick (Gregory *et al.*, 1984), presumably via non-enzymatic reduction *in vivo*.

In patients with severely impaired gastric function, the stomach pH will be high and endogenous gastric ascorbate levels will be low. Under these conditions, 5-methyl-5,6-DHF is stable within the stomach, but it will not be reduced to 5-methyl-THF. 5-Methyl-5,6-DHF, although rapidly absorbed, persists in the bloodstream without being metabolized and does not therefore appear to enter the folate metabolic pool (Rata-nasthien *et al.*, 1977). It follows that such patients are likely to have a reduced bioavailability of dietary folate if 5-methyl-5,6-DHF is a major constituent. There will be no problem with supplemental folic acid, which is stable within a broad pH range.

Analytical methods for determining folate in food or plasma invariably use ascorbic acid or other antioxidant to prevent degradation of labile folates during heat treatment. Any 5-methyl-5,6-DHF that might be present in the food or plasma will therefore be converted to 5-methyl-THF and escape detection. For individuals who cannot salvage dietary folate in the manner described, the assay result will suggest a bioavailability value higher than it actually is.

Bioavailability

It is impossible to determine the extent of inherent difference in bioavailability among the various forms of monoglutamyl folate, but little or no difference would be predicted on the basis of their total interconversion in mammalian metabolism (Gregory, 1997).

Tamura and Stokstad (1973) studied the bioavailability of folate in various foods by measuring the urinary excretion of folate in human subjects kept in a metabolic unit. All subjects were presaturated with folate in order to obtain linear response curves relating the oral dose to the urinary excretion. The apparent bioavailability for each food was calculated from the dose–response curve for each subject. A similar technique was adopted in another study (Babu and Srikantia, 1976). Data from the two studies are summarized in Table 13.4. The wide variation among subjects could reflect a varying ability to digest the unusually large amounts of test food administered. The large amounts were necessary to effect significant changes in urinary folate levels. Despite inherent limitations of these studies and the recording of bioavailability values of > 100%, the results strongly suggest that substantial

Table 13.4 Apparent bioavailability of endogenous folate in various foods as determined by urinary excretion measurement in preloaded human subjects

Food supplement	Total amount of test food consumed (g)	Apparent bioavailability (%)		Reference
		Mean	Range	
Banana	890	82	0–148	(1)
Lima beans (dried, cooked)	560	70	0–138	(1)
Lima beans (frozen, cooked)	360	96	48–181	(1)
Beef liver (cooked)	63–94	50	22–103	(1)
Brewer's yeast	30	60	55–67	(1)
Cabbage (cooked)	500–700	47	0–127	(1)
Cabbage (raw)	500	47	0–93	(1)
Defatted soybean meal	200	46	0–83	(1)
Wheat germ	170	30	0–64	(1)
Orange juice concentrate	600	35	29–40	(1)
Egg yolk (cooked)	250	82	61–100	(1)
Tomato	1000	37	24–71	(2)
Spinach	200	63	26–99	(2)
Banana	533–700	46	0–148	(2)
Egg, whole	300–500	72	35–137	(2)
Goat liver	100	70	9–125	(2)

(1) Tamura and Stokstad (1973); (2) Babu and Srikantia (1976).

variation in the bioavailability of naturally occurring folate exists among common foods.

The low folate bioavailability reported in orange juice (Tamura and Stokstad, 1973) was later attributed to the effects of low intraluminal pH produced by large amounts of ingested citric acid (Tamura *et al.*, 1976). Other studies show that the bioavailability of orange juice folate is high in humans. Using intraluminal perfusion of the human small intestine, Nelson, Streiff and Cerda (1975) compared the absorption of folate from orange juice and from a solution of synthetic folic acid, and found no significant difference. Rhode, Cooper and Farmer (1983) reported similar serum folate levels in women taking orange juice or synthetic folic acid during 9 weeks of a folate-restricted diet. The availability of folate in orange juice was unaffected by oral contraceptive medication. Gregory (1989) opined that any effects of orange juice components on folate bioavailability of other foods in mixed diets would be negligible.

Grossowicz, Rachmilewitz and Izak (1975) reported that when small amounts of yeast containing 300 µg of folate per dose were fed to healthy human subjects, the yeast folate was fully utilized. Other investigators, who used much higher amounts of yeast in their trials, observed incomplete bioavailability of folate, probably because the larger doses exceeded the capacity of intestinal deconjugation and/or absorption mechanisms.

Bioavailability data for various foods have also been obtained by a rat bioassay using serum and liver folate levels and body weight gain as response criteria (Clifford, Jones and Bills, 1990; Clifford *et al.*, 1991). The data presented in Table 13.5 show that estimated folate bioavailability of

Table 13.5 Relative bioavailability of endogenous folate in various foods as determined by rat bioassay

Food	Relative bioavailability (%)			Reference
	Serum folate	*Liver folate*	*Weight gain*	
Broccoli, cooked	82	74	95	(1)
raw	88	58	103	(1)
Cabbage, cooked	58	83	74	(1)
Cantaloupe	66	116	81	(1)
Beans, refried	nd	82	113	(1)
Beef liver	94	101	91	(2)
Lima beans	75	85	111	(2)
Peas	68	75	97	(2)
Spinach	65	71	75	(2)
Mushrooms	123	89	73	(2)
Collards	83	74	85	(2)
Orange juice	107	99	91	(2)
Wheat germ	63	68	77	(2)

(1) Clifford, Jones and Bills (1990); (2) Clifford *et al.* (1991).

the foods tested exceeded 50% by all three response criteria. Folates of beef liver and orange juice were as available as folic acid, while folates of spinach and wheat germ were significantly less available than folic acid. The relevance of these bioavailability estimates to human nutrition is uncertain because of the significant differences in physical characteristics and mode of action between intestinal conjugases of the rat and human (Wang, Reisenauer and Halsted, 1985).

Colman, Green and Metz (1975) compared the efficiency of absorption of folic acid from fortified whole wheat bread, rice, maize meal (porridge) and an aqueous solution of folic acid in human subjects. Each helping of the food contained 1 mg of added folic acid, the same amount as in the aqueous dose. The absorption profile for each subject was determined as the summated increases in serum folate 1 and 2 h after administration of the test materials. The absorption efficiencies of the rice and maize were similar (about 55%), relative to the response observed with aqueous folic acid, while the absorption efficiencies of the bread were lower (about 30%). The folic acid was found to resist destruction by the conventional cooking conditions used to prepare the test foods. The long-term rate of change of erythrocyte folate during daily administration of fortified maize meal (Colman *et al.*, 1975) or fortified bread (Margo *et al.*, 1975) corroborated these findings.

The results of Colman and associates suggested that interactions of the added folic acid with food components, possibly during cooking, caused impaired absorption. To investigate this possibility, Ristow, Gregory and Damron (1982a) prepared lactose-casein liquid model food systems fortified with folic acid or 5-methyl-THF and subjected them to retort processing (121 °C for 20 min). Microbiological and HPLC analysis indicated that folic acid was very stable during thermal processing, while 5-methyl-THF was *c.*75% degraded. A chick bioassay showed that the bioavailability of the remaining folate was complete for both fortified model food systems, indicating that no complexes were formed during processing that inhibited folate utilization.

Folic acid is completely absorbed when administered under fasting conditions and there is little nutritionally significant difference in the *in vivo* absorption and utilization among the various monoglutamyl folates under normal dietary conditions (Gregory, 1995). In the same review, various studies indicated effective, but frequently incomplete, bioavailability of polyglutamyl folate. For example, using an intestinal perfusion technique, Bailey *et al.* (1984) reported a bioavailability of 59% for [^{14}C]PteGlu$_7$ relative to [^{3}H]PteGlu on the basis of urinary recovery in young and elderly healthy human subjects. In the same study, the luminal disappearance (absorption) of [^{14}C]PteGlu$_7$ compared with that of [^{3}H]PteGlu was 81% and 72% (no significant difference) for elderly and young subjects, respectively. Gregory *et al.* (1991) reported a bioavailability of *c*.50% for PteGlu₆ relative to PteGlu as judged by urinary excretion when stable isotopes of these compounds (deuterium-labelled folates) were administered orally to human subjects. Semchuk, Allen and O'Connor (1994) found that the relative bioavailability of PteGlu₆ added to rat diets containing human, cow or goat milk solids was 49–71% that of PteGlu added to the same milk-containing diets. In a 92-day metabolic study in women, dietary folate appeared to be no more than 50% bioavailable when compared with synthetic folic acid (Sauberlich *et al.*, 1987).

The bioavailability of folate in vegetables may be affected by the extent of polyglutamyl folate deconjugation by endogenous conjugase. Thus raw cabbage, which had undergone enzymatic deconjugation during diet preparation and storage, was found to be completely biologically available by a chick bioassay (Ristow, Gregory and Damron, 1982a). In the same study *c*.60% of the folate in cooked cabbage, which corresponded to the polyglutamate fraction, was not biologically available. The heat treatment would have inactivated the endogenous conjugase in the cooked cabbage, leaving the majority of the folate content in the polyglutamyl form. Endogenous conjugase will also be deactivated by the blanching of cabbage.

A possible explanation for the incomplete bioavailability of polyglutamyl folate is that the intestinal deconjugation process is rate-limiting. Reisenauer and Halsted (1987) calculated that human intestinal brush-border conjugase is present in sufficient quantity as not to limit folate absorption. However, changes in the pH of the luminal contents and the presence of dietary conjugase inhibitors and folate binders could adversely affect the rate of deconjugation and absorption of folate. These factors may account for the variation in folate bioavailability in individual foods.

Bhandari and Gregory (1990) showed that extracts of many foods can inhibit brush-border conjugase activity from human and porcine intestine *in vitro*. Such inhibition may be a factor affecting the bioavailability of polyglutamyl folates in diets containing these foods, although further investigation is needed. As shown in Table 13.6, beans, banana and spinach caused a moderate (14–36%) inhibition, but more dramatic inhibition was caused by tomato (46%) and orange juice (73–80%). Foods that exhibited no significant inhibition included whole-wheat flour, wheat bran, whole egg, milk, cabbage, cauliflower and lettuce. The observed inhibition was not accompanied by binding of enzyme to the substrate, and pH was not a factor because all the food extracts were neutralized before testing. Citrate was not the causative factor of inhibition for tomato or orange juice. Although citrate significantly inhibited the enzyme activity, the effect was far less pronounced than that caused by orange juice or tomato extract when calculated on the basis of equivalent citrate concentration. Inhibition was specific for the conjugase; two other brush-border

Table 13.6 *In vitro* effect of extracts from different food substances on the conjugase activity of human and porcine jejunal brush-border membrane vesicles (from Bhandari and Gregory, 1990)

Food substance	Relative activity (% of control)	
	Porcine	Human
Red kidney beans	64.5*	84.1*
Pinto beans	64.9*	66.8*
Green lima beans	64.4*	64.8*
Cut green beans	74.1*	80.7*
Black-eyed peas	64.7*	71.3*
Yellow-corn meal	102.5	100
Wheat bran	91.9	85.8*
Tomato	54.1*	54.0*
Banana	74.8*	74.8*
Cauliflower	85.8	84.3*
Spinach	78.9*	86.1*
Orange juice	20.0*	26.6*
Whole egg	88.5	94.7
Evaporated milk	86.3	nd
Cabbage	87.9	nd
Lettuce	93.8	nd
Whole-wheat flour	99.7	nd
Medium rye flour	100.2	nd

Control = the enzyme assayed in the absence of any food extract; nd, not determined.
* Significantly different from control, $P < 0.01$.

enzymes, alkaline phosphatase and sucrase, were not affected by the foods tested. The nature of the inhibitory substance(s) is not known. No inhibitory effect was shown by dietary folate-binding protein, soluble anionic polysaccharides, sulphated compounds, phytohaemagglutinins or trypsin inhibitors at nutritionally relevant concentrations.

Butterworth, Newman and Krumdieck (1974) reported that a wide variety of beans and other legumes contained a substance which inhibited the activity of hog kidney conjugase *in vitro*. The conjugases of human plasma, chicken pancreas and rat liver were similarly inhibited (Krumdieck, Newman and Butterworth, 1973). The significance of these findings is not clear with respect to the conjugase from the intestinal brush-border membrane.

Keagy, Shane and Oace (1988) reported that the bioavailability of monoglutamyl and polyglutamyl folates did not differ significantly when consumed with Californian white beans. It was also observed that the bioavailability of PteGlu was lower when added to beans than when added to wheat bran. A possible explanation for this effect is the binding of folate to slowly digested legume starch, as observed by Luther *et al.* (1965).

The presence of blanched cabbage and orange juice solids significantly ($P < 0.05$) retarded the utilization of folic acid and bacterial polyglutamyl folates in rats as measured by hepatic retention of tritiated folates (Abad and Gregory, 1988). The magnitude of the effect was not sufficient to cause a difference in urinary tritium excretion. Because both mono- and polyglutamyl folates were affected, it appears that, in this animal model, certain dietary components exert an adverse effect on the absorption, rather than the deconjugation of folates.

Bioavailability of folate in milk
The physiological function of the milk folate-binding protein (FBP) remains controversial, although it may be implicated in the absorption of folate from milk in the suckling animal (Mason and Selhub, 1988), as previously discussed.

Smith, Picciano and Deering (1985) monitored the folate status of infants by measuring serum and erythrocyte folate concentrations from birth to 1 year. When solid foods were introduced after 3 months of life, blood concentrations of folate were significantly correlated with the total amount of folate ingested from milk (either human milk or bovine milk-based formula) but not with the total amount of folate in the diet. This strong correlation between folate status and milk folate intake suggests that folate present in the milk component of the diet (whether from human milk, bovine milk or bovine milk-based formula) is more available to the infant than is folate ingested from the remainder of the diet.

The role of milk FBP in the absorption and retention of dietary folate appears to be beneficial. Tani and Iwai (1984) showed that, in the adult rat, urinary excretion of folate when administered bound to bovine milk FBP was some four-fold lower than when administered as free folate. This decreased urinary loss, together with the slower rate of absorption of bound folate, suggests that FBP might act to regulate the bioavailability of folate in milk.

Swiatlo *et al.* (1990) assessed the relative bioavailability of folate from rat diets formulated to contain 20% by weight of unpasteurized human, bovine and goat milk solids. Using plasma and kidney folate concentrations as the response criteria, the bioavailability of folate from human and bovine milk-containing diets was significantly greater than that from milk-free diets. There was no enhancement of bioavailability from the diet containing goat's milk. It was later demonstrated (Semchuk, Allen and O'Connor (1994) that the superior folate bioavailability from rat diets containing human milk is due, at least in part, to changes in the intestinal environment that promote a net increase in microbial folate biosynthesis and availability to the host. The binding capacity of goat's milk solids is greater than that of either human or bovine milk solids (Swiatlo *et al.*,

1990) and therefore these results do not support the suggestion that the presence of FBP in the diet enhances folate bioavailability. The results underscore the important contribution of gut microbial biosynthesis of folate in rats and the impact that diet has on altering microbial production of folate. The relevance of these rat studies to folate bioavailability in humans is a matter for conjecture.

Wigertz *et al.* (1996) reported significant ($P < 0.05$) losses of 5-methyl-THF in milk as a result of pasteurization (62.5 °C for 30 min), ultra-heat treatment (UHT) (140 °C for 5 s) and fermentation (90 °C for 10 min to produce yoghurt). The respective losses were 8%, 19% and 20%. Pasteurization caused a significant ($P < 0.05$) 20% decrease in FBP concentration as measured by ELISA, while the protein-binding capacity was unaffected. UHT processing and fermentation reduced the FBP concentration by 97% and 100% respectively, with accompanying losses of folate-binding capacity. Even if the heating step had been omitted in the yoghurt production, folate in fermented milk would occur in the free form, since low pH, such as that in yoghurt, is known to cause dissociation between FBP and the folate. The results indicated that all folate in unprocessed milk and pasteurized milk is protein-bound, while folate in UHT-processed milk and yoghurt occurs in the free form. According to these data, pasteurized cow's milk, like natural human milk, contains FBP which facilitates folate absorption in the infant. Gregory (1982) also reported that the pasteurization of milk had little effect on the folate-binding capacity, although minor alterations of binding characteristics were observed. UHT milk does not afford infants the potential benefits of protein-bound folate, although the infant has the ability to absorb unbound folate.

Wigertz and Jägerstad (1993) found no significant difference in the relative bioavailability of folate between processed milk (pasteurized, UHT-treated or fermented milk) and raw milk in a rat bioassay, suggesting that the loss of protein-binding capacity does not affect the bioavailability of milk folate in the adult rat.

Proprietary formulas marketed in the United States are supplemented with unbound folic acid and contain two to three times the amount of folate present in human milk (Smith, Picciano and Deering, 1985). The nutritional adequacy of infant formulas that contain no active FBP is well documented (Gregory, 1995).

Effects of dietary fibre

The possible influence of dietary fibre on folate bioavailability is of interest, since many of the food products previously reported to have low folate bioavailability are also high in dietary fibre.

The bioavailability of monoglutamyl folate does not appear to be inhibited by high-fibre foods. Neither spinach nor bran cereal impaired the absorption of folic acid in humans (Bailey *et al.*, 1988), and high-fibre Iranian breads were also without effect (Russell, Ismail-Beigi and Reinhold, 1976).

Ristow, Gregory and Damron (1982b) investigated the ability of purified dietary fibre components to sequester folic acid *in vitro* using equilibrium dialysis under neutral isotonic conditions. The recovery of essentially 100% of the folic acid from each material (cellulose, pectin, lignin, sodium alginate and wheat bran) indicated the absence of any binding or entrapment. *In vivo* effects were evaluated by a chick bioassay with graded levels of folic acid in semi-purified diets containing the fibre material at 3% (w/w). There were no differences between any of the materials with respect to the rise in plasma and liver folate as a function of dietary folic acid. The results of this study suggest that the fibre component of human diets would have a negligible effect on the bioavailability of folic acid employed in food fortification. A similar conclusion can be drawn from data produced by Keagy and Oace (1984), who showed that cellulose, xylan, pectin or wheat bran had no detectable effect on the utilization of folic acid added to rat diets.

In contrast to the lack of effect of dietary fibre on the bioavailability of monoglutamyl folate, a negative effect of wheat bran on the bioavailability of polyglutamyl folate has been reported in several studies. Wheat bran decreased the folate liver response to added PteGlu$_7$ relative to PteGlu in rats (Keagy, 1985). In a human study the addition of 30 g of wheat bran to the formulated meal delayed the absorption of PteGlu$_7$ relative to PteGlu (Keagy, 1979). The serum folate response of human subjects to PteGlu$_7$ when added to bran cereal was significantly less than the response to PteGlu added to bran cereal (Bailey *et al.*, 1988). The addition of 30 g of wheat bran to the formula meal accelerated PteGlu absorption, but did not significantly alter the absorption of PteGlu$_7$ (Keagy, Shane and Oace, 1988). The latter authors presented possible factors that may explain the lower bioavailability of polyglutamyl folate compared with monoglutamyl folate when consumed with wheat bran. The six additional alpha carboxyl groups in the PteGlu$_7$ molecule increase the potential for anionic interactions. Wheat bran has cation exchange properties and can decrease intragastric concentrations of hydrogen ions and pepsin. Thus it has the potential for direct interaction with folate compounds, it may interfere with folate binding to other diet components, may alter the pH of the medium and may alter the rate and extent of digestion of other diet components. Wheat bran also reaches the colon faster than other sources of fibre.

Effects of alcohol

The majority of alcoholics are clinically folate deficient or at least in negative folate balance. Aside from the poor diet of alcoholics, ethanol adversely affects folate absorption, hepatic uptake and urinary excretion. Destruction of the folate molecule as a consequence of ethanol metabolism may also contribute to the vitamin deficiency. The aetiology of folate deficiency in alcoholism has been reviewed by Halsted (1995).

Dietary inadequacy Except for certain beers, little or no folate is present in alcoholic beverages. Thus chronic alcoholics who substitute ethanol for other sources of calories typically deprive themselves of dietary folate and are potential cases of megaloblastic anaemia. The well-nourished alcoholic rarely manifests this condition.

Intestinal malabsorption Both hydrolysis of polyglutamyl folate by brush-border jejunal conjugase and jejunal uptake of folate are affected by chronic alcoholism. Naughton *et al.* (1989) demonstrated a significant (50%) decrease in conjugase activity in jejunal brush-border vesicles obtained from miniature pigs that had been fed ethanol as 50% of dietary energy for one year. Conjugase activity was also decreased by the addition of physiological concentrations of ethanol to brush-border vesicles from control-fed pigs. These data suggest that polyglutamyl folate hydrolysis is inhibited by chronic and acute exposure to alcohol.

In a clinical study conducted by Halsted, Robles and Mezey (1973), jejunal uptake of orally administered radioactive folic acid was decreased in two hospitalized alcoholic patients who developed megaloblastic bone marrow while consuming a folate-deficient diet supplemented with ethanol. Uptake was unchanged in one patient who received the folate-deficient diet without added ethanol, and in one other patient who received an adequate diet with added ethanol. These findings suggest that folate malabsorption in alcoholics is the result of a dietary inadequacy and chronic alcohol ingestion acting synergistically.

Hepatic uptake Acute exposure of isolated rat hepatocytes to ethanol increased the uptake of 5-methyl-THF as a consequence of ethanol metabolism (Horne, Briggs and Wagner, 1979). On the other hand, monkeys fed ethanol as 50% of dietary energy for two years showed significantly decreased liver folate levels (Tamura *et al.*, 1981). The hepatic processes of reduction, methylation and formylation of reduced folate and the synthesis of polyglutamyl folate were not affected in the ethanol-fed monkeys, suggesting that chronic ethanol ingestion somehow caused a decreased ability to retain folate in the liver. A 30-day folate turnover study in the same alcoholic monkeys showed increased faecal excretion of folate, consistent with an increase in biliary excretion and/or a decrease in intestinal reabsorption of folate (Tamura and Halsted, 1983).

Enterohepatic cycle Short-term feeding appeared to decrease biliary folate excretion in rats (Hillman, McGuffin and Campbell, 1977); however, this study did not include alcohol-treated but folate-replete animals. When such a group was included in a similar study, ethanol had no effect on the biliary excretion of folate (Weir, McGing and Scott, 1985). Thus, the precise influence of ethanol on folate enterohepatic circulation is still open to question.

Renal excretion The feeding of acute doses of ethanol to fasting rats produced a marked increase in urinary folate levels, in amounts that accounted for a subsequent decrease in plasma folate levels (McMartin, 1984). Eisenga, Collins and McMartin (1989) presented evidence that acute ethanol treatment decreases proximal tubular reabsorption of 5-methyl-THF in the rat, thus enhancing the urinary excretion of this compound. McMartin, Collins and Bairnsfather (1986) showed that repeated daily doses of ethanol produced a cumulative increase in urinary folate excretion in fed and fasted rats, although the effect in fed rats was less marked. There was no adaptation to the loss of folate during the subacute treatment. Chronic ethanol treatment (McMartin *et al.*, 1989) also led to increased urinary folate excretion in rats fed folate-containing diets, but the rats did not become folate deficient. In contrast, when ethanol-treated rats were fed folate-deficient diets for two weeks, only very small amounts of folate were excreted in the urine. Thus, in the absence of a dietary supply of folate, the rat adapts its renal processing of folate to conserve the vitamin. The adaptive mechanism evidently opposes and overcomes the inherent effect of ethanol on the renal processing of folate.

Urinary folate excretion was moderately increased in human volunteers given folic acid supplements and alcohol for two weeks (Russell *et al.*, 1983).

Direct destruction of the folate molecule The metabolism of ethanol by its major metabolic pathway via alcohol dehydrogenase leads to the formation of acetaldehyde, which is metabolized to acetate. The oxidation of acetaldehyde, however, can also occur by reaction with the ubiquitous enzyme xanthine oxidase, during which superoxide radicals are generated. Shaw *et al.* (1989) demonstrated that superoxide generated from acetaldehyde/xanthine oxidase *in vitro* cleaves folates. Thus superoxide-mediated folate cleavage during ethanol metabolism may contribute to the increased requirement for folate seen in the alcoholic.

A possible unifying explanation for the effect of ethanol on folate metabolism and bioavailability is its known ability to alter lipid composition and physical properties of cellular membranes (Halsted, 1995). As discussed previously, folate homeostasis requires its interaction with many membrane-bound proteins and therefore any changes in membrane structure would have potential effects on folate binding and transport. For example, the decreased conjugase activity observed in alcoholic pigs (Naughton *et al.*, 1989) could have been caused by a change in the physicochemical composition of the jejunal brush-border membrane.

13.6.3 Effects of food processing and cooking

Little is known about the extent and mechanisms of loss of folate in processed foods compared with other water-soluble vitamins. The folic acid that is added to fortified foods is retained to a greater extent than the less stable reduced derivatives that are naturally present. Thus baking destroyed one-third of the folate naturally present in wheat, but only 11% of added folate (Keagy, Stokstad and Fellers, 1975).

Heat processing of canned and sealed foods results in a markedly higher folate retention than domestic cooking methods. The addition of ascorbic acid to canned vegetables before retorting provides additional stability to the folate content during subsequent storage, but little protective effect takes place during the heat process itself (Hawkes and Villota, 1989). Lin, Luh and Schweigert (1975) reported a 30% loss of folate in canned garbanzo beans regardless of whether or not 0.2% ascorbic acid was added to the curing brine. Folate analysis was performed on the homogenized contents of the whole can, i.e. beans plus brine. No further loss of folate took place when the process time at 118 °C was extended from 30 min to 53 min. The pH of canned garbanzo beans is between 5.8 and 6.2, a range at which folate is quite stable toward heat.

It has been known for many years that cooking lowers the folate content of foods (Cheldelin, Woods and Williams, 1943; Schweigert, Pollard and Elvehjem, 1946; Hurdle, Barton and Searles, 1968). Boiling, steaming and pressure-cooking green vegetables results in a loss of folate of some 90%, the majority of the loss occurring in the first minute. Foods of animal origin show smaller losses or are unaffected by boiling or frying.

Leichter, Switzer and Landymore (1978) examined the effect of cooking (vigorous boiling for 10 min) on the folate content of vegetables which are good sources of the vitamin. Folate was determined in both the raw and cooked vegetables as well as the water in which they were cooked. With the exception of spinach, the amount of folate retained in the cooked vegetable plus the amount in the cooking water was in excess of 100% of the amount found in the raw vegetable. Thus the loss of folate during the cooking of vegetables is caused mostly by leaching into the surrounding cooking water and not by actual destruction of the vitamin. In the case of broccoli, cabbage, cauliflower and spinach, the cooking water contained higher quantities of folate than the cooked vegetables. Brussels sprouts and asparagus were exceptions, probably because of their lower surface-to-weight ratios. Folate losses increase as the ratio of cooking water to food increases. Discarding the cooking water would deprive the consumer of a significant amount of readily available folate.

13.7 DIETARY INTAKE

The average folate intake in the United Kingdom has been estimated to be *c*.190 μg per person per day (Spring, Robertson and Buss, 1979; Bates *et al.*, 1982; Poh Tan, Wenlock and Buss, 1984). This estimate approximates to the RNI of 200 μg/day for adult males and nonpregnant, nonlactating females. The corresponding US RDA is 200 μg/day for males and 180 μg/day for females (National Research Council, 1989). Populations with low or marginal intakes of folate have a high incidence of folate deficiency, leading to megaloblastic anaemia. Folate body stores are depleted within a period of a few months; therefore it is important to maintain an adequate dietary intake, especially during pregnancy, lactation, infancy and adolescence, which are associated with rapid growth. The pregnant adolescent is at particular risk of developing folate deficiency because increased requirements for foetal growth are superimposed on those for maternal growth. Folate inadequacy among the aged is mainly confined to low income groups in which socioeconomic factors lead to suboptimal health care (Bailey, 1990a,b). In addition to physiological factors associated with phases of the life cycle, environmental factors – including use of alcohol, drugs and tobacco – increase the risk of developing folate deficiency (Bailey, 1990a,b).

All manifestations of folate deficiency in women who start pregnancy with moderate folate stores could probably be prevented by diets containing the equivalent of 200 μg folic acid/day (Herbert, 1977). In women receiving a normal diet in the UK, a daily oral supplement of 100 μg of folic acid prevented any fall in the mean concentration of erythrocyte folate during pregnancy (Chanarin *et al.*, 1968). The US RDA of 400 μg/day during pregnancy is higher than needed for most women and is intended to meet the needs of those with poor folate stores, essentially no other dietary folate, and multiple or twin pregnancies. This level of intake cannot usually be met without oral supplementation or food fortification.

The maternal requirement for dietary folate increases during lactation and folate is taken up by the mammary gland in preference to other maternal tissues. On the basis of a milk volume of 750 ml/day and 50% absorption of dietary folate, the RDA for folate during lactation is 280 μg/day during the first six months and 260 μg/day during the second six months (National Research Council, 1989). The National Academy of Science's Subcommittee on Nutrition During Lactation predicted in 1991 that the dietary recommendations of folate during lactation would not be met by women with nutrient intakes of fewer than 2000 kcal/day (Picciano, Green and O'Connor, 1994).

An adequate supply of folate is essential to support the rapid growth during infancy. These needs are met by human or bovine milk, but not by goat's milk. Although blood concentrations of folate in the new-born are about three times higher than those of maternal blood folate, liver stores at birth are small and are rapidly depleted, especially in premature infants. By two weeks of age, blood folate levels fall below adult levels and remain low during the first year of life. This reduction in infant blood folate coincides with the ingestion of other foods, in which the folate is less bioavailable than it is in milk.

The US Food and Drug Administration (FDA) has announced that with effect from January 1, 1998, folic acid must be added to most enriched flour, breads, corn meals, rice, noodles, macaroni and other grain products. These foods were chosen for folate fortification because they are staple products for most of the US population, and because they are suitable vehicles for this purpose (Anon., 1996). Enriched flour will be fortified with 140 µg folic acid per 100 g flour, giving a folate content nearly four times that in whole-grain flour (Hine, 1996). This public health measure was designed to reduce the number of infants born with neural tube defects.

In the event of vitamin B_{12} deficiency, folate supplements can delay and prevent the haematological changes associated with pernicious anaemia. However, folate cannot prevent the irreversible neurological damage caused by vitamin B_{12} deficiency. There is a real danger, therefore, that the obvious symptoms of pernicious anaemia will be alleviated by the folate treatment, but the concomitant nerve degeneration will be undetected. For this reason, prophylactic vitamin supplementation always includes vitamin B_{12} with folate.

It has been proposed by Milne *et al.* (1984) and several other research groups that oral folate supplementation might have an adverse effect on zinc status. On the basis of these reports, zinc supplements have to be considered if folate supplements are to be given routinely before and during pregnancy (Halsted, 1993). Kauwell *et al.* (1995) presented data from a short-term folic acid supplementation study in young men that did not support the hypothesis that supplemental folic acid adversely affects zinc status. No differences in plasma zinc, erythrocyte zinc or urinary zinc due to supplemental folic acid were detected at two levels of zinc intake. Data from the same study also suggested that zinc intake does not impair folate utilization.

Effects of high intakes

Folate is generally considered to have a low acute and chronic toxicity for humans. In adults no adverse effects were noted after 400 mg/day for 5 months or after 10 mg/day for 5 years (Brody, 1991). Folic acid appears to require binding to polypeptides before it can be stored: amounts in excess of the limited binding capacity in serum and tissues tend to be excreted in the urine rather than retained (Herbert and Das, 1994).

Assessment of nutritional status

Measurement of blood folate
Measurement of plasma folate concentration provides a sensitive indicator of folate status long before clinical signs of deficiency develop. Normal plasma levels of 6 to 20 ng/ml drop to less than 3 ng/ml in about one week on diets very low in folate. Therefore folate deficiency can only be ascribed to those cases in which plasma folate levels remain low over a period of time. Monoglutamyl 5-methyl-THF is the predominant folate in plasma and this vitamer can be rapidly and reliably determined by radioassay.

Erythrocyte folate levels, which drop after 15 to 20 weeks of deficient intake, reflect body folate stores at the time of erythrocyte formation; they therefore represent a more accurate and less variable indication of folate status than plasma folate levels. Folates in the erythrocytes are polyglutamyl and must be hydrolysed to monoglutamates before microbiological assay or radioassay. As most of the whole-blood folate is located in the erythrocytes, hydrolysis can be simply accomplished by lysing whole blood and allowing plasma conjugase access to the polyglutamates. Erythrocyte folate levels of < 140 ng/ml packed cells, measured by microbiological assay, are indicative of folate deficiency; levels between 140 and 160 ng/ml suggest marginal status; and levels > 160 ng/ml normal folate status (Brody, Shane and Stokstad, 1984).

It should be understood that a lowered plasma or erythrocyte folate level does not necessarily point to folate deficiency as the causative factor. For instance, vitamin B_{12} deficiency causes a functional folate deficiency that would be indistinguishable in these tests from the deficiency caused by inadequate folate intake or defective folate metabolism, unless specific tests for B_{12} status are also carried out.

Deoxyuridine suppression test
This functional test can be used to distinguish between folate and vitamin B_{12} deficiency. It examines the ability of deoxyuridine (dU) to suppress the incorporation of tritiated deoxythymidine ($[^3H]dT$) into the DNA of rapidly dividing cells *in vitro* and is best applied to bone marrow biopsy

$$\text{Deoxyuridine monophosphate} + 5,10\text{-CH}_2\text{-THF} \xrightarrow[\text{synthetase}]{\text{thymidylate}} \text{deoxythymidine monophosphate} + \text{DHF}$$

Figure 13.3 The role of 5,10-methylene-THF as a one-carbon donor. (Note: Deoxythymidine monophosphate is synonymous with thymidine monophosphate.)

samples. The test is centred around the formation of deoxythymidine monophosphate (dTMP) from deoxyuridine monophosphate (dUMP) through the catalytic action of thymidylate synthetase. dTMP is the nucleotide of thymine, one of the base constituents of DNA. The folate dependency of thymidylate synthetase activity is due to the role of 5,10-methylene-THF as the one-carbon donor (Figure 13.3). Formation of dTMP also depends on the presence of vitamin B_{12} which, in the form of methylcobalamin, is a coenzyme for methionine synthetase. As shown in Figure 13.2, this enzyme is involved in the regeneration of THF, which is needed to synthesize 5,10-methylene-THF.

In normal cells, the incorporation of [3H]dT into DNA after pre-incubation with dU is 1.4–1.8% of that without pre-incubation. This is because the labelled precursor is diluted in the larger intracellular pool of newly synthesized unlabelled dTMP. By contrast, cells which are deficient in folate form little or no dTMP from dU, and hence incorporate into their DNA nearly as much of the [3H]dT after incubation with dU as they do without pre-incubation (Bender, 1992).

In folate deficiency, the addition of any biologically active form of folate to the *in vitro* system will increase the dU suppression of [3H]dT incorporation, but the addition of vitamin B_{12} will not. In vitamin B_{12} deficiency, addition of methylcobalamin or any folate, with the exception of 5-methyl-THF, will increase the suppression (Brody, Shane and Stokstad, 1984).

13.8 SYNOPSIS

Approximately 75% of the folate in mixed American diets is estimated to be present in the form of polyglutamates. Polyglutamyl folate is deconjugated by brush-border conjugase in the intestinal lumen and the monoglutamates are absorbed by a carrier-mediated system that is markedly influenced by the luminal pH. At present, a great deal of uncertainty exists concerning the bioavailability of folate, particularly with respect to the polyglutamyl forms. Folate absorption could be impaired by inhibition of intestinal conjugase, intralumenal entrapment of folates by the food matrix, or by inhibition of transport processes. Because of these dietary factors, differences exist in the bioavailability of the folates in individual foods.

A variety of experimental approaches have been used to determine the bioavailability of food folate in humans and in rat or chick models. They include measuring changes in haematological values, tissue folate levels, urinary folate excretion and growth in response to known intakes of folic acid or folate-containing foods. The inconsistencies of reported bioavailability estimates are probably due in part to the many different experimental procedures used. Studies on a limited number of food items indicate that the mean bioavailability of folate is about 50%.

REFERENCES

Abad, A.R. and Gregory, J.F. III (1988) Assessment of folate bioavailability in the rat using extrinsic dietary enrichment with radiolabeled folates. *J. Agric. Food Chem.*, **36**, 97–104.

Anon. (1992) Folate supplements prevent recurrence of neural tube defects. *Nutr. Rev.*, **50**, 22–4.

Anon. (1996) Folic acid fortification. *Nutr. Rev.*, **54**, 94–5.

Babu, S. and Srikantia, S.G. (1976) Availability of folates from some foods. *Am. J. Clin. Nutr.*, **29**, 376–9.

Bailey, L.B. (1990a) The role of folate in human nutrition. *Nutrition Today*, **25**, 12–19.

Bailey, L.D. (1990b) Folate status assessment. *J. Nutr.*, **120**, 1508–11.

Bailey, L.B., Cerda, J.J., Bloch, B.S *et al.* (1984) Effect of age on poly- and monoglutamyl folacin absorption in human subjects. *J. Nutr.*, **114**, 1770–6.

Bailey, L.B., Barton, L.E., Hillier, S.E. and Cerda, J.J. (1988) Bioavailability of mono and polyglutamyl folate in human subjects. *Nutr. Rep. Int.*, **38**, 509–18.

Bates, C.J., Black, A.E., Phillips, D.R. *et al.* (1982) The discrepancy between normal folate intakes and the folate RDA. *Human. Nutr. Appl. Nutr.*, **36A**, 422–9.

Baugh, C.M. and Krumdieck, C.L. (1971) Naturally occurring folates. *Ann. NY Acad. Sci.*, **186**, 7–28.

Bell, J.G. (1974) Microbiological assay of vitamins of the B group in foodstuffs. *Lab. Pract.*, **23**, 235–42, 252.

Bender, D.A. (1992) *Nutritional Biochemistry of the Vitamins*, Cambridge University Press, Cambridge.

Bhandari, S.D. and Gregory, J.F. III (1990) Inhibition by selected food components of human and porcine intestinal pteroylpolyglutamate hydrolase activity. *Am. J. Clin. Nutr.*, **51**, 87–94.

Bhandari, S.D., Gregory, J.F. III, Renuart, D.R. and Merritt, A.M. (1990) Properties of pteroylpolyglutamate hydrolase in pancreatic juice of the pig. *J. Nutr.*, **120**, 467–75.

Bird, O.D., McGlohon, V.M. and Vaitkus, J.W. (1969) A microbiological assay system for naturally occurring folates. *Can. J. Microbiol.*, **15**, 465–72.

Birn, H., Selhub, J. and Christensen, E.I. (1993) Internalization and intracellular transport of folate-binding protein in rat kidney proximal tubule. *Am. J. Physiol.*, **264**, C302–10.

Brody, T. (1991) Folic acid. In *Handbook of Vitamins*, 2nd edn (ed. L.J. Machlin), Marcel Dekker, Inc., New York, pp. 453–89.

Brody, T., Shane, B. and Stokstad, E.L.R. (1979) Separation and identification of pteroylpolyglutamates by polyacrylamide gel chromatography. *Analyt. Biochem.*, **92**, 501–9.

Brody, T., Shane, B. and Stokstad, E.L.R. (1984) Folic acid. In *Handbook of Vitamins. Nutritional, Biochemical, and Clinical Aspects* (ed. L.J. Machlin), Marcel Dekker, Inc., New York, pp. 459–96.

Butterworth, C.E. Jr, Baugh, C.M. and Krumdieck, C. (1969) A study of folate absorption and metabolism in man utilizing carbon-14-labeled polyglumates synthesized by the solid phase method. *J. Clin. Invest.*, **48**, 1131–42.

Butterworth, C.E., Newman, A.J. and Krumdieck, C.L. (1974) Tropical sprue: a consideration of possible etiologic mechanisms with emphasis on pteroylpolyglutamate metabolism. *Trans. Am. Clin. Climatol. Ass.*, **86**, 11–22.

Černá, J. and Káš, J. (1983) New conception of folacin assay in starch or glycogen containing food samples. *Nahrung*, **27**, 957–64.

Chanarin, I. and Perry, J. (1969) Evidence for reduction and methylation of folate in the intestine during normal absorption. *Lancet*, **II** (No. 7624), 776–8.

Chanarin, I., Rothman, D., Ward, A. and Perry, J. (1968) Folate status and requirement in pregnancy. *Br. Med. J.*, **2**, 390–4.

Chandler, C.J., Harrison, D.A., Buffington, C.A. *et al.* (1991) Functional specificity of jejunal brush-border pteroylpolyglutamate hydrolase in pig. *Am. J. Physiol.*, **260**, G865–72.

Cheldelin, V.H., Woods, A.M. and Williams, R.J. (1943) Losses of B vitamins due to cooking of foods. *J. Nutr.*, **26**, 477–85.

Chen, T.-S. and Cooper, R.G. (1979) Thermal destruction of folacin: effect of ascorbic acid, oxygen and temperature. *J. Food Sci.*, **44**, 713–16.

Clifford, A.J., Jones, A.D. and Bills, N.D. (1990) Bioavailability of folates in selected foods incorporated into amino acid-based diets fed to rats. *J. Nutr.*, **120**, 1640–7.

Clifford, A.J., Heid, M.K., Peerson, J.M. and Bills, N.D. (1991) Bioavailability of food folates and evaluation of food matrix effects with a rat bioassay. *J. Nutr.*, **121**, 445–53.

Colman, N., Green, R. and Metz, J. (1975) Prevention of folate deficiency by food fortification. II. Absorption of folic acid from fortified staple foods. *Am. J. Clin. Nutr.*, **28**, 459–64.

Colman, N., Barker, E.A., Barker, M. *et al.* (1975) Prevention of folate deficiency by food fortification. IV. Identification of target groups in addition to pregnant women in an adult rural population. *Am. J. Clin. Nutr.*, **28**, 471–6.

Cooper, B.A. (1977) Physiology of absorption of monoglutamyl folates from the gastrointestinal tract. In *Folic Acid. Biochemistry and Physiology in Relation to the Human Nutrition Requirement* (Food and Nutrition Board/National Research Council), National Academy of Sciences, Washington, DC, pp. 188–97.

Darcy-Vrillon, B., Selhub, J. and Rosenberg, I.W. (1988) Analysis of sequential events in intestinal absorption of folylpolyglutamate. *Am. J. Physiol.*, **255**, G361–6.

Day, B.P. and Gregory, J.F. III (1981) Determination of folacin derivatives in selected foods by high-performance liquid chromatography. *J. Agric. Food Chem.*, **29**, 374–7.

Day, B.P.F. and Gregory, J.F. III (1983) Thermal stability of folic acid and 5-methyltetrahydrofolic acid in liquid model food systems. *J. Food Sci.*, **48**, 581–7, 599.

De Souza, S. and Eitenmiller, R. (1990) Effects of different enzyme treatments on extraction of total folate from various foods prior to microbiological assay and radioassay. *J. Micronutr. Anal.*, **7**, 37–57.

Donaldson, K.O. and Keresztesy, J.C. (1962) Naturally occurring forms of folic acid. III. Characterization and properties of 5-methyldihydrofolate, an oxidation product of 5-methyltetrahydrofolate. *J. Biol. Chem.*, **237**, 3815–9.

Eisenga, B.H., Collins, T.D. and McMartin, K.E. (1989) Effects of acute ethanol on urinary excretion of 5-methyltetrahydrofolic acid and folate derivatives in the rat. *J. Nutr.*, **119**, 1498–1505.

Engelhardt, R. and Gregory, J.F. III (1990) Adequacy of enzymatic deconjugation in quantification of folate in foods. *J. Agric. Food Chem.*, **38**, 154–8.

Ford, J.E. (1974) Some observations on the possible nutritional significance of vitamin B$_{12}$- and folate-binding proteins in milk. *Br. J. Nutr.*, **31**, 243–57.

Ford, J.E., Salter, D.N. and Scott, K.J. (1969) The folate-binding protein in milk. *J. Dairy Res.*, **36**, 435–46.

Gauch, R., Leuenberger, U. and Müller, U. (1993) The determination of folic acid (pteroyl-L-glutamic acid) in food by HPLC. *Mitt. Gebiete Lebensm. Hyg.*, **84**, 295–302 (in German).

Ghitis, J., Mandelbaum-Shavit, F. and Grossowicz, N. (1969) Binding of folic acid and derivatives by milk. *Am. J. Clin. Nutr.*, **22**, 156–62.

Goli, D.M. and Vanderslice, J.T. (1989) Microbiological assays of folacin using a CO_2 analyzer system. *J. Micronutr. Anal.*, **6**, 19–33.

Graham, D.C., Roe, D.A. and Ostertag, S.G. (1980) Radiometric determination and chick bioassay of folacin in fortified and unfortified frozen foods. *J. Food Sci.*, **45**, 47–51.

Gregory, J.F. III (1982) Denaturation of the folacin-binding protein in pasteurized milk products. *J. Nutr.*, **112**, 1329–38.

Gregory, J.F. III (1984) Determination of folacin in foods and other biological materials. *J. Ass. Off. Analyt. Chem.*, **67**, 1015–19.

Gregory, J.F. III (1985a) Chemical changes of vitamins during food processing. In *Chemical Changes of Food during Processing* (eds T. Richardson and J.W. Finley), Van Nostrand Reinhold Co., New York, pp. 373–408.

Gregory, J.F. III (1985b) Folacin. Chromatographic and radiometric assays. In *Methods of Vitamin Assay*, 4th edn (eds J. Augustin, B.P. Klein, D. Becker and P.B. Venugopal), John Wiley & Sons, New York, pp. 473–96.

Gregory, J.F. III (1989) Chemical and nutritional aspects of folate research: analytical procedures, methods of folate synthesis, stability, and bioavailability of dietary folates. *Adv. Food Nutr. Res.*, **33**, 1–101.

Gregory, J.F. III (1995) The bioavailability of folate. In *Folate in Health and Disease* (ed L.B. Bailey), Marcel Dekker, Inc., New York, pp. 195–235.

Gregory, J.F. III (1997) Bioavailability of folate. *Eur. J. Clin. Nutr.*, **51**, Suppl. 1, S54–9.

Gregory, J.F. III, Day, B.P.F. and Ristow, K.A. (1982) Comparison of high performance liquid chromatographic, radiometric, and *Lactobacillus casei* methods for the determination of folacin in selected foods. *J. Food Sci.*, **47**, 1568–71.

Gregory, J.F. III, Sartain, D.B., and Day, B.P.F. (1984) Fluorometric determination of folacin in biological materials using high performance liquid chromatography. *J. Nutr.*, **114**, 341–53.

Gregory, J.F. III, Ristow, K.A., Sartain, D.B. and Damron, B.L. (1984) Biological activity of the folacin oxidation products 10-formylfolic acid and 5-methyl-5,6-dihydrofolic acid. *J. Agric. Food Chem.*, **32**, 1337–42.

Gregory, J.F. III, Engelhardt, R., Bhandari, S.D. et al. (1990) Adequacy of extraction techniques for determination of folate in foods and other biological materials. *J. Food Comp. Anal.*, **3**, 134–44.

Gregory, J.F. III, Bhandari, S.D. Bailey, L.B. et al. (1991) Relative bioavailability of deuterium-labeled monoglutamyl and hexaglutamyl folates in human subjects. *Am. J. Clin. Nutr.*, **53**, 736–40.

Grossowicz, N., Rachmilewitz, M. and Izak, G. (1975) Utilization of yeast polyglutamate folates in man. *Proc. Soc. Exp. Biol. Med.*, **150**, 77–9.

Halsted, C.H. (1990) Intestinal absorption of dietary folates. In *Folic Acid Metabolism in Health and Disease* (eds M.F. Picciano, E.L.R. Stokstad and J.F. Gregory III), Wiley-Liss, Inc., New York, pp. 23–45.

Halsted, C.H. (1993) Water-soluble vitamins. In *Human Nutrition and Dieletics*, 9th edn (eds J.S. Garrow, W.P.T. James and A. Ralph), Churchill Livingstone, Edinburgh, pp. 239–63.

Halsted, C.H. (1995) Alcohol and folate interactions: clinical implications. In *Folate in Health and Disease* (ed. L.B. Bailey), Marcel Dekker, Inc., New York, pp. 313–27.

Halsted, C.H., Robles, E.A. and Mezey, E. (1973) Intestinal malabsorption in folate-deficient alcoholics. *Gastroenterology*, **64**, 526–32.

Hawkes, J.G. and Villota, R. (1989) Folates in foods: reactivity, stability during processing, and nutritional implications. *Crit. Rev. Food Sci. Nutr.*, **28**, 439–539.

Henderson, G.B. (1990) Folate-binding proteins. *Ann. Rev. Nutr.*, **10**, 319–35.

Herbert, V. (1977) Folate requirement in adults (including pregnant and lactating females). In *Folic Acid. Biochemistry and Physiology in Relation to the Human Nutrition Requirement* (Food and Nutrition Board/National Research Council), National Academy of Sciences, Washington, D.C., pp. 247–55.

Herbert, V. and Das, K.C. (1994) Folic acid and vitamin B_{12}. In *Modern Nutrition in Health and Disease*, 8th edn, Vol. 1 (eds M.E. Shils, J.A. Olson and M. Shike), Lea & Febiger, Philadelphia, pp. 402–25.

Herbert, V., Larrabee, A.R. and Buchanan, J.M. (1962) Studies on the identification of a folate compound of human serum. *J. Clin. Invest.*, **41**, 1134–8.

Hillman, R.S., McGuffin, R. and Campbell, C. (1977) Alcohol interference with the folate enterohepatic cycle. *Trans. Ass. Am. Physicians*, **90**, 145–56.

Hine, R.J. (1996) What practitioners need to know about folic acid. *J. Am. Dietetic Ass.*, **96**, 451–2.

Hjelle, J.T., Christensen, E.I., Carone, F.A. and Selhub, J. (1991) Cell fractionation and electron microscope studies of kidney folate-binding protein. *Am. J. Physiol.*, **260**, C338–46.

Holt, D.L., Wehling, R.L. and Zeece, M.G. (1988) Determination of native folates in milk and other dairy products by high-performance liquid chromatography. *J. Chromat.*, **440**, 271–9.

Hoppner, K. and Lampi, B. (1982) The determination of folic acid (pteroylmonoglutamic acid) in fortified products by reversed phase high pressure liquid chromatography. *J. Liquid Chromat.*, **5**, 953–66.

Horne, D.W. (1986) Studies on the mechanism of folate transport in isolated hepatocytes. In *Chemistry and Biology of Pteridines 1986. Pteridines and Folic Acid Derivatives* (eds B.A. Cooper and V.M. Whitehead), Walter de Gruyter, New York, pp. 559–62.

Horne, D.W. (1990) Na^+ and pH dependence of 5-methyltetrahydrofolic acid and methotrexate transport in freshly isolated hepatocytes. *Biochim. Biophys. Acta*, **1023**, 47–53.

Horne, D.W., Briggs, W.T. and Wagner, C. (1978) Transport of 5-methyltetrahydrofolic acid and folic acid in freshly isolated hepatocytes. *J. Biol. Chem.*, **253**, 3529–35.

Horne, D.W., Briggs, W.T. and Wagner, C. (1979) Studies on the transport mechanism of 5-methyltetrahydrofolic acid in freshly isolated hepatocytes: effect of ethanol. *Arch. Biochem. Biophys.*, **196**, 557–65.

Horne, D.W., Reed, K.A. and Said, H.M. (1992) Transport of 5-methyltetrahydrofolate in basolateral membrane vesicles of rat liver. *Am. J. Physiol.*, **262**, G150–8.

Horne, D.W., Reed, K.A., Hoefs, J. and Said, H.M. (1993) 5-Methyltetrahydrofolate transport in basolateral membrane vesicles from human liver. *Am. J. Clin. Nutr.*, **58**, 80–4.

Hurdle, A.D.F., Barton, D. and Searles, I.H. (1968) A method for measuring folate in food and its application to a hospital diet. *Am. J. Clin. Nutr.*, **21**, 1202–7.

Iwai, K., Tani, M. and Fushiki, T. (1983) Electrophoretic and immunological properties of folate-binding protein isolated from bovine milk. *Agric. Biol. Chem.*, **47**, 1523–30.

Jacoby, B.T. and Henry, F.T. (1992) Liquid chromatographic determination of folic acid in infant formula and adult medical nutritionals. *J. AOAC Int.*, **75**, 891–8.

Kamen, B.A. and Capdevila, A. (1986) Receptor-mediated folate accumulation is regulated by the cellular folate content. *Proc. Natl Acad. Sci USA*, **83**, 5983–7.

Kauwell, G.P.A., Bailey, L.B., Gregory, J.F. III *et al.* (1995) Zinc status is not adversely affected by folic acid supplementation and zinc intake does not impair folate utilization in human subjects. *J. Nutr.*, **125**, 66–72.

Keagy, P.M. (1979) Dietary fiber effects on folacin bioavailability. *Cereal Foods World*, **24**, 461 (abstr).

Keagy, P.M. (1985) Rat bioassay of folate monoglutamate (PG1), heptaglutamate (PG7) and wheat bran. *Fed. Proc.*, **44**, 777 (abstr).

Keagy, P.A. and Oace, S.M. (1984) Folic acid utilization from high fiber diets in rats. *J. Nutr.*, **114**, 1252–9.

Keagy, P.M., Shane, B. and Oace, S.M. (1988) Folate bioavailability in humans: effects of wheat bran and beans. *Am. J. Clin. Nutr.*, **47**, 80–8.

Keagy, P.M., Stokstad, E.L.R. and Fellers, D.A. (1975) Folacin stability during bread processing and family flour storage. *Cereal Chem.*, **52**, 348–56.

Klein B.P. and Kuo, C.H.Y. (1981) Comparison of microbiological and radiometric assays for determining total folacin in spinach. *J. Food Sci.*, **46**, 552–4.

Krumdieck, C.L., Newman, A.J. and Butterworth, C.E. Jr (1973) A naturally occurring inhibitor of folic acid conjugase (pteroyl-polyglutamyl hydrolase) in beans and other pulses. *Am. J. Clin. Nutr.*, **26**, 460 (abstr).

Krumdieck, C.L., Tamura, T. and Eto, I. (1983) Synthesis and analysis of the pteroylpolyglutamates. *Vitams Horm.*, **40**, 45–104.

Laskowski, M. and Laskowski, M. (1951) Crystalline trypsin inhibitor from colostrum. *J. Biol. Chem.*, **190**, 563–73.

Laurence, K.M., James, N., Miller, M.H. *et al.* (1981) Double-blind randomised controlled trial of folate treatment before conception to prevent recurrence of neural-tube defects. *Br. Med. J.*, **282**, 1509–11.

Leichter, J., Landymore, A.F. and Krumdieck, C.L. (1979) Folate conjugase activity in fresh vegetables and its effect on the determination of free folate content. *Am. J. Clin. Nutr.*, **32**, 92–5.

Leichter, J., Switzer, V.P. and Landymore, A.F. (1978) Effect of cooking on folate content of vegetables. *Nutr. Rep. Int.*, **18**, 475–9.

Leslie, G.I. and Rowe, P.B. (1972) Folate binding by the brush-border membrane proteins of small intestinal epithelial cells. *Biochemistry*, **11**, 1696–1703.

Lewis, G.P. and Rowe, P.B. (1979) Oxidative and reductive cleavage of folates – a critical appraisal. *Analyt. Biochem.*, **93**, 91–7.

Lin, K.C., Luh, B.S. and Schweigert, B.S. (1975) Folic acid content of canned garbanzo beans. *J. Food Sci.*, **40**, 562–5.

Lucock, M.D., Hartley, R. and Smithells, R.W. (1989) A rapid and specific HPLC-electrochemical method for the determination of endogenous 5-methyltetrahydrofolic acid in plasma using solid phase sample preparation with internal standardization. *Biomed. Chromat.*, **3**, 58–63.

Lucock, M.D., Green, M., Hartley, R. and Levene, M.I. (1993) Physicochemical and biological factors influencing methylfolate stability: use of dithiothreitol for HPLC analysis with electrochemical detection. *Food Chem.*, **47**, 79–86.

Lucock, M.D., Nayeemuddin, F.A., Habibzadeh, N. *et al.* (1994) Methylfolate exhibits a negative in-vitro interaction with important dietary metal cations. *Food Chem.*, **50**, 007–10.

Lucock, M.D., Green, M., Priestnall, M. *et al.* (1995a) Optimisation of chromatographic conditions for the determination of folates in foods and biological tissues for nutritional and clinical work. *Food Chem.*, **53**, 329–38.

Lucock, M.D., Priestnall, M., Daskalakis, I. *et al.* (1995b) Nonenzymatic degradation and salvage of dietary folate: physicochemical factors likely to influence bioavailability. *Biochem. Mol. Med.*, **55**, 43–53.

Luther, L., Santini, R., Brewster, C. *et al.* (1965) Folate binding by insoluble components of American and Puerto Rican diets. *Alabama J. Med. Sci.*, **2**, 389–93.

Malin, J.D. (1975) Folic acid. *World Rev. Nutr. Dietetics*, **21**, 198–223.

Margo, G., Barker, M., Fernandes-Costa, F. *et al.* (1975) Prevention of folate deficiency by food fortification. VII. The use of bread as a vehicle for folate supplementation. *Am. J. Clin. Nutr.*, **28**, 761–3.

Martin, J.I., Landen, W.O. Jr, Soliman, A.-G.M. and Eitenmiller, R.R. (1990) Application of a tri-enzyme extraction for total folate determination in foods. *J. Ass. Off. Analyt. Chem.*, **73**, 805–8.

Maruyama, T., Shiota, T. and Krumdieck, C.L. (1978) The oxidative cleavage of folates – a critical study. *Analyt. Biochem.*, **84**, 277–95.

Mason, J.B. and Rosenberg, I.H. (1994) Intestinal absorption of folate. In *Physiology of the Gastrointestinal Tract*, 3rd edn, Vol. 2 (ed. L.R. Johnson), Raven Press, New York, pp. 1979–95.

Mason, J.B. and Selhub, J. (1988) Folate-binding protein and the absorption of folic acid in the small intestine of the suckling rat. *Am. J. Clin. Nutr.*, **48**, 620–5.

McMartin, K.E. (1984) Increased urinary folate excretion and decreased plasma folate levels in the rat after acute ethanol treatment. *Alcoholism Clin. Exp. Res.*, **8**, 172–8.

McMartin, K.E., Collins, T.D. and Bairnsfather, L. (1986) Cumulative excess urinary excretion of folate in rats after repeated ethanol treatment. *J. Nutr.*, **116**, 1316–25.

McMartin, K.E., Collins, T.D., Eisenga, B.H. *et al.* (1989) Effects of chronic ethanol and diet treatment on urinary folate excretion and development of folate deficiency in the rat. *J. Nutr.*, **119**, 1490–7

McPartlin, J., Halligan, A., Scott, J.M. *et al.* (1993) Accelerated folate breakdown in pregnancy. *Lancet*, **341**, 148–9.

Milne, D.B., Canfield, W.K., Mahalko, J.R. and Sandstead, H.H. (1984) Effect of oral folic acid supplements on zinc, copper, and iron absorption and excretion. *Am. J. Clin. Nutr.*, **39**, 535–9.

Milunsky, A., Jick, H., Jick, S.S. *et al.* (1989) Multivitamin/folic acid supplementation in early pregnancy reduces the prevalence of neural tube defects. *J. Am. Med. Assoc.*, **262**, 2847–52.

MRC Vitamin Study Research Group (1991) Prevention of neural tube defects: results of the Medical Research Council Vitamin Study. *Lancet*, **338**, 131–7.

National Research Council (1989) Water-soluble vitamins. In *Recommended Dietary Allowances*, 10th edn, National Academy Press, Washington, DC, pp. 115–73.

Naughton, C.A., Chandler, C.J., Duplantier, R.B. and Halsted, C.H. (1989) Folate absorption in alcoholic pigs: in vitro hydrolysis and transport at the brush border membrane. *Am. J. Clin. Nutr.*, **50**, 1436–41.

Nelson, E.W., Streiff, R.R. and Cerda, J.J. (1975) Comparative bioavailability of folate and vitamin C from a synthetic and a natural source. *Am. J. Clin. Nutr.*, **28**, 1014–9.

O'Broin, J.D., Temperley, I.J., Brown, J.P. and Scott, J.M. (1975) Nutritional stability of various naturally occurring monoglutamate derivatives of folic acid. *Am. J. Clin. Nutr.*, **28**, 438–44.

Olinger, E.J., Bertino, J.R. and Binder, H.J. (1973) Intestinal folate absorption. II. Conversion and retention of pteroylmonoglutamate by jejunum. *J. Clin. Invest.*, **52**, 2138–45.

Österdahl, B.-G. and Johansson, E. (1989) Comparison of radiometric and microbiological assays for the determination of folate in fortified gruel and porridge. *Int. J. Vitam. Nutr. Res.*, **59**, 147–50.

Paine-Wilson, B. and Chen, T.-S. (1979) Thermal destruction of folacin: effect of pH and buffer ions. *J. Food Sci.*, **44**, 717–22.

Pedersen, J.C. (1988) Comparison of γ-glutamyl hydrolase (conjugase; EC 3.4.22.12) and amylase treatment procedures in the microbiological assay for food folates. *Br. J. Nutr.*, **59**, 261–71.

Pedersen, T.G., Svendsen, I., Hansen, S.I. *et al.* (1980) Aggregation of a folate-binding protein from cow's milk. *Carlsberg Res. Commun.*, **45**, 161–6.

Phillips, D.R. and Wright, A.J.A. (1982) Studies on the response of *Lactobacillus casei* to different folate monoglutamates. *Br. J. Nutr.*, **47**, 183–9.

Phillips, D.R. and Wright, A.J.A. (1983) Studies on the response of *Lactobacillus casei* to folate vitamin in foods. *Br. J. Nutr.*, **49**, 181–6.

Picciano, M.F., Green, T. and O'Connor, D.L. (1994) The folate status of women and health. *Nutrition Today*, **29**, 20–9.

Poh Tan, S., Wenlock, R.W. and Buss, D.H. (1984) Folic acid content of the diet in various types of British household. *Human Nutr. Appl. Nutr.*, **38A**, 17–22.

Ratanasthien, K., Blair, J.A., Leeming, R.J. *et al.* (1977) Serum folates in man. *J. Clin. Path.*, **30**, 438–48.

Reed, L.S. and Archer, M.C. (1979) Action of sodium nitrite on folic acid and tetrahydrofolic acid. *J. Agric. Food Chem.*, **27**, 995–9.

Reed, L.S. and Archer, M.C. (1980) Oxidation of tetrahydrofolic acid by air. *J. Agric. Food Chem.*, **28**, 801–5.

Reingold, R.N. and Picciano, M.F. (1982) Two improved high-performance liquid chromatographic separations of biologically significant forms of folate. *J. Chromat.*, **234**, 171–9.

Reisenauer, A.M. (1990) Affinity labelling of the folate-binding protein in pig intestine. *Biochem. J.*, **267**, 249–52.

Reisenauer, A.M. and Halsted, C.H. (1987) Human folate requirements. *J. Nutr.*, **117**, 600–602.

Reisenauer, A.M., Chandler, C.J. and Halsted, C.H. (1986) Folate binding and hydrolysis by pig intestinal brush-border membranes. *Am. J. Physiol.*, **251**, G481–6.

Rhode, B.M., Cooper, B.A. and Farmer, F.A. (1983) Effect of orange juice, folic acid, and oral contraceptives on serum folate in women taking a folate-restricted diet. *J. Am. Coll. Nutr.*, **2**, 221–30.

Ristow, K.A., Gregory, J.F. III and Damron, B.L. (1982a) Thermal processing effects on folacin bioavailability in liquid model food systems, liver, and cabbage. *J. Agric. Food Chem.*, **30**, 801–806.

Ristow, K.A., Gregory, J.F. III and Damron, B.L. (1982b) Effects of dietary fiber on the bioavailability of folic acid monoglutamate. *J. Nutr.*, **112**, 750–8.

Robinson, D.R. (1971) The nonenzymatic hydrolysis of N^5, N^{10}-methenyltetrahydrofolic acid and related reactions. *Methods Enzymol.*, **18B**, 716–25.

Rosenberg, I.H. (1977) Role of intestinal conjugase in the control of the absorption of polyglutamyl folates. In *Folic Acid. Biochemistry and Physiology in Relation to the Human and Nutrition Requirement* (Food and Nutrition Board/National Research Council), National Academy of Sciences, Washington, DC, pp. 136–46.

Ruddick, J.E., Vanderstoep, J. and Richards, J.F. (1978) Folate levels in food – a comparison of microbiological assay and radioassay methods for measuring folate. *J. Food Sci.*, **43**, 1238–41.

Russell, R.M., Ismail-Beigi, F. and Reinhold, J.G. (1976) Folate content of Iranian breads and the effect of their fiber content on the intestinal absorption of folic acid. *Am. J. Clin. Nutr.*, **29**, 799–802.

Russell, R.M., Dhar, G.J., Dutta, S.K. and Rosenberg, I.H. (1979) Influence of intraluminal pH on folate absorption: studies in control subjects and in patients with pancreatic insufficiency. *J. Lab. Clin. Med.*, **93**, 428–36.

Russell, R.M., Rosenberg, I.M., Wilson, P.D. *et al.* (1983) Increased urinary excretion and prolonged turnover time of folic acid during ethanol ingestion. *Am. J. Clin. Nutr.*, **38**, 64–70.

Said, H.M. and Redha, R. (1987) A carrier-mediated transport for folate in basolateral membrane vesicles of rat small intestine. *Biochem. J.*, **247**, 141–6.

Said, H.M., Horne, D.W. and Wagner, C. (1986) Effect of human milk folate binding protein on folate intestinal transport. *Arch. Biochem. Biophys.*, **251**, 114–20.

Salter, D.N. and Blakeborough, P. (1988) Influence of goat's-milk folate-binding protein on transport of 5-methyltetrahydrofolate in neonatal-goat small intestinal brush-border-membrane vesicles. *Br. J. Nutr.*, **59**, 497–507.

Salter, D.N. and Mowlem, A. (1983) Neonatal role of milk folate-binding protein: studies on the course of digestion of goat's milk folate binder in the 6-d old kid. *Br. J. Nutr.*, **50**, 589–96.

Sauberlich, H.E., Kretsch, M.J., Skala, J.H. *et al.* (1987) Folate requirement and metabolism in nonpregnant women. *Am. J. Clin. Nutr.*, **46**, 1016–28.

Schieffer, G.W., Wheeler, G.P. and Cimino, C.O. (1984) Determination of folic acid in commercial diets by anion-exchange solid-phase extraction and subsequent reversed-phase HPLC. *J. Liquid Chromat.*, **7**, 2659–69.

Schron, C.M., Washington, C. Jr and Blitzer, B.L. (1985) The transmembrane pH gradient drives uphill folate transport in rabbit jejunum. *J. Clin. Invest.*, **76**, 2030–3.

Schron, C.M., Washington, C. Jr and Blitzer, B.L. (1988) Anion specificity of the jejunal folate carrier: effects of reduced folate analogues on folate uptake and efflux. *J. Membrane Biol.*, **102**, 175–83.

Schweigert, B.S., Pollard, A.E. and Elvehjem, C.A. (1946) The folic acid content of meats and the retention of this vitamin during cooking. *Arch. Biochem.*, **10**, 107–11.

Selhub, J., Brin, H. and Grossowicz, N. (1973) Uptake and reduction of radioactive folate by everted sacs of rat small intestine. *Eur. J. Biochem.*, **33**, 433–8.

Semchuk, G.M., Allen, O.B. and O'Connor, D.L. (1994) Folate bioavailability from milk-containing diets is affected by altered intestinal biosynthesis of folate in rats. *J. Nutr.*, **124**, 1118–25.

Shane, B., Tamura, T. and Stokstad, E.L.R. (1980) Folate assay: a comparison of radioassay and microbiological methods. *Clin. Chim. Acta*, **100**, 13–19.

Shaw, S., Jayatilleke, E., Herbert, V. and Colman, N. (1989) Cleavage of folates during ethanol metabolism. Role of acetaldehyde/xanthine oxidase-generated superoxide. *Biochem. J.*, **257**, 277–80.

Shin, H.-C., Takakuwa, F., Shimoda, M. and Kokue, E. (1995) Enterohepatic circulation kinetics of bile-active folate derivatives and folate homeostasis in rats. *Am. J. Physiol.*, **269**, R421–5.

Shoda, R., Mason, J.B., Selhub, J. and Rosenberg, I.H. (1990) Folate binding in intestinal brush border membranes: evidence for the presence of two binding activities. *J. Nutr. Biochem.*, **1**, 257–61.

Smith, A.M., Picciano, M.F. and Deering, R.H. (1985) Folate intake and blood concentrations of term infants. *Am. J. Clin. Nutr.*, **41**, 590–8.

Smithells, R.W., Sheppard, S., Schora, C.J. *et al.* (1981) Apparent prevention of neural tube defects by periconceptional vitamin supplementation. *Arch. Disease Childhood.*, **56**, 911–18.

Spring, J.A., Robertson, J. and Buss, D.H. (1979) Trace nutrients. 3. Magnesium, copper, zinc, vitamin B_6, vitamin B_{12} and folic acid in the British household food supply. *Br. J. Nutr.*, **41**, 487–93.

Steinberg, S.E. (1984) Mechanisms of folate homeostasis. *Am. J. Physiol.*, **246**, G319–24.

Steinberg, S.E., Campbell, C.L. and Hillman, R.S. (1979) Kinetics of the normal folate enterohepatic cycle. *J. Clin. Invest.*, **64**, 83–8.

Stokstad, E.L.R. and Koch, J. (1967) Folic acid metabolism. *Physiol. Rev.*, **47**, 83–116.

Strum, W.B. (1979) Enzymatic reduction and methylation of folate following pH-dependent, carrier-mediated transport in rat jejunum. *Biochim. Biophys. Acta*, **554**, 249–57.

Strum, W., Nixon, P.F., Bertino, J.B. and Binder, H.J. (1971) Intestinal folate absorption. I. 5-Methyltetrahydrofolic acid. *J. Clin. Invest.*, **50**, 1910–16.

Subar, A.F., Block, G. and James, L.D. (1989) Folate intake and food sources in the US population. *Am. J. Clin. Nutr.*, **50**, 508–16.

Svendsen, I., Martin, B, Pedersen, T.G. *et al.* (1979) Isolation and characterization of the folate-binding protein from cow's milk. *Carlsberg Res. Commun.*, **44**, 89–99.

Swiatlo, N., O'Connor, D.L., Andrews, J. and Picciano, M.F. (1990) Relative folate bioavailability from diets containing human, bovine and goat milk. *J. Nutr.*, **120**, 172–7.

Tamura, T. and Halsted, C.H. (1983) Folate turnover in chronically alcoholic monkeys. *J. Lab. Clin. Med.*, **101**, 623–8.

Tamura, T. and Stokstad, E.L.R. (1973) The availability of food folate in man. *Br. J. Haematol.*, **25**, 513–32.

Tamura, T., Shin, Y.S., Buehring, K.U. and Stokstad, E.L.R. (1976) The availability of folates in man: effect of orange juice supplement on intestinal conjugase. *Br. J. Haematol.*, **32**, 123–33.

Tamura, T., Romero, J.J., Watson, J.E. *et al.* (1981) Hepatic folate metabolism in the chronic alcoholic monkey. *J. Lab. Clin. Med.*, **97**, 654–61.

Tani, M. and Iwai, K. (1984) Some nutritional effects of folate-binding protein in bovine milk on the bioavailability of folate to rats. *J. Nutr.*, **114**, 778–85.

Tani, M., Fushiki, T. and Iwai, K. (1983) Influence of folate-binding protein from bovine milk on the absorption of folate in gastrointestinal tract of rat. *Biochim. Biophys. Acta*, **757**, 274–81.

Tannenbaum, S.R., Young, V.R. and Archer, M.C. (1985) Vitamins and minerals. In *Food Chemistry*, 2nd edn (ed. O.R. Fennema), Marcel Dekker, Inc., New York, pp. 477–544.

Temple, C. Jr and Montgomery, J.A. (1984) Chemical and physical properties of folic acid and reduced derivatives. In *Folates and Pterins, Vol. 1. Chemistry and Biochemistry of Folates* (eds R.L. Blakley and S.J. Benkovic), John Wiley & Sons, New York, pp. 61–120.

Uyeda, K. and Rabinowitz, J.C. (1963) Fluorescence properties of tetrahydrofolate and related compounds. *Analyt. Biochem.*, **6**, 100–8.

Vahteristo, L., Ollilainen, V. and Varo, P. (1996) HPLC determination of folate in liver and liver products. *J. Food Sci.*, **61** 524 6.

Vahteristo, L. T., Ollilainen, V., Koivistoinen, P.E. and Varo, P. (1996) Improvements in the analysis of reduced folate monoglutamates and folic acid in food by high-performance liquid chromatography. *J. Agric. Food Chem.*, **44**, 477–82.

Wagner, C. (1985) Folate-binding proteins. *Nutr. Rev.*, **43**, 293–9.

Wang, T.T.Y., Reisenauer, A.M. and Halsted, C.H. (1985) Comparison of folate conjugase activities in human, pig, rat and monkey intestine. *J. Nutr.*, **115**, 814–9.

Waxman, S. and Schreiber, C. (1980) Determination of folate by use of radioactive folate and binding proteins. *Methods Enzymol.*, **66**, 468–83.

Waxman, S., Schreiber, C. and Herbert, V. (1971) Radioisotopic assay for measurement of serum folate levels. *Blood*, **38**, 219–28.

Weir, D.G., McGing, P.G. and Scott, J.M. (1985) Folate metabolism, the enterohepatic circulation and alcohol. *Biochem. Pharmacol.*, **34**, 1–7.

White, D.R. Jr (1990) Determination of 5-methyltetrahydrofolate in citrus juice by reversed-phase high-performance liquid chromatography with electrochemical detection. *J. Agric. Food Chem.*, **38**, 1515–18.

White, D.R. Jr, Lee, H.S. and Krüger, R.E. (1991) Reversed-phase HPLC/EC determination of folate in citrus juice by direct injection with column switching. *J. Agric. Food Chem.*, **39**, 714–7.

Whitehead, V.M. (1986) Pharmacokinetics and physiological disposition of folate and its derivatives. In *Folates and Pterins*, Vol. 3, *Nutritional, Pharmacological, and Physiological Aspects* (eds R.L. Blakley and V.M. Whitehead), John Wiley & Sons, New York, pp. 177–205.

Wigertz, K. and Jägerstad, M. (1993) Analysis and characterization of milk folates from raw; pasteurized; UHT-treated and fermented milk related to availability *in vivo*. In *Bioavailability '93. Nutritional, Chemical and Food Processing Implications of Nutrient Availability*. Conference proceedings, Part 2 (ed. U. Schlemmer), Ettlingen, May 9–12, 1993, Bundesforschungsanstalt für Ernährung, pp. 431–5.

Wigertz, K. and Jägerstad, M. (1995) Comparison of a HPLC and radioproteinbinding assay for the determination of folates in milk and blood samples. *Food Chem.*, **54**, 429–36.

Wigertz, K., Hansen, I., Høier-Madsen, M. *et al.* (1996) Effect of milk processing on the concentration of folate-binding protein (FBP), folate-binding capacity and retention of 5-methyltetrahydrofolate. *Int. J. Food Sci. Nutr.*, **47**, 315–22.

Wilson, S.D. and Horne, D.W. (1983) Evaluation of ascorbic acid in protecting labile folic acid derivatives. *Proc. Natl Acad. Sci. USA*, **80**, 6500–4.

Wilson, S.D. and Horne, D.W. (1984) High-performance liquid chromatographic determination of the distribution of naturally occurring folic acid derivatives in rat liver. *Analyt. Biochem.*, **142**, 529–35.

Wright, A.J.A. and Phillips, D.R. (1985) The threshold growth response of *Lactobacillus casei* to 5-methyl-tetrahydrofolic acid: implications for folate assays. *Br. J. Nutr.*, **53**, 569–73.

Zimmerman, J. (1990) Folic acid transport in organ-cultured mucosa of human intestine. *Gastroenterology*, **99**, 964–72.

Zimmerman, J., Selhub, J. and Rosenberg, I.H. (1986) Role of sodium ion in transport of folic acid in the small intestine. *Am. J. Physiol.*, **251**, G218–22.

14

Vitamin B$_{12}$

14.1 INTRODUCTION

A type of anaemia attributed to a digestive disorder was reported by Combe in 1822 and later recognized as pernicious anaemia by Addison in 1849. It was not until 1926 that Minot and Murphy started to cure patients suffering from pernicious anaemia by feeding them with large amounts of raw liver. In 1929, Castle showed that the intestinal absorption of the 'antipernicious anaemia principal' required prior binding to a specific protein (intrinsic factor) secreted by the stomach. Research into isolating the active principal from liver was hampered by the difficulty of the only known bioassay, which was the haemopoietic response of patients with pernicious anaemia. Eventually, in 1948, a red crystalline substance having the clinical activity of liver and designated as vitamin B$_{12}$ was isolated almost simultaneously by two independent groups led by Folkers (Merck Company, USA) and E.L. Smith (Glaxo Company, UK). The success of Folkers' group was largely attributable to a microbiological assay developed by Shorb in 1947; Smith used the human bioassay procedure. The complicated structure of vitamin B$_{12}$ was established by Hodgkin using X-ray crystallography in 1955. Its complete chemical synthesis was achieved in 1973, but because of the large number of stages required (over 70) the procedure is of no commercial interest.

Vitamin B$_{12}$ is essential for the regeneration of THF in folate metabolism through the coenzymatic action of methylcobalamin. Adenosylcobalamin is a coenzyme in the conversion of L-methylmalonyl-CoA to succinyl-CoA, an important intermediary in the tricarboxylic acid cycle.

14.2 CHEMICAL STRUCTURE AND NOMENCLATURE

In accordance with the literature on nutrition and pharmacology, the term vitamin B$_{12}$ is used in this text as the generic descriptor for all cobalamins that exhibit anti-pernicious anaemia activity. Individual cobalamins will be referred to by their specific names (e.g. cyanocobalamin).

The cobalamin molecule is a six coordination cobalt complex containing a corrin ring system substituted with numerous methyl, acetamide and propionamide radicals (Figure 14.1). Methylene bridges link the pyrrole rings A to B, B to C and C to D but not A to D, which are linked directly. The cobalt atom, which may assume an oxidation state of (I), (II) or (III), is linked by four equatorial coordinate bonds to the four pyrrole nitrogens, and by an axial coordinate bond to a 5,6-dimethylbenzimidazole (DMB) moiety, which extends in α-glycosidic linkage to ribose-3-phosphate. The phosphate group is linked to the D ring of the corrin structure via a substituted propionamide chain. The pseudonucleotide (DMB base plus sugar phosphate) is oriented perpendicularly to the corrin structure. The remaining axial coordinate bond at the X position links the cobalt atom to a cyano (—CN) group in the case of cyanocobalamin $(C_{63}H_{88}O_{14}N_{14}PCo; MW = 1355.4)$ or, depending on the chemical environment, to some other group (e.g.—OH in hydroxocobalamin and- HSO$_3$ in sulphitocobalamin).

There are two vitamin B$_{12}$ coenzymes with known metabolic activity in humans, namely methylcobalamin and 5'-deoxyadenosylcobalamin (frequently abbreviated to adenosylcobalamin and also known as coenzyme B$_{12}$). The methyl or adenosyl ligands of the coenzymes occupy the X position in the corrin structure (Herbert and Colman, 1988). The coenzymes are bound intracellularly to their protein apoenzymes through a covalent peptide link, or in milk and plasma to specific transport proteins.

14.3 DEFICIENCY SYNDROMES

Vitamin B$_{12}$ deficiency in humans is manifested by an anaemia and a neuropathy (Scott, 1992). The anaemia is clinically indistinguishable from

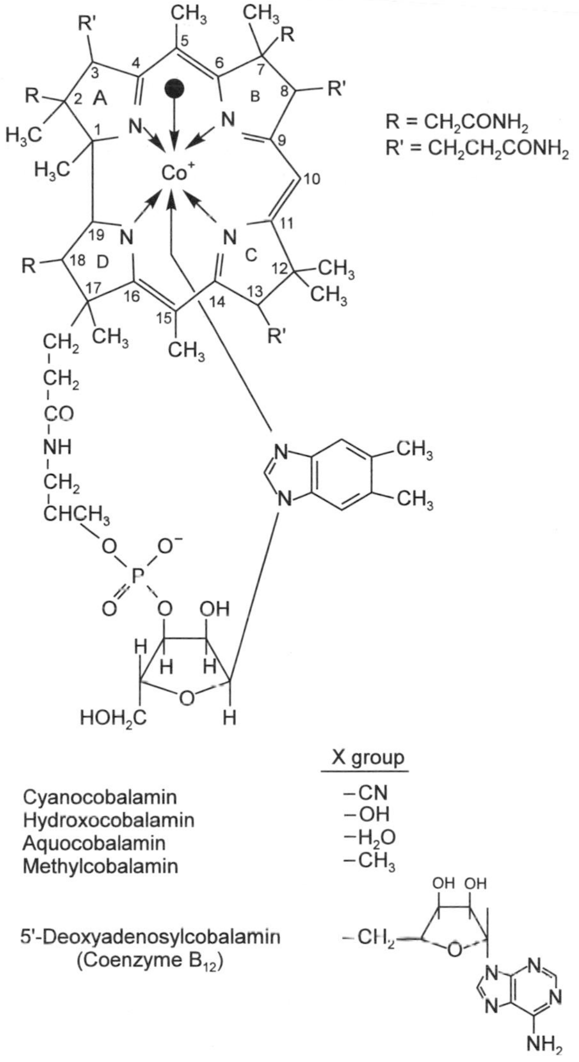

Figure 14.1 Structures of vitamin B_{12} compounds.

that induced by folate deficiency and is due to the inactivity of a vitamin B_{12}-dependent enzyme in folate metabolism. When this type of megaloblastic anaemia arises from a deficiency of vitamin B_{12} it is called pernicious or Addisonian anaemia. The neurological changes are unrelated to folate deficiency and are caused by the inability to manufacture the lipid component of myelin, which results in a generalized demyelinization of nerve tissue. Neuropathy begins in the peripheral nerves, affecting first the feet and fingers, and then progressing to the spinal cord and brain.

In the event of vitamin B$_{12}$ deficiency, there is a danger of irreversible neurological damage if folic acid supplements are taken without including vitamin B$_{12}$. This is because the obvious symptoms of anaemia will be alleviated by the folate treatment, but the concomitant nerve degeneration will be undetected. For this reason, prophylactic vitamin supplementation always includes vitamin B$_{12}$ with folic acid.

14.4 PHYSICOCHEMICAL PROPERTIES

Solubility

Cyanocobalamin is an artificial product used in pharmaceutical preparations because of its stability. It is a tasteless dark red crystalline hygroscopic powder which can take up appreciably more than the 12% of moisture permitted by the British Pharmacopoeia. The anhydrous material can be obtained by drying under reduced pressure at 105 °C. Cyanocobalamin is soluble in water (1.25 g/100 ml at 25 °C) and in lower alcohols, phenol and other hydroxylated solvents such as ethylene diol; it is insoluble in acetone, chloroform, ether and most other organic solvents. Aqueous solutions of cyanocobalamin are of neutral pH.

Stability

Cyanocob (III)alamin is the most stable of the vitamin B$_{12}$-active cobalamins and is the one mostly used in pharmaceutical preparations and food supplementation. Aqueous solutions of cyanocobalamin are stable in air at room temperature if protected from light. The pH region of optimal stability is 4.5–5 and solutions of pH 4–7 can be autoclaved at 120 °C for 20 min with negligible loss of vitamin activity. The addition of ammonium sulphate increases the stability of cyanocobalamin in aqueous solution. Heating with dilute acid deactivates the vitamin owing to hydrolysis of the amide substituents or further degradation of the molecule. Mild alkaline hydrolysis at 100 °C promotes cyclization of the acetamide side-chain at C-7 to form a biologically inactive γ-lactam or, in the presence of an oxidizing agent, a γ-lactone (Wong, 1989). The vitamin B$_{12}$ activity in aqueous solutions is destroyed in the presence of strong oxidizing agents and high concentrations of reducing agents, such as ascorbic acid, sulphite and iron(II) salts. On exposure to light, the cyano group dissociates from cyanocobalamin and hydroxocob(III)-alamin is formed. In neutral and acid solution hydroxocobalamin exists in the form of aquocobalamin (Gräsbeck and Salonen, 1976). This photolytic reaction does not cause a loss of activity. Adenosylcobalamin and methylcobalamin are reduced cob(I)alamin derivatives which are easily oxidized by light to the hydroxo compound (Hoffbrand, 1979).

14.5 ANALYSIS

14.5.1 Scope of analytical techniques

Vitamin B_{12} occurs intracellularly in the living tissues of animals in the form of the two coenzymes, adenosylcobalamin and methylcobalamin, which are covalently bound to their protein apoenzymes. In milk, these coenzymes are bound noncovalently to specific transport proteins. Hydroxocobalamin is present in animal-derived foods, especially in milk, as a result of the photo-oxidation of the coenzyme forms. Cyanocobalamin is a synthetic stable form of the vitamin and is used in fortification. The potential vitamin B_{12} activity of a food sample is represented by the total cobalamin content, regardless of the ligand attached. The determination of total vitamin B_{12} may be performed by microbiological assay or by radioassay. Supplemental vitamin B_{12} (cyanocobalamin) may be determined by a nonisotopic protein-binding assay.

No IU for vitamin B_{12} activity has been defined, and the assay results are expressed in milligrams, micrograms or nanograms of pure crystalline cyanocobalamin. The measurement of biological activity in preparations containing vitamin B_{12} relies on microbiological assays, there being no animal bioassay.

14.5.2 Extraction techniques

Extraction procedures generally have the dual purpose of liberating protein-bound cobalamins and converting the labile naturally occurring forms to a single, stable form – cyanocobalamin or sulphitocobalamin. Conversion to the sulphitocobalamin by reaction with metabisulphite avoids the use of lethally toxic cyanide solutions required to form cyanocobalamin.

The extraction procedure employed in the AOAC microbiological method for determining vitamin B_{12} activity in vitamin preparations (AOAC, 1990a) is also applicable to foods, having been found satisfactory by interlaboratory collaborative analysis of a crude liver paste, condensed fish solubles and a crude vitamin B_{12} fermentation product (Krieger, 1954). The procedure entails homogenizing the sample with 0.1 M phosphate-citrate buffer at pH 4.5 containing freshly prepared sodium metabisulphite ($Na_2S_2O_5$), and then autoclaving the mixture for 10 min at 121 °C. The heat treatment denatures the proteins, inactivates the enzymes and accelerates the conversion of liberated cobalamins to sulphitocobalamin.

For the determination of vitamin B_{12} activity in milk-based infant formula (AOAC, 1990b), protein is removed by filtration after adjustment of the autoclaved extract to the point of maximum precipitation (*c*.pH 4.5).

Methods in which the sample is heated on a boiling water bath, rather than autoclaved, may not completely extract all of the bound vitamin (Casey *et al.*, 1982).

14.5.3 Microbiological assay

Most current procedures for the microbiological determination of vitamin B$_{12}$ activity in foods use *L. delbrueckii* subsp. *lactis* (formerly called *L. leichmannii*) (ATCC No. 4797), which gives a response range of 1–10 pg/ml assay medium (Skeggs, 1967). Either *L. delbrueckii* (ATCC No. 4797) or the 313 strain (ATCC No. 7830) may be used, although the latter requires a shorter time to reach nearly maximum growth (20 h versus 48 h) (Hoffmann *et al.*, 1949). The growth response of the 313 strain to cyanocobalamin is similar to its growth response to hydroxocobalamin, sulphitocobalamin, dicyanocobalamin and nitritocobalamin but lower than that to adenosylcobalamin and higher than that to methylcobalamin. Accurate measurement of vitamin B$_{12}$ in foods can be made using *L. delbrueckii* (ATCC No. 7830) and cyanocobalamin as the calibration standard if the sample extracts are exposed to light before analysis. Complete conversion of adenosylcobalamin and methylcobalamin to hydroxocobalamin takes place in 20 min after exposure to white light from a 15 W bulb at a distance of 20 cm (Muhammad, Briggs and Jones, 1993a).

L. delbrueckii does not respond specifically to vitamin B$_{12}$, as biologically inactive analogues, in which the 5,6-dimethylbenzimidazole moiety is substituted by a purine base or a purine derivative, can replace the vitamin as a growth factor (Shaw and Bessell, 1960). Such analogues are found mainly in natural material that has undergone microbial fermentation, and they do not occur to any significant extent in foods. Examples of potentially interfering analogues include so-called pseudovitamin B$_{12}$ and Factor A, in which the base substituents are adenine and 2-methyladenine, respectively. DNA, deoxyribonucleotides and deoxyribonucleosides can also replace vitamin B$_{12}$ for the growth of *L. delbrueckii* (Kitay, McNutt and Snell, 1949). The nucleotides and nucleosides are less active than vitamin B$_{12}$ by a factor of 10^4, while DNA is less active by a factor of 10^6 (Hoff-Jørgensen, 1954). The activity attributable to deoxyribonucleosides can be determined by *L. delbrueckii* assay after the cobalamins have been destroyed by heating to 100 °C at pH 11 for 30 min (Hoffmann *et al.*, 1949). Deoxyribonucleosides do not occur in usual foods at levels likely to constitute a significant interference (Rosenthal, 1968), and they only substitute for B$_{12}$ at concentrations above 1 µg/ml of assay solution (Skeggs, 1963). Any such interference can therefore be eliminated by simply diluting to an inactive concentration.

Ford (1953) developed an assay using the protozoan *Ochromonas malhamensis* (ATCC No. 11532), which is as sensitive as the *L. delbrueckii* assay, but is more specific in that it responds only slightly, if at all, to clinically inactive cobalamins. A disadvantage of the *O. malhamensis* assay is the long incubation period of 72 h, during which the culture must be shaken continously in the dark. A comparative study between the *O. malhamensis* and *L. delbrueckii* methods applied to a cross-section of 27 different foods did not reveal any problems of interference with the latter method. On the contrary, the *O. malhamensis* results were statistically higher at the 5% level of significance than the *L. delbrueckii* results, suggesting the presence in certain foods of unknown noncobalamin substances which stimulate the growth of the protozoan (Lichtenstein, Beloian and Reynolds, 1959). Ford's method has been recommended for the determination of vitamin B_{12} in animal feedstuffs (Analytical Methods Committee, 1964), although Shrimpton (1956) reported that *O. malhamensis* was no better than *L. delbrueckii* in predicting the vitamin B_{12} activity of feeding stuffs for chicks.

Chen and McIntyre (1979) applied the radiometric–microbiological assay to the determination of vitamin B_{12} in foods by measuring the $^{14}CO_2$ generated from the metabolism of L-[guanido-^{14}C]arginine by *L. delbrueckii*. The assay allowed immediate inoculation with previously lyophilized cultures of *L. delbrueckii*, thus eliminating the routine subculturing required in the standard maintenance procedure.

14.5.4 Radioassay

Published radioassay techniques for determining vitamin B_{12} in foods (Richardson *et al.*, 1978; Beck, 1979; Casey *et al.*, 1982; Österdahl *et al.*, 1986; Andersson, Lundqvist and Öste, 1990) are based on the original method of Lau *et al.* (1965), which was developed for measurement of serum B_{12}. Food extracts prepared for radioassay contain 20–1000 pg vitamin B_{12}/ml, which is well within the 50–2000 pg/ml range of commercially available assay kits. The radioassay utilizes [^{57}Co]cyanocobalamin as the tracer, and hog intrinsic factor as the binding protein. Cyanocobalamin has a binding affinity for hog intrinsic factor equal to that of methylcobalamin, dicyanocobalamin and nitrotocobalamin but not to that of hydroxocobalamin, sulphitocobalamin and adenosylcobalamin. For an accurate assay, it is therefore necessary to extract foods in the presence of excess cyanide, in order to convert the latter three cobalamins to dicyanocobalamin (Muhammad, Briggs and Jones, 1993b).

In one of two commercial assay kits evaluated by Richardson *et al.* (1978), the binding agent is supplied in the form of a Sephadex (dextran)–intrinsic factor complex, which simplifies the analysis and was found to function more satisfactorily with food extracts than the separate use of intrinsic factor and albumin-coated charcoal. The results obtained by radioassay for the determination of vitamin B_{12} in food were compared with those obtained by the *L. delbrueckii* (ATCC No. 7830) microbiological assay using the same sample extracts (Muhammad, Briggs and Jones, 1993c). The agreement between the two methods was very good in eight out of the 10 foods analysed. In the case of pork, the radioassay gave the higher figure, whilst the opposite was observed for yoghurt. The radioassay is supposed to be more specific for vitamin B$_{12}$, as intrinsic factor binds with a very narrow range of corrinoids.

14.6 BIOAVAILABILITY

14.6.1 Physiological aspects

The absorption and transport of the physiological amounts of vitamin B$_{12}$ present in natural foods take place by specialized mechanisms that accumulate the vitamin and deliver it to cells that require it. Structurally similar but biologically inactive analogues, which may be harmful, are rejected.

Intestinal absorption

All forms of vitamin B$_{12}$ are absorbed and transported in the same way. Ingested cobalamins are released from their peptide bonding to proteins by the combined action of hydrochloric acid and pepsin in the stomach. Gastric juice contains two cobalamin-binding proteins known as R protein and intrinsic factor. In humans, the former originates in saliva and the latter is synthesized and secreted directly into gastric juice by the parietal cells of the stomach. At the acid pH of the stomach, cobalamins have a greater affinity for R protein than for intrinsic factor, and therefore cobalamins leave the stomach and enter the duodenum bound to R protein accompanied by free intrinsic factor. In the mildly alkaline environment of the jejunum, pancreatic proteases, particularly trypsin, partially degrade both free R protein and R protein complexed with cobalamins. Intrinsic factor, which is resistant to proteolysis by pancreatic enzymes, then binds avidly to the released cobalamins.

The intrinsic factor–cobalamin complex is carried down to the ileum, where it binds avidly to specific receptors on the brush border of the ileal enterocyte. The presence of calcium ions and a pH above 5.5 are necessary to induce the appropriate configuration of the receptor for binding (Donaldson, 1987). The intrinsic factor–cobalamin complex is absorbed intact, but little is known about the mechanism of absorption and subsequent events within the enterocyte. Seetharum (1994) presented evidence that the intrinsic factor–cobalamin complex enters the cell by receptor-mediated endocytosis, i.e. engulfment by the cell membrane. The cobalamin is subsequently released by the action of acid within a non-lysosomal cellular compartment. The intrinsic factor may be degraded by proteolysis after releasing its bound cobalamin, there being no apparent recycling of intrinsic factor to the brush-border membrane.

The structural specificity for the binding of cobalamin by intrinsic factor is rather complex, but the 5,6-dimethylbenzimidazole moiety is absolutely essential (Seetharam and Alpers, 1991). Cyanocobalamin, hydroxocobalamin, methylcobalamin, or adenosylcobalamin all bind to intrinsic factor with approximately the same high affinity (Donaldson, 1987). R proteins bind a wide variety of cobalamin analogues in addition to cobalamin itself and receptors for these proteins appear to be present only on the surface of liver cells. R proteins may therefore function to bind excess cobalamin or unwanted analogues for storage and ultimate excretion via the liver.

The intrinsic factor-mediated system is capable of handling between 1.5 to 3.0 µg of vitamin B_{12}. The limited capacity of the ileum to absorb vitamin B_{12} can be explained by the limited number of receptor sites, there being only about one receptor per microvillus. The number of ileal receptor sites is doubled during pregnancy in the mouse (Robertson and Gallagher, 1979). Saturation of the system at one meal does not preclude absorption of normal amounts of the vitamin some hours later. The entire absorptive process, from ingestion of the vitamin to its appearance in the portal vein, takes 8–12 h.

Absorption may also occur rapidly by simple diffusion across the entire small intestine. This process probably accounts for the absorption of only 1% to 3% of the vitamin consumed in ordinary diets, but can provide a physiologically significant source of the vitamin when it is administered as free cobalamin in pharmacological doses of 30 µg or more.

Schilling test for defective absorption

The classical procedure for defining the nature of vitamin B_{12} malabsorption in adult humans is the urinary excretion (Schilling) test. In the basic procedure, a physiological (0.5 to 2.0 µg) quantity of cyanocobalamin labelled with [57]Co or [58]Co is given orally to the fasting subject and within 2 h a 'flushing dose' of nonradioactive cyanocobalamin (1000 µg) is administered by intramuscular injection. Under these conditions all the plasma cobalamin binders will be saturated with nonradioactive vitamin B_{12} so that absorbed radioactive B_{12} enters the plasma in free (unbound) form and is filtered through the glomerulus before excretion in the urine. Normally, about 10–20% of the labelled dose is absorbed and excreted in a 24 h collection of the urine. If < 5% of the labelled dose is excreted, malabsorption is indicated and the test must be repeated to determine if a lack of endogenous intrinsic factor is the cause. In the re-test, labelled cyanocobalamin is fed together with an excess of intrinsic factor, ensuring that residual radioactivity from the first test will not be excreted in the urine. If the excretion is restored to normal by the co-administration of intrinsic factor, pernicious anaemia is indicated (Donaldson, 1987).

The Schilling test can be shortened by using two different isotopes of cobalt to label free and intrinsic factor-bound cyanocobalamin and administering the mixture simultaneously. Absorption is determined by measuring the relative amounts of each isotope excreted in the urine.

The Schilling test does not provide evidence of vitamin B$_{12}$ deficiency. For example, strict vegetarians, who have low plasma B$_{12}$ levels, will produce a normal result. On the other hand, patients with pernicious anaemia will produce an abnormal result, even though their plasma B$_{12}$ levels are normal from treatment with the vitamin. Because vitamin B$_{12}$ in food is protein-bound, the Schilling test only reflects true absorption if proteolysis in the subject is adequate. Subjects whose gastric acid and enzyme production are inadequate cannot absorb protein-bound cobalamins but, because there is still substantial intrinsic factor secretion, will be able to absorb the dose of free cyanocobalamin and thus produce a normal Schilling test.

Other metabolic considerations

About 90% of circulating vitamin B$_{12}$ (mainly in the form of methylcobalamin) is bound to transcobalamin I, a glycoprotein which appears to have largely a storage function and does not seem to be involved in tissue uptake. Vitamin B$_{12}$ destined for tissue uptake is bound to transcobalamin II, a β-globulin that binds very few cobalamin analogues and hence limits the cellular uptake of any such analogues that may have gained access to the circulation. Hepatocytes and the cells of many other tissues, including bone marrow cells, take up the protein–cobalamin complex by receptor-mediated endocytosis. Following lysosomal proteolysis, the vitamin is released into the cytoplasm in the form of hydroxocobalamin. This compound is either converted directly to methylcobalamin in the cytoplasm or eventually to adenosylcobalamin in the mitochondria.

The body is extremely efficient at conserving vitamin B$_{12}$. Unlike the other water-soluble vitamins, vitamin B$_{12}$ is stored in the liver, primarily in the form of adenosylcobalamin. In an adult man, the total body store of vitamin B$_{12}$ is estimated to be 2000 to 5000 μg, of which about 80% is in the liver. The amount stored in the liver increases with age, more than doubling between the ages of 20 and 60. The remainder of the stored vitamin is located in muscle, skin and blood plasma. Only 2 to 5 μg of vitamin B$_{12}$ are lost daily through metabolic turnover, regardless of the amount stored in the body.

Approximately 0.5–5 µg of cobalamin is secreted into the jejunum daily, mainly in the bile. Of this, at least 65–75% is reabsorbed in the ileum along with dietary sources of the vitamin (Ellenbogen, 1984). This enterohepatic cycle allows the excretion of unwanted nonvitamin B_{12} analogues, which constitute about 60% of the corrinoids secreted in bile, and returns vitamin B_{12} relatively free of analogues.

The binding of vitamin B_{12} with transcobalamin-II in plasma prevents the vitamin molecule from being excreted in the urine as it passes through the kidney. The efficient conservation by the kidney and entero-hepatic circulation, together with the slow rate of turnover, explains why strict vegetarians, whose absorptive capacity is intact, take 20 years or more to develop signs of deficiency. People with absorptive malfunction develop deficiency signs within 2–3 years.

Pernicious anaemia

The deficiency disease of pernicious anaemia is due to the lack of absorp-tion of vitamin B_{12} caused by the inability to synthesize intrinsic factor or by some other digestive malfunction. It is rarely, if ever, caused by a lack of dietary vitamin B_{12}. Malabsorption syndromes also interfere with enterohepatic circulation, thereby increasing the amount of vitamin B_{12} required to meet body needs. In many cases of pernicious anaemia the cause is an auto-immune reaction with formation of antibodies against intrinsic factor in certain genetically predisposed individuals. Two types of antibody are involved: type I prevents intrinsic factor from binding to cobalamin and type II blocks the binding of the intrinsic factor–cobalamin complex to the ileal receptor (Seetharum, 1994). The deficiency symptoms are relieved quickly when cobalamin is injected into the bloodstream to bypass the absorption process.

14.6.2 Dietary sources and their bioavailability

Dietary sources

Naturally occurring vitamin B_{12} originates solely from synthesis by bac-teria and other microorganisms growing in soil or water, in sewage, and in the rumen and intestinal tract of animals. Any traces of the vitamin that may be detected in plants are due to microbial contamination from the soil or manure or, in the case of certain legumes, to bacterial synthesis in the root nodules.

Vitamin B_{12} is ubiquitous in foods of animal origin and is derived from the animal's ingestion of cobalamin-containing animal tissue or micro-biologically contaminated plant material, in addition to vitamin absorbed from the animal's own digestive tract. The vitamin B_{12} contents of some

Table 14.1 Vitamin B$_{12}$ content of various foods (from Holland, 1991)

Food	µg Vitamin B$_{12}$/100 g
Milk, pasteurized (whole, semi-skimmed or skimmed)	0.4
Cream, single, pasteurized	0.3
Cheese	
Brie	1.2
Camembert	1.1
Cheddar, average	1.1
Cottage, plain	0.7
Yoghurt, whole milk, plain	0.2
Egg	
whole, raw	2.5
white, raw	0.1
yolk, raw	6.9
whole, boiled	1.1
Beef, rump steak, lean and fat (raw, fried or grilled)	2.0
Lamb, leg or shoulder, lean and fat (raw or roast)	2.0
Pork, chop, loin, lean and fat, grilled	1.0
Chicken, meat only (raw or roast)	trace
Turkey, meat only (raw or roast)	2.0
Rabbit, meat only, stewed	12.0
Heart, ox, stewed	15.0
Kidney, lamb, fried	79.0
Liver, lamb, fried	81.0
Cod, fillets, raw	2
Pilchards, canned in tomato sauce	12.0
Salmon, canned	4.0
Sardines, canned in oil, drained	28
Tuna, canned in oil, drained	4.8

foods in which the vitamin is found are listed in Table 14.1. Liver is the outstanding dietary source of the vitamin, followed by kidney and heart. Muscle meats, fish, eggs, cheese and milk are other important food sources. Vitamin B$_{12}$ activity has been reported in yeast, but this has since been attributed to the presence of noncobalamin corrinoids or vitamin B$_{12}$ originating from the enriching medium (Herbert, 1988). Spirulina, a type of seaweed, is claimed to be a source of vitamin B$_{12}$, but in fact is practically devoid of the vitamin. The so-called vitamin B$_{12}$ is actually inactive analogues, two of which have been shown to block vitamin B$_{12}$ metabolism in human cell cultures (Herbert, 1988). About 5 to 30% of the reported vitamin B$_{12}$ in foods may be microbiologically active noncobalamin corrinoids rather than true B$_{12}$ (National Research Council, 1989).

Vitamin B_{12} in foods exists in several forms as reported by Farquharson and Adams (1976). Meat and fish contain mostly adenosyl- and hydroxocobalamins; these compounds, accompanied by methylcobalamin, also occur in dairy products, with hydroxocobalamin predominating in milk. Sulphitocobalamin is found in canned meats and fish. Cyanocobalamin was not detected, apart from small amounts in egg white, cheese and boiled haddock. In bovine milk, naturally occurring vitamin B_{12} is bound to proteins, with a high proportion being present in whey proteins (Hartman and Dryden, 1974).

Bioavailability

Humans appear to be entirely dependent on a dietary intake of vitamin B_{12}. Although microbial synthesis of the vitamin occurs in the human colon, it is apparently not absorbed (Shinton, 1972). Strict vegetarians may obtain limited amounts of vitamin B_{12} through ingestion of the vitamin-containing root nodules of certain legumes and from plant material contaminated with microorganisms.

The percentage of ingested vitamin B_{12} that is absorbed decreases as the actual amount in the diet increases. At intakes of 0.5 µg or less, about 70% of the available vitamin B_{12} is absorbed. At an intake of 5.0 µg, a mean of 28% is absorbed (range 2–50%) while at a 10 µg intake the mean absorption is 16% (range 0–34%) (Herbert, 1987). When 100 µg or more of crystalline vitamin B_{12} is taken, the absorption efficiency drops to 1%, and the excess vitamin is excreted in the urine.

In normal human subjects, the vitamin B_{12} in lean mutton (Heyssel *et al.*, 1966), chicken meat (Doscherholmen, McMahon and Ripley, 1978) and fish (Doscherholmen, McMahon and Economon, 1981) is absorbed as efficiently as a comparable amount of crystalline cyanocobalamin administered orally in aqueous solution. In contrast, the vitamin B_{12} in eggs is poorly absorbed (Doscherholmen, McMahon and Ripley, 1975) owing to the presence of distinct vitamin B_{12} binding proteins in egg white and egg yolk (Levine and Doscherholmen, 1983).

Megadosing with vitamin C (500 mg or more daily) may adversely affect the bioavailability of vitamin B_{12} from food (Herbert, 1990).

Effects of dietary fibre
Vitamin B_{12} depletion occurs more rapidly in the presence than in the absence of intestinal microorganisms, presumably due to competition for available B_{12} between the gut flora and the host. Cullen and Oace (1978) investigated the possibility that dietary fibre, by stimulating the growth of intestinal bacteria, might increase the rate of vitamin B_{12} utilization by the rat. The results showed that cellulose or pectin added to purified vitamin B_{12}-deficient diets increased the faecal excretion of radioactive

B$_{12}$ that was injected after several weeks of depletion. Thus, both cellulose and pectin may have bound or adsorbed biliary B$_{12}$ and carried it past the ileal absorptive sites. In addition, pectin, which is hydrolysed to an appreciable extent by intestinal microorganisms, might have served as a substrate for the growth of vitamin B$_{12}$-requiring bacteria.

Lewis, Kies and Fox (1986) found no significant difference in the urinary excretion of vitamin B$_{12}$ in human subjects receiving controlled diets supplemented with or without wheat bran.

Effects of alcohol

Although alcohol ingestion has been shown to decrease vitamin B$_{12}$ absorption in volunteers after several weeks of intake, alcoholics do not commonly suffer from vitamin B$_{12}$ deficiency, probably because of the large body stores of the vitamin and the reserve capacity for absorption (Lieber, 1988).

Effects of smoking

Cigarette smokers, but not nonsmokers, showed a high urinary thiocyanate excretion, which was associated with increased vitamin B$_{12}$ excretion and a relatively low serum B$_{12}$ concentration (Linnell *et al.*, 1968). Because urinary thiocyanate excretion is an index of cyanide detoxication, it appears possible that a high plasma cyanide concentration caused through smoking disturbs the equilibrium between plasma and urinary vitamin B$_{12}$.

14.6.3 Effects of food processing and cooking

The cobalamins present in food are generally resistant to thermal processing and cooking in a nonalkaline medium. Vitamin B$_{12}$ losses incurred during the normal oven heating of frozen convenience dinners prior to serving ranged from 0% to 21% in products containing fish, fried chicken, turkey and beef (De Ritter, 1976). A 27–33% loss of vitamin B$_{12}$ expressed per unit of nitrogen occurred during the cooking of beef due to the loss of moisture and fat; the vitamin content of raw and cooked beef was, however, similar on a moisture basis (Bennink and Ono, 1982). Thus there is little loss of the vitamin in the cooking of meat provided that the meat juices are utilized.

In the processing of milk, the following losses of available vitamin B$_{12}$ have been reported: boiling for 2–5 min, 30% (Herbert, 1990); pasteurization for 2–3 s, 7%; sterilization in the bottle at 120 °C for 13 min, 77%; rapid sterilization at 143 °C for 3–4 s with superheated steam, 10%; evaporation, 70–90%; and spray-drying, 20–35% (Tannenbaum, Young and Archer, 1985). The light-sensitive coenzyme forms of vitamin B$_{12}$ are largely converted to hydroxocobalamin in light-exposed milk, but with no loss of vitamin activity (Farquharson and Adams, 1976).

14.7 DIETARY INTAKE

The average omnivorous American diet probably supplies 5–15 µg of vitamin B_{12} daily, but the amount can range from as low as 1 to as high as 100 µg/day (Herbert, 1987). The human need for vitamin B_{12} is extremely small, and for normal people does not exceed 1 µg daily. To ensure normal serum concentrations and adequate body stores, the RDA for adults is set at 2.0 µg/day (National Research Council, 1989). This allowance is based on the amount of the vitamin required to elicit a haematological response in strict vegetarians.

The human placenta concentrates vitamin B_{12} and newborns often have twice or more the serum B_{12} concentrations of their mothers. Although maternal body stores are normally sufficient to meet the needs of pregnancy, an additional allowance of 0.2 µg/day is recommended. An increment of 0.6 µg/day during lactation ensures an adequate supply for breast milk.

Effects of high intakes

Vitamin B_{12} is nontoxic, even in doses that exceed the maximum daily adult human requirement by 10 000 times. Excessive intakes above the limited serum- and tissue-available binding capacity tend to be excreted in the urine rather than retained (Herbert and Colman, 1988).

Assessment of nutritional status

Functional deficiency of vitamin B_{12} is signalled by the appearance of hypersegmented neutrophils in the circulating blood. The requirement of vitamin B_{12} for the conversion of circulating 5-methyl-THF to metabolically active THF results in normal or elevated serum folate levels and a decreased red-cell folate level to below 140 ng/ml. Plasma vitamin B_{12} levels fall to below 200 pg/ml and the deoxyuridine suppression test (see previous chapter) becomes abnormal (Halsted, 1993).

The Schilling test will indicate the existence of pernicious anaemia and the detection of circulating antibodies to intrinsic factor will confirm auto-immune pernicious anaemia.

14.8 SYNOPSIS

Naturally occurring vitamin B_{12} originates solely from microbial synthesis and dietary sources are confined mainly to foods of animal origin. Humans cannot absorb the vitamin B_{12} produced by colonic microorganisms and hence depend entirely on a dietary intake to meet their requirement. The healthy body is extremely efficient at conserving the small amounts of dietary vitamin B_{12}. The vitamin B_{12} present in meat and fish is completely absorbed, in contrast to the poor absorption of the vitamin in eggs.

For many years the microbiological assay has been the only practical method with adequate sensitivity and specificity to measure the low concentrations of vitamin B$_{12}$ in foods. Using the conventional turbidimetric procedure, the maintenance of *L. delbrueckii*, the generally preferred assay organism, requires special techniques to overcome problems of lack of growth in subcultures or loss of sensitivity to B$_{12}$. The radiometric–microbiological assay developed by Chen and McIntyre allows immediate inoculation with previously lyophilized cultures of *L. delbrueckii*, and hence does not suffer from the problems encountered in subculturing. *Ochromonas malhamensis* has been proposed as an alternative assay organism, on the grounds of its non-response to clinically inactive cobalamins. However, the long (3–4 days) growth requirement and the need for a large surface for growing are practical disadvantages for assays using protozoa.

Radioassays have been routinely used by clinical laboratories to monitor serum concentrations of vitamin B$_{12}$, and investigated for their possible application to food analysis. The investigations have led to the general conclusion that, provided the extraction technique is sufficiently rigorous to liberate the cobalamins completely from their bound forms, the radioassay yields B$_{12}$ values that are comparable with those obtained using the *L. delbrueckii* assay.

REFERENCES

Analytical Methods Committee (1964) The determination of water-soluble vitamins in compound feeding stuffs. *Analyst, Lond.*, **89**, 1–6.

Andersson, I., Lundqvist, R. and Öste, R. (1990) Analysis of vitamin B$_{12}$ in milk by a radioisotope dilution assay. *Milchwissenschaft*, **45**, 507–9.

AOAC (1990a) Cobalamin (vitamin B$_{12}$ activity) in vitamin preparations. Microbiological methods. Final action 1960. In *AOAC Official Methods of Analysis*, 15th edn (ed.K. Helrich), Association of Official Analytical Chemists, Inc., Arlington, VA, 952.20.

AOAC (1990b) Cobalamin (vitamin B$_{12}$ activity) in milk-based infant formula. Turbidimetric method. Final action 1988. In *AOAC Official Methods of Analysis*, 15th edn (ed. K. Helrich), Association of Official Analytical Chemists, Inc., Arlington, VA, 986.23.

Beck, R.A. (1979) Comparison of two radioassay methods for cyanocobalamin in seafoods. *J. Food Sci.*, **44**, 1077–9.

Bennink, M.R. and Ono, K. (1982) Vitamin B$_{12}$, E and D content of raw and cooked beef. *J. Food Sci.*, **47**, 1786–92.

Casey, P.J., Speckman, K.R., Ebert, F.J. and Hobbs, W.E. (1982) Radioisotope dilution technique for determination of vitamin B$_{12}$ in foods. *J. Ass. Off. Analyt. Chem.*, **65**, 85–8.

Chen, M. and McIntyre, P.A. (1979) Measurement of the trace amounts of vitamin B$_{12}$ present in various foods by a new radiometric microbiologic technique. In *Trace Organic Analysis: a New Frontier in Analytical Chemistry* (eds H.S. Hertz and S.N. Chesler), National Bureau of Standards, Washington, DC, pp. 257–65.

Cullen, R.W. and Oace, S.M. (1978) Methylmalonic acid and vitamin B_{12} excretion of rats consuming diets varying in cellulose and pectin. *J. Nutr.*, **108**, 640–7.

De Ritter, E. (1976) Stability characteristics of vitamins in processed foods. *Food Technol.*, **30**(1), 48–51, 54.

Donaldson, R.M. (1987) Intrinsic factor and the transport of cobalamin. In *Physiology of the Gastrointestinal Tract*, 2nd edn (ed. L.R. Johnson), Raven Press, New York, pp. 959–73.

Doscherholmen, A., McMahon, J. and Economon, P. (1981) Vitamin B_{12} absorption from fish (41201). *Proc. Soc. Exp. Biol. Med.*, **167**, 480–4.

Doscherholmen, A., McMahon, J. and Ripley, D. (1975) Vitamin B_{12} absorption from eggs (38940). *Proc. Soc. Exp. Biol. Med.*, **149**, 987–90.

Doscherholmen, A., McMahon, J. and Ripley, D. (1978) Vitamin B_{12} assimilation from chicken meat. *Am. J. Clin. Nutr.*, **31**, 825–30.

Ellenbogen, L. (1984) Vitamin B_{12}. In *Handbook of Vitamins. Nutritional, Biochemical, and Clinical Apects* (ed. L.J. Machlin), Marcel Dekker, Inc., New York, pp. 497–547.

Farquharson, J. and Adams, J.F. (1976) The forms of vitamin B-12 in foods. *Br. J. Nutr.*, **36**, 127–36.

Ford, J.E. (1953) The microbiological assay of 'vitamin B_{12}'. The specificity of the requirement of *Ochromonas malhamensis* for cyanocobalamin. *Br. J. Nutr.*, **7**, 299–306.

Gräsbeck, R. and Salonen, E.-M. (1976) Vitamin B-12. *Prog. Food Nutr. Sci.*, **2**, 193–231.

Halsted, C.H. (1993) Water-soluble vitamins. In *Human Nutrition and Dietetics*, 9th edn (eds J.S. Garrow, W.P.T. James and A. Ralph), Churchill Livingstone, Edinburgh, pp. 239–63.

Hartman, A.M. and Dryden, L.P. (1974) Vitamins in milk and milk products. In *Fundamentals of Dairy Chemistry*, 2nd edn (eds B.H. Webb, A.H. Johnson and J.A. Alford), AVI Publishing Co., Inc., Westport, CT, pp. 325–441.

Herbert, V. (1987) Recommended dietary intakes (RDI) of vitamin B-12 in humans. *Am. J. Clin. Nutr.*, **45**, 671–8.

Herbert, V. (1988) Vitamin B-12: plant sources, requirements, and assays. *Am. J. Clin. Nutr.*, **48**, 852–8.

Herbert, V. (1990) Vitamin B-12. In *Present Knowledge in Nutrition*, 6th edn (ed. M.L. Brown), International Life Sciences Institute, Nutrition Foundation, Washington, DC, pp. 170–8.

Herbert, V.D. and Colman, N. (1988) Folic acid and vitamin B-12. In *Modern Nutrition in Health and Disease*, 7th edn (eds M.E. Shils and V.R. Young), Lea & Febiger, Philadelphia, pp. 388–416.

Heyssel, R.M., Bozian, R.C., Darby, W.J. and Bell, M.C. (1966) Vitamin B_{12} turnover in man. The assimilation of vitamin B_{12} from natural foodstuff by man and estimates of minimal daily dietary requirements. *Am. J. Clin. Nutr.*, **18**, 176–84.

Hoffbrand, A.V. (1979) Nutritional and biochemical aspects of vitamin B_{12}. In *The Importance of Vitamins to Human Health* (ed. T.G. Taylor), MTP Press Ltd, Lancaster, pp. 41–6.

Hoff-Jørgensen, E. (1954) Microbiological assay of vitamin B_{12}. In *Methods of Biochemical Analysis*, Vol. I. (ed. D. Glick), Interscience Publishers, Inc., New York, pp. 81–113.

Hoffmann, C.E., Stokstad, E.L.R., Hutchings, B.L. *et al.* (1949) The microbiological assay of vitamin B_{12} with *Lactobacillus leichmannii*. *J. Biol. Chem.*, **181**, 635–44.

Holland, B., Welch, A.A., Unwin, I.D. *et al.* (1991) *McCance and Widdowson's The Composition of Foods*, 5th edn, Royal Society of Chemistry and Ministry of Agriculture, Fisheries and Foods.

Kitay, E., McNutt, W.S. and Snell, E.E. (1949) The non-specificity of thymidine as a growth factor for lactic acid bacteria. *J. Biol. Chem.*, **177**, 993–4.

Krieger, C.H. (1954) Report on vitamin B$_{12}$. Microbiological method. *J. Ass. Off. Agric. Chem.*, **37**, 781–92.

Lau, K.-S., Gottlieb, C., Wasserman, L.R. and Herbert, V. (1965) Measurement of serum vitamin B$_{12}$ level using radioisotope dilution and coated charcoal. *Blood*, **26**, 202–14.

Levine, A.S. and Doscherholmen, A. (1983) Vitamin B$_{12}$ bioavailability from egg yolk and egg white: relationship to binding proteins. *Am. J. Clin. Nutr.*, **38**, 436–9.

Lewis, N.M., Kies, C. and Fox, H.M. (1986) Vitamin B12 status of humans as affected by wheat bran supplements. *Nutr. Rep. Int.*, **34**, 495–9.

Lichtenstein, H., Beloian, A. and Reynolds, H. (1959) Comparative vitamin B$_{12}$ assay of foods of animal origin by *Lactobacillus leichmannii* and *Ochromonas malhamensis*. *J. Agric. Food Chem.*, **7**, 771–4.

Lieber, C.S. (1988) The influence of alcohol on nutritional status. *Nutr. Rev.*, **46**, 241–54.

Linnell, J.C., Smith, A.D.M., Smith, C.L. *et al.* (1968) Effects of smoking on metabolism and excretion of vitamin B$_{12}$. *Br. Med. J.*, **II**, 215–6.

Muhammad, K., Briggs, D. and Jones, G. (1993a) The appropriateness of using cyanocobalamin as calibration standards in *Lactobacillus leichmannii* A.T.C.C. 7830 assay of vitamin B$_{12}$. *Food Chem.*, **48**, 427–9.

Muhammad, K., Briggs, D. and Jones, G. (1993b) The appropriateness of using cyanocobalamin as calibration standards in competitive binding assays of vitamin B$_{12}$. *Food Chem.*, **48**, 423–5.

Muhammad, K., Briggs, D. and Jones, G. (1993c) Comparison of a competitive binding assay with *Lactobacillus leichmannii* A.T.C.C. 7830 assay for the determination of vitamin B$_{12}$ in foods. *Food Chem.*, **48**, 431–4.

National Research Council (1989) Water-soluble vitamins. In *Recommended Dietary Allowances*, 10th edn, National Academy Press, Washington, DC, pp. 115–73.

Österdahl, B.-G., Janné, K., Johansson, E. and Johnsson, H. (1986) Determination of vitamin B$_{12}$ in gruel by a radioisotope dilution assay. *Int. J. Vitam. Nutr. Res.*, **56**, 95–9.

Richardson, P.J., Favell, D.J., Gidley, G.C. and Jones, G.H. (1978) Application of a commercial radioassay test kit to the determination of vitamin B$_{12}$ in food. *Analyst, Lond.*, **103**, 865–8.

Robertson, J.A. and Gallagher, N.D. (1979) Effect of placental lactogen on the number of intrinsic factor receptors in the pregnant mouse. *Gastroenterology*, **77**, 511–17.

Rosenthal, H.L. (1968) Vitamin B$_{12}$. IV. Estimation in foods and food supplements. In *The Vitamins. Chemistry, Physiology, Pathology, Methods*, 2nd edn, Vol. II (eds W.H. Sebrell, Jr and R.S. Harris), Academic Press, New York, pp. 145–70.

Scott, J.M. (1992) Folate–vitamin B$_{12}$ interrelationships in the central nervous system. *Proc. Nutr. Soc.*, **51**, 219–24.

Seetharam, B. (1994) Gastrointestinal absorption and transport of cobalamin (vitamin B$_{12}$). In *Physiology of the Gastrointestinal Tract*, 3rd edn (ed. L.R. Johnson), Raven Press, New York, pp. 1997–2026.

Seetharam, B. and Alpers, D.H. (1991) Gastric intrinsic factor and cobalamin absorption. In *Handbook of Physiology, Section 6: The Gastrointestinal System*,

Vol. IV. Intestinal Absorption and Secretion (volume eds M. Field and R.A. Frizzell), American Physiological Society, Bethesda, MD, pp. 437–61.

Shaw, W.H.C. and Bessell, C.J. (1960) The determination of vitamin B_{12}. A critical review. *Analyst, Lond.*, **85**, 389–409.

Shinton, N.K. (1972) Vitamin B-12 and folate metabolism. *Br. Med. J.*, **1**, 556–9.

Shrimpton, D.H. (1956) The estimation of vitamin-B_{12} activity in feeding stuffs with *Lactobacillus leichmannii* and *Ochromonas malhamensis*. *Analyst, Lond.*, **81**, 94–9.

Skeggs, H.R. (1963) *Lactobacillus leichmannii* assay for vitamin B_{12}. In *Analytical Microbiology* (ed. F. Kavanagh), Academic Press, New York, pp. 551–65.

Skeggs, H.R. (1967) Vitamin B_{12}. In *The Vitamins. Chemistry, Physiology, Pathology, Methods*, 2nd edn, Vol. VII (eds P. György and W.N. Pearson), Academic Press, New York, pp. 277–93.

Tannenbaum, S.R., Young, V.R. and Archer, M.C. (1985) Vitamins and minerals. In *Food Chemistry*, 2nd edn (ed. O.R. Fennema), Marcel Dekker, Inc., New York, pp. 477–544.

Wong, D.W.S. (1989) *Mechanism and Theory in Food Chemistry*, Van Nostrand Reinhold Co., New York, pp. 335–99.

15

Vitamin C

15.1 INTRODUCTION

The first scientific account of the role of citrus fruits in the prevention and cure of scurvy was published by James Lind, a Scottish naval physician, in 1753. The concept of an antiscorbutic vitamin was postulated by Casimir Funk in 1912 after Holz and Fröhlich in 1907 induced the disease in guinea pigs. By 1924, Zilva had obtained concentrates of the antiscorbutic factor from lemons and showed that the factor possessed strong reducing properties. In 1928 Szent-Györgyi isolated in crystalline form from adrenal glands and from cabbage a strongly reducing acidic substance designated as a 'hexuronic acid', which gave colour tests characteristic of the sugars. In 1932, Waugh and King isolated an apparently identical substance from lemon juice which was identified as the anti-scorbutic factor, vitamin C. The chemical structure of the vitamin was established by Haworth's group in 1933 and in the same year Haworth, in England, and Reichstein, in Switzerland, accomplished its synthesis.

Vitamin C serves a key role in hydroxylation reactions that prevent scurvy and is thought to be a major component of the defence against oxidative damage from free radical reactions. Many animal species synthesize ascorbic acid from glucose and have no need for dietary vitamin C. Humans and other primates, guinea pigs and fruit-eating bats lack the enzyme gulonolactone oxidase, which catalyses the final step in the biosynthetic pathway, and thus rely on their diet to provide the vitamin.

15.2 CHEMICAL STRUCTURE AND NOMENCLATURE

The term vitamin C is used as the generic descriptor for all compounds exhibiting qualitatively the biological activity of ascorbic acid. The principal natural compound with vitamin C activity is L-ascorbic acid (Figure 15.1), which is systematically named 2-oxo-L-*threo*-hexono-1,4-lactone-2,3-ene-diol ($C_6H_8O_6$; MW = 176.13). There are two enantiomeric pairs; namely, L- and D-ascorbic acid, and L- and D-isoascorbic acid. D-Ascorbic acid and L-isoascorbic acid are devoid of vitamin C activity and do not occur in nature. D-Isoascorbic acid is also not found in natural products, apart from its occurrence in certain microorganisms. It possesses similar reductive properties to L-ascorbic acid, but exhibits only 5% of the anti-scorbutic activity of L-ascorbic acid in guinea pigs (Hornig and Weiser, 1976; Pelletier and Godin, 1969).

Ascorbic acid is used extensively in food technology as a stabilizer for the processing of beverages, wines and meat products. D-Isoascorbic

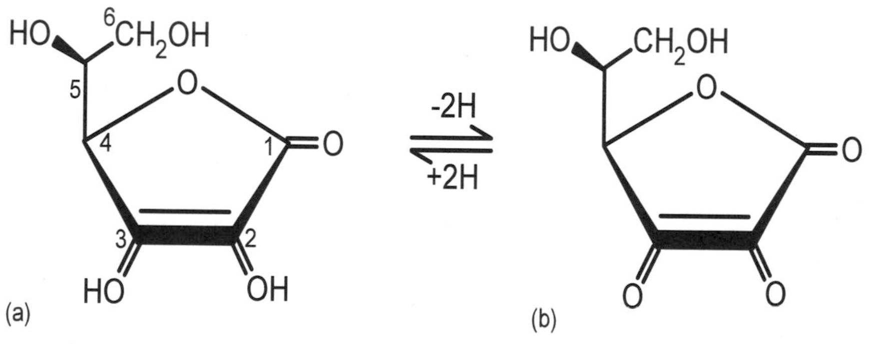

Figure 15.1 Structures of vitamin C compounds: (a) ascorbic acid; (b) dehydro-L-ascorbic acid

acid, known more commonly as erythorbic acid, is cheaper to manufacture on a commercial scale than L-ascorbic acid. In some countries, it is legal to substitute erythorbate for ascorbate when, for technological reasons, the antioxidant or reducing properties and not the vitamin C activity of the additive is required. In the United Kingdom and certain other countries, erythorbate is not permitted as an antioxidant, and it is also prohibited for use with raw and unprocessed meats.

L-Ascorbic acid is easily and reversibly oxidized to dehydro-L-ascorbic acid (Figure 15.1), which is equally effective in preventing scurvy (Todhunter, McMillan and Ehmke, 1950). Dehydroascorbic acid is a toxic compound that disrupts leucocytes and erythrocytes, and increases the permeability of pancreatic B-cells, resulting in loss of insulin (Rose, Choi and Koch, 1988).

15.3 DEFICIENCY SYNDROMES

A deficiency of vitamin C results in scurvy, the primary symptoms of which are haemorrhages in the gums, skin, bones and joints, and the failure of wound healing. These symptoms are accompanied by listlessness, malaise and other behavioural effects.

15.4 PHYSICOCHEMICAL PROPERTIES

Solubility and other properties

Ascorbic acid is an almost odourless white or very pale yellow crystalline powder with a pleasant sharp taste and a melting point of about 190 °C with decomposition. It is readily soluble in water (33 g/ 100 ml at 25 °C), less soluble in 95% ethanol (3.3 g/100 ml), absolute ethanol (2 g/100 ml), acetic acid (0.2 g/100 ml) and acetonitrile (0.05 g/ 100 ml) and insoluble in fat solvents (Seib, 1985). A 5% aqueous solution of ascorbic acid has a pH of 2.2–2.5, the acidic nature being due to the facile ionization of the hydroxyl group on C-3 ($pK_{a1} = 4.17$); the hydroxyl group on C-2 is much more resistant to ionization ($pK_{a2} = 11.79$) (Crawford and Crawford, 1980). Sodium ascorbate is freely soluble in water (62 g/100 ml at 25 °C; 78 g/100 ml at 75 °C) and practically insoluble in ethanol, diethyl ether and chloroform; the pH of an aqueous solution is 5.6–7.0.

Ascorbyl palmitate is a stable fat-soluble form of the vitamin which is practically insoluble at 25 °C in water (0.2 g/100 ml), soluble in ethanol (20 g/100 ml) and slightly soluble in diethyl ether. The solubility in vegetable oils at room temperature is very low (30 mg/100 ml) but increases sharply with increasing temperature.

The carbonyl enediol group of ascorbic acid confers strong reducing properties to the molecule, as indicated by its ability to reduce Fehling's or Tollen's solution at room temperature. The redox potential of the first stage at pH 5.0 is $\epsilon_0^1 = +0.127$ V

Spectroscopic properties

Ascorbic acid displays an absorption spectrum which is attributable to its enediol structure. The absorption maximum depends on the ionic state of the molecule and is therefore influenced by the pH of the medium. For the nondissociated form (pH 2, with dilute hydrochloric acid) the $A_{1\,cm}^{1\%}$ at λ_{max} 245 nm is 695; for the mono-dissociated form (pH 6.4, with phosphate buffer) the $A_{1cm}^{1\%}$ at λ_{max} 265 nm is 940 (Lawendel, 1957). Dehydroascorbic acid has a λ_{max} at 223 nm (Karayannis, Samios and Gousetis, 1977). The spectra of D-isoascorbic acid at pH 2.0 and 6.0 are virtually identical to those of L-ascorbic acid.

Stability

Pure dry crystalline ascorbic acid is stable on exposure to air and daylight at normal room temperature for long periods of time. Commercial vitamin C tablets possess virtually their original potency even after 8 years' storage at 25 °C (Killeit, 1986). Crystalline sodium ascorbate is also very stable, apart from a tendency to turn yellow.

The aerobic oxidation of ascorbic acid in the presence of transition metals is the most important reaction responsible for loss of vitamin C in foods. In the presence of molecular oxygen and trace amounts of transition metals, particularly copper(II) and iron(III), an intermediate ternary complex is formed which undergoes a single two-electron transfer to yield dehydroascorbic acid and hydrogen peroxide (Seib, 1985). A high hydrogen ion concentration (pH about 1) suppresses the ionization of ascorbic acid, and the fully protonated molecule is relatively slowly attacked by oxygen. Consequently, the rate of oxidation of ascorbic acid accelerates as acidity is decreased from pH 1.5 to 3.5. The photochemical oxidation of ascorbic acid proceeds through the generation of free radicals and is accelerated in the presence of metal ions and oxygen (Hay, Lewis and Smith, 1967). Other reactions of ascorbic acid related to foods have been discussed by Liao and Seib (1987, 1988).

Dehydroascorbic acid is not a true organic acid, as it contains no readily ionizable protons. It exists in aqueous solution as the bicyclic hemiketal hydrate rather than as the 2,3-diketo compound and is stable for several days at 4 °C at pH 2.5–5.5. In buffered solution at neutral or alkaline pH, dehydroascorbic acid undergoes a nonreversible oxidation in which the two rings open to give 2,3-diketogulonic acid in a straight-chain structure. Upon standing for 14 h in water at pH 7 and 27 °C, the diketogulonic acid undergoes decarboxylation to give L-*threo*-2-pentulose (Seib, 1985). Dehydroascorbic acid is thermolabile, and has a thermal half-life at pH 6 of less than 1 min at 100 °C and 2 min at 70 °C, irrespective of the presence of air (Bender, 1979).

15.5 ANALYSIS

15.5.1 Scope of analytical techniques

In food analysis, a method for determining vitamin C should ideally account for both ascorbic acid and its reversible oxidation product, dehydroascorbic acid, to give a value for total vitamin C. In addition, the ability to distinguish ascorbic acid from its epimer D-isoascorbic acid (erythorbic acid) is useful in the analysis of processed foods.

The classic titrimetric method using 2,6-dichlorophenolindophenol accounts for ascorbic acid, but not dehydroascorbic acid. Nonchromatographic methods for determining total vitamin C include colorimetry and fluorimetry, in which ascorbic acid is oxidized to dehydroascorbic acid and then reacted with a chemical reagent to form a coloured or fluorescent compound. Total vitamin C can be determined by HPLC using absorbance or electrochemical detection after reduction of dehydroascorbic acid to ascorbic acid, or using fluorescence detection after oxidation of ascorbic acid and derivatization of the dehydroascorbic acid formed. Capillary electrophoresis offers an alternative to HPLC and eliminates the need for organic mobile phases and expensive chromatography columns. Flow-injection analysis coupled with immobilized enzyme and electrochemical detection confers high specificity and provides a rapid automated procedure using relatively simple apparatus.

The IU of vitamin C activity is 0.05 mg of L-ascorbic acid: hence, by definition, 1 g of L-ascorbic acid is equivalent to 20 000 IU (Association of Vitamin Chemists, Inc., 1966). Dehydroascorbic acid possesses full vitamin C activity because it is readily reduced to ascorbic acid in the animal body. Ascorbyl palmitate exhibits the full antiscorbutic activity of ascorbic acid on a molar basis: i.e. 1 g of ascorbyl palmitate is equivalent to the potency of 0.425 g of ascorbic acid. Results obtained by chemical analysis are usually expressed in milligrams of pure L-ascorbic acid.

15.5.2 Bioassay

The most precise and objective bioassay technique for vitamin C is the odontoblast assay, which is based on the linear relationship between the length of the mature odontoblast cells in longitudinally sectioned incisors and the log dose of ascorbic acid. Guinea pigs are placed on a basal scurvy-producing diet supplemented with graded levels of the test sample in some groups, and of the standard ascorbic acid in others. Another group on the basal ration alone serves as a negative control. The length of the odontoblasts in histological preparations of the decalcified incisors is measured, and the results of the standard ascorbic acid are compared with those of the test samples (Bliss and György, 1967).

15.5.3 Extraction techniques

An effective means of extracting vitamin C from foods is homogenization with a solution of 3% w/v metaphosphoric acid dissolved in 8% glacial acetic acid. This extracting solution denatures and precipitates proteins (thereby inactivating all enzymes) and provides a medium below pH 4, which favours the stability of ascorbic acid and dehydroascorbic acid. Furthermore, metaphosphoric acid prevents catalysis of the oxidation of ascorbic acid by copper(II) or iron(III) ions (Ponting, 1943). Extracting solutions should be deoxygenated by bubbling oxygen-free nitrogen through the solution before use.

15.5.4 AOAC titrimetric method

This method for determining ascorbic acid is based upon the reduction of the dye 2,6-dichlorophenolindophenol (DCPIP) with ascorbic acid in an acid solution (Figure 15.2). Dehydroascorbic acid does not participate in the redox reaction, so the method does not yield the total vitamin C activity of a sample if this compound is present in significant quantities. The method has been adopted as Final Action for the determination of ascorbic acid in vitamin preparations and juices by the AOAC (AOAC,

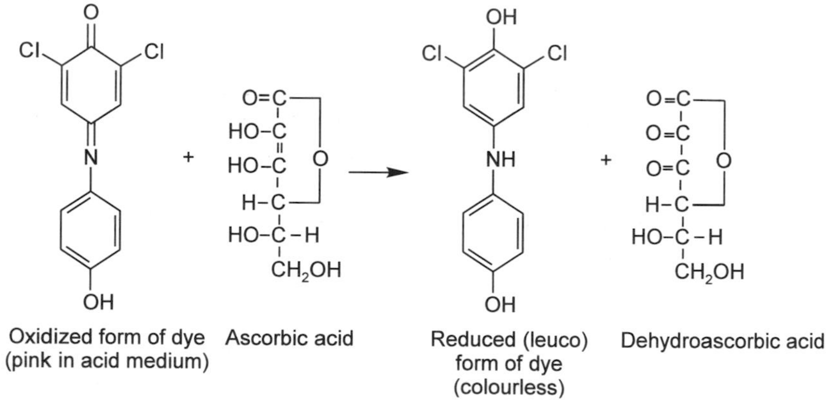

Figure 15.2 Reduction of 2,6-dichlorophenolindophenol (DCPIP) dye with ascorbic acid in an acid medium.

1990a). In the oxidized form, DCPIP is purplish-blue in neutral or alkaline solution, and pink in acid solution; the reduced leuco compound is colourless. The procedure entails titrating a standardized solution of the dye into an acid extract of the sample. The pink end-point signals the presence of excess unreduced dye. The titration should be performed rapidly (within 1–2 min) in the pH range 3–4, taking the first definite end-point. In the absence of interfering substances, the capacity of the extract to reduce the dye is directly proportional to the ascorbic acid content. The assay may be performed using potentiometric titration instead of visual titration as a means of overcoming the end-point difficulty encountered with coloured solutions.

Unless suitable measures are taken, substances other than ascorbic acid, which have reduction potentials lower than that of the DCPIP indicator, will react with the dye and give a falsely high result for the vitamin C content of the sample. Substances known to interfere in the assay include sulphydryl compounds (e.g. glutathione and cysteine), phenols, sulphites, metal ions (copper(I), iron(II) and tin(II)) and reductones. Sulphites are a common cause of difficulty because of their use as a preservative of foodstuffs. Provided that the titration is performed rapidly, sulphydryl and phenolic compounds should not cause interference, as the reduction of DCPIP by these compounds is relatively slow (Roe, 1967). Metal ions are not normally present in sufficient concentration to cause a significant interference. However, iron(II) and tin(II) ions can be leached from nonlacquered cans containing fruit drinks and juices and, in combination with the traces of naturally occurring oxalic acid, can produce a measurable interference (Pelletier and Morrison, 1966). Reductones are only likely to be found in processed foods after prolonged boiling or in canned foods after long standing at elevated temperatures (Roe, 1967). Various modifications of the AOAC titrimetric method have been proposed to eliminate the interference from other reducing compounds which may be present (Ball, 1994).

The titrimetric method gives results that generally agree with the biological estimation of vitamin C in raw and canned fruit and vegetables and their juices, which usually contain negligible amounts of dehydro-ascorbic acid. Analytical results for fresh fruits and vegetables showed good general agreement between the ascorbic acid values obtained by the AOAC titrimetric method and by HPLC (Wills, Wimalasiri and Greenfield, 1983). Albrecht and Schafer (1990) found that the titrimetric method and HPLC gave comparable results for fresh and three-week stored broccoli, cauliflower, green beans, turnips and three-week stored Brussels sprouts and spinach. Ascorbic acid contents, when measured by the HPLC method, were higher for fresh Brussels sprouts and spinach possibly because of co-eluting compounds that were oxidized during storage.

15.5.5 Direct spectrophotometry

Direct spectrophotometry has been applied to the determination of ascorbic acid in soft drinks, fruit juices and cordials after correction for background absorption in the UV region (Lau, Luk and Wong, 1986). The method of background correction was to measure the absorbance of the sample solution before and after the catalytic oxidation of ascorbic acid with copper(II) sulphate, and then to calculate the concentration of ascorbic acid from the difference. The sample blank was prepared by adding copper(II) sulphate to an aliquot of diluted sample and heating at 50 °C for 15 min. The heating step was necessary to overcome the inhibitory effect of citrate upon the copper-catalysed oxidation. To correct for the absorption due to Cu(II), ethylenediaminetetraacetic acid (EDTA) was added after the oxidation. Samples and standard solutions were prepared to contain the same concentration of the Cu(II)–EDTA complex, which does not catalyse the oxidation of ascorbic acid at room temperature. The absorption due to the Cu(II)–EDTA complex constituted part of the reagent blank against which the ascorbic acid standard solutions were read. Absorbance measurements were made at 267 nm and at pH 6. The calibration graph was linear within the range 0–20 µg ascorbic acid/ ml. The precision was 0.1–0.5% for ascorbic acid in the concentration range 5–13 µg/ml.

15.5.6 AOAC microfluorimetric methods

A fluorimetric method for determining microgram quantities of total vitamin C in pharmaceutical preparations, beverages and special dietary foods has been described by Deutsch and Weeks (1965). The method involves the oxidation of ascorbic acid to dehydroascorbic acid with active charcoal, followed by the reaction of dehydroascorbic acid with 1,2-phenylenediamine dihydrochloride (OPDA) to form the fluorescent quinoxaline derivative 3-(1,2-dihydroxyethyl)furo[3,4-*b*]quinoxaline-1-one (DFQ) (Figure 15.3). The blank reveals any fluorescence due to interfering substances, and is prepared by complexing the oxidized vitamin with boric acid to prevent the formation of the quinoxaline derivative.

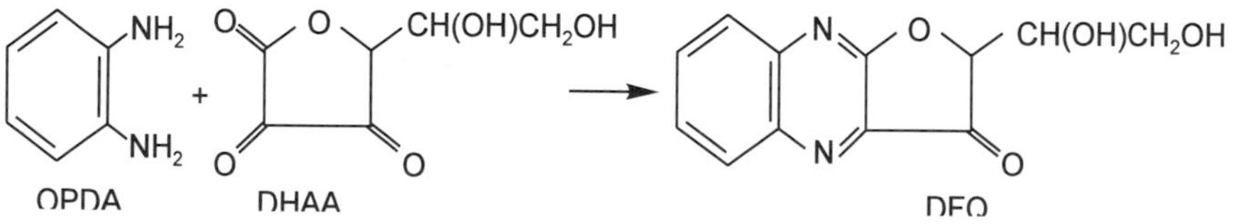

Figure 15.3 Reaction between dehydroascorbic acid and *o*-phenylenediamine to form the quinoxaline derivative (DFQ).

The fluorimetric method has been adopted as Final Action by the AOAC (AOAC, 1990b) for vitamin preparations.

The method shows a high degree of specificity. Deutsch and Weeks (1965) ascertained that a substance will only interfere in the assay if all of the following conditions are satisfied.

1. The substance must have α-diketo groups, which react with OPDA under the assay conditions.
2. The excitation and emission wavelengths of the quinoxaline derivative must be within the regions prescribed for the assay.
3. It must contain adjacent *cis* hydroxyl groups, which react with the boric acid solution to form a complex.

Additionally, the substance must be present in the sample assay solution in sufficient quantity to have an effect. Of a large number of possibly interfering substances tested (Deutsch and Weeks, 1965; Deutsch, 1967), no individual compound was found which satisfied all of the above criteria. The procedure was therefore judged to be suitable for samples containing large amounts of reducing substances. An additional advantage is its ability to cope with highly coloured materials.

Christie (1975) reported from the Laboratory of the Government Chemist in the United Kingdom that the fluorimetric method could be applied successfully to a wide range of foodstuffs, including liver, milk, fresh and canned fruit, raw and cooked vegetables and potato powder. However, Wills, Wimalasiri and Greenfield (1983) found that total vitamin C values for green leafy vegetables were higher when measured by the fluorimetric assay than when measured by HPLC with UV detection, suggesting a pigment-related interference in the fluorescence measurement. Also, Augustin, Beck and Marousek (1981) obtained unrealistically high vitamin C values with the fluorimetric method in the analysis of processed potato products.

Segmented-flow analysis has been utilized for modifying the AOAC fluorimetric procedure for total vitamin C using DCPIP (Kirk and Ting, 1975) or N-bromosuccinimide (Roy, Conetta and Salpeter, 1976) as the oxidizing agent instead of activated charcoal. Roy, Conetta and Salpeter (1976) reported that cocoa and chocolate, known to contain significant amounts of reductic acids, reductones and alkaloids, required prior filtration through charcoal to remove interfering fluorescent compounds. On the basis of this observation, Egberg, Potter and Heroff (1977) proposed a semi-automated method involving simultaneous oxidation and clean-up with charcoal prior to automated analysis. Dunmire *et al.* (1979) compared the methods of Roy, Conetta and Salpeter (1976) and Egberg, Potter and Heroff (1977) and concluded that the latter method, with its potential for sample clean-up, was more appropriate for the majority of food commodities analysed. Following a collaborative study (DeVries, 1983), the semi-automated microfluorimetric procedure was adopted as Final Action for the determination of total vitamin C in foods by the AOAC (AOAC, 1990c).

15.5.7 High-performance liquid chromatography

General considerations

The ability of an HPLC technique to separate ascorbic acid from erythorbic acid, and to detect these epimers, provides the opportunity to use erythorbic acid as an internal standard for the quantification of ascorbic acid, provided it is shown that erythorbic acid is absent in the food sample extract presented for analysis. HPLC, using a weak anion exchange mechanism, has the potential to separate isocratically not only ascorbic acid from its oxidation products, dehydroascorbic acid and diketogulonic acid, but, simultaneously, erythorbic acid from its analogous oxidation products (Doner and Hicks, 1981). Other chromatographic modes, in addition to weak anion exchange, facilitate useful separations and are utilized in accordance with the particular analytical requirements (Table 15.1).

Detection

Ascorbic acid can be detected directly by either absorbance or electrochemical monitoring, and fluorescence detection can be utilized after chemical derivatization to a fluorescent compound. The detection of dehydroascorbic acid, however, poses a problem as the molar absorptivity of this compound is relatively very weak and it is electrochemically inactive.

Hydrodynamic voltammograms for ascorbic acid show that the peak current reaches a plateau at $+0.8\,V$ using either a platinum or glassy carbon electrode (Huang, Duda and Kissinger, 1987). An oxidative potential of $+0.8\,V$ is relatively high, and the many compounds with lower or very close potentials that are present in biological materials will also be detected. Thus the specificity of the assay relies largely on the chromatographic separation. In practice, a potential of $+0.6\,V$ is frequently chosen which, though not as sensitive as $+0.8\,V$, results in improved selectivity.

Dehydroascorbic acid, being electrochemically inactive, must be reduced chemically to ascorbic acid if it is to be taken into account. These circumstances also apply to erythorbic acid and its oxidation product, dehydroerythorbic acid.

The minimum detection limit of an amperometric detector towards ascorbic acid, using an oxidation potential of $+0.6\,V$, was reported to be

0.3 ng on-column (based on a signal/noise ratio of 3 : 1) and was an order of magnitude lower than the limit obtained with a photometric detector used for comparison (Karp, Ciambra and Miklean, 1990). The higher sensitivity gives electrochemical detection the advantage over absorbance detection for the analysis of foods that contain relatively small amounts of vitamin C.

Doner and Hicks (1981) used a refractive index monitor to detect dehydroascorbic acid, dehydroerythorbic acid, diketogulonic acid and diketogluconic acid, which were all transparent at the wavelength (268 nm) employed for detecting ascorbic acid and erythorbic acid. Refractive index detection is nonselective and only moderately sensitive, and hence it has no utility in the accurate measurement of the small amounts of dehydroascorbic acid and other oxidation products that might be present in foods.

Some HPLC methods for analysing fruits, fruit juices and vegetables have employed dual absorbance detection, ascorbic acid being measured at 254 nm and dehydroascorbic acid at 210 nm (Finley and Duang, 1981; Bradbury and Singh, 1986) or 214 nm (Wimalasiri and Wills, 1983). This dual detector technique is simple to perform and does not involve the analytes in any chemical reactions. However, the accuracy of dehydroascorbic acid measurement is expected to be poor at such low detection wavelengths.

A more popular approach is to reduce the dehydroascorbic acid in the sample extract to ascorbic acid by pre-column reaction with either homocysteine, dithiothreitol or L-cysteine, and then to determine the ascorbic acid (representing total vitamin C) by HPLC using absorbance or electrochemical detection. The dehydroascorbic acid value can be calculated by subtracting the ascorbic acid value (obtained without reduction) from the total vitamin C value, but the error will be appreciable if the concentration of dehydroascorbic acid is very low relative to that of ascorbic acid. Diop *et al.* (1988) referred to previous work that reported an incomplete reduction of dehydroascorbic acid by homocysteine (55% instead of 97.8% claimed by Dennison, Brawley and Hunter, 1981), and which led to a low (50%) recovery of dehydroascorbic acid as measured by spiking.

Ziegler, Meier and Sticher (1987) determined dehydroascorbic acid directly, in addition to ascorbic acid, by separating the two compounds using reversed-phase HPLC with ion suppression, and reducing dehydroascorbic acid to ascorbic acid with dithiothreitol in a post-column in-line reaction system. This technique enabled the dehydroascorbic acid to be measured photometrically (267 nm) at the same sensitivity as ascorbic acid with a detection limit of 1.4 ng per 10 µl injection volume (signal/noise ratio of 2 : 1). Attempts to use electrochemical detection were unsuccessful, owing to a high background signal and

Table 15.1 HPLC methods used for the determination of vitamin C

Food	Sample preparation	Column	Mobile phase	Compounds separated	Detection	Reference
		Weak anion exchange chromatography				
Nonfat dry milk	Dissolve in water containing 1 mg/ml dithiothreitol (DTT), stir, add 12% HPO₃, stir, add 2 ml MeCN, stir, centrifuge	μBondapak-NH₂ 300 × 7.8 mm i.d. Column temperature 21 °C	MeCN/0.005M KH₂PO₄(70 + 30) + 0.125 ml/1 of mercaptoethanol	AA (representing total vitamin C)	UV 268 nm	Margolis and Black (1987)
Fruits, vegetables	Homogenize with 3% citric acid, dilute to volume and filter. Pass through C₁₈ Sep-Pak solid-phase extraction cartridge	μBondapak-NH₂ 10 μm 300 × 3.9 mm i.d.	MeCN/0.01M NH₄ H₂PO₄ buffer (pH 4.3) (70:30)	AA, DHAA	Dual UV 254 nm (AA) 214 nm (DHAA)	Vimalasiri and Wills (1983)
Tropical root crops	Homogenize with 5% HPO₃ and filter	μBondapak-NH₂ in Z-module cartridge	MeCN/0.005M KH₂PO₄ buffer (pH 4.6) (70:30)	AA, DHAA	Dual UV 210 nm (DHAA) 254 nm (AA)	Bradbury and Singh (1986)
Orange juice, soft drinks, protein-fortified drinks	*For AA,* deproteinize (if necessary) with 12.5% TCA, centrifuge, filter *For total vitamin C,* adjust filtrate to pH 7.0 with 45% K₂HPO₄, add 0.8% homocysteine solution (allow 15 min reaction time), filter	μBondapak-NH₂ 300 × 3.9 mm i.d.	MeOH/0.25% KH₂PO₄ buffer (pH 3.5) (50:50)	AA (representing total vitamin C if DHAA is reduced with homocysteine)	UV 244 nm	Dennison, Brawley and Hunter 1981)
Orange juice	*For AA,* dilute juice with extraction solution (1% citric acid containing 0.05% EDTA.2Na in 50% MeOH) and filter *For total vitamin C,* add 0.8% DL-homocysteine and adjust pH to 7 with 45% K₂HPO₄ (allow 15 min reaction time). Dilute with extraction solution and filter	Zorbax-NH₂ 5 μm 250 × 4.6 mm i.d.	0.25% KH₂PO₄ buffer (pH 3.5)/MeOH (60:40)	AA Determine DHAA as the difference between total vitamin C and AA contents	UV 244 nm	Hare, Jones and Lindsay (1993)

Sample	Procedure	Column	Mobile phase	Detection	Compounds	Reference
Fresh apple and potato	*For AA*, blend with 2.5% HPO₃, and mobile phase, filter through Celite, refilter, pass through C₁₈ Sep-Pak solid-phase extraction cartridge. *For total vitamin C*, treat second filtrate with dithiothreitol (DTT) after adjusting to pH 6 (allow 30 min reaction time at ambient temperature), pass through C₁₈ Sep-Pak cartridge	Dynamax-60A-NH₂ 8 μm (Rainin) 250 × 4.6 mm i.d.	MeCN/0.05 M KH₂PO₄ (75:25)	UV 254 nm	AA (representing total vitamin C if DHAA is reduced with DTT), ascorbic acid-2-phosphate	Sapers *et al.* (1990)
Frozen apples, potato products, concentrated fruit and vegetable juices, frozen juices, natural and artificially flavoured drink mixes, Hi-C drinks, cured meat products	Suspend blended sample in water or EtOH (potato products), filter, pass through C₁₈ Sep-Pak solid-phase extraction cartridge. Add dithiothreitol (DTT) to one portion of each sample to obtain total vitamin C and total isovitamin C	LiChrosorb-NH₂ 250 × 4 mm i.d.	MeCN/0.05 M KH₂PO₄ buffer (pH 5.95) (75:25)	UV 268 nm	AA and EA (representing total vitamin C and isovitamin C, respectively, if DHAA and DHEA are reduced with DTT)	Tuan, Wyatt and Anglemier (1987)
Fresh fruit and vegetables, fruit drinks	Homogenize with 3% HPO₃, filter, pass through C₁₈ Sep-Pak solid-phase extraction cartridge	NH₂-phase (Alltech)	MeCN/0.05 M KH₂PO₄ buffer (pH 5.9) (75:25)	UV 254 nm (AA) Fluorescence (DFQ) through post-column derivatization of DHAA with OPDA	AA, DHAA	Kacem *et al.* (1986)

Ion-exclusion chromatography

Sample	Procedure	Column	Mobile phase	Detection	Compounds	Reference
Citrus fruit juices, fruits, vegetables	Dilute or extract with 0.05% EDTA.2NA in 0.1 M H₂SO₄, filter or centrifuge	Aminex HPX-87 SCX sulphonated PS-DVB resin 300 × 7.8 mm i.d.	4.5 mM H₂SO₄	UV 245 nm	AA	Ashoor, Monte and Welty (1984)

Table 15.1 *Continued*

Food	Sample preparation	Column	Mobile phase	Compounds separated	Detection	Reference
Strawberries (fresh), potatoes (raw and cooked)	Blend samples with 62.5 mM HPO_3, centrifuge, filter, dilute with 62.5 mM HPO_3. *For AA*, analyse extract directly. *For total vitamin C*, add 30 mM DL-homocysteine to extract and adjust pH to 6.8–7.0 with 2.6 M K_2HPO_4. After 30 min, stop reduction by addition of 6.25 M HPO_3	Aminex HPX-87H sulphorated PS-DVB resin 300 × 7.8 mm i.d.	4.5 mM H_2SO_4	AA. Determine DHAA as the difference between total vitamin C after DHAA reduction and AA content of orginal sample	UV 245 nm	Graham and Annette (1992)
Potatoes	Homogenize with 95% EtOH/water containing 0.2% conc. H_2SO_4 (60:40), filter	Aminex HPX-87 SCX sulphonated PS-DVB resin 9 μm 300 × 7.8 mm i.d.	9 mM H_2SO_4	AA, oxalic acid, citric acid, malic acid, fumaric acid	UV 260 nm (AA) UV 210 nm (other acids)	Eashway, Bureau and McGann (1984)
Milk (fluid and powdered)	Powdered milk: dilute with 1% HPO_3, centrifuge. Fluid milk: centrifuge	SCX sulphonated PS-DVB resin, H^+-form (Bio-Rad) 9 μm 100 × 7.8 mm i.d.	1 mM H_2SO_4	AA	Amperometric: glassy carbon electrode, +0.70 V vs Ag/AgCl	Mannino and Pagliarini (1988)
Fresh fruits, fruit juices, dry mixed fruits, potatoes	Homogenize solid samples or dilute liquid samples with 0.005 M H_2SO_4 (pH 2) + 0.01 M mannitol, centrifuge, filter (0.45 μm) or (for some fresh fruits) pass through C_{18} Sep-Pak solid-phase extraction cartridge	Polypore H 10 μm (Bioanalytical Systems) 100 × 4.6 mm i.d.	10 mM H_2SO_4 + 10 mM mannitol	AA, free sulphite	Amperometric dual electrode Free sulphite: Pt, +0.70 V AA: glassy carbon, +0.80 V vs Ag/AgCl	Huang, Duda and Eissinger (1987)
Fruits, fruit drinks, vegetables	Dilute or homogenize with 0.02 M H_2SO_4, centrifuge, filter (0.45 μm). Mix with 0.05 M phosphate buffer (pH 7) and 0.01 M dithiothreitol (2 min reaction time at ambient temperature)	SCX sulphonated PS-DVB resin (Wescan) 100 × 4.6 mm i.d.	20 mM H_2SO_4	AA (representing total vitamin C)	Amperometric: Pt electrode, +0.60 V or +0.80 V vs Ag/AgCl	Kim (1989)

Reversed-phase chromatography

Sample	Preparation	Column	Mobile phase	Analytes	Detection	Reference
Green beans	Stir homogenized sample with 4.5% HPO$_3$ for 15 min. Filter, dilute with 4.5% HPO$_3$, filter (0.22 μm)	Spherisorb ODS-2 5 μm 250 × 4.6 mm i.d.	Water adjusted to pH 2.2 with H$_2$SO$_4$	AA (separated from oxalic, malic, citric, succinic and fumaric acids)	Dual UV: 245 nm for AA 215 nm for other organic acids	Vazquez Oderiz et al. (1994)
Fresh fruits	Shake homogenized sample with extracting solution (8% acetic acid and 3% HPO$_3$ in water), dilute with the same solution and filter	Spherisorb ODS-2 5 μm 250 × 4.6 mm i.d.	Water acidified to pH 2.2 with H$_2$SO$_4$	AA (separated from quinic, malic and citric acids)	Dual UV: 254 nm for AA 214 nm for organic acids	Romero Rodriguez et al. (1992)
Fresh citrus fruit juices	Dilute with mobile phase, filter	PLRP-S 100 Å, 5 μm (Polymer Laboratories) 250 × 4.6 mm i.d.	0.2 M NaH$_2$PO$_4$ buffer (pH 2.14)	AA, EA	Dual UV: UV 244 nm (AA) UV 220 nm (AA, EA)	Lloyd et al. (1988a, b)
Solid and liquid foods	Add extraction solution containing 30 g HPO$_3$, 0.5 g EDTA.2Na and 80 ml glacial acetic acid/1 plus EA (internal standard) and blend. *For non-fat and low starch samples,* centrifuge and filter. *For low-starch samples with fat,* add hexane and vortex-mix before centrifugation, filter the aqueous layer. *For high-starch samples,* add n-butanol and vortex-mix before centrifugation, filter the aqueous layer	Two PLRP-S (100 Å, 5 μm) columns in series: 150 × 4.6 mm + 250 × 4.6 mm i.d.	0.2 M NaH$_2$PO$_4$ adjusted to pH 2.14 with H$_3$PO$_4$	DHEA, DHAA, AA, EA	Fluorescence: ex. 350 nm em. 430 nm Post-column derivatization (oxidation of AA to DHAA and EA to DHEA using CuCl$_2$ followed by reaction with OPDA to form the quinoxaline derivatives	Vanderslice and Higgs (1993)
Fruits, vegetables, juices	Homogenize with 0.2 M phosphate buffer (pH 2.0) and extract with 3% HPO$_3$, centrifuge	PLRP-S 100 Å, 5 μm 250 × 4.6 mm i.d.	1.8% THF and 0.3% HPO$_3$ in water	AA	UV 244 nm	Bushway et al. (1988)
Fruits, vegetables	Extract with 6% HPO$_3$ containing 1 μM EDTA and 0.1 μM diethylthiocarbamate	μBondapak C$_{18}$ 10 μm in a 100 × 8 mm radial compression module (Waters)	1.5% NH$_4$H$_2$PO$_4$ buffer (pH 3)	AA	UV 254 nm	Watada (1982)
Citrus juices	Mix with 5% HPO$_3$, centrifuge, pass through C$_{18}$ solid-phase extraction cartridge, add quinic acid as internal standard	Zorbax ODS 250 × 4.6 mm i.d.	2% KH$_2$PO$_4$ buffer (pH 2.4)	AA, quinic acid (internal standard)	UV 245 nm	Lee and Coates (1987)

Table 15.1 *Continued*

Food	Sample preparation	Column	Mobile phase	Compounds separated	Detection	Reference
Fruits, vegetables, commercial orange juices	Blend sample with 0.05 N H_3PO_4, centrifuge and dilute. Pass through preconditioned Sep-Pak C_{18} solid-phase extraction cartridge and filter (0.45 μm)	Spheri-5 RP-18 5 μm 110 × 4.6 mm i.d. and two Polypore H organic acid columns (110 × 4.6 and 220 × 4.6 mm i.d.) connected in series	2% KH_2PO_4 buffer (pH 2.3)	AA (separated from DHAA and oxalic, malic, succinic and citric acids)	UV 260 nm	Visperos-Carriedo, Buslig and Shaw (1992)
Rose hips	Extract with 1% HPO_3, pass through C_{18} solid-phase extraction cartridge	LiChrosorb RP-18 5 μm 250 × 4 mm i.d.	0.25% HPO_3	AA, DHAA	UV 267 nm for both peaks after post-column reduction of DHAA to AA with dithiothreitol at 50°C	Ziegler, Meier and Sticher (1987)
Orange juice	Centrifuge, mix 1:1 with 6% HPO_3, filter	Brownlee RP-18 5 μm 220 × 4.6 mm i.d. (or 100 × 4.6 mm i.d.)	2% $NH_4H_2PO_4$ (pH 2.8)	AA	Amperometric: glassy carbon electrode, +0.6 V vs Ag/AgCl	Wilson and Shaw (1987)
Fruit juices	Dilute juice with water and filter *For AA* add α-methyl-L-DOPA (internal standard) and 2% HPO_3 *For total vitamin C* add internal standard and L-cysteine diluted in 0.01 M phosphate buffer, pH 6.8 (allow 15 min reaction time)	Inertsil ODS-2 5 μm 150 × 4.6 mm i.d.	0.1 M KH_2PO_4 buffer (pH 3) containing 1 mM EDTA.2Na	AA, α-methyl-L-DOPA (internal standard) Determine DHAA as the difference between total vitamin C and AA contents	Amperometric: +0.3V vs Ag/AgCl	Iwase and Ono 1993)
Citrus juices, vegetables	*For total vitamin C and total isovitamin C:* blend with 0.3 M TCA, dilute to volume and filter. Add 4.5 M acetate buffer (pH 6.2), incubate at 37°C for 5 min with ascorbate oxidase. Add 0.1% OPDA, react at 37°C for 30 min *For DHAA and DHEA:* as above, but omit enzymatic oxidation	Hypersil – ODS 3 μm 125 × 4.6 mm i.d.	0.08 M KH_2PO_4/MeOH (80:20) final pH 7.8	Quinoxaline derivatives of DHAA and DHEA representing total vitamin C (AA + DHAA) and total isovitamin C (EA + DHEA), respectively	Fluorescence: ex. 365 nm (filter) em. 418 nm (filter)	Speek, Schrijver and Schreurs (1984)

Sample	Extraction	Column	Mobile phase	Analyte	Detection	Reference
Processed meats, skimmed and condensed milk, cherry cheesecake, chocolate diet product, fruit preservative	Homogenize in cold 17% HPO$_3$, centrifuge, filter. *For AA and EA*: dilute with phosphate buffer (pH 9.8) to give a final pH of 7.1. After 30 min at 25°C further dilute with cold 0.85% HPO$_3$. Dilute an aliquot with mobile phase buffer. *For total vitamin C (AA+DHAA) and total isovitamin C (EA + DHEA)*: dilute with 1% homocysteine in phosphate buffer (pH 9.8) to give a final pH of 7.1. Then as for AA and EA	FLRP-S 100 Å, 5 μm 250 × 4.6 mm i.d. Two columns connected in series	0.02 M sodium phosphate monobasic monohydrate containing 0.17% HPO$_3$ (final pH 2.2)	AA, EA (determine DHAA and DHEA by subtracting AA and EA values from respective total vitamin C and isovitamin C values)	Amperometric: glassy carbon electrode, +0.7V vs Ag/AgCl	Behrens and Madère (1994)
Infant formulas, potato crisps, cereal (fortified), canned corn (maize), canned green beans, canned potatoes, cranberry juice, cranapple juice	Mix homogenized or blended sample with extraction solution (3% HPO$_3$ in 8% acetic acid) and centrifuge. Re-extract and centrifuge. *Conversion to DHAA*: shake with acid-washed Norit (carbon) and filter. Clarify by mixing with sodium acetate solution and MeOH, filter or centrifuge. *Derivatization*: add OPDA solution and dilute with HPLC mobile phase (allow 60 min reaction time)	μBondapak C$_{18}$ 10 μm 300 × 3.9 mm i.d.	MeOH/H$_2$O (55 + 45)	DFQ, representing total vitamin C	Fluorescence: ex. 350 nm em. 430 nm	Dodson, Young and Soliman (1992)

Ion interaction chromatography

Sample	Extraction	Column	Mobile phase	Analyte	Detection	Reference
Fresh tomatoes	Homogenize with 6% HPO$_3$, mix aliquot of slurry with MeOH, add 0.05% pentanophenone as internal standard, centrifuge, dilute with MeOH containing 0.0015 M pyrogallol	Vydac 201-HS C$_{18}$ 10 μm 250 × 4.6 mm i.d.	MeOH/H$_2$O/MeCN (60 + 40 + 1) containing 0.5 mM tridecylammonium formate, adjusted to pH 4.25	AA	UV 247 nm	Russell (1986)
Fresh fruits and vegetables	Mix macerated sample with 2% HPO$_3$, shake, filter	Spherisorb ODS-2 10 μm 250 × 4.6 mm i.d.	0.01 M KH$_2$PO$_4$/20% tetrabutylammonium hydroxide/MeOH, 970 + 1 + 30 adjusted to pH 2.75 using 85% H$_3$PO$_4$	AA (separated from succinic, citric, fumaric and gallic acids)	UV 244 nm	Daood et al. (1994)

Table 15.1 *Continued*

Food	Sample preparation	Column	Mobile phase	Compounds separated	Detection	Reference
Fruits, fruit juice products, vegetables	Dilute or extract with 6% HPO₃/EtOH (10:90), filter (paper), centrifuge, filter (0.45 µm membrane), pass through Sep-Pak C₁₈ solid-phase extraction cartridge	Spherisorb ODS 10 µm 250 × 4 mm i.d.	(i) 0.1 M phosphate buffer (pH 4.2) containing 5 mM tetrabutyl-ammonium hydroxide (ii) 0.2 M phosphate buffer (pH 4.2) containing 10 mM hexadecyltrimethyl-ammonium bromide/MeOH (1-1)	AA	UV 254 nm	Moledina and Flink (1982)
Cured meats	Homogenize with 6% HPO₃, filter through glass wool and pass through Sep-Pak C₁₈ solid-phase extraction cartridge	µBondapak C₁₈ 300 × 3.9 mm i.d.	5 mM tetrabutyl-ammonium formate (pH 4.77)	EA	UV 254 nm	Lee and Marder (1983)
Fruits, fruit juices, vegetables	Extract with HPO₃-acetic acid solution, filter, pass through C₁₈ solid-phase extraction cartridge	Two 300 × 3.9 mm i.d. µBondapak C₁₈ columns connected in series	0.008 M phosphate buffer, pH 5.3 containing 0.7 ml/l tri-n-butylamine	AA, DHAA, DKGulA	UV 210 nm and 254 nm	Finley and Duang (1981)
Wine and beer	Membrane-filter (0.2 µm)	Nucleosil 120 C₁₈ 7 µm 250 × 4 mm i.d.	0.5% aqueous MeOH containing 0.05 M acetate buffer and 5 mM n-octylamine	AA, EA	UV 266 nm	Seifert, Swaczyna and Schaefer (1992)
Fruits, infant foods, milk	Fruits, infant foods: extract with 3% HPO₃–8% acetic acid, centrifuge, dilute to volume with cold 0.05 M perchloric acid Milk: dilute to volume with cold 0.05 M perchloric acid, centrifuge	LiChrosorb RP-18 150 × 4.6 mm i.d.	0.08 M acetate buffer containing 1 mM tridecylamine and 15% MeOH (final pH 4.5)	AA	Amperometric: carbon paste electrode, +0.7 V vs Ag/AgCl	Pachla and Kissinger (1979)
Bread	Grind freeze-dried bread, extract with 3% HPO₃, centrifuge, dilute to volume with cold 0.05 M perchloric acid	Alltech C-18 5 µm 250 × 4.6 mm i.d. Column temperature 25 °C	0.08 M acetate buffer, pH 4.2 containing 0.1 mM EDTA.2Na and 1 mM octyltriethylammonium phosphate	AA	Amperometric: glassy carbon electrode, +0.72 V vs Ag/AgCl	Hong, Seib and Kramer (1987)
Beers	Ultrasonic degassing	µBondapak C₁₈ 300 × 3.9 mm i.d.	Citrate buffer pH 4.4 containing 0.5 mM EDTA and 1 mM N-methyldodecylamine	AA	Amperometric: glassy carbon electrode, +0.60 V vs Ag/AgCl	McL and Joly (1987)

Sample	Sample preparation	Column	Mobile phase	Compounds	Detection	Reference
Orange juice, milk	Dilute with cold 0.05M perchloric acid and centrifuge	Ultrasphere ODS 5 μm 250 × 4.6 mm i.d.	0.04 M acetate buffer containing 1 mM decylamine and 15% MeOH	AA, EA	Amperometric: carbon paste electrode, +0.7 V vs Ag/AgCl	Tsao and Salimi (1982)
Beer	Ultrasonic degassing	Zorbax ODS 250 × 4.6 mm i.d.	0.1 M acetate buffer pH 5.5 containing 200 mg/l EDTA.2Na and 1 mV octylamine	AA or EA	Amperometric: glassy carbon electrode, +0.60 V vs Ag/AgCl	Knudson and Siebert (1987)
Cured meats	Grind with 5% HPO_3 containing 0.1 mg/ml EDTA.2Na, centrifuge and dilute with mobile phase	Ultrasphere ODS 5 μm 250 × 4.6 mm i.d. Column temperature 30°C	0.04 M acetate buffer containing 5 mM tetrabutyl-ammonium phosphate and 0.2 mg/ml EDTA.2Na (final pH 5.25)	AA, EA	Amperometric: glassy carbon electrode, +0.6 V vs Ag/AgCl	Kutnink and Omaye (1987)
Fresh oranges, canned orange juice, orange soft drink, beer, fresh kiwi fruit, fresh tomato	Dilute homogenized sample with MeOH/H_2O (5:95), add EA (internal standard) and centrifuge. Adjust pH to between 2.20 and 2.45 if necessary, bring final volume to 10 ml and pass through preconditioned Sep-Pak C_{18} solid-phase extraction cartridge. Discard first 5 ml of eluate and collect the next 3 ml. Add OPDA solution and filter (0.22 μm). Allow 37 min reaction time before HPLC analysis	μBondapak C_{18} 10 μm 300 × 3.9 mm i.d.	MeOH/H_2O (5:95) containing 5 mM hexadecyltrimethyl-ammonium bromide (cetrimide) and 0.05M KH_2PO_4 (final pH 4.59)	DFQ, AA, EA (internal standard)	UV 348 nm for DFQ and 261 nm for AA and EA	Zapata and Dufour (1992)
Processed meat	Homogenize with 0.5% HPO_3 and centrifuge. Dilute with 0.5% HPO_3, add MeOH to give a 4% final concentration	Hypersil ODS 5 μm 250 × 4 mm i.d.	H_2O/MeOH/acetate buffer/1,5-dimeth/lhexylamine (945 + 40 + 15 + 1.5)	AA, EA	UV 254 nm	Schüep and Keck (1990)
Fresh fruits	Homogenize with 0.4% HPO_3 and dilute to volume. Filter (0.2 μm), pass through C_{18} Sep-Pak solid-phase extraction cartridge (if coloured)	μBondapak C_{18} 10 μm 300 × 3.9 mm i.d.	0.1 M KH_2PO_4/MeOH (9:1) containing 5 mM cetyltrimethyl-ammonium bromide	AA	UV 265 nm	Diop et al. (1988)

Abbreviations: see footnote to Table 2.4.

HPO_3, metaphosphoric acid; OPDA, o-phenylenediamine; DFQ, quinoxaline derivative.

electrode poisoning that resulted from the excess dithiothreitol. Karp, Ciambra and Miklean (1990) extended the system of Ziegler, Meier and Sticher (1987) by reacting the excess dithiothreitol with N-ethylmalei- mide, thereby permitting the electrochemical detection of both ascorbic acid and dehydroascorbic acid with a detection limit of 0.3 ng per 20 μl injection volume.

An alternative approach is to oxidize the ascorbic acid in the sample extract to dehydroascorbic acid and then to react the latter compound with OPDA to form DFQ. HPLC analysis of the resulting solution, with fluorescence detection of the DFQ, gives the total vitamin C value, while dehydroascorbic acid can be determined directly in a separate analysis by omitting the oxidation step. Speek, Schrijver and Schreurs (1984) determined total vitamin C and total isovitamin C (erythorbic acid + dehydroerythorbic acid) in citrus juices and vegetables, after enzymatic oxidation of ascorbic acid and erythorbic acid to the respective dehydro compounds, through fluorimetric detection of the corresponding quinoxaline derivatives separated by reversed-phase HPLC. Stability studies showed that the quinoxaline derivatives deteriorated rapidly under daylight exposure and in the dark at 22 °C, whereas they were fairly stable in the dark at 4 °C or –20 °C for at least 12 h. Reductones react slowly with OPDA and do not interfere (Jaffe, 1984). Vanderslice and Higgs (1993) determined total vitamin C and total isovitamin C in solid and liquid foods through post-column in-line chemistry involving oxidation of ascorbic acid and erythorbic acid with copper(II) sulphate, followed by reaction with OPDA and fluorimetric detection.

Applications

The principal chromatographic modes that have been utilized for the determination of vitamin C and related compounds in foods are weak anion exchange, ion exclusion, reversed-phase with ion suppression, and ion interaction.

Weak anion exchange (WAX) chromatography
Aminopropyl-bonded stationary phases operated in the WAX mode have the potential to separate, isocratically, not only ascorbic acid from its oxidation products, dehydroascorbic acid and diketogulonic acid, but, simultaneously, erythorbic acid from its analogous oxidation products, dehydroerythorbic acid and diketogluconic acid. Since there is no significant difference in pK_a values of the ionizable protons between ascorbic acid and erythorbic acid, one must assume that factors other than degree of ionization affect interactions with the stationary phase. One possible mechanism is hydrogen bonding between hydroxyl protons in

the compounds with the neutral amino group in the stationary phase (Doner and Hicks, 1981).

Kacem *et al.* (1986) employed tandem absorbance and fluorimetric detection for the simultaneous determination of ascorbic acid and dehydroascorbic acid, respectively, after post-column derivatization of the latter compound to DFQ.

In order to obtain a quantitative recovery of dehydroascorbic acid using WAX chromatography and absorbance detection, it is essential to neutralize sample extracts to pH 6 if metaphosphoric acid is used as the extractant. Recovery of dehydroascorbic acid was only 14% in non-neutralized extracts (pH 2.9) and 45% at pH 4 (Sapers *et al.*, 1990). Wimalasiri and Wills (1983) also experienced interference from metaphosphoric acid in the measurement of dehydroascorbic acid at 214 nm and used, instead, citric acid as the extractant for fruit and vegetables.

For the determination of total vitamin C and/or total isovitamin C after pre-column reduction of the oxidized forms, the reducing agent dithiothreitol elutes near the solvent front, and hence does not interfere in the determination (Margolis and Black, 1987). An alternative reducing agent, homocysteine, produced a strongly absorbing doublet peak in front of the ascorbic acid peak, and so would mask the erythorbic acid peak. Tuan, Wyatt and Anglemier (1987) reported that phenylalanine could be used as an internal standard for determining total vitamin C in fruit and vegetable samples, but could not be used for determining total isovitamin C in meat products, owing to the presence of interfering peaks.

The column life of aminopropyl-bonded phases is somewhat limited compared with that of ODS-bonded reversed phases, owing to the loss of the amine function when it reacts irreversibly with the carbonyl groups of reducing sugars or other compounds to form Schiff bases. Bui-Nguyên (1985) reported that the resolution of ascorbic acid and erythorbic acid decreased after 50–100 injections using an aminopropyl-bonded column. The resolution was not significantly improved by increasing the percentage of methanol in the mobile phase (decreasing the polarity), but increasing the column temperature from 35 to 60 °C restored the separation.

Ion exclusion chromatography
Kim and Kim (1988) employed ion exclusion chromatography with amperometric detection to determine ascorbic acid in foods and beverages. The method was rapid (total analysis time, 5 min), sensitive (< 0.1 mg/100 g food sample detectable) and selective. A useful feature of this technique was its ability to determine simultaneously the sulphite present in fruit juices and instant mashed potato. The results obtained for these commodities using the AOAC titrimetric method were

higher than those obtained chromatographically, due to the interference by free sulphite. The method was extended to include the determination of dehydroascorbic acid after reduction to ascorbic acid with dithiothreitol (Kim, 1989). Dithiothreitol is electrochemically active, but its peak appears well after the ascorbic acid peak in the chromatogram, so it does not hinder the quantitative determination. A disadvantage of ion exclusion chromatography is that ascorbic acid and erythorbic acid are inseparable.

Reversed-phase chromatography

Lloyd *et al.* (1988a,b) examined the effects of changing ionic strength and pH on the separation of ascorbic acid and erythorbic acid on a PS-DVB copolymer column operated in the reversed-phase mode. The optimum mobile phase composition was found to be 0.2 M NaH_2PO_4, with the pH adjusted to 2.14 using 0.5 N HCl. This pH value is 2 pH units below the pK_a of 4.17 for ascorbic acid, in accordance with the principle of ion suppression. Under these conditions small changes in mobile phase buffer salt composition and pH would have a negligible effect on the solute capacity factor *(k)* and selectivity (α). This optimized separation was applied to the routine determination of ascorbic acid in fresh citrus fruit juices using absorbance detection at 220 nm. Vanderslice and Higgs (1993) used two PLRP-S copolymer columns connected in series and a pH 2.14 phosphate buffer mobile phase to separate dehydroerythorbic acid, dehydroascorbic acid, ascorbic acid and erythorbic acid. Post-column in-line chemistry involving oxidation and reaction with OPDA enabled total vitamin C and total isovitamin C in solid and liquid foods to be determined using fluorimetric detection.

Reversed-phase chromatography at around neutral pH (i.e. without ion suppression) can be utilized if DFQ is chromatographed after pre-column oxidation of ascorbic acid and derivatization (Speek, Schrijver and Schreurs, 1984).

Ion interaction chromatography

The tetrabutylammonium salts that are generally popular as ion interaction agents cause a relatively poor retention of ascorbic acid, but they have been used without an organic modifier to determine ascorbic acid and erythorbic acid in meat products (Kutnink and Omaye, 1987). Carnevale (1980) reported that these reagents are unsuitable for the analysis of citrus juices, because the relatively high concentration of ionizable material (mostly citric acid) exceeds their ion interaction capacity, resulting in loss of resolution, distortion of peak shapes and changes in elution volume.

Most published ion interaction chromatographic methods for determining vitamin C and related compounds use a more hydrophobic reagent with longer aliphatic chains, and methanol as organic modifier to achieve the desired separation and to impart the required solubility. The reagents represent a variety of primary, secondary or tertiary amines. Some ion interaction agents, namely n-octylamine, n-decylamine, hexadecyltrimethylammonium bromide (Kutnink and Omaye, 1987) and tridecylammonium formate (Watada, 1982; Lee and Marder, 1983), reportedly form precipitates (presumably phosphate complexes) when mixed with dilute (5% or 6%) metaphosphoric acid used in the extraction procedure. This precipitation causes the problem of pressure build-up within the HPLC system, which is not prevented by the addition of methanol to the mobile phase.

Moledina and Flink (1982) showed that buffering of the mobile phase to pH 4.2 (the pK_{a1} value of ascorbic acid) was essential for obtaining a sharp, well defined ascorbic acid peak. Maintaining the pH between 4.2 and 4.5 with a buffer salt would ensure that the ascorbic acid and the ion interaction reagent are in the ionic form, whilst also ensuring a sufficiently acidic medium in the interest of ascorbic acid stability. Potatoes and green bell peppers were the only samples among the fruits, juices and vegetables tested where the ascorbic acid values obtained by HPLC and the AOAC titrimetric method were essentially equal. The agreement obtained with these samples could be due to a lack of reducing compounds or to co-eluting substances, which absorb at the detection wavelength of 254 nm.

Zapata and Dufour (1992) used absorbance detection for the simultaneous determination of ascorbic acid and DFQ in selected foods and beverages after pre-column derivatization of dehydroascorbic acid. The detector wavelength was initially set at 348 nm and, after elution of DFQ, it was manually changed to 261 nm for ascorbic acid and erythorbic acid (internal standard) detection.

15.5.8 Capillary electrophoresis

Both capillary zone electrophoresis (CZE) and micellar electrokinetic capillary chromatography (MECC) have been applied to the determination of ascorbic acid in fruit juices and plant tissues (Table 15.2). Dehydroascorbic acid, being nonionizable, cannot itself be directly determined electrophoretically. However, total vitamin C can be determined by first reducing the dehydroascorbic acid to ascorbic acid by treatment with homocysteine (Chiari *et al.*, 1993) or dithiothreitol (Marshall, Trenerry and Thompson, 1995; Thompson and Trenerry, 1995). Davey, Bauw and van Montagu (1996), using an uncoated capillary, found that CZE gave superior resolution to comparable HPLC separations and a comparable analysis time (Figure 15.4).

Table 15.2 Capillary electrophoresis methods for the determination of vitamin C

Food	Sample preparation	Sample loading	Capillary	Separation buffer	Separation voltage	Compound separated	Detection	Reference
Capillary zone electrophoresis								
Lemon and orange juices	Filter (0.2 μm) and dilute in buffer solution	On cathodic side by electromigration at 8 kV for 6 s	Fused-silica, coated 20 cm × 25 μm i.d. Bio-Rad	0.1 M phosphate buffer, pH 5.0	8 kV	AA	UV 265 nm	Lin Ling et al. (1992)
Fruit beverages	Add EA internal standard. Dilute with HPO$_3$ (100 g/l) and filter (0.45 μm) by centrifugation	Pressure injection	Fused-silica, uncoated 37 cm × 75 μm i.d. (30 cm to detector window)	0.1 M tricine buffer, pH 8.8	11 kV (297 V/cm)	AA, EA (internal standard)	UV 254 nm	Koh, Bissell and Ito (1993)
Orange juice	*For AA*, add 12.5% TCA, centrifuge, filter *For total vitamin C*, adjust filtrate pH to 7 and add 0.8% DL-homocysteine (allow 15 min reaction time), filter	Hydrostatic pressure injection for 6 s	Fused-silica coated internally with polyacrylamide 40 cm × 100 μm i.d.	0.02 M phosphate buffer, pH 7.0	6 kV	AA (representing total vitamin C after treatment of sample extract with DL-homocysteine)	UV 254 nm	Chiari et al. (1993)
Plant tissues	Pulverize in liquid N$_2$. Extract twice with 3% HPO$_3$/1 nM EDTA, centrifuge. Pass through C$_{18}$ Sep-Pak solid-phase extraction cartridge	Hydrostatic (N$_2$) pressure injection	Fused-silica, uncoated 57 cm × 75 μm i.d. (50 cm to detector window) T = 25 °C	0.2 M borate buffer, pH 9	25 kV	AA, EA	UV 260 nm	Davey, Bauw and van Montagu (1996)
Micellar electrokinetic capillary chromatography								
Fruits and vegetables	Blend with 3% HPO$_3$, filter. Add 0.2% DL-dithiothreitol containing EA (internal standard). Pass through C$_{18}$ Sep-Pak solid-phase extraction cartridge, filter'(0.8 μm)	Under vacuum	Fused-silica *For fruits* 65 cm ×75 μm i.d. (40 cm to detector window) *For vegetables* 75 cm × 75 μm i.d. (50 cm to detector window) T = 28 °C	0.05 M sodium deoxycholate (surfactant), 0.01 M sodium borate and 0.01 M KH$_2$PO$_4$	25 kV	AA (representing total vitamin C), EA (internal standard)	UV 254 nm	Thompson and Trenerry (1995)

Abbreviations: see footnote to Table 2.4.
HPO$_3$, metaphosphoric acid.

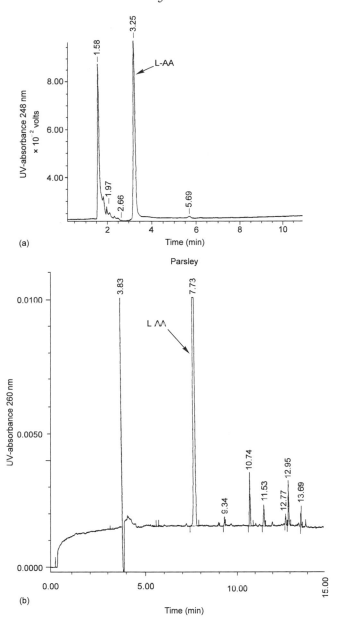

Figure 15.4 (a) HPLC and (b) capillary zone electrophoresis (CZE) analyses of a parsley extract. Reversed-phase HPLC was performed on a Bio-Rad C$_{18}$, 3 μm, 250 × 4.6 mm i.d. column with phosphoric acid/0.1 nM EDTA (pH 2.5) as the mobile phase. Conditions for CZE as given in Table 15.2. (Reproduced with permission of the authors and publisher from Davey, Bauw and van Montagu, (1996) *Anal. Biochem.*, **239**, 8–19. Published by Academic Press Inc. Florida, USA.)

15.5.9 Enzymatic method

Tsumura *et al.* (1993) developed a method of ascorbic acid determination which directly measures the change in absorbance of ascorbic acid during oxidation by guaiacol peroxidase. In contrast to ascorbate peroxidase, which is unstable and commercially unavailable in purified form, guaiacol peroxidase, extracted from horseradish, is stable and available at reasonable cost. In the procedure described, food samples were extracted with 2% metaphosphoric acid, then centrifuged and filtered. Into a quartz spectrophotometer cuvette were mixed an aliquot of the sample filtrate or ascorbic acid standard solution, M/30 phosphate buffer (pH 7.0) and guaiacol peroxidase in phosphate buffer containing 1.81 mM EDTA and 0.13 mM 2-mercaptoethanol. The initial absorbance at 265 nm was recorded with a spectrophotometer and the reaction was initiated by adding 50 mM hydrogen peroxide. Temperature was controlled at 37 °C by circulating water around the cuvette. The decrease in absorbance at 265 nm due to oxidation of ascorbic acid to dehydroascorbic acid was recorded until absorbance reached the final value. The difference between initial and final absorbance corresponding to ascorbic acid was then calculated. Quantitation was carried out by reference to a calibration graph of ascorbic acid concentration to absorbance at 265 nm using pH 7.0 phosphate buffer and 0.05 mg/ml peroxidase at 37 °C. No interference was seen for any of 30 different food samples tested, including vegetables, fruits, ham, liver and jam. Erythorbic acid also acts as a substrate for guaiacol peroxidase and, if present in a processed food, will interfere with the assay. The method was applicable to coloured and/or sugar-rich samples, and was more precise than officially adopted chemical methods.

15.5.10 Flow-injection analysis

Vanderslice and Higgs (1989) combined a flow-injection system with a robotic extraction procedure to automate the determination of total vitamin C in fruits, juices and vegetables from after sample weighing to final quantification. The procedure was based on the AOAC (1990b) microfluorimetric method, except that mercury(II) chloride was used as the oxidizing agent.

Lázaro, Luque de Castro and Valcárel (1987) proposed a flow-injection method for the simultaneous determination of ascorbate and sulphite in soft drinks based on the reaction with chloramine-T. The sample, dissolved in an acidic medium, is injected into the chloramine-T stream after merging with a starch–iodide stream. Iodide forms HI in an acidic medium, which subsequently reacts with chloramine-T to liberate iodine. The iodine oxidizes ascorbate and sulphite and, as soon as these reducing analytes have disappeared, the remaining iodine binds to starch to form the starch–iodine complex, whose absorbance is monitored at 581 nm. Maximum absorbance is obtained with the blank sample because no iodide is consumed when the reducing analytes are absent; thus the difference in absorbance between the blank and sample solutions is related to the concentration of analyte. The absorbance due to sulphite alone is determined by simultaneously injecting a sample previously mixed with NaOH to destroy the ascorbate. The ascorbic acid content can then be calculated by subtracting the result of the NaOH-treated sample from the untreated sample.

Using simple flow-injection apparatus, Elgilany Elbashier and Greenway (1990) determined ascorbic acid in beverages by monitoring fluorimetrically the reduction of cerium(IV) to cerium(III) by the action of ascorbic acid in acidic conditions.

Greenway and Ongomo (1990) devised a flow-injection system for determining ascorbic acid in fruit and vegetable juices using an amperometric detector and an on-line column containing ascorbate oxidase immobilized on activated Sepharose 4B. On passage through the immobilized enzyme, a fraction of the ascorbic acid was converted into the electrochemically inactive dehydroascorbic acid, and the decrease in signal compared with that obtained using a blank (no enzyme) column was measured. The signal from the blank column represented total oxidizable material, while the signal from the enzyme column represented total oxidizable material less ascorbic acid. The difference between the two signals therefore gave the ascorbic acid content, with the enzyme reaction providing the selectivity. Glucose, oxalic acid, citric acid and tartaric acid did not seriously interfere, even in great excess, while interference from copper was eliminated by addition of EDTA. The immobilized enzyme was used continuously each day for a period of 3 weeks without significant loss in activity.

In a flow-injection method for determining total vitamin C (Daily *et al.*, 1991), food samples were extracted with 15 mM phosphate buffer (pH 5.0) containing 1 mM dithiothreitol, which reduced dehydroascorbic acid present in the sample to ascorbic acid whilst also stabilizing the ascorbic acid. Initially, an aliquot of each sample extract was taken and divided into two. The first aliquot was passed through an enzyme packed-bed (ascorbate oxidase immobilized on aminopropyl controlled-pore glass beads) that had been previously heat-denatured. An amperometric signal, proportional to the amount of ascorbic acid plus other electro-oxidizable species (interferents) present in the sample, was produced at the electrochemical detector. The second aliquot was then passed through a similar bed containing active enzyme, where ascorbic acid was selectively removed giving a second signal attributable only to the interfering species. The difference between the two signals generated at the detector electrode could be related to the concentration of ascorbic acid, representing total vitamin C. With the automated instrumentation, the method provided high sample throughout (15 samples per hour).

15.6 BIOAVAILABILITY

15.6.1 Physiological aspects

Intestinal and renal transport of vitamin C are similar in several respects and the model shown in Figure 15.5 can be applied to both situations.

Intestinal absorption

The enterocytes of animal species that require dietary vitamin C possess separate transport mechanisms for taking up L-ascorbate and dehydro-L-ascorbic acid at the brush border. Inside the cell, dehydroascorbic acid is enzymatically reduced and the accumulated L-ascorbic acid is transported across the basolateral membrane to the bloodstream. In addition to brush-border uptake, dehydroascorbic acid from the bloodstream can be taken up at the basolateral membrane, reduced within the cell, and returned to the circulation in the form of the useful and nontoxic L-ascorbic acid. The serosal uptake of dehydroascorbic acid from the bloodstream and intracellular reduction to ascorbic acid take place in animal species which do not have a dietary vitamin C requirement as well as those species that do. The ability of the enterocyte to absorb dehydroascorbic acid efficiently is important because, apart from the indigenous dehydroascorbic acid content of the diet, additional oxidation of ascorbic acid occurs in the gastrointestinal tract as the vitamin functions to maintain other nutrients such as iron in the reduced state. The intestinal uptake and reduction of dehydroascorbic acid explains why this compound, orally administered, maintains plasma concentrations of ascorbic acid and prevents scurvy. The overall system of intestinal transport and

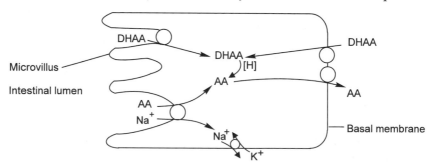

Figure 15.5 Model of transport and metabolism of vitamin C in enterocyte of vitamin C-dependent animals.

metabolism is designed to maximize the conservation of vitamin C and also to maintain the vitamin in its nontoxic reduced state whether it is derived from the diet or from the circulation.

The transport mechanisms for the intestinal absorption of physiological intakes of L-ascorbic acid in humans and other vitamin C-dependent species are similar to those involved in the transport of sugars. Absorption takes place mainly in the ileum and occurs as a result of specific carrier-mediated mechanisms in the brush-border and basolateral membranes of enterocytes (Bianchi, Wilson and Rose, 1986). Ascorbic acid is highly ionized within the pH range of intestinal chyme, and therefore it is the ascorbate anion that is transported. Sodium-dependent transport across the brush-border membrane is electrically neutral, as opposed to the electrogenic transport of glucose, which indicates that the anionic charge is balanced by the simultaneous transfer of a sodium cation, i.e. a $1:1$ Na^+/ascorbate$^-$ transport stoichiometry (Siliprandi *et al.*, 1979). The brush-border transport system is competitively inhibited by D-isoascorbic acid (Toggenburger, Landoldt and Semenza, 1979). The immediate source of energy for the uptake of L-ascorbate by the cell is the sum of the concentration gradients of ascorbate and sodium, rather than the sodium electrochemical potential gradient that drives glucose uptake. The sodium concentration gradient is maintained by continuous action of the basolateral sodium pump. L-Ascorbate leaves the enterocyte by sodium-independent, facilitated diffusion at the basolateral membrane. Because the anionic charge is not balanced by simultaneous transfer of a sodium cation, ascorbate transport at the basolateral membrane is sensitive to the membrane potential.

Rose and Nahrwold (1978) found that high oral doses (5 or 25 times normal) of ascorbic acid for 14 or 28 days significantly reduced the influx of ascorbate from the lumen into the ileal mucosa of guinea pigs. The influx rate was reduced by intramuscular injection of ascorbic acid, suggesting that the transport mechanism may respond to circulating levels of the vitamin. Guinea pigs which were fed an ascorbic acid-free diet had a lower rate of ascorbate influx after 28 days compared with control animals, at which time they developed scurvy. If the reduction in intestinal absorption following high ascorbate intake were irreversible, then this could contribute to systemic conditioning, also known as rebound scurvy. In this condition, humans previously ingesting high levels of ascorbate display deficiency symptoms when later consuming lower normal levels of ascorbate (Rivers, 1987). Karasov, Darken and Bottum (1991) reported that the suppression of ascorbic acid absorption in guinea pigs was reversible within 7 days. They concluded that decreased bioavailability of dietary ascorbate is probably not a major factor in the development of rebound scurvy in humans, at least in adults. A more likely cause of rebound scurvy is an induction of ascorbate-metabolizing enzymes.

Dehydro-L-ascorbic acid enters the enterocyte at both the brush-border and basolateral membranes by means of sodium-independent facilitated diffusion (Rose, Choi and Koch, 1988). Dehydroascorbic acid, lacking the dissociable protons at the carbon 2 and 3 positions, is electrically neutral under physiological conditions. The transport processes at both sites are therefore insensitive to the membrane potential, the driving force for cellular uptake being the steep concentration gradient maintained by the intracellular reduction of dehydroascorbic acid.

Renal reabsorption

The kidney participates with the intestine in maximizing vitamin C conservation in the body. The kidneys of all mammals handle vitamin C in a similar manner. Renal reabsorption of vitamin C is an essential process for humans as urinary loss would far exceed the average daily intake of the vitamin. Although species that have the ability to synthesize ascorbic acid might be able to replace that lost in the urine, the metabolic costs would be high.

Uptake of the L-ascorbate anion at the brush-border membrane of the absorptive cell of the proximal convoluted tubule, in common with intestinal uptake of ascorbate in the guinea pig or human, is a sodium-coupled secondary active transport system. D-Isoascorbate is a competitive inhibitor. The velocity of renal uptake far exceeds that of intestinal uptake, which is logical considering that the transit time of ascorbate in the proximal tubule is only about 10 s. Whereas the intestinal transport system is electroneutral owing to a $1:1$ Na^+/ascorbate$^-$ flux stoichiometry, the renal transport system is electrogenic and displays a $2:1$ Na^+/ascorbate$^-$ stoichiometry (Toggenburger *et al.*, 1981). In this respect, the renal transport system resembles that for the intestinal uptake of glucose; that is, both the membrane potential and the sodium concentration gradient provide the immediate driving force.

Dehydro-L-ascorbic acid is taken up at both brush-border and basolateral membranes by an electroneutral, sodium-independent facilitated diffusion mechanism. A favourable gradient for continued reabsorption is maintained by intracellular reduction of recently reabsorbed dehydroascorbic acid. Recently absorbed or regenerated L-ascorbic acid leaves the absorptive cell by sodium-independent facilitated diffusion at the basolateral membrane (Bianchi and Rose, 1985a,b; Rose, 1989).

Other metabolic considerations

Vitamin C circulates in the bloodstream mainly in the free (nonprotein-bound) and reduced form; c.5% of the circulating vitamin is dehydroascorbic acid. About 70% of blood-bound ascorbate is present in plasma and erythrocytes. The remainder is in leucocytes (white cells) which have the ability to concentrate the vitamin from the plasma. Uptake of ascorbic acid by human leucocytes is a stereospecific, sodium-dependent, active transport process that shows different transport kinetics for different cell types (Moser, 1987).

Vitamin C is widely distributed to the cells of the body, which all take up and utilize the vitamin. Leucocytes and the adrenal and pituitary glands have the highest tissue concentrations. High concentrations are also found in the liver, spleen and brain. Skeletal muscle contains much of the body's pool of ascorbate, although the concentration is relatively low. In humans, the metabolic fate of vitamin C is urinary excretion of ascorbic acid or metabolites.

15.6.2 Dietary sources and their bioavailability

Dietary sources

The ascorbic acid and dehydroascorbic acid contents of some vegetables and fruits are listed in Table 15.3. The values shown are typical of the observed concentrations found in these samples, but they can vary greatly (Table 15.4) and should not be taken as absolute. Genetic variation, maturity, climate, sunlight, method of harvesting and storage can all affect the levels of vitamin C.

Fresh fruits (especially citrus fruits and blackcurrants) and green vegetables constitute rich sources of vitamin C. Potatoes contain moderate amounts but, because of their high consumption, represent the most important source of the vitamin in the British diet. Liver (containing 10–40 mg/100 g), kidney and heart are good sources, but muscle meats and cereal grains do not contain the vitamin. Human milk provides enough ascorbic acid to prevent scurvy in breast-fed infants, but preparations of cow's milk are a poor source owing to oxidative losses incurred during processing. Cabbage and other brassica contain a bound form of ascorbic acid known as ascorbigen, which exhibits 15–20% bioavailability in guinea pigs (Matano and Kato, 1967).

Bioavailability

The efficiency of vitamin C absorption of foods over a range of usual intakes (20–120 mg/day) is c.90% (Olson and Hodges, 1987). Absorption by the brush-border carrier mechanism reaches its maximum rate at a relatively low luminal concentration. Higher pharmacological concentrations are absorbed additionally by passive diffusion, which proceeds at a very low rate. Absorption therefore becomes progressively less efficient with increasing dose levels, the efficiency falling from 50% of a

Table 15.3 Vitamin C content of some vegetables and fruits (from Vanderslice *et al.*, 1990)

Food	Concentration (mg/100 g)[ab]		
	AA	DHAA	Total
Vegetables			
Broccoli			
fresh, raw	89.0 ± 2.0	7.7 ± 0.6	97 ± 2
boiled	37.0 ± 1.0	2.6 ± 0.6	40 ± 1
microwaved	111.0 ± 2.0	4.7 ± 0.6	116 ± 2
Cabbage			
fresh, raw	42.3 ± 3.4	–	42 ± 3
boiled	24.4 ± 1.6	–	24 ± 2
Cauliflower, fresh, raw	54.0 ± 1.0	8.7 ± 0.6	63 ± 1
Spinach			
fresh, raw	52.4 ± 2.5	–	52 ± 3
boiled	19.6 ± 1.0	–	20 ± 1
microwaved (4 min)	48.3 ± 3.7	5.8 ± 0.6	54 ± 4
Peppers			
red, fresh	151.0 ± 3.0	4.0 ± 1.0	155 ± 4
green, fresh	129.0 ± 1.0	5.0 ± 0.0	134 ± 1
Potatoes (without skin)			
raw	8.0 ± 0.0	3.0 ± 0.0	11 ± 0
boiled	7.0 ± 1.0	1.3 ± 0.6	9 ± 1
Tomatoes, fresh	10.6 ± 0.6	3.0 ± 0.0	14 ± 1
Fruits			
Bananas	15.3 ± 2.5	3.3 ± 0.6	19 ± 3
Grapefruit, fresh	21.3 ± 0.6	2.3 ± 0.6	24 ± 1
Oranges			
Florida	54.7 ± 2.5	8.3 ± 1.2	63 ± 3
California navel	75 ± 4.5	8.2 ± 1.6	83 ± 5

[a] Values reported are mean ± standard deviation based on three measurements. When no values are listed, the concentration was < 1 mg/100 g sample.
[b] Foods analysed by HPLC (Vanderslice and Higgs, 1988, 1990) together with robotic extraction procedures (Vanderslice and Higgs, 1985).

single 1.5 g dose to 16% of a single 12 g dose (Kübler and Gehler, 1970). Ingestion of single doses greater than 200 mg results in post-absorptive degradation of ascorbic acid in the intestine to carbon dioxide, which is expired in the breath (Kallner, Hornig and Pellikka, 1985). Absorption efficiency is increased by the ingestion of several spaced doses throughout the day, rather than ingestion of a single megadose (Mayersohn, 1972). Ingesting the vitamin in a sustained release form also improves absorption efficiency. For example, synthetic ascorbic acid in a natural citrus extract containing bioflavonoids, proteins and carbohydrates was found to be 35% more bioavailable ($P < 0.001$) than ascorbic acid alone in human subjects (Vinson and Bose, 1988).

Table 15.4 Range of vitamin C values in some vegetables and fruits (from Vanderslice and Higgs, 1991)

Sample	Total vitamin C content (AA + DHAA) (mg/100 g)
Broccoli, raw	97–163
Cabbage, raw	42–83
Spinach, fresh	25–70
Potatoes, Idaho	11–13
Tomatoes	14–19
Bananas	12–19
Grapefruit, red	21–31
Oranges	
Florida	53–63
California navel	52–78

In a human study involving 68 adult male nonsmokers, Mangels *et al.* (1993) examined the relative bioavailability of ascorbic acid from several sources. Subjects underwent two 8-week ascorbic acid depletion–repletion cycles. In repletion, subjects were randomized to receive 108 mg of ascorbic acid per day as a tablet taken with or without an iron tablet (63 mg of ferrous fumarate to release 20 mg of elemental iron), as orange segments or juice, or as raw or cooked broccoli. The experiment was designed with a crossover within each major treatment group (e.g. cooked to raw broccoli) for the second repletion. The response to the tested source of ascorbic acid was quantified as the rate of change of plasma ascorbate concentration over the first 3 weeks of each repletion period. The relative bioavailability of ascorbic acid from the various sources was determined by comparing these responses. Statistical analysis of the data showed no significant overall differences in ascorbic acid bioavailability among the three main sources of the vitamin (i.e. tablets, orange and broccoli). In addition, the bioavailability of ascorbic acid in the tablet alone did not differ from that in the tablet plus iron, and there was no difference in bioavailability between orange segments and juice. The bioavailability of ascorbic acid from raw broccoli was about 20% lower than from cooked broccoli. The lack of a significant effect of the iron supplement on the bioavailability of synthetic ascorbic acid suggests that the presence of iron has no influence on ascorbate absorption or its stability in the intestinal lumen prior to absorption. This is rather surprising, as ascorbic acid is known to enhance the absorption of nonhaem iron when the two nutrients are ingested together (Hallberg, 1981). In summary, the observation that ascorbic acid bioavailability from the fruit and vegetable sources examined was not significantly different from that from synthetic ascorbic acid indicates that the bioavailability of vitamin C in a typical American diet is high.

Effects of dietary fibre

Keltz, Kies and Fox (1978) studied the effects of dietary fibre constituents (pectin, cellulose and hemicellulose) on the urinary excretion of ascorbic acid in human subjects. Diets supplemented with hemicellulose material (14.2 g/day) resulted in increased excretion, suggesting an enhanced absorption of the vitamin. Conversely, pectin supplementation (14.2 g/day) resulted in decreased excretion, reflecting a less efficient absorption. Supplementation with cellulose had no significant effect.

Effects of alcohol

In a study of the acute effects of alcohol on plasma ascorbic acid in healthy subjects, Fazio, Flint and Wahlqvist (1981) observed the increase in plasma ascorbate concentrations over fasting levels after the ingestion of 2.0 g of ascorbic acid and breakfast. When 35 g of ethanol was ingested with ascorbic acid and breakfast, plasma concentrations were significantly lower for at least 24 h. This effect was probably due to an impairment in absorption of ascorbic acid by ethanol. These findings indicated that ethanol may reduce the availability of ascorbic acid from food and predispose to deficiency of the vitamin. Whether ethanol increases the excretion, catabolism or utilization of ascorbic acid is not known (Lieber, 1988).

15.6.3 Effects of food processing, storage and cooking

Ascorbic acid is very susceptible to chemical and enzymatic oxidation during the processing, cooking and storage of food. Thus a significant proportion of vitamin C in our diet is present in the form of dehydroascorbic acid. Ascorbic acid oxidase, which promotes the direct oxidation of ascorbic acid in plant tissues, exhibits maximum activity at 40 °C and is almost completely inactivated at 65 °C (Bender, 1979). Hence rapid heating, such as the blanching of fruits and vegetables or the pasteurization of fruit juices, prevents the action of this enzyme during post-process storage. Other plant enzymes, including phenolase (formerly called polyphenol oxidase), cytochrome oxidase and peroxidase, are indirectly responsible for ascorbic acid loss (Erdman and Klein, 1982). The ascorbic acid in plant tissue is protected from these enzymes by cellular compartmentation. However, when tissues are disrupted after bruising, wilting, rotting, or during advanced stages of senescence, the enzymes gain access to the vitamin and begin to destroy it.

Chemical oxidation of ascorbic acid is lowered during processing by carrying out vacuum deaeration and inert gas treatment where feasible. The headspace in containers should be minimized and hermetically sealed systems used. Ascorbic acid is very stable in canned or bottled foods after the oxygen in the headspace has been used up, and provided the food is not subjected to high-temperature storage or exposed to light. The vitamin also keeps well in frozen storage. In contrast to glass containers, plastic bottles and cardboard cartons are permeable to oxygen, so a lowered vitamin C retention is to be expected. Bronze, brass, copper and iron equipment should be avoided, whilst sequestering agents such as EDTA, polyphosphates and citrates prevent the catalytic action of traces of copper and iron. The sulphites and metabisulphites, which are added to juices or beverages as a source of SO_2, exert a stabilizing effect on ascorbic acid in addition to their role as antimicrobial agents. The addition of food-grade antioxidants such as butylated hydroxyanisole (BHA), butylated hydroxytoluene (BHT) and propyl gallate also protect the vitamin.

Vitamin C in freshly secreted cow's milk is predominantly in the form of ascorbic acid, but this is rapidly oxidized by the dissolved oxygen content. If milk is subjected to light, the oxidation process is greatly accelerated due to the catalytic effect of lumichrome and lumiflavin produced by the photochemical destruction of riboflavin. Losses of vitamin C activity during HTST and ultra-HTST treatment of milk average 20%, but losses incurred during in-bottle sterilization of milk are higher, at 40–79% (Cremin and Power, 1985).

Ascorbic acid can leach away from fruit and vegetables during processing or cooking. This is of little importance with canned, bottled or stewed fruits where the juice is eaten with the tissue, but may represent a serious loss with vegetables, where the liquor is drained away before serving. If vegetables are steamed or pressure-cooked instead of boiled, the leaching effect is greatly reduced, but a greater loss of ascorbic acid is to be expected from oxidation. Cold water washing or steeping does not normally leach out a significant amount of the vitamin in whole undamaged fruits and vegetables. In jam making, when the fruit is boiled with sugar, ascorbic acid is remarkably stable (Olliver, 1967).

15.7 DIETARY INTAKE

In human adults, 10 mg/day of vitamin C is sufficient to prevent overt symptoms of scurvy, but this level of intake is insufficient to provide measurable plasma concentrations of the vitamin. Vitamin C begins to appear in plasma at an intake of about 30 mg/day and reaches a maximum concentration at an intake of about 70 mg/day. At 40 mg/day, measurable amounts of ascorbic acid are present in the plasma and this intake has been selected as the Reference Nutrient Intake for adults of both sexes in the United Kingdom. A 40 mg/day intake will maintain an exchangeable body pool of about 900 mg and prevent signs of scurvy for at least 4 weeks, even on a zero intake during this period (Department of Health, 1991).

In the United States, the 1989 RDA for vitamin C of 60 mg/day for adults is based on maintenance of a body pool of 1500 mg (National Research Council, 1989). At this level, saturation of tissue binding and maximal rates of metabolic and renal tubular absorption seem to be approached. The 1989 RDA of 100 mg/day for smokers is higher because the plasma concentration of vitamin C is lowered by the use of cigarettes, apparently due to increased vitamin turnover (Kallner, Hartmann and Hornig, 1981). Plasma ascorbate levels of women decrease during pregnancy and the RDA includes a 10 mg/day increment for pregnant women. Plasma ascorbate levels of the foetus and neonate are some 50% higher than maternal levels, indicating active transplacental transport and a relatively larger body pool of the vitamin. A daily increment of 35 mg is recommended during the first 6 months of lactation, and 30 mg thereafter.

The amount of vitamin C that prevents deficiency with a margin of safety may or may not be similar to the amount needed for optimal human health. Levine *et al.* (1995) defined optimal vitamin C requirements operationally based on the following: dose–function relations, availability for ingestion in the diet, steady-state concentrations in plasma and tissues achieved at each dose of vitamin C, urinary excretion, bioavailability, toxicity, and epidemiological observations in populations.

Estimates of vitamin C intake based on amounts of the vitamin in the diet may be considerably higher than actual intakes because of oxidative destruction and cooking losses.

Effects of high intakes

Ascorbic acid is generally regarded as being nontoxic (Rivers, 1987). When excessive pharmacological doses are taken, almost all of the ingested amount is excreted in the urine. In healthy individuals daily doses of up to at least 1 g may be taken over a period of weeks without the manifestations of toxic symptoms (Hornig and Moser, 1981). Massive oral doses (several grams) of vitamin C daily are likely to produce diarrhoea from the osmotic effects of unabsorbed vitamin in the intestinal lumen. Excessive daily amounts can also cause an increased production of oxalic acid in some individuals, leading to an increased risk of kidney stone formation.

Assessment of nutritional status

The measurement of plasma ascorbic acid concentrations reflects recent dietary intake of vitamin C and is the most practical test. Since > 95% of plasma vitamin C exists as ascorbic acid, it is valid to measure only this reduced form; thus the assay is simple. In general, plasma or serum ascorbic acid values $< 0.20\,mg/100\,ml$ are considered to represent frank deficiency (biochemical and/or clinical symptoms). Values between 0.2 and $0.4\,mg/100\,ml$ represent marginal status, i.e. moderate risk of developing clinical signs of vitamin C deficiency due to low vitamin intakes or depleted body pool (Jacob, 1990).

15.8 SYNOPSIS

Vitamin C is considered the most labile of the vitamins in our food supply. Oxidation can occur in the presence of metal catalysts or plant oxidase enzymes, particularly following cell damage, or as a result of food processing. Vitamin C is easily leached from foods during processing and is discarded with washing, soaking or cooking water. Ascorbic acid losses begin with harvesting and continue through handling, industrial or home preparation, cooking, and storage of plant foods.

The efficiency of vitamin C absorption of foods in typical Western diets is high (c.90%), indicating a high bioavailability.

Vitamin C activity is less of a problem to determine than that of the B-group vitamins, as it occurs mainly in unbound water-extractable form, and in milligram amounts per $100\,g$ in fruits and vegetables. Ascorbic acid and its immediate oxidation product, dehydroascorbic acid, possess equal biological activity and are readily interconvertible, thus permitting the measurement of only one compound.

The AOAC dye titration method is simple and rapid, but it does not account for dehydroascorbic acid. Moreover, the end-point is difficult or impossible to observe when sample extracts are coloured, and the method is subject to interference from sulphites and many naturally occurring reducing substances present in foods. The AOAC microfluorimetric method provides a value for total vitamin C and is less susceptible to interference, but it is a lengthy and tedious method when performed manually. A semi-automated microfluorimetric method has been adopted by the AOAC using segmented-flow analysis. Other titrimetric methods and colorimetric methods have fallen into disuse.

Many HPLC methods have been published, reflecting the popularity of this technique for vitamin C analysis. HPLC is accurate and sensitive for the determination of ascorbic acid, but it is less than ideal for determining total vitamin C. The poor optical absorptivity of dehydroascorbic acid and its lack of electrochemical activity is a drawback, and necessitates the chemical reduction of this compound to ascorbic acid. Fluorescence detection can be used, but the lack of inherent fluorescence necessitates the chemical oxidation of ascorbic acid to dehydroascorbic acid, and subsequent derivatization. These chemical manipulations somewhat defeat the purpose of HPLC, as one of its principal advantages over gas chromatography is that derivatization is not usually necessary. Capillary electrophoresis offers an alternative to HPLC and eliminates the need for organic solvents and expensive chromatography columns.

Flow-injection systems provide rapid sample analysis with high sample throughput and could find useful application for the commercial analysis of fruit juices. The use of immobilized enzyme and amperometric detection confers high specificity. Sample and mobile phase purity are less important than they are when HPLC is used and an automated system can be constructed using relatively simple apparatus.

REFERENCES

Albrecht, J.A. and Schafer, H.W. (1990) Comparison of two methods of ascorbic acid determination in vegetables. *J. Liquid Chromat*, **13**, 2633–41.

AOAC (1990a) Vitamin C (ascorbic acid) in vitamin preparations and juices. 2, 6-Dichloroindophenol titrimetric method. Final action 1968. In *AOAC Official Methods of Analysis*, 15th edn (ed. K. Helrich), Association of Official Analytical Chemists, Inc., Arlington, VA, 967.21.

AOAC (1990b) Vitamin C (ascorbic acid) in vitamin preparations. Microfluorometric method. Final action 1968. In *AOAC Official Methods of Analysis*, 15th edn (ed. K. Helrich), Association of Official Analytical Chemists, Inc., Arlington, VA, 967.22.

AOAC (1990c) Vitamin C (total) in food. Semiautomated fluorometric method. Final action 1985. In *AOAC Official Methods of Analysis*, 15th edn (ed. K. Helrich), Association of Official Analytical Chemists, Inc., Arlington, VA, 984.26.

Ashoor, S.H., Monte, W.C. and Welty, J. (1984) Liquid chromatographic determination of ascorbic acid in foods. *J. Ass. Off. Analyt. Chem.*, **67**, 78–80.

Association of Vitamin Chemists, Inc. (1966) *Methods of Vitamin Assay*, 3rd edn, Interscience Publishers, New York.

Augustin, J., Beck, C. and Marousek, G.I. (1981) Quantitative determination of ascorbic acid in potatoes and potato products by high performance liquid chromatography. *J. Food Sci.*, **46**, 312–3, 316.

Ball, G.F.M. (1994) *Water-soluble Vitamin Assays in Human Nutrition*, Chapman & Hall, London.

Behrens, W.A. and Madère, R. (1994) A procedure for the separation and quantitative analysis of ascorbic acid, dehydroascorbic acid, isoascorbic acid, and dehydroascorbic acid in food and animal tissue. *J. Liquid Chromat.*, **17**, 2445–55.

Bender, A.E. (1979) The effects of processing on the stability of vitamins in foods. In *Proceedings of the Kellogg Nutrition Symposium*, London, 14–15th Dec., 1978 (ed. T.G. Taylor), MTP Press, Lancaster, pp. 111–25.

Bianchi, J. and Rose, R.C. (1985a) Na^+-independent dehydro-L-ascorbic acid uptake in renal brush-border membrane vesicles. *Biochim. Biophys. Acta*, **819**, 75–82.

Bianchi, J. and Rose, R.C. (1985b) Transport of L-ascorbic acid and dehydro-L-ascorbic acid across renal cortical basolateral membrane vesicles. *Biochim. Biophys. Acta*, **820**, 265–73.

Bianchi, J., Wilson, F.A. and Rose, R.C. (1986) Dehydroascorbic acid and ascorbic acid transport systems in the guinea pig ileum. *Am. J. Physiol.*, **250**, G461–8.

Bliss, C.I. and György, P. (1967) Ascorbic acid. III. Animal assay. In *The Vitamins. Chemistry, Physiology, Pathology, Methods*, 2nd edn, Vol. VII (eds P. György and W.N. Pearson), Academic Press, New York, pp. 49–51.

Bradbury, J.H. and Singh, U. (1986) Ascorbic acid and dehydroascorbic acid content of tropical root crops from the South Pacific. *J. Food Sci.*, **51**, 975–8, 987.

Bui-Nguyên, M.H. (1985) Ascorbic acid and related compounds. In *Modern Chromatographic Analysis of the Vitamins* (eds A.P. De Leenheer, W.E. Lambert and M.G.M. De Ruyter), Marcel Dekker, Inc., New York, pp. 267–301.

Bushway, R.J., Bureau, J.L. and McGann, D.F. (1984) Determinations of organic acids in potatoes by high performance liquid chromatography. *J. Food Sci.*, **49**, 75–7, 81.

Bushway, R.J., King, J.M., Perkins, B. and Krishnan, M. (1988) High performance liquid chromatographic determination of ascorbic acid in fruits, vegetables and juices. *J. Liquid Chromat.*, **11**, 3415–23.

Carnevale, J. (1980) Determination of ascorbic, sorbic and benzoic acids in citrus juices by high-performance liquid chromatography. *Food Technol. Aust.*, **32**, 302, 304–5.

Chiari, M., Nesi, M., Carrea, G. and Righetti, P.G. (1993) Determination of total vitamin C in fruits by capillary zone electrophoresis. *J. Chromat.*, **645**, 197–200.

Christie, A.A. (1975) Analysis for selected vitamins in a nutritional labelling programme. *Inst. Food Sci. Technol. Proc.*, **8**, 163–8.

Crawford, T.C. and Crawford, S.A. (1980) Synthesis of L-ascorbic acid. *Adv. Carbohydr. Chem.*, **37**, 79–155.

Cremin, F.M. and Power, P. (1985) Vitamins in bovine and human milks. In *Developments in Dairy Chemistry*, Vol. III (ed. P.F. Fox), Elsevier Applied Science Publishers, London and New York, pp. 337–98.

Daily, S., Armfield, S.J., Haggett, B.G.D. and Downs, M.E.A. (1991) Automated enzyme packed-bed system for the determination of vitamin C in foodstuffs. *Analyst, Lond.*, **116**, 569–72.

Daood, H.G., Biacs, P.A., Dakar, M.A. and Hajdu, F. (1994) Ion-pair chromatography and photodiode-array detection of vitamin C and organic acids. *J. Chromatogr. Sci.*, **32**, 481–7.

Davey, M.W., Bauw, G. and van Montagu, M. (1996) Analysis of ascorbate in plant tissues by high-performance capillary zone electrophoresis. *Analyt. Biochem.*, **239**, 8–19.

Dennison, D.B., Brawley, T.G. and Hunter, G.L.K. (1981) Rapid high-performance liquid chromatographic determination of ascorbic acid and combined ascorbic acid–dehydroascorbic acid in beverages. *J. Agric. Food Chem.*, **29**, 927–9.

Department of Health (1991) *Dietary Reference Values for Food Energy and Nutrients for the United Kingdom.* Report on Health and Social Subjects, No. 41, HM Stationery Office, London.

Deutsch, M.J. (1967) Assay for vitamin C: a collaborative study. *J. Ass. Off. Analyt. Chem.*, **50**, 798–806.

Deutsch, M.J. and Weeks, C.E. (1965) Microfluorimetric assay for vitamin C. *J. Ass. Off. Analyt. Chem.*, **48**, 1248–56.

DeVries, J.W. (1983) Semiautomated fluorometric method for the determination of vitamin C in foods: a collaborative study. *J. Ass. Off. Analyt. Chem.*, **66**, 1371–6.

Diop, P.A., Franck, D., Grimm, P. and Hasselmann, C. (1988) High-performance liquid chromatographic determination of vitamin C in fresh fruits from West Africa. *J. Food Comp. Anal.*, **1**, 265–9.

Dodson, K.Y., Young, E.R. and Soliman, A.-G.M. (1992) Determination of total vitamin C in various food matrixes by liquid chromatography and fluorescence detection. *J. AOAC Int.*, **75**, 887–91.

Doner, L.W. and Hicks, K.B. (1981) High-performance liquid chromatographic separation of ascorbic acid, erythorbic acid, dehydroascorbic acid, dehydroerythorbic acid, diketogulonic acid, and diketogluconic acid. *Analyt. Biochem.*, **115**, 225–30.

Dunmire, D.L., Reese, J.D., Bryan, R. and Seegers, M. (1979) Automated fluorometric determination of vitamin C in foods. *J. Ass. Off. Analyt. Chem.*, **62**, 648–52.

Egberg, D.C., Potter, R.H. and Heroff, J.C. (1977) Semiautomated method for the fluorometric determination of total vitamin C in food products. *J. Ass. Off. Analyt. Chem.*, **60**, 126–31.

Elgilany Elbashier, E. and Greenway, G.M. (1990) Determination of vitamin C in karkady using flow injection analysis. *J. Micronutr. Anal.*, **8**, 311–16.

Erdman, J.W. Jr and Klein, B.P. (1982) Harvesting, processing, and cooking influences on vitamin C in foods. In *Ascorbic Acid: Chemistry, Metabolism, and Uses* (eds P.A. Seib and B.M. Tolbert), American Chemical Society, Washington, DC, pp. 499–532.

Fazio, V., Flint, D.M. and Wahlqvist, M.L. (1981) Acute effects of alcohol on plasma ascorbic acid in healthy subjects. *Am. J. Clin. Nutr.*, **34**, 2394–6.

Finley, J.W. and Duang, E. (1981) Resolution of ascorbic, dehydroascorbic and diketogulonic acids by paired-ion reversed-phase chromatography. *J. Chromat.*, **207**, 449–53.

Graham, W.D. and Annette, D. (1992) Determination of ascorbic and dehydroascorbic acid in potatoes (*Solanum tuberosum*) and strawberries using ion-exclusion chromatography. *J. Chromat.*, **594**, 187–94.

Greenway, G.M. and Ongomo, P. (1990) Determination of L-ascorbic acid in fruit and vegetable juices by flow injection with immobilised ascorbate oxidase. *Analyst, Lond.*, **115**, 1297–9.

Hallberg, L. (1981) Bioavailability of dietary iron in man. *Ann. Rev. Nutr.*, **1**, 123–47.

Hay, G.W., Lewis, B.A. and Smith, F. (1967) Ascorbic acid. II. Chemistry. In *The Vitamins. Chemistry, Physiology, Pathology, Methods*, 2nd edn, Vol. I (eds W.H. Sebrell, Jr and R.S. Harris), Academic Press, New York, pp. 307–36.

Hoare, M., Jones, S. and Lindsay, J. (1993) Total vitamin C analysis of orange juice. *Food Australia*, **45**, 341–5.

Hornig, D.H. and Moser, U. (1981) The safety of high vitamin C intakes in man. In *Vitamin C (Ascorbic Acid)* (eds J.N. Counsell and D.H. Hornig), Applied Science Publishers, London and New Jersey, pp. 225–48.

Hornig, D. and Weiser, H. (1976) Interaction of erythorbic acid with ascorbic acid catabolism. *Int. J. Vitam. Nutr. Res.*, **46**, 40–7.

Huang, T., Duda, C. and Kissinger, P.T. (1987) LCEC determination of sulphite in food. *Current Separations* (Bioanalytical Systems, Inc., West Lafayette, IN), **8**, 49–52.

Hung, T.-H.T., Seib, P.A. and Kramer, K.J. (1987) Determination of L-ascorbyl 6-palmitate in bread using reverse-phase high-performance liquid chromatography (HPLC) with electrochemical (EC) detection. *J. Food Sci.*, **52**, 948–53, 974.

Iwase, H. and Ono, I. (1993) Determination of ascorbic acid and dehydroascorbic acid in juices by high-performance liquid chromatography with electrochemical detection using L-cysteine as precolumn reductant. *J. Chromat. A*, **654**, 215–20.

Jacob, R.A. (1990) Assessment of human vitamin C status. *J. Nutr.*, **120**, 1480–5.

Jaffe, G.M. (1984) Vitamin C. In *Handbook of Vitamins. Nutritional, Biochemical, and Clinical Aspects* (ed. L.J. Machlin), Marcel Dekker, Inc., New York, pp. 199–244.

Kacem, B., Marshall, M.R., Matthews, R.F. and Gregory, J.F. III (1986) Simultaneous analysis of ascorbic acid and dehydroascorbic acid by high-performance liquid chromatography with postcolumn derivatization and UV absorbance. *J. Agric. Food Chem.*, **34**, 271–4.

Kallner, A.B., Hartmann, D. and Hornig, D.H. (1981) On the requirements of ascorbic acid in man: steady-state turnover and body pool in smokers. *Am. J. Clin. Nutr.*, **34**, 1347–55.

Kallner, A., Hornig, D. and Pellikka, R. (1985) Formation of carbon dioxide from ascorbate in man. *Am. J. Clin. Nutr.*, **41**, 609–13.

Karasov, W.H., Darken, B.W. and Bottum, M.C. (1991) Dietary regulation of intestinal ascorbate uptake in guinea pigs. *Am. J. Physiol.*, **260**, G108–18.

Karayannis, M.I., Samios, D.N. and Gousetis, Ch.P. (1977) A study of the molar absorptivity of ascorbic acid at different wavelengths and pH values. *Analyt. Chim. Acta*, **93**, 275–9.

Karp, S., Ciambra, C.M. and Miklean, S. (1990) High-performance liquid chromatographic post-column reaction system for the electrochemical detection of ascorbic acid and dehydroascorbic acid. *J. Chromat.*, **504**, 434–40.

Keltz, F.R., Kies, C. and Fox, H.M. (1978) Urinary ascorbic acid excretion in the human as affected by dietary fiber and zinc. *Am. J. Clin. Nutr.*, **31**, 1167–71.

Killeit, U. (1986) The stability of vitamins. *Food Europe*, No. 3, March/April, 21–4.

Kim, H.-J. (1989) Determination of total vitamin C by ion exclusion chromatography with electrochemical detection. *J. Ass. Off. Analyt. Chem.*, **72**, 681–6.

Kim, H.-J. and Kim, Y.-K. (1988) Analysis of ascorbic acid by ion exclusion chromatography with electrochemical detection. *J. Food Sci.*, **53**, 1525–7.

Kirk, J.R. and Ting, N. (1975) Fluorometric assay for total vitamin C using continuous flow analysis. *J. Food Sci.*, **40**, 463–6.

Knudson, E.J. and Siebert, K.J. (1987) The determination of ascorbates in beer by liquid chromatography with electrochemical detection. *J. Am. Soc. Brewing Chem.*, **45**, 33–7.

Koh, E.V., Bissell, M.G. and Ito, R.K. (1993) Measurement of vitamin C by capillary electrophoresis in biological fluids and fruit beverages using a stereoisomer as an internal standard. *J. Chromat.*, **633**, 245–50.

Kübler, W. and Gehler, J. (1970) On the kinetics of the intestinal absorption of ascorbic acid: a contribution to the calculation of an absorption process that is not proportional to the dose. *Int. J. Vitam. Nutr. Res.*, **40**, 442–43 (in German).

Kutnink, M.A. and Omaye, S.T. (1987) Determination of ascorbic acid, erythorbic acid, and uric acid in cured meats by high performance liquid chromatography. *J. Food Sci.*, **52**, 53–6.

Lau, O.-W., Luk, S.-F. and Wong, K.-S. (1986) Background correction method for the determination of ascorbic acid in soft drinks, fruit juices and cordials using direct ultra-violet spectrophotometry. *Analyst, Lond.*, **111**, 665–70.

Lawendel, J.S. (1957) Ultra-violet absorption spectra of L-ascorbic acid in aqueous solutions. *Nature*, **180**, 434–5.

Lázaro, F., Luque de Castro, M.D. and Valcárcel, M. (1987) Simultaneous determination of ascorbic acid and sulphite in soft drinks by flow injection analysis. *Analusis*, **15**, 183–7.

Lee, H.S. and Coates, G.A. (1987) Liquid chromatographic determination of vitamin C in commercial Florida citrus juices. *J. Micronutr. Anal.*, **3**, 199–209.

Lee, K. and Marder, S. (1983) HPLC determination of erythorbate in cured meats. *J. Food Sci.*, **48**, 306, 308.

Levine, M., Dhariwal, K.R., Welch, R.W. *et al.* (1995) Determination of optimal vitamin C requirements in humans. *Am. J. Clin. Nutr.*, **62** (suppl.) 1347S–56S.

Liao, M.-L. and Seib, P.A. (1987) Selected reactions of L-ascorbic acid related to foods. *Food Technol.*, **41**(11), 104–7.

Liao, M.-L. and Seib, P.A. (1988) Chemistry of L-ascorbic acid related to foods. *Food Chem.*, **30**, 289–312.

Lieber, C.S. (1988) The influence of alcohol on nutritional status. *Nutr. Rev.*, **46**, 241–54.

Lin Ling, B., Baeyens, W.R.G., van Acker, P. and Dewaele, C. (1992) Determination of ascorbic acid and isoascorbic acid by capillary zone electrophoresis: application to fruit juices and to a pharmaceutical formulation. *J. Pharm. Biomed. Anal.*, **10**, 717–21.

Lloyd, L.L., Warner, F.P., Kennedy, J.F. and White, C.A. (1988a) Quantitative analysis of vitamin C (L-ascorbic acid) by ion suppression reversed phase chromatography. *Food Chem.*, **28**, 257–68.

Lloyd, L.L., Warner, F.P., Kennedy, J.F. and White, C.A. (1988b) Ion suppression reversed-phase high-performance liquid chromatography method for the separation of L-ascorbic acid in fresh fruit juice. *J. Chromat.*, **437**, 447–52.

Mangels, A.R., Block, G., Frey, C.M. *et al.* (1993) The bioavailability to humans of ascorbic acid from oranges, orange juice and cooked broccoli is similar to that of synthetic ascorbic acid. *J. Nutr.*, **123**, 1054–61.

Mannino, S. and Pagliarini, E. (1988) A rapid HPLC method for the determination of vitamin C in milk. *Lebensm.-Wiss. u.-Technol.*, **21**, 313–4.

Margolis, S.A. and Black, I. (1987) Stabilization of ascorbic acid and its measurement by liquid chromatography in nonfat dry milk. *J. Ass. Off. Analyt. Chem.*, **70**, 806–9.

Marshall, P.A., Trenerry, V.C. and Thompson, C.O. (1995) The determination of total ascorbic acid in beers, wines, and fruit drinks by micellar electrokinetic capillary chromatography. *J. Chromatogr., Sci.*, **33**, 426–32.

Matano, K. and Kato, N. (1967) Studies on synthetic ascorbigen as a source of vitamin C for guinea pigs. *Acta Chem. Scand.*, **21**, 2886–7.

Mayersohn, M. (1972) Ascorbic acid absorption in man – pharmacokinetic implications. *Eur. J. Pharmacol.*, **19**, 140–2.

Moledina, K.H. and Flink, J.M. (1982) Determination of ascorbic acid in plant food products by high performance liquid chromatography. *Lebensm.-Wiss. u.-Technol.*, **15**, 351–8.

Moll, N. and Joly, J.P. (1987) Determination of ascorbic acid in beers by high-performance liquid chromatography with electrochemical detection. *J. Chromat.*, **405**, 347–56.

Moser, U. (1987) Uptake of ascorbic acid by leukocytes. *Ann. NY Acad. Sci.*, **498**, 200–15.

National Research Council (1989) Water-soluble vitamins. In *Recommended Dietary Allowances*, 10th edn, National Academy Press, Washington, DC, pp. 115–73.

Nisperos-Carriedo, M.O., Buslig, B.S. and Shaw, P.E. (1992) Simultaneous detection of dehydroascorbic, ascorbic, and some organic acids in fruits and vegetables by HPLC. *J. Agric. Food Chem.*, **40**, 1127–30.

Olliver, M. (1967) Ascorbic acid. V. Occurrence in foods. In *The Vitamins. Chemistry, Physiology, Pathology, Methods*, 2nd edn, Vol. I (eds W.H. Sebrell, Jr and R.S. Harris), Academic Press, New York, pp. 359–67.

Olson, J.A. and Hodges, R.E. (1987) Recommended dietary intakes (RDI) of vitamin C in humans. *Am. J. Clin. Nutr.*, **45**, 693–703.

Pachla, L.A. and Kissinger, P.T. (1979) Analysis of ascorbic acid by liquid chromatography with amperometric detection. *Methods Enzymol.*, **62D**, 15–24.

Pelletier, O. and Godin, C. (1969) Vitamin C activity of D-isoascorbic acid for the guinea pig. *Can. J. Physiol. Pharmacol.*, **47**, 985–91.

Pelletier, O. and Morrison, A.B. (1966) Determination of ascorbic acid in the presence of ferrous and stannous salts. *J. Ass. Off. Analyt. Chem.*, **49**, 800–3.

Ponting, J.D. (1943) Extraction of ascorbic acid from plant materials. *Ind. Engng Chem., Analyt. Edn*, **15**, 389–91.

Rivers, J.M. (1987) Safety of high-level vitamin C ingestion. *Ann. NY Acad. Sci.*, **498**, 445–54.

Roe, J.H. (1967) Ascorbic acid. In *The Vitamins. Chemistry, Physiology, Pathology, Methods*, 2nd edn, Vol. VII (eds P. György and W.N. Pearson), Academic Press, New York, pp. 27–51.

Romero Rodriguez, M.A., Vazquez Oderiz, M.L., Lopez Hernandez, J. and Simal Lozano, J. (1992) Determination of vitamin C and organic acids in various fruits by HPLC. *J. Chromatogr. Sci.*, **30**, 433–7.

Rose, R.C. (1989) Renal metabolism of the oxidized form of ascorbic acid (dehydro-L-ascorbic acid). *Am. J. Physiol.*, **256**, F52–6.

Rose, R.C. and Nahrwold, D.L. (1978) Intestinal ascorbic acid transport following diets of high or low ascorbic acid content. *Int. J. Vitam. Nutr. Res.*, **48**, 382–6.

Rose, R.C., Choi, J.-L. and Koch, M.J. (1988) Intestinal transport and metabolism of oxidized ascorbic acid (dehydroascorbic acid). *Am. J. Physiol.*, **254**, G824–8.

Roy, R.B., Conetta, A. and Salpeter, J. (1976) Automated fluorometric method for the determination of total vitamin C in food products. *J. Ass. Off. Analyt. Chem.*, **59**, 1244–50.

Russell, L.F. (1986) High performance liquid chromatographic determination of vitamin C in fresh tomatoes. *J. Food Sci.*, **51**, 1567–8.

Sapers, G.M., Douglas, F.W. Jr, Ziolkowski, M.A. *et al.* (1990) Determination of ascorbic acid, dehydroascorbic acid and ascorbic acid-2-phosphate in infiltrated apple and potato tissue by high-performance liquid chromatography. *J. Chromat.*, **503**, 431–6.

Schüep, W. and Keck, E. (1990) Measurement of ascorbic acid and erythorbic acid in processed meat by HPLC. *Z. Lebensmittelunters u.-Forsch.*, **191**, 290–2.

Seib, P.A. (1985) Oxidation, monosubstitution and industrial synthesis of ascorbic acid. *Int. J. Vitam. Nutr. Res.*, Suppl. 27, 259–306.

Seiffert, B., Swaczyna, H. and Schaefer, I. (1992) Simultaneous determination of L-ascorbic acid and D-isoascorbic acid by HPLC in wine and beer. *Deutsche Lebensm.-Rundschau*, **88**, 38–40.

Siliprandi, L., Vanni, P., Kessler, M. and Semenza, G. (1979) Na^+-dependent, electroneutral L-ascorbate transport across brush border membrane vesicles from guinea pig small intestine. *Biochim. Biophys. Acta*, **552**, 129–42.

Speek, A.J., Schrijver, J. and Schreurs, W.H.P. (1984) Fluorometric determination of total vitamin C and total isovitamin C in foodstuffs and beverages by high-performance liquid chromatography with precolumn derivatization. *J. Agric. Food Chem.*, **32**, 352–5.

Thompson, C.O. and Trenerry, V.C. (1995) A rapid method for the determination of total L-ascorbic acid in fruits and vegetables by micellar electrokinetic capillary chromatography. *Food Chem.*, **53**, 43–50.

Todhunter, E.N., McMillan, T. and Ehmke, D.A. (1950) Utilization of dehydroascorbic acid by human subjects. *J. Nutr.*, **42**, 297–308.

Toggenburger, G., Landoldt, M. and Semenza, G. (1979) Na^+-dependent, electroneutral L-ascorbate transport across brush border membrane vesicles from human small intestine. *FEBS Letters*, **108**, 473–6.

Toggenburger, G., Häusermann, M., Mütsch *et al.* (1981) Na^+-dependent, potential-sensitive L-ascorbate transport across brush border membrane vesicles from kidney cortex. *Biochim. Biophys. Acta*, **646**, 433–43.

Tsao, C.S. and Salimi, S.L. (1982) Differential determination of L-ascorbic acid and D-isoascorbic acid by reversed-phase high-performance liquid chromatography with electrochemical detection. *J. Chromat.*, **245**, 355–8.

Tsumura, F., Ohsako, Y., Haraguchi, Y. *et al.* (1993) Rapid enzymatic assay for ascorbic acid in various foods using peroxidase. *J. Food Sci.*, **58**, 619–22, 687.

Tuan, S., Wyatt, J. and Anglemier, A.F. (1987) The effect of erythorbic acid on the determination of ascorbic acid levels in selected foods by HPLC and spectrophotometry. *J. Micronutr. Anal.*, **3**, 211–28.

Vanderslice, J.T. and Higgs, D.J. (1985) Robotic extraction of vitamin C from food samples. *J. Micronutr. Anal.*, **1**, 143–54.

Vanderslice, J.T. and Higgs, D.J. (1988) Chromatographic separation of ascorbic acid, isoascorbic acid, dehydroascorbic acid and dehydroisoascorbic acid and their quantitation in food products. *J. Micronutr. Anal.*, **4**, 109–18.

Vanderslice, J.T. and Higgs, D.J. (1989) Automated analysis of total vitamin C in food. *J. Micronutr. Anal.*, **6**, 109–17.

Vanderslice, J.T. and Higgs, D.J. (1990) Separation of ascorbic acid, isoascorbic acid, dehydroascorbic acid and dehydroisoascorbic acid in food and animal tissue. *J. Micronutr. Anal.*, **7**, 67–70.

Vanderslice, J.T. and Higgs, D.J. (1991) Vitamin C content of foods: sample variability. *Am. J. Clin. Nutr.*, **54**, 1323S–7S.

Vanderslice, J.T. and Higgs, D.J. (1993) Quantitative determination of ascorbic, dehydroascorbic, isoascorbic, and dehydroascorbic acids by HPLC in foods and other matrices. *J. Nutr. Biochem.*, **4**, 184–90.

Vanderslice, J.T., Higgs, D.J., Hayes, J.M. and Block, G. (1990) Ascorbic acid and dehydroascorbic acid content of foods-as-eaten. *J. Food Comp. Anal.*, **3**, 105–18.

Vazquez Oderiz, M.L., Vazquez Blanco, M.E., Lopez Hernandez, J. *et al.* (1994) Simultaneous determination of organic acids and vitamin C in green beans by liquid chromatography. *J. AOAC Int.*, **77**, 1056–9.

Vinson, J.A. and Bose, P. (1988) Comparative bioavailability to humans of ascorbic acid alone or in a citrus extract. *Am. J. Clin. Nutr.*, **48**, 601–4.

Watada, A.E. (1982) A high-performance liquid chromatography method for determining ascorbic acid content of fresh fruits and vegetables. *Hort. Sci.*, **17**, 334–5.

Wills, R.B.H., Wimalasiri, P. and Greenfield, H. (1983) Liquid chromatography, microfluorometry, and dye-titration determination of vitamin C in fresh fruit and vegetables. *J. Ass. Off. Analyt. Chem.*, **66**, 1377–9.

Wilson, C.W. and Shaw, P.E. (1987) High-performance liquid chromatographic determination of ascorbic acid in aseptically packaged orange juice using ultraviolet and electrochemical detectors. *J. Agric. Food Chem.*, **35**, 329–31.

Wimalasiri, P. and Wills, R.B.H. (1983) Simultaneous analysis of ascorbic acid and dehydroascorbic acid in fruit and vegetables by high-performance liquid chromatography. *J. Chromat.*, **256**, 368–71.

Zapata, S. and Dufour, J.-P. (1992) Ascorbic, dehydroascorbic and isoascorbic acid simultaneous determinations by reverse phase ion interaction HPLC. *J. Food Sci.*, **57**, 506–11.

Ziegler, S.J., Meier, B. and Sticher, O. (1987) Rapid and sensitive determination of dehydroascorbic acid in addition to ascorbic acid by reversed-phase high performance liquid chromatography using a post-column reduction system. *J. Chromat.*, **391**, 419–26.

Index

Page numbers in **bold** type refer to figures and those in *italics* refer to tables.

Index